Studies in Computational Intelligence

Volume 777

Series editor

Janusz Kacprzyk, Polish Academy of Sciences, Warsaw, Poland
e-mail: kacprzyk@ibspan.waw.pl

The series "Studies in Computational Intelligence" (SCI) publishes new developments and advances in the various areas of computational intelligence—quickly and with a high quality. The intent is to cover the theory, applications, and design methods of computational intelligence, as embedded in the fields of engineering, computer science, physics and life sciences, as well as the methodologies behind them. The series contains monographs, lecture notes and edited volumes in computational intelligence spanning the areas of neural networks, connectionist systems, genetic algorithms, evolutionary computation, artificial intelligence, cellular automata, self-organizing systems, soft computing, fuzzy systems, and hybrid intelligent systems. Of particular value to both the contributors and the readership are the short publication timeframe and the world-wide distribution, which enable both wide and rapid dissemination of research output.

More information about this series at http://www.springer.com/series/7092

Witold Pedrycz · Shyi-Ming Chen
Editors

Computational Intelligence for Pattern Recognition

Editors
Witold Pedrycz
University of Alberta
Edmonton, AB
Canada

Shyi-Ming Chen
National Taiwan University of Science and Technology
Taipei
Taiwan

ISSN 1860-949X ISSN 1860-9503 (electronic)
Studies in Computational Intelligence
ISBN 978-3-030-07819-5 ISBN 978-3-319-89629-8 (eBook)
https://doi.org/10.1007/978-3-319-89629-8

Softcover re-print of the Hardcover 1st edition 2018

Printed on acid-free paper

This Springer imprint is published by the registered company Springer International Publishing AG part of Springer Nature
The registered company address is: Gewerbestrasse 11, 6330 Cham, Switzerland

Preface

Since the inception of fuzzy sets, fuzzy pattern recognition, including its methodology, algorithms, and applications, has been at the center of the developments of the technology of fuzzy sets. One can refer here to the seminal paper entitled *Abstraction and pattern classification* authored by Bellman, Kalaba, and Zadeh, which has opened uncharted research areas and offered a new attractive insight into the principles of pattern classification. As of now, pattern recognition augmented by the methodology and algorithms of fuzzy sets has established itself as a mature, well-developed research discipline with a variety of advanced applications. Pattern recognition comes with a great deal of challenges, exhibits a continuous paradigm shift (quite often dictated by new applications), and becomes vividly manifested through a growing diversity of areas of its usage. All of those call for substantial enhancements of the existing fundamentals or the formation of new paradigms.

Computational Intelligence (CI) with its impressive armamentarium of methodologies and tools is positioned in a unique way to address the growing needs of pattern recognition. As a matter of fact, this can be accomplished in several tangible ways realized both at the methodological and algorithmic level. There are at least five dominant manifestations of CI in the realm of pattern recognition. They are associated with: (i) coping with a large volume of data and their diversity, (ii) setting a suitable level of abstraction, (iii) dealing with a distributed nature of data along with associated requirements of privacy and security, (iv) building efficient feature spaces, and (v) building interpretable findings of classification at a suitable level of abstraction.

The key objective of the proposed volume is to provide the community with a comprehensive and up-to-date treatise in the area of pattern recognition and computational intelligence. It covers a spectrum of methodological and algorithmic issues, discusses implementations and case studies, identifies the best design practices, and assesses business models and practices of pattern recognition in industry, health care, administration, and business. The collection of contributions forming the edited volume offers the reader a representative view at the progress and accomplishments of the area with a timely, in-depth, and comprehensive material on the conceptually appealing and practically sound methodology and practices of CI-based pattern recognition.

The book engages a wealth of methods of CI, brings new concepts, architectures and practice of fuzzy sets, neurocomputing, and biologically inspired optimization. The chapters cover a wealth of ideas, algorithms, and applications and are a testimony to the synergistic linkages within the CI area and CI and pattern recognition.

Given the leading theme of this undertaking, the book is aimed at a broad audience of researchers and practitioners. Thanks to the nature of the material being covered and the way the main threads have been organized, the volume will appeal to the well-established communities including those active in various disciplines in which pattern recognition plays a central role and serves as an efficient vehicle to produce solutions to numerous classification problems and augments solutions constructed with the aid of the "standard" methodology and algorithms of pattern recognition.

With the required prerequisites covered, the book caters to the broad readership. Those involved in operations research, management, various branches of engineering, sciences, data science, medicine, and bioinformatics will benefit from the exposure to the subject matter.

We would like to take this opportunity to express our sincere thanks to the contributors to the volume for sharing results of their advanced, far-reaching, and original research, and delivering their views at the rapidly expanding areas of fundamental and applied research. The reviewers deserve our thanks for their constructive and timely input. We greatly appreciate a continuous support and encouragement coming from the Editor-in-Chief, Prof. Janusz Kacprzyk whose leadership and vision has helped us arrive at the successful completion of this project. The editorial staff at Springer has done a meticulous job and working with them was a pleasant experience.

We hope that the readers will find this volume interesting and the variety of ideas put forward in this volume will become instrumental in fostering the progress in research, education, and numerous practical endeavors in the CI-oriented pattern recognition.

Edmonton, Canada — Witold Pedrycz
Taipei, Taiwan — Shyi-Ming Chen

Contents

Fuzzy Choquet Integration of Deep Convolutional Neural Networks for Remote Sensing

Derek T. Anderson, Grant J. Scott, Muhammad Aminul Islam, Bryce Murray and Richard Marcum

Abstract What deep learning lacks at the moment is the heterogeneous and dynamic capabilities of the human system. In part, this is because a single architecture is not currently capable of the level of modeling and representation of the complex human system. Therefore, a heterogeneous set of pathways from sensory stimulus to cognitive function needs to be developed in a richer computational model. Herein, we explore the learning of multiple pathways–as different deep neural network architectures–coupled with appropriate data/information fusion. Specifically, we explore the advantage of data-driven optimization of fusing different deep nets–GoogleNet, CaffeNet and ResNet–at a per class (neuron) or shared weight (single data fusion across classes) fashion. In addition, we explore indices that tell us the importance of each network, how they interact and what aggregation was learned. Experiments are provided in the context of remote sensing on the UC Merced and WHU-RS19 data sets. In particular, we show that fusion is the top performer, each network is needed across the various target classes, and unique aggregations (i.e., not common operators) are learned.

Keywords Fuzzy integral · Convolutional neural network · Remote sensing

---

D. T. Anderson (✉) · G. J. Scott · R. Marcum
Electrical Engineering and Computer Science, University of Missouri, Columbia, MO, USA
e-mail: andersondt@missouri.edu

G. J. Scott
e-mail: grantscott@missouri.edu

R. Marcum
e-mail: ram7cd@missouri.edu

M. A. Islam · B. Murray
Electrical and Computer Engineering, Mississippi State University, Starkville, MS, USA
e-mail: mi160@msstate.edu

B. Murray
e-mail: bjm260@msstate.edu

W. Pedrycz and S.-M. Chen (eds.), *Computational Intelligence for Pattern Recognition*, Studies in Computational Intelligence 777,
https://doi.org/10.1007/978-3-319-89629-8_1

1 Introduction

We humans excel at many robust pattern recognition tasks in which computational systems can only perform well when limited in scope and constrained in operating environment. The human visual system is no exception. Humans develop at an early age a comprehensive visual processing and pattern recognition ability. Our vision allows us to process our physical environment (navigation) and facilitates many higher-level cognitive functions such as object classification and entity resolution. We accomplish this via a complex multistage visual system that begins with basic lightness and color receptors, then builds upon the perceived edges to derive shapes, spatial relationships, and eventually to organization of components into objects of interest – and this is before any higher level cognitive processing.

Deep neural network models follow a similar paradigm conceptually, extracting first edges and other simple geometric primitives in the lowest levels, then later mid-level assemblies of these primitives into visual concepts, which are then combined in higher-level layers as object components (blobs), that are eventually agglomerated into objects. These visual objects are agglomerated within fully connected neural layers for eventual classifications, which is an informational (cognitive) output. What deep architectures lack at the moment is the heterogeneous and dynamic capabilities of the human system, which is in part because a single architecture is not capable of the level of modeling and representation of the complex human system. Therefore, a heterogeneous set of pathways from sensory stimulus to cognitive function needs to be developed in a richer computational model. The model proposed in this chapter represents the learning of multiple pathways–as deep neural networks–coupled with appropriate information fusion. We feel fusion of the cognitive outputs (information) from multiple heterogeneous models (pathways) is the next step towards robust computational cognitive processing of visual, and visual-like, sensory data.

In general, *computational intelligence* (CI) is a branch of mathematics inspired by nature. Specifically, CI is associated with *neural networks* (NNs), *evolutionary algorithms* (EA) and *fuzzy set theory* (FST). NNs were established in 1943 by McCulloch and Pitts [1], FST was established in 1965 by Zadeh [2] and EAs were made popular by Holland in the early 1970s [3] (but arguably have roots going back as far as Turing in 1950). The point is, CI has existed in one form or another since the advent of *artificial intelligence* (AI). In this chapter, we focus on the intersection of NNs and FST for pattern recognition. In the last decade, substantial interest and effort has gone into *deep learning* (DL), a re-branding of NNs. This shift has forced us to re-address fundamental questions like; should humans design features (the classical approach to pattern recognition) or is a machine better at this task? Empirically, DL has more-or-less unanimously topped the charts in many domains (e.g., natural language processing [4, 5], vision [6–10], remote sensing [11–14]). However, while DL has generated great excitement, much remains to be explored and explained. In this chapter, we focus on the specific question of how to perform decision-level fusion of DL networks.

DL can be viewed as a generalization of the classical pattern recognition pipeline–e.g., pre-processing, feature extraction (selection and/or reduction), classification and post-processing. In some settings this is now being called *shallow learning* because there are only a few "layers" in the pattern recognition pipeline. In the context of computer vision, DL can also be decomposed into levels; "low" (e.g., signal/image analysis via convolution), "mid" and "high" (more AI than signal processing, e.g., MLP classification). In the extreme, DL is nothing more than a series of operations that transform data to decisions. The point is, fusion can (and often does) take place at different levels in pattern recognition/DL. For example, keeping with the fusion nomenclature of the *Joint Directors of Laboratories* (JDL) [15], some fusion algorithms do *signal-in-signal-out* (SISO), whereas others do *feature-in-feature-out* (FIFO) and *decision-in-decision-out* (DIDO). If we regard DL as a SIDO process (e.g., SI = image and DO = class label), then it can be decomposed into its corresponding SISO, SIFO, FIFO, FIDO, DIDO (and combinations therein). In summary, fusion is not as simple as "cram data into a DL and let it do its thing".

Herein, we restrict our analysis to *deep convolutional neural networks* (DCNNs) [6–10, 16, 17], versus *auto encoders* (AEs) [18–22], *deep belief nets* (DBNs) [23, 24], etc., for sake of discussion tractability. The reality is, we still know little-to-nothing about fundamental DL fusion questions such as; (i) how/where is fusion currently happening, (ii) based on our current set of neurons/transformations, what is mathematically expressible and what is not (but should be), (iii) how should we be performing fusion at different levels, (iv) how do we address heterogeneity with respect to semantics and/or uncertainty across data/information sources, and (v) how do we explain what fusion is doing (aka *explainable AI* (XAI)), to list a few. Independent of DL, fusion is a complicated topic that often means different things to different people in different fields (and even within the same field). Fusion is a wealth of challenges wrapped up into one term. Fusion ranges from data association (e.g., finding a one-to-one mapping between pixels in one sensor to pixels in another) to the mathematics of aggregation (specific functions/operators). In general, the idea of fusion is to obtain a "better" result than if we only used the individual inputs. However, better is not a well defined concept. In some applications, better might mean taking a set of inputs and reducing them into a single result that can be more efficiently or effectively used for visualization. Better could also refer to obtaining more desirable properties such as higher information content or lower conflict. In areas like pattern recognition, better often refers to some desirable property like more robust and generalizable solutions (e.g., classifiers). Regardless of the task at hand or the particular application, fusion is a core tool at the heart of numerous modern scientific thrusts.

In this chapter, we make the following contributions. First, we discuss two approaches for heterogeneous DCNN architecture fusion; density-based imputation and full *Choquet integral* (ChI) learning (per neuron and "shared weight"). Second, we outline indices for introspection and information theoretic indices to understand the capacity and integral (moving us closer to a so-called XAI solution versus black box solution). Third, we demonstrate and analyze these ideas on remotely sensed data. Last, we provide open source code at www.derektanderson.com/FuzzyLibrary and www.github.com/scottgs/fi_library.

2 Deep Convolutional Neural Networks

To date, the AE [18–20], CNN [6–10, 16, 17], DBN [23, 24] and *recurrent NNs* (RNNs) [25, 26] are the most mainstream DLs. However, other DL approaches exist, e.g., deep inference nets ([27] Verma et al. Takagi-Sugeno-Kang deep net), deconvolution CNNs (specifically transpose matrix convolution) [28–30] and morphological shared weight neural networks [31, 32]. Herein, we focus on the CNN, which is by far the most employed and often the highest performer. With respect to the CNN, a number of architectures have been explored to date, e.g., AlexNet [7], GoogLeNet [17], VGGNet [33] and their derivatives. These architectures can be downloaded and extended (training, evaluation, visualization) via open source libraries like TensorFlow [34], CaffeNet [35], and MatConvNet [36]. The fundamental challenges of which architecture, how deep versus wide, hyperparameter tuning, what neuron types, how to transfer a DL from one domain to another (transfer learning [37]), and other questions are unanswered. Also, numerous challenges exist; e.g., lack of training data volume (and variety), class imbalance, dimensionality (spatial, temporal and spectral), explainable DL (what did the DL learn, versus a black box), to name a few. While DL has sparked a revolution in computer vision, pattern recognition and AI in general, an overwhelming number of theoretical and applied questions remain ripe for exploration.

In general, most CNNs consist of combinations of the following operations (see Fig. 1). First, let the input to the system, O_0, be a three dimensional data cube of size $N_0 \times M_0 \times D_0$; where N_0 and M_0 are spatial dimensions and D_0 is the temporal or spectral dimensionality (e.g., RGB imagery has $D_0 = 3$). **(Convolution)** The backbone of a CNN is filtering via convolution. Filtering can take a number of meanings, e.g., enhancement, denoising or detection. Convolution specifics include factors like (i) stride (spatial and/or spectral/temporal "jumps") and (ii) padding (if no padding is used then the spatial dimension shrinks). **(Pooling)** Pooling is often applied to reduce spatial dimensionality–and combat challenges related to affine variation, noise, etc. Most often, average and max pooling are used. **(Activation)** Nonlinearity is also typically applied, in the form of a function like hyperbolic tangent (*tanh*), sigmoid, or ReLU ($ReLU(x) = \max(0, x)$). **(Training Techniques)** In order to combat factors like sensitivity to parameter selection and overtraining, methods like dropout [38], regularization [39] and/or batch normalization [40] (addresses internal covariate shift and vanishing gradients) are often used. Beyond architecture, there are factors like GPU acceleration [41], training (e.g., *stochastic gradient decent* (SGD) [42], SGD with momentum [43, 44], AdaGrad [45], RMSProp [46] and ADAM [47]). The reader can refer to [39] for additional mathematical and algorithm

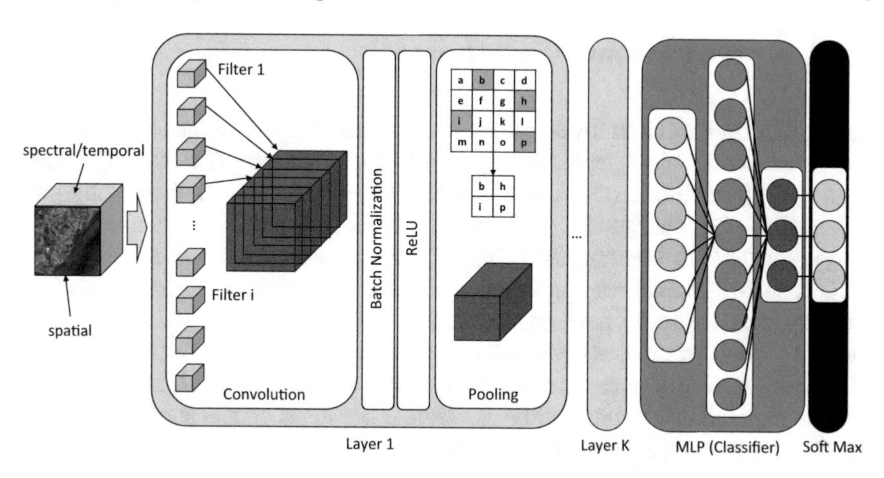

Fig. 1 Example CNN. Input is a 3D cube (x-y spatial, z spectral), green layers consist of subset of convolution (morphology, etc.), pooling (average, max, etc.), batch normalization (or other method to mitigate overfitting, vanishing gradients, internal covariate shift, etc.) and nonlinear function (e.g., ReLU activation). The output of the green layers are typically fed to a MLP and optional post-processing steps (e.g., soft max normalization)

details related to CNNs. The reader can also refer to [48] for a recent survey of DL in remote sensing (theory, applications and open challenges).

The idea of FST in NNs is not new. The reader can refer to the work of Pal and Mitra [49] for neuro-fuzzy pattern recognition. Pal, Mitra, and others (e.g., Keller and the fuzzy perceptron [50]), explored a variety of topics such as fuzzy min-max networks, fuzzy MLPs, and fuzzy Kohonen networks. In terms of aggregation, a few FST works have been explored to date. In 1992 [51], Yager put forth the *ordered weighted average* (OWA) [52]–which technically is a *linear combination of order statistics* (LCOS) since the weights are real-valued numbers (versus sets)–neuron. In 1995, Sung-Bae utilized the OWA for NN aggregation (at the decision/output level) [53]. In 1995, Sung-Bae et al. also explored the fuzzy integral, the Sugeno fuzzy integral not Sugeno's fuzzy ChI, for NN aggregation [54]. Specifically, they used the Sugeno λ-*fuzzy measure* (FM) and the densities were derived using their respective accuracy rates on training data. In 2017 [55], we (Scott et al.) used the Sugeno and ChIs for DCNN fusion. Specifically, Scott et al. used transfer learning to adapt GoogLeNet, AlexNet and ResNet50 from camera imagery to remote sensing imagery. Scott then applied different aggregations–fuzzy integral, voting, arrogance, and weighted sum–to these DCNNs. Scott's fusion was based on the Sugeno λ FM and the densities were (i) set to the normalized classifier accuracies and (ii) a GA learned the densities (which led to higher performance).

3 Fuzzy Measure and Fuzzy Integrals

The ChI has been successfully demonstrated in numerous applications; e.g., explosive hazard detection [56–58], computer vision [59], pattern recognition [60–64], remote sensing [65], multi-criteria decision making [66, 67], forensic anthropology [68–70], control [71], multiple kernel learning [56, 72–75], multiple instance learning [76], ontologies [77], missing data [78], and most relevant to the current chapter, DL [55]. The ChI is a nonlinear aggregation function that is parameterized by the FM (aka capacity). Countless mathematical variations of the fuzzy integral have been put forward for different reasons; e.g., address different types (i.e., real-valued, interval-valued, set-valued) of uncertainty in the integrand and/or measure, limit the number of input interactions for tractability, etc. Herein, we focus on and succinctly review just the real-valued discrete (finite X) ChI for DCNN fusion.

3.1 Discrete (Finite X) Fuzzy Measure

Let $X = \{x_1, x_2, \ldots, x_N\}$ be N sources, e.g., experts, sensors, or in the case of this chapter, DCNNs. The first action we face is how to assign "worth/utility" to different subsets of DCNNs. For example, the well-known backbone of calculus on real-valued domains is the Lebesgue measure; which coincides with length, area and hypervolume. However, when X is a discrete domain, e.g., set of DCNNs, what is the corresponding "measure"? In [59], Keller et al. first investigated the idea of using the fuzzy integral for pattern recognition. A FM is a function, μ, on the power set of X, 2^X, which satisfies (1) (boundary condition) $\mu(\emptyset) = 0$ (often $\mu(X) = 1$[1]) and (2) (monotonicity) if $A, B \subseteq X$ and $A \subseteq B$, then $\mu(A) \leq \mu(B)$.

3.2 Discrete (Finite X) Fuzzy Integral

The FM models important "interactions" (e.g., subjective worth, statistical correlation, etc.) between different source subsets. The input provided by our sources is $\{h(\{x_1\}), h(\{x_2\}), \ldots, h(\{x_N\})\}$. The fuzzy integral is a way to combine the integrand (h) information relative to the FM (μ). Let $h(\{x_i\}) \in \Re^{\geq 0}$ be the data from source i. The discrete (finite X) Sugeno FI is[2]

[1]If $\mu(X) < 1$, properties like idempotency and boundedness are not guaranteed.

[2]Due to the maximum (t-conorm) and minimum operators (t-norm), the Sugeno FI does not actually generate any possible number between the minimum and maximum of the inputs. Instead, it selects one of the FM or input values, i.e., at most one of $2^N + N$ values.

$$\int_S h \circ \mu = S_\mu(h) = \bigvee_{i=1}^{N} \left(h(\{x_{\pi(i)}\}) \wedge \mu(A_i) \right), \quad (1)$$

where π is the permutation $h(\{x_{\pi(1)}\}) \geq h(\{x_{\pi(2)}\}), \ldots, \geq h(\{x_{\pi(N)}\})$, $A_i = \{x_{\pi(1)}, \ldots, x_{\pi(i)}\}$ and $\mu(A_0) = 0$. The discrete (finite X) ChI is[3]

$$\int_C h \circ \mu = C_\mu(h) = \sum_{i=1}^{N} h(\{x_{\pi(i)}\}) \left[\mu(A_i) - \mu(A_{i-1}) \right]. \quad (2)$$

Since the ChI is a parametric aggregation function, once the FM is determined the ChI turns into a specific operator. For example: if $\mu(A) = 1, \forall A \in 2^X \setminus \emptyset$, the ChI becomes the maximum operator; if $\mu(A) = 0, \forall A \in 2^X \setminus X$, we recover the minimum; and if $\mu(A) = \frac{|A|}{N}$, we recover the mean. Each of these cases can be viewed as constraints or simplifications on the FM (and therefore the ChI). In general, the discrete ChI has $N!$ unique input sortings and each yields a linear convex sum operator.

3.3 *Data-Driven Optimization*

The first challenge we must confront is, where do we get the FM (μ) from? One option is to have an expert specify it. However, this is not practical (assuming the expert could even meaningfully assign values to the interactions) as the number of inputs (e.g., DLs) increases. Another option is we can specify or try to learn the worth of just the singletons (the densities). From there, a number of formulas can be used to impute (fill in) the missing variable values. Popular approaches include the Sugeno λ-FM and the S-Decomposable FM [79]. However, while convenient, most often we do not obtain the desired values for variables that we need. With respect to pattern recognition, the focus of this chapter, another route is to learn it from data. Next, we review one way to learn the FM, and therefore the ChI, in the context of DIDO for DL. However, the reader can refer to [80] for an efficient learning method with only data-supported variables and [81] for a review of alternative FM/ChI learning methods.

We quickly summarize one way to learn the full FM/ChI (see [82] for full mathematical explanation). Let $O = \{\mathbf{h}_j, y_j\},\ j = 1, 2, \ldots, M$, be M training examples; where $\mathbf{h}_j$ is the j-th instance with data from N inputs and y_j is the ground-truth for $\mathbf{h}_j$. The sum of squared error for training dataset O is

[3]The ChI is used frequently for various reasons; e.g., it is differentiable [62], for an additive (probability) measure it recovers the Lebesgue integral, it yields a wider spectrum of values between the minimum and maximum (versus the discrete and relatively small number of values that the Sugeno FI selects from), etc.

$$E(O, \mathbf{u}) = \sum_{j=1}^{M} e^j = \sum_{j=1}^{M} (\mathbf{c}_j^T \mathbf{u} - y_j)^2 = ||D\mathbf{u} - \mathbf{y}||_2^2, \tag{3}$$

where $\mathbf{u} = [\mu(\{x_1\}), \ldots, \mu(\{x_N\}), \mu(\{x_1, x_2\}), \mu(\{x_1, x_3\}), \ldots, \mu(X)]$ (lexicographic vector of size $2^N - 1$), $\mathbf{c}_j$ holds the coefficients of $\mathbf{u}$ for observation $\mathbf{h}_j$, e.g., for $N = 3$ and $h(\{x_2\}) \geq h(\{x_1\}) \geq h(\{x_3\})$,

$$c = [0, h(\{x_2\}) - h(\{x_1\}), 0, h(\{x_1\}) - h(\{x_3\}), 0, 0, h(\{x_3\})]\,,$$

$D = [\mathbf{c}_1\ \mathbf{c}_2\ \ldots\ \mathbf{c}_M]^T$ (full dataset), $\mathbf{y} = [y_1\ y_2\ \ldots\ y_M]^T$, and $||\cdot||_2$ is norm-2 operation, $\mathbf{u}$. The regularized SSE optimization problem is

$$\min_{\mathbf{u}} f(\mathbf{u}) = ||D\mathbf{u} - \mathbf{y}||^2 + \beta v(\mathbf{u}), \tag{4}$$

where $\beta \in \Re^{\geq 0}$ (regularization constant, which balances the "cost" (or penalty) of obtaining minimum function error relative to our desire to have minimal model complexity) and $v(\mathbf{u})$ is an index of model complexity (e.g., k-additive and Mobius, Gini-Simpson, ℓ_p-norm, etc. [83]), subject to the FM boundary and monotonicity conditions (see [82] for how to pack the constraints into a linear algebra expression), which can be solved via quadratic programming. Full code (including how to build C) can be found at www.derektanderson.com/FuzzyLibrary and www.github.com/scottgs/fi_library.

3.4 Explainable AI (XAI) Fusion

It is one thing to train a network and another to understand it! In this subsection, we highlight FM and ChI indices for the purpose of *explainable AI* (XAI).[4] XAI is an attempt to explain the inner operations of pattern recognition for purposes like describing it to others for domain knowledge transfer, trust, etc. The Shapley index addresses the importance or worth of each input (aka DL),

$$\Phi_\mu(i) = \sum_{K \subseteq X \setminus \{i\}} \zeta_{X,1}(K)\left(\mu(K \cup \{i\}) - \mu(K)\right), \tag{5}$$

where $\zeta_{X,1}(K) = \frac{(|X|-|K|-1)!|K|!}{|X|!}$, $K \subseteq X\backslash\{i\}$ denotes all proper subsets from X that do not include source i. The Shapley value of μ is the vector $\Phi_\mu = (\Phi_\mu(1), \ldots, \Phi_\mu(N))^t$ and $\sum_{i=1}^{N} \Phi_\mu(i) = 1$. The Shapley index can be interpreted as the average amount of *contribution* of source i across all coalitions. The next index informs us

[4] Go to www.derektanderson.com/FuzzyLibrary and www.github.com/scottgs/fi_library.

about how two inputs interact with one another (aka what advantage is there in combining DLs). The interaction index (Murofushi and Soneda [84]) between i and j is

$$\mathcal{I}_{\mu}(i, j) = \sum_{K \subseteq X \setminus \{i,j\}} \zeta_{X,2}(K)(\mu(K \cup \{i, j\}) - \mu(K \cup \{i\}) - \mu(K \cup \{j\}) + \mu(K)), \tag{6}$$

where $\zeta_{X,2}(K) = \frac{(|X|-|K|-2)!|K|!}{(|X|-1)!}$, $\mathcal{I}_{\mu}(i, j) \in [-1, 1], \forall i, j \in \{1, 2, \ldots, N\}$. A value of 1 (respectively, -1) represents the maximum complementary (respective redundancy) between i and j. Refer to [85] for further details about the interaction index, its connections to game theory and interpretations. Grabisch later extended the interaction index to the general case of any coalition [86],

$$\mathcal{I}_{\mu}(A) = \sum_{K \subseteq X \setminus A} \zeta_{X,3}(K, A) \sum_{C \subseteq A} (-1)^{|A \setminus C|} \mu(C \cup K), \quad i = 1, \ldots, N, \tag{7}$$

where $\zeta_{X,3}(K, A) = \frac{(|X|-|K|-|A|)!|K|!}{(|X|-|A|+1)!}$. Equation (7) is a generalization of both the Shapley index and Murofushi and Soneda's interaction index as $\Phi_{\mu}(i)$ corresponds with $\mathcal{I}_{\mu}(\{i\})$ and $\mathcal{I}_{\mu}(i, j)$ with $\mathcal{I}_{\mu}(\{i, j\})$.

The above indices are focused strictly on the FM. A different fundamental type of question is what "type" of aggregation is the ChI performing? Answering this question helps us understand how the DL information is being combined (e.g., in an optimistic, pessimistic, expected value like fashion, etc.). In [87], we established an index (D_1) to measure the degree to which a given FM/ChI is an maximum operator. In the following, we discuss the FM in terms of its underlying lattice structure. Let "layer k" (measure defined on sets of cardinality k) be denoted by $L(k)$, e.g., $L(1) = \{\mu(\{x_1\}), \mu(\{x_2\}), \mu(\{x_3\})\}$ for $N = 3$. Next, let $\mathbf{W} = \frac{[\frac{1}{N}, \ldots, 1]}{\sum_{i=1}^{N} \frac{i}{N}}$ be weights (penalty or costs) for each layer and

$$D_1(\mu) = \sum_{k=1}^{1} \frac{\mathbf{W}(k)}{2}(T_1 + T_4) + \left[\sum_{k=2}^{N} \frac{\mathbf{W}(k)}{3}(T_1 + T_2 + T_4)\right], \tag{8}$$

$T_1 = 1 - \left(\frac{\sum_{I \in L(k)} \mu(I)}{|L(k)|}\right)$, $T_2 = \left(\frac{\sum_{I \in L(i)} \mu(I)}{|L(k)|} - \frac{\sum_{J \in L(k-1)} \mu(J)}{|L(k-1)|}\right)$, $T_3 = \frac{\sum_{I \in L(k)} \mu(I)}{|L(k)|}$ and $T_4 = \frac{\sum_{I \in L(k)} (\mu(I) - T_3)^2}{|L(k)|-1}$. A value of $D_1 = 0$ means the ChI is the maximum operator. The distance of a learned capacity to a minimum operator (D_2), mean (D_3) and LCOS (D_4), for $\mathbf{W}_2 = \frac{[1, \ldots, \frac{1}{N-1}]}{\sum_{i=1}^{N-1} \frac{i}{N-1}}$, is

$$D_2(\mu) = \sum_{k=1}^{1} \frac{\mathbf{W}_2(k)}{2}(T_3 + T_4) + \left[\sum_{k=2}^{N-1} \frac{\mathbf{W}_2(i)}{3}(T_3 + T_2 + T_4)\right], \tag{9}$$

$$D_3(\mu) = \frac{1}{2^N - 2} \sum_{k=1}^{N-1} \sum_{I \in L(k)} \left| \mu(I) - \frac{k}{N} \right|, \tag{10}$$

$$D_4(\mu) = \frac{1}{N-1} \sum_{k=1}^{N-1} \sqrt{T_4}. \tag{11}$$

4 DCNN Fusion Based on Fuzzy Integration

The focus of this chapter is fusing different state-of-the-art DCNN architectures. However, the procedures outlined are applicable to other neural inputs (see Fig. 2).

4.1 DCNN Architectures Used for Fusion

The first NN used herein for fusion is CaffeNet [35], which is a derivative of AlexNet with similar structure, except that the order of pooling and normalization is reversed to

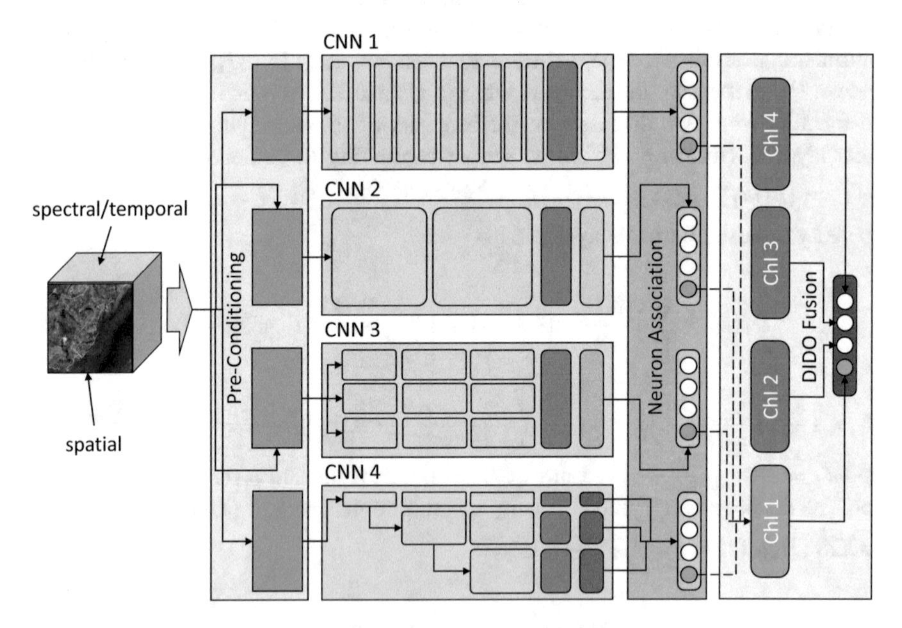

Fig. 2 Illustration of DIDO DCNN fusion. Note, many possibilities exist; e.g., variations in architecture, pre-conditioning/transforms (e.g., conversion to frequency analysis versus spatial domain, band selection or grouping, etc.), training data, etc. Next, neuron mapping/association is required followed by aggregation. Herein, a different fusion operator is learned per output neuron (versus shared fusions/weights)

reduce learnable parameters. CaffeNet contains five convolutional feature extraction steps and three fully connected layers for classification. Classification is performed with two fully connected inner product layers and a final soft-max layer for the network output. The output of soft-max layer is effectively a classification vector. CaffeNet represents the most simple and shallow of our DL investigated herein.

GoogLeNet [17] is a much deeper NN than CaffeNet–it has 27 parameterized layers. Because of this network depth, GoogLeNet has three classification outputs at various stages of the network to facilitate error back propagation. GoogLeNet's novel *inception layer* processes the input with max-pooling, 1×1, 3×3, and 5×5 convolutions simultaneously in a feature extraction step, and the outputs are concatenated as the layer output to achieve a multi-scale feature extraction. Using multiple convolutions at each stage follows the intuition that features from different kernel scales can be extracted and processed at the same time, thereby extracting multi-scale visual features. GoogLeNet is from a family of networks commonly referred to as *inception networks*.

ResNet [88] is a collection of DCNN architectures inspired by VGGNet [33]. In both ResNet and VGGNet, the primary kernels used to construct the convolution layers are 3×3. The architecture design incorporates the following rules to govern their structure. First, if the output of the feature map is the same, then the same number of 3×3 convolutional layers will be used. Second, if the output of the feature map is halved, then it will use twice as many 3×3 convolutional kernels The ResNet architectures employ residual connections that bypass two or more convolution layers at a time, allowing error to better propagate backward through the network. These are commonly referred to as *residual networks*, and here the ResNet50 and ResNet101 architectures are used within our experimental design. These networks have 50 and 101 feature extraction steps, respectively.

4.2 Transfer Learning, Neuron Association and Conditioning

If we design a set of custom DCNNs then it is trivial to ensure a bijection (one-to-one and onto) output neuron mapping. However, if existing community pretrained DCNNs (GoogLeNet, AlexNet, etc.) are leveraged–a task encountered frequently in practice–then this is not guaranteed. One way to resolve the one-to-one mapping task is to replace and retrain the DCNN classification layers per the labels for the task at hand. This is a type of transfer learning that keeps the feature layers intact. In [89], we (i) replaced and retrained the classification layers and we also (ii) updated the feature weights (e.g., convolution layers). Thus, we built custom classifiers for remote sensing of aerial imagery based on a network initialized by ground-perspective RGB imagery. In addition, data augmentation via rotation and image flipping was applied as well. However, we remark that other avenues exist; e.g., one could manually resolve the mapping or use an automated method based on an ontology. Regardless, using multiple custom or pretrained networks of different architectures raises another question; are the outputs numerically (e.g., all in [0, 1]) and semantically "to scale" (e.g., does a (e.g., $a = 0.5$) in domain i map to a in the other domains). One way to mitigate this issue in practice is to add a soft max normalization (aka normalized exponent function) layer after the raw neuronal output layer. For example, if η is the soft max output for neuron o_j then the soft max function is $\eta(o_j) = \frac{e^{o_j}}{\sum_{n=1}^{N} e^{o_n}}$. Thereby, we bound the domain of input for the subsequent fusion layer of our pattern recognition system, ensure the data across networks and neurons is well conditioned.

4.3 Imputation: λ-FM ChI

The first fusion approach explored here is to exploit our knowledge about the performance of the individual DCNNs on training data [54, 55]. A classical approach to obtaining the remaining $2^N - 2 - N$ FM values (beyond the densities) is the Sugeno λ-measure. For sets $A, B \subseteq X$, such that $A \cap B = \phi$,

$$\mu_\lambda(A \cup B) = \mu_\lambda(A) + \mu_\lambda(B) + \lambda\mu_\lambda(A)\mu_\lambda(B), \tag{12}$$

for some $\lambda > -1$. In particular, Sugeno showed that λ can be found by solving

$$\lambda + 1 = \prod_{i=1}^{N} (1 + \lambda\mu(x_i)), \lambda > -1, \tag{13}$$

where there exists exactly one real solution such that $\lambda > -1$. Some advantages of the Sugeno λ-measure include its simplicity, the N densities can be more tractable to acquire, fewer number of parameters can help address overfitting (versus using the full 2^N variables), and it is a probability measure when $\lambda = 0$. However, there is no guarantee in practice that the values that it imputes are what we actually need. Simply speaking, more information or a different imputation formula may be required; e.g., the S-Decomposable imputation formula, $\mu(A) = \bigvee_{i \in A}(\mu(x_i))$ (where $\bigvee$ is a t-conorm). Algorithm (1) describes how to use the Sugeno λ-measure to fuse a set of pretrained DLs based on individual performance for density.

Algorithm 1 λ-FM Based Imputation of ChI from Pre-Trained DCNNs

INPUT: DL_i - N DCNNs (B neurons each); $\bar{O}$ - labeled training data
1. Run each DCNN on $\bar{O}$, get *overall accuries* (OA); $a_{b,i} \in [0, 1]$ (i.e., performance of DL i on class b)
2. Assign the ith density its corresponding OA; i.e., $\mu_{\lambda_b}(x_i) = a_{b,i}$
3. Find λ_b (using $\{\mu_{\lambda_b}(\{x_1\}), ..., \mu_{\lambda_b}(\{x_N\})\}$) for Sugeno λ-FM (solve Eq. (13))
4. Recursively calculate $\mu_{\lambda_b}(A)$, $\forall A \in 2^X \setminus \{\{x_1\}, ..., \{x_N\}\}$, using the densities and λ_b (Eq. (12))
OUTPUT: B full fuzzy measures - $\{\mu_{\lambda_1}, ..., \mu_{\lambda_B}\}$

4.4 Optimization Approach: Learning the Full ChI

As stated in Sect. 4.3, there is no guarantee that imputation from densities results in the input interactions that we desire (and thus results in an appropriate aggregation operator). Algorithm (2) shows how to use quadratic programming for acquisition of the full FM for DIDO fusion of DCNNs (Algorithm (3) is how to learn a single "shared" FM to be applied to all neurons). Thus, training data is directly used to learn these crucial interactions–which means better selection of appropriate aggregation operator. However, as we discuss in [80], this process can lead to a big boost in performance but it is not without flaw. Specifically, in [80] we show that training data only typically supports a subset of FM variables. In return, we put forth an extended optimization of the ChI by (1) identifying which variables are supported by data, (2) optimizing just those variables and then (3) looking at imputation methods to infer the value of data unsupported variables based on application specific criteria. We do not have space to go into depth about the extension here, the reader can refer to [80] for full details.

Algorithm 2 Learn a Full FM/ChI Per Class for a Set of Pre-Trained DCNNs

INPUT: DL_i - N DCNNs (B neurons each); $\bar{O}$ - training data; β - regularization
1. Per class/output neuron (b), run each instance ($1 \leq j \leq |\bar{O}|$) through each DCNN (i); get $h_j^b(x_i)$ terms
2. Per neuron (b), construct the individual D_b from the $h_j^b(x_i)$ terms
3. Run B independent QPs (on the D_b respectively); yielding $\{\mu_1, ..., \mu_B\}$
OUTPUT: B full fuzzy measures - $\{\mu_1, ..., \mu_B\}$

Algorithm 3 Learn a Single "Shared Weight" Full FM/ChI for Pre-Trained DCNNs

INPUT: DL_i - N DCNNs (B neurons each); $\bar{O}$ - training data; β - regularization
1. Per class/output neuron (b), run each instance ($1 \leq j \leq |\bar{O}|$) through each DCNN (i); get $h_j^b(x_i)$ terms
2. Per neuron (b), construct the individual D_b from the $h_j^b(x_i)$ terms
3. Use QP to solve $\left(||D_1\mathbf{u} - \mathbf{y_1}||_2^2 + ... + ||D_B\mathbf{u} - \mathbf{y_B}||_2^2 + \beta v(\mathbf{u})\right)$; yields μ
OUTPUT: Full fuzzy measure - μ

5 Experiments

In this chapter, two benchmark remote sensing datasets suitable for classification tasks of objects or land-cover/land-use are used. Remote sensing data represents a significant pattern recognition challenge. As can be seen in Figs. 3 and 9 below, the variability and complexity of overhead imagery is immense. The visual cues exist at multiple levels: fine-scale (e.g. airplane shapes, vehicle presence, etc.) to

Fig. 3 Sample image chips from the 21 class UCM benchmark dataset, each 256×256 pixels approximately 0.3m *ground sampling distance* (GSD) spatial resolution. Classes in left-to-right, top-down order: 1 *agricultural*, 2 *airplane*, 3 *baseball diamond*, 4 *beach*, 5 *buildings*, 6 *chaparral*, 7 *dense residential*, 8 *forest*, 9 *freeway*, 10 *golf course*, 11 *harbor*, 12 *intersection*, 13 *medium residential*, 14 *mobile home park*, 15 *overpass*, 16 *parking lot*, 17 *river*, 18 *runway*, 19 *sparse residential*, 20 *storage tanks*, and 21 *tennis court*. In Sect. 5.1, neuron indices are used instead of text descriptions for sake of compactness

large-scale (e.g., road way configurations in overpasses versus intersections versus freeway). In fact the entire field of photo-interpretation revolves around developing human expertise in this pattern recognition task. For each of the datasets herein DCNNs were trained using the techniques in [89], including transfer learning and data augmentation via rotation and flipping. The trained DCNNs are then used in a locked state, i.e., no further learning happens in DL during the fusion stage. The training of the DCNNs are done in five-fold, cross validation manner; such that we have 5 sets of 80% training and 20% testing for both datasets. Per DCNN fold, three-fold CV fusion is used (due to limited data).

5.1 UC Merced (UCM) Dataset

The UC Merced (UCM) benchmark dataset [90, 91] has been used in a wide range of remote sensing research, including prior work in classification of objects and land-cover such as [55, 89, 92]. Figure 3 shows exemplar image chips from each class of the UCM dataset. The dataset includes 21 classes that are a mix of objects (airplane, baseball diamond, etc.) and landcover (beach, chaparral, etc.). We see that some classes, e.g., harbor and parking lot, are complex compositions of sub-entities (boats and vehicles); while others are general structural patterns of shapes (e.g., intersection and baseball diamonds).

Table 1 is the result of fusion on the UCM dataset. First, we see that aggregation outperforms no aggregation (i.e., the individual DCNNs) in four out of five folds.

Table 1 Fusion results for the UCM dataset

	Method									
	ChI Per Neuron	ChI Shared	SLFM Shared	CNet	GNet	RNet 50	RNet 101	Max	Avg	Min
Fold 1	0.979	0.977	0.984	0.957	0.957	0.985	0.973	0.978	0.981	0.976
Fold 2	0.991	0.994	0.993	0.964	0.983	0.978	0.988	0.993	0.994	0.993
Fold 3	0.994	0.990	0.996	0.971	0.985	0.992	0.988	0.996	0.996	0.998
Fold 4	0.992	0.996	0.996	0.988	0.980	0.983	0.988	0.996	0.992	0.998
Fold 5	0.989	0.985	0.989	0.976	0.973	0.983	0.980	0.989	0.989	0.986

	Max	Min	Mean	OWA
1	0.26	0.14	0.29	0.24
2	0.17	0.20	0.10	0.00
3	0.23	0.29	0.43	0.52
4	0.17	0.20	0.10	0.00
5	0.25	0.19	0.34	0.32
6	0.17	0.20	0.10	0.00
7	0.12	0.33	0.35	0.29
8	0.23	0.29	0.43	0.51
9	0.24	0.21	0.39	0.35
10	0.20	0.19	0.27	0.14
11	0.24	0.16	0.24	0.23
12	0.21	0.20	0.29	0.27
13	0.23	0.27	0.41	0.49
14	0.23	0.16	0.29	0.20
15	0.26	0.16	0.34	0.29
16	0.25	0.15	0.24	0.23
17	0.18	0.23	0.32	0.16
18	0.15	0.31	0.31	0.31
19	0.25	0.16	0.34	0.25
20	0.17	0.23	0.28	0.15
21	0.21	0.16	0.18	0.12

Fig. 4 Color coded matrix showing the distances obtained using the four reported indices of introspection ($D_1(\mu)$ to $D_4(\mu)$) relative to the learned full ChI per neuron on fold 1 of the UCM dataset. y-axis is the neuron index (see Fig. 3) and x-axis is the distance measure. Neurons two, four and six are OWA operators (but not min, max or mean like)

Second, we see that min, max and average (basic aggregation operators) do well in comparison to the ChI. However, these three operators are specific instances of the ChI, which informs us that there are challenges with variety and thus generalizability of this particular data set (otherwise they should have been selected). Next, it is interesting to see that the shared weight fusion solutions do as well as they do. It is our suspicion–something to be explored in future work–that a shared FM for the ChI helps combat overfitting. It is also our suspicion–again, subject of future work–that while the Sugeno λ-FM would not be our first choice, it might also help combat overfitting as it has just N parameters versus the otherwise $2^N - 1$. However, the performance of the individual DCNNs (which were used as the densities) are so high that ultimately this forces the Sugeno λ-FM to more-or-less be the maximum operator.

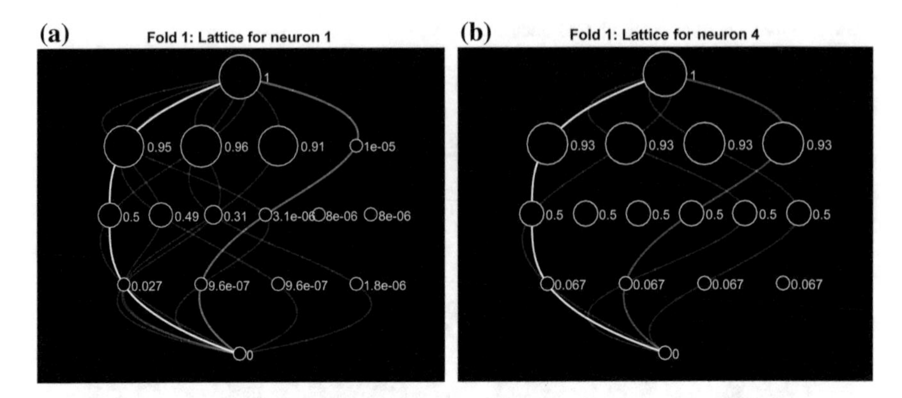

Fig. 5 Example of two full FMs for the **a** first and **b** fourth neuron in fold 1 of the UCM dataset. "Layer" l (from bottom to top) in the image denotes FM variables with cardinally l. Thus, layer 0 (bottom node) is the empty set, the next layer is the singletons, top is $\mu(X)$, etc. Each variable is presented in lexicographic order, i.e., layer 2 is $\{x_1, x_2\}$, $\{x_1, x_3\}$, $\{x_1, x_4\}$, $\{x_2, x_3\}$, $\{x_2, x_4\}$ and $\{x_3, x_4\}$. The nodes are also drawn size-wise proportional to their value (a minimum size and maximum was specified to make them still show up for 0 valued variables). In addition, the "paths" drawn indicate the visitation frequency (the brighter the line, the higher the visitation) for the test data in fold 1. Furthermore, the fourth neuron learned an OWA with weights $(0.067, 0.433, 0.43, 0.07)^t$– a trimmed mean operator. Conversely, neuron one is more complex to decode. It does not reduce into a single compact description like an OWA. However, we can view it in terms of the $N!$ walks (possible sorts). Since the $h(\{x_1\}) \geq h(\{x_2\}) \geq h(\{x_3\}) \geq h(\{x_4\})$ is encountered frequently, we decode and analyze its weights. The linear convex sum weights for the ChI of this walk (sorting) are $(0.027, 0.473, 0.45, 0.05)$ respectively. Thus, it is a weighted average of GoogLeNet and ResNet50. This analytic process can be repeated for the other $N! - 1$ walks if desired

Next, Fig. 4 gives us a feel for what type of aggregation strategy is being used for the 21 classes. Again, the max, min and mean are all OWAs, so we can start first with analyzing column four. There are three neurons (2, 4 and 6–i.e., airplane, beach and chaparral) that learned an OWA. The other neurons have learned something more unique, which helps justify the inclusion of the ChI versus say a simpler operator (see Fig. 5(a)). At that, none of the learned OWAs are that similar to our extreme markers of max (a t-conorm or union like operator), min (a t-conorm or intersection like operator) or average (an expected value like operator). For example, Fig. 5(b) shows one of these OWA operators, which breaks down into a trimmed mean operator.

Last, Fig. 6 shows the FM and Shapley values. While it is more-or-less impossible to read individual values in these plots, they show that there is no consensus in values nor importance of DCNNs. Meaning, different output neurons (classes) appears to use these different DCNNs in different ways. Furthermore, Fig. 7 shows the corresponding interaction index values. These values also reinforce the complex interplay and back-and-forth exchange of complementary, independent and redundant information between DCNNs across output neurons (classes). In total, the combination of

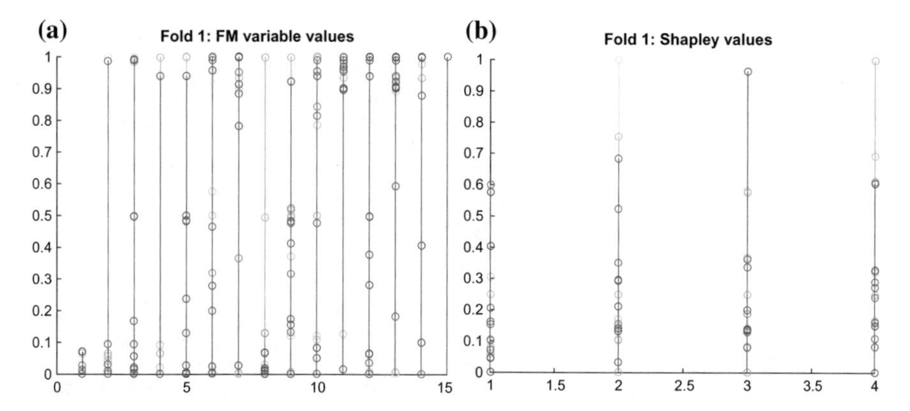

Fig. 6 Learned full ChI per neuron on fold 1 of the UCM dataset. **a** Plot of the $2^4 - 1$ binary encoded FM variables, i.e., for $N = 3$ the order is ($x1$, $x2$, $x12$, $x3$, $x13$, $x23$, $x123$, $x4$, $x14$, $x24$, $x124$, $x34$, $x134$, $x234$, $x1234$). This plot shows the agreement/disagreement of variable values across the 21 neurons. If all neurons required the same fusion then each x-axis location would have a single convergent set of circles (FM variable values). However, we can see that each x-axis location (FM variable) has for the most part significant variation (outside the CaffeNet density). **b** Plot of the 4 neuron Shapley index values across the 21 neurons. Again, this plot shows the variety of values learned. With respect to individual output neurons, some NNs could be eliminated. However, across the 21 neurons, no NN can be eliminated (we would expect to see approximately all zero values for that Shapley if so)

analysis of underlying aggregation function, importance of individual DCNNs and their pair-wise interaction behavior help the claim that performance appears to be improving due to diversity in the way these DCNNs operate. This is in line with our intuition about these DCNNs based on the ways their architectures were created.

Last, Fig. 8 shows example images *missed* by our fusion approach. As the reader can visually verify, these examples are extreme and represent incorrectly labeled or fundamentally ambiguous labels. We would not expect fusion to be able to fix this type of problem. At that, it is hard to say that the DCNNs should have got these, as a human might just as well mistaken them.

5.2 WHU-RS19 (RSD) Dataset

The WHU-RS19 (RSD) dataset is composed of 600×600 pixel, JPEG compressed images [93]. This class includes 19 classes, and approximately 50 chips per class. This imagery was screen scrapped from Google Earth, and therefore they are of variable spatial resolutions. Figure 9 shows exemplar image chips from each class of the RSD benchmark dataset. Similar to the UCM dataset, this dataset is a mixture of landcover and objects within the image chips. Table 2 shows the result of fusion, Fig. 10 are the indices of introspection, Fig. 11 are example lattices, Fig. 12 are the FM and the Shapley indices and Fig. 13 are example interaction indices. Overall, we see

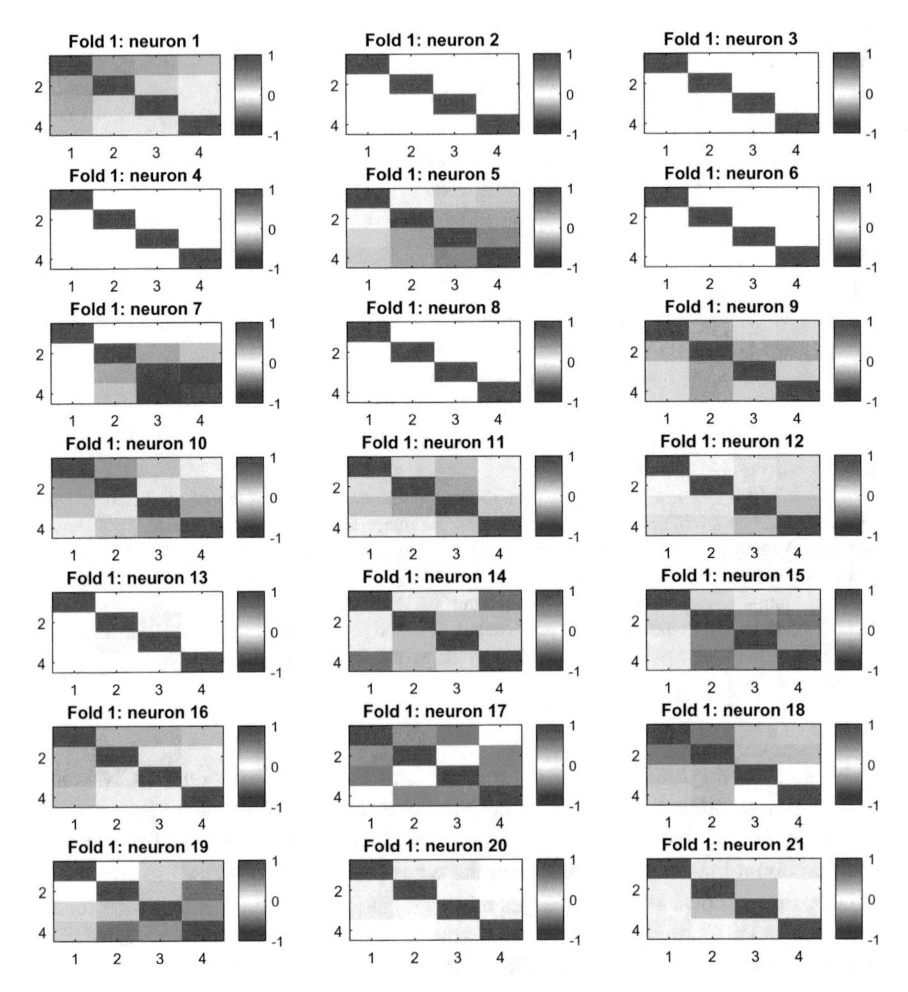

Fig. 7 Interaction index values for the learned full ChI per neuron on fold 1 of the UCM dataset. Index 1 is CaffeNet, 2 is GoogLeNet, 3 is ResNet50 and 4 is ResNet101. Consider neuron 1. CaffeNet has positive interactions (complementary information) with the other three NNs (0.37, 0.34 and 0.3 respectively). On the other hand, GoogLeNet has negative interaction values (redundancy) with the ResNet NNs (–0.19 and –0.1 respectively). The two ResNet NNs have a negative interaction index of –0.12. Also, in neuron 7, CaffeNet has approximately a zero interaction index with the other NNs (independence), whereas GoogLeNet has a value of –0.29 with ResNet50 and a positive interaction value of 0.22 with ResNet101. Last, ResNet50 and ResNet101 have a large negative interaction index of –0.72 with each other

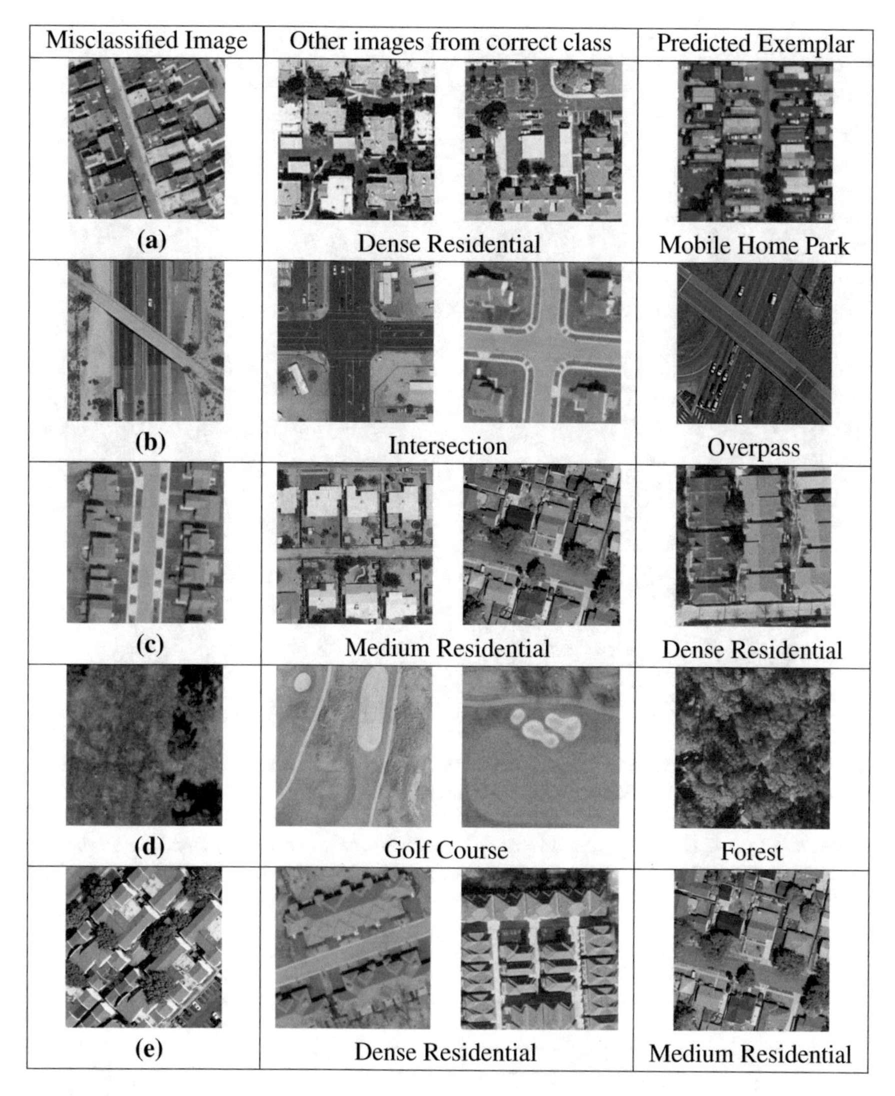

Fig. 8 Five images *missed* by the fusion framework; **a** dense residential misclassified as mobile home park, **b** (incorrectly labeled) intersection misclassified as overpass (correct label), **c** medium residential misclassified as dense residential, **d** (incorrectly labeled) golf course misclassified as forest, and **e** dense residential misclassified as medium residential

Fig. 9 Sample image chips from the 19 class RSD benchmark dataset, each 600×600 pixels of various spatial resolution. Classes in left-to-right, top-down order: 1 *airport*, 2 *beach*, 3 *bridge*, 4 *commercial area*, 5 *desert*, 6 *farmland*, 7 *football field*, 8 *forest*, 9 *industrial area*, 10 *meadow*, 11 *mountain*, 12 *park*, 13 *parking lot*, 14 *pond*, 15 *port*, 16 *railway station*, 17 *residential area*, 18 *river*, and 19 *viaduct*. In Sect. 5.2, neuron indices are used instead of text descriptions for sake of compactness

Table 2 Fusion results for the RSD dataset

	Method								
	ChI Per Neuron	ChI Shared	SLFM Shared	CNet	GNet	RNet50	Max	Avg	Min
Fold 1	0.989	0.991	0.991	0.982	0.977	0.988	0.991	0.991	0.991
Fold 2	0.992	0.984	0.992	0.978	0.994	0.989	0.987	0.992	0.987
Fold 3	0.984	0.992	0.979	0.955	0.988	0.966	0.979	0.979	0.979
Fold4	0.983	0.983	0.983	0.983	0.960	0.971	0.983	0.988	0.987
Fold 5	0.998	1.00	1.00	0.977	0.994	0.994	1.00	1.00	1.00

the same general trend (as the UCM dataset). Namely, (i) aggregation outperforms no aggregation in general and (ii) there are challenges with variety (and therefore generalizability) in the RSD data set as well.

	Max	Min	Mean	OWA
1	0.20	0.17	0.24	0.26
2	0.16	0.18	0.22	0.21
3	0.24	0.19	0.44	0.58
4	0.10	0.21	0.37	0.24
5	0.18	0.15	0.09	0.00
6	0.18	0.15	0.09	0.00
7	0.20	0.16	0.16	0.12
8	0.29	0.14	0.34	0.25
9	0.15	0.23	0.39	0.29
10	0.18	0.15	0.09	0.00
11	0.15	0.21	0.32	0.34
12	0.27	0.17	0.35	0.28
13	0.33	0.11	0.43	0.28
14	0.18	0.19	0.33	0.45
15	0.15	0.23	0.39	0.29
16	0.19	0.18	0.23	0.23
17	0.29	0.15	0.34	0.26
18	0.21	0.18	0.26	0.16
19	0.26	0.13	0.23	0.24

Fig. 10 Color coded matrix showing the distances obtained using the four reported indices of introspection ($D_1(\mu)$ to $D_4(\mu)$) relative to the learned full ChI per neuron on fold 1 of the RSD dataset. y-axis is the neuron index (see Fig. 9) and x-axis is the distance measure

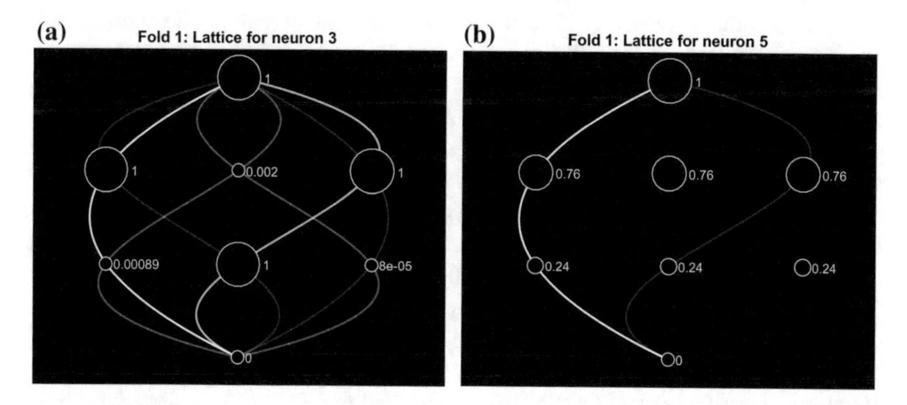

Fig. 11 Example FMs for fold 1 of the RSD dataset. Neuron three is for all intents and purposes a binary FM (see [94] for a formal characterization of binary FMs, the resultant FI and efficient ways of representing and learning such a function). For binary FMs, the Sugeno FI and the ChI are mathematically equivalent [94]. The FI is acting like a "dynamic maximum operator" with respect to FM variables that have a value one–or conversely a "dynamic minimum" with respect to zero valued FM variables. For example, if $h(\{x_1\}) \geq h(\{x_2\}) \geq h(\{x_3\})$ (aka CaffeNet is more confident than GoogLeNet followed by ResNet) then we take the output of GoogLeNet. However, if $h(\{x_2\})$ (GoogLeNet) is the most confident then we take its input. This reasoning can be followed to get similar stories for the other $N! - 2$ walks. Another interesting observation of neuron 3, versus neuron 5, is a slightly more diverse visitation (walk) pattern

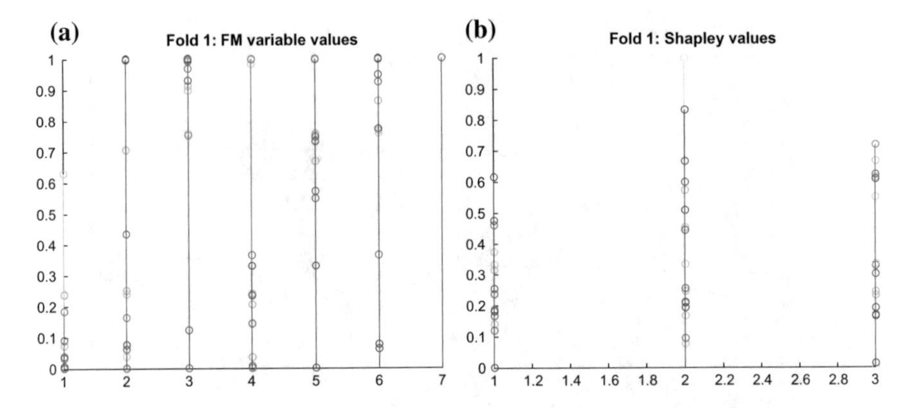

Fig. 12 Binary encoded $2^3 - 1$ FM variables and the 3 Shapley index values for the nineteen output neurons in the RSD dataset. As demonstrated in the UCM dataset, great variability exists in FM variable and Shapley values for these nineteen output neurons

6 Conclusion and Future Work

In summary, this chapter outlined a data-driven method for optimizing Choquet integral-based fusion of heterogeneous deep convolutional neural networks for pattern recognition in remotely sensed data. To the best of our knowledge, no one has previously learned the full fuzzy Choquet integral for fusing neural networks, just density-based fuzzy measures. This chapter brought together state-of-the-art advancements in two important parts of computational intelligence; fuzzy set theory and neural networks. Specifically, CaffeNet, GoogLeNet, ResNet50 and ResNet101 were fused at the per-output-neuron and with respect to a single "shared weight" solution. In a strive for explainable AI, versus black box solutions, different indices of introspection of the Choquet integral and information theoretic indices of the fuzzy measure were highlighted for analysis of the final deep learning solution. These indices showed us that there does appear to be diversity in these different heterogeneous DCNNs. Two benchmark remote sensing datasets were used, UCM and RSD, and our fused results showed improvement over the individual deep learners. However, this data set and DCNNs were saturated and therefore limited data (both volume and variety) existed for training fusion. Last, analysis of mislabeled imagery from fusion revealed incorrectly labeled data and ambiguous image chips that would lead to a human mislabeling imagery.

While encouraging, more research (theory and application) is needed. In future work, we will migrate our Choquet integral solution into a strictly neural representation for optimization and speed. Furthermore, we will move away from DIDO and explore fusion neurons at various layers in the network. We will also investigate what types of neural inputs should be fed to DIDO fusion; e.g., combinations of deep and shallow, different convolution map scales, etc. Future work will also include simultaneously learning the DCNNs and our fusion operators (they are learned inde-

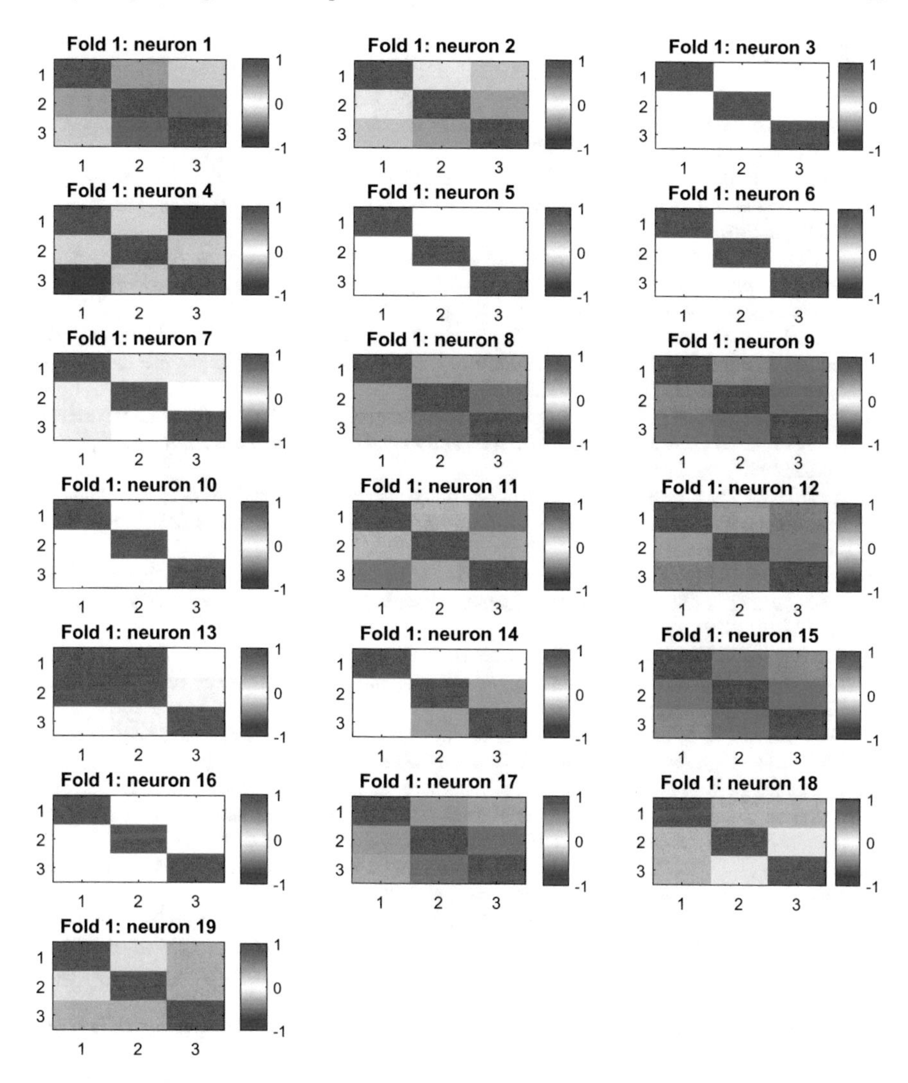

Fig. 13 Interaction index values for nineteen outputs in the RSD dataset. Index 1 is CaffeNet, 2 is GoogLeNet and 3 is ResNet50

pendently herein). Last, we will look to use our explainable AI methods to make improvements to the fusion and DCNNs, manually as well as possibly using them directly computationally to promote diversity and/or aid in the design of our networks.

References

1. W.S. McCulloch, W. Pitts, A logical calculus of the ideas immanent in nervous activity. Bull. Math. Biophys. **5**(4), 115–133 (1943)
2. L.A. Zadeh, Fuzzy sets. Inf. Control **8**(3), 338–353 (1965)
3. J.H. Holland, *Adaptation in Natural and Artificial Systems, 1992* (Ann Arbor, University of Michigan Press, MI, 1975)
4. R. Collobert, J. Weston, A unified architecture for natural language processing: deep neural networks with multitask learning, in *Proceedings of the 25th International Conference on Machine Learning* (ACM, New York, 2008), pp. 160–167
5. R. Socher, C.C. Lin, C. Manning, A.Y. Ng, Parsing natural scenes and natural language with recursive neural networks, in *Proceedings of the 28th International Conference on Machine Learning (ICML-11)* (2011), pp. 129–136
6. K. Fukushima, S. Miyake, Neocognitron: a self-organizing neural network model for a mechanism of visual pattern recognition, in *Competition and Cooperation in Neural Nets* (Springer, Berlin, 1982), pp. 267–285
7. A. Krizhevsky, I. Sutskever, G.E. Hinton, Imagenet classification with deep convolutional neural networks, in *Advances in Neural Information Processing Systems* (2012), pp. 1097–1105
8. D. Ciregan, U. Meier, J. Schmidhuber, Multi-column deep neural networks for image classification, in *2012 IEEE Conference on Computer Vision and Pattern Recognition (CVPR)* (IEEE, New York, 2012), pp. 3642–3649
9. C. Szegedy, V. Vanhoucke, S. Ioffe, J. Shlens, Z. Wojna, Rethinking the inception architecture for computer vision, in *Proceedings of the IEEE Conference on Computer Vision and Pattern Recognition* (2016), pp. 2818–2826
10. D.C. Ciresan, U. Meier, L.M. Gambardella, J. Schmidhuber, Deep, big, simple neural nets for handwritten digit recognition. Neural Comput. **22**(12), 3207–3220 (2010)
11. C. Bentes, D. Velotto, S. Lehner, Target classification in oceanographic sar images with deep neural networks: architecture and initial results, in *2015 IEEE International Geoscience and Remote Sensing Symposium (IGARSS)* IEEE, New York, 2015), pp. 3703–3706
12. W. Huang, L. Xiao, Z. Wei, H. Liu, S. Tang, A new pan-sharpening method with deep neural networks. IEEE Geosci. Remote Sens. Lett. **12**(5), 1037–1041 (2015)
13. X. Chen, S. Xiang, C.L. Liu, C.H. Pan, Vehicle detection in satellite images by hybrid deep convolutional neural networks. IEEE Geosci. Remote Sens. Lett. **11**(10), 1797–1801 (2014)
14. J. Yue, W. Zhao, S. Mao, H. Liu, Spectral-spatial classification of hyperspectral images using deep convolutional neural networks. Remote Sens. Lett. **6**(6), 468–477 (2015)
15. A.N. Steinberg, C.L. Bowman, F.E. White, Revisions to the JDL data fusion model, in *Handbook of Data Fusion* (1999)
16. M.D. Zeiler, R. Fergus, Visualizing and understanding convolutional networks, in *European Conference on Computer Vision* (Springer, Berlin, 2014), pp. 818–833
17. C. Szegedy, W. Liu, Y. Jia, P. Sermanet, S. Reed, D. Anguelov, D. Erhan, V. Vanhoucke, A. Rabinovich, Going deeper with convolutions, in *Proceedings of the IEEE Conference on Computer Vision and Pattern Recognition* (2015), pp. 1–9
18. G.E. Hinton, R.R. Salakhutdinov, Reducing the dimensionality of data with neural networks. Science **313**(5786), 504–507 (2006)
19. P. Vincent, H. Larochelle, Y. Bengio, P.-A. Manzagol, Extracting and composing robust features with denoising autoencoders, in *Proceedings of the 25th International Conference on Machine learning* (ACM, New York, 2008), pp. 1096–1103
20. M. Chen, Z. Xu, K. Weinberger, F. Sha, *Marginalized denoising autoencoders for domain adaptation* (2012). arXiv preprint arXiv:1206.4683
21. Q. Fu, X. Yu, X. Wei, Z. Xue, Semi-supervised classification of hyperspectral imagery based on stacked autoencoders, in *Eighth International Conference on Digital Image Processing (ICDIP 2016)*, 100332B-100332B. International Society for Optics and Photonics (2016)

22. J. Geng, J. Fan, H. Wang, X. Ma, B. Li, F. Chen, High-resolution sar image classification via deep convolutional autoencoders. IEEE Geosci. Remote Sens. Lett. **12**(11), 2351–2355 (2015)
23. G.E. Hinton, Deep belief networks. Scholarpedia **4**(5), 5947 (2009)
24. H. Lee, R. Grosse, R. Ranganath, A.Y. Ng, Convolutional deep belief networks for scalable unsupervised learning of hierarchical representations, in *Proceedings of the 26th Annual International Conference on Machine Learning* (ACM, New York, 2009), pp. 609–616
25. T. Mikolov, M. Karafiát, L. Burget, J. Cernock'y, S. Khudanpur, Recurrent neural network based language model, in *Interspeech*, vol. 2 (2010), 3 p
26. K. Funahashi, Y. Nakamura, Approximation of dynamical systems by continuous time recurrent neural networks. Neural Netw. **6**(6), 801–806 (1993)
27. S. Rajurkar, N.K. Verma, Developing deep fuzzy network with takagi sugeno fuzzy inference system, in *2017 IEEE International Conference on Fuzzy Systems (FUZZ-IEEE)* (2017), pp. 1–6. https://doi.org/10.1109/FUZZ-IEEE.2017.8015718
28. L. Xu, J.S. Ren, C. Liu, J. Jia, Deep convolutional neural network for image deconvolution, in *Advances in Neural Information Processing Systems* (2014), pp. 1790–1798
29. M.D. Zeiler, D. Krishnan, G.W. Taylor, R. Fergus, Deconvolutional networks, in *2010 IEEE Conference on Computer Vision and Pattern Recognition (CVPR)* (IEEE, New York, 2010), pp. 2528–2535
30. M.D. Zeiler, G.W. Taylor, R. Fergus, Adaptive deconvolutional networks for mid and high level feature learning, in *2011 IEEE International Conference on Computer Vision (ICCV)* (IEEE, New York, 2011), pp. 2018–2025
31. Y. Won, P.D. Gader, P.C. Coffield, Morphological shared-weight networks with applications to automatic target recognition. IEEE Trans. Neural Netw. **8**(5), 1195–1203 (1997)
32. X. Jin, C.H. Davis, Vehicle detection from high-resolution satellite imagery using morphological shared-weight neural networks. Image Vis. Comput. **25**(9), 1422–1431 (2007)
33. K. Simonyan, A. Zisserman, *Very Deep Convolutional Networks for Large-scale Image Recognition* (2014). arXiv preprint arXiv:1409.1556
34. M. Abadi, A. Agarwal, P. Barham, E. Brevdo, Z. Chen, C. Citro, G.S. Corrado, A. Davis, J. Dean, M. Devin, et al., *Tensorflow: Large-scale Machine Learning on Heterogeneous Distributed Systems* (2016). arXiv preprint arXiv:1603.04467
35. Y. Jia, E. Shelhamer, J. Donahue, S. Karayev, J. Long, R. Girshick, S. Guadarrama, T. Darrell, Caffe: convolutional architecture for fast feature embedding, in *Proceedings of the 22nd ACM International Conference on Multimedia* (ACM, New York, 2014), pp. 675–678
36. A. Vedaldi, K. Lenc, Matconvnet: convolutional neural networks for matlab, in *Proceedings of the 23rd ACM International Conference on Multimedia* (ACM, New York, 2015), pp. 689–692
37. J. Yosinski, J. Clune, Y. Bengio, H. Lipson, How transferable are features in deep neural networks? in *Advances in Neural Information Processing Systems* (2014), pp. 3320–3328
38. N. Srivastava, G.E. Hinton, A. Krizhevsky, I. Sutskever, R. Salakhutdinov, Dropout: a simple way to prevent neural networks from overfitting. J. Mach. Learn. Res. **15**(1), 1929–1958 (2014)
39. I. Goodfellow, Y. Bengio, A. Courville, *Deep Learning* (MIT Press, 2016)
40. S. Ioffe, C. Szegedy, Batch normalization: accelerating deep network training by reducing internal covariate shift, in *International Conference on Machine Learning* (2015), pp. 448–456
41. L. Brown, *Deep Learning with GPUs*, http://www.nvidia.com/content/events/geoInt2015/
42. L. Bottou, Stochastic gradient learning in neural networks. Proc. Neuro-Names **91**(8) (1991)
43. B.T. Polyak, Some methods of speeding up the convergence of iteration methods. USSR Comput. Math. Math. Phys. **4**(5), 1–17 (1964)
44. I. Sutskever, J. Martens, G. Dahl, G. Hinton, On the importance of initialization and momentum in deep learning, in *International Conference on Machine Learning* (2013), pp. 1139–1147
45. J. Duchi, E. Hazan, Y. Singer, Adaptive subgradient methods for online learning and stochastic optimization. J. Mach. Learn. Res. **12**(Jul), 2121–2159 (2011)
46. T. Tieleman, G. Hinton, Lecture 6.5-rmsprop: divide the gradient by a running average of its recent magnitude. Coursera: Neural Netw. Mach. Learn. **4**(2), 26–31 (2012)
47. D. Kingma, J. Ba, Adam: a method for stochastic optimization, in *3rd International Conference for Learning Representations* (2015)

48. J.E. Ball, D.T. Anderson, C.S. Chan, A comprehensive survey of deep learning in remote sensing: theories, tools and challenges for the community. J. Appl. Remote Sens. (2017)
49. S.K. Pal, S. Mitra, *Neuro-fuzzy Pattern Recognition: Methods in Soft Computing* (Wiley Inc, New Jersey, 1999)
50. J.M. Keller, D.J. Hunt, Incorporating fuzzy membership functions into the perceptron algorithm. IEEE Trans. Pattern Anal. Mach. Intell. **6**, 693–699 (1985)
51. R.R. Yager, Applications and extensions of owa aggregations. Int. J. Man Mach. Stud. **37**(1), 103–122 (1992)
52. R.R. Yager, On ordered weighted averaging aggregation operators in multicriteria decision-making. IEEE Trans. Syst. Man Cybern. **18**(1), 183–190 (1988)
53. C. Sung-Bae, Fuzzy aggregation of modular neural networks with ordered weighted averaging operators. English. Int. J. Approx. Reas. **13**(4), 359–375 (1995)
54. S.B. Cho, J.H. Kim, Combining multiple neural networks by fuzzy integral for robust classification. IEEE Trans. Syst. Man Cybern. **25**(2), 380–384 (1995)
55. G.J. Scott, R.A. Marcum, C.H. Davis, T.W. Nivin, Fusion of deep convolutional neural networks for land cover classification of high-resolution imagery. IEEE Geosci. Remote Sens. Lett. (2017)
56. S.R. Price, B. Murray, L. Hu, D.T. Anderson, T.C. Havens, R.H. Luke, J.M. Keller, Multiple kernel based feature and decision level fusion of IECO individuals for explosive hazard detection in flir imagery, in *SPIE*, vol. 9823 (2016), pp. 98231G-98231G-11. https://doi.org/10.1117/12.2223297
57. R.E. Smith, D.T. Anderson, A. Zare, J.E. Ball, B. Alvey, J.R. Fairley, S.E. Howington, Genetic programming based Choquet integral for multi-source fusion, in *IEEE International Conference on Fuzzy Systems (FUZZ-IEEE* (2017)
58. R.E. Smith, D.T. Anerson, J.E. Ball, A. Zare, B. Alvey, Aggregation of Choquet integrals in GPR and EMI for handheld platform-based explosive hazard detection, in *Proceedings of the SPIE 10182, Detection and Sensing of Mines, Explosive Objects, and Obscured Targets XXII* (2017)
59. H. Tahani, J. Keller, Information fusion in computer vision using the fuzzy integral. IEEE Trans. Syst. Man Cybern. **20**, 733–741 (1990)
60. M. Grabisch, J.-M. Nicolas, Classification by fuzzy integral: performance and tests. Fuzzy Sets Syst. **65**(2–3), 255–271 (1994)
61. M. Grabisch, M. Sugeno, Multi-attribute classification using fuzzy integral, in *IEEE International Conference on Fuzzy Systems, 1992* (IEEE, New York, 1992), pp. 47–54
62. A. Mendez-Vazquez, P. Gader, J.M. Keller, K. Chamberlin, Minimum classification error training for Choquet integrals with applications to landmine detection. IEEE Trans. Fuzzy Syst. **16**(1), 225–238 (2008). https://doi.org/10.1109/TFUZZ.2007.902024. ISSN: 1063-6706
63. J.M. Keller, P. Gader, H. Tahani, J. Chiang, M. Mohamed, Advances in fuzzy integration for pattern recognition. Fuzzy Sets Syst. **65**(2–3), 273–283 (1994)
64. P.D. Gader, J.M. Keller, B.N. Nelson, Recognition technology for the detection of buried land mines **9**(1), 31–43 (2001)
65. G.J. Scott, D.T. Anderson, Importance-weighted multi-scale texture and shape descriptor for object recognition in satellite imagery, in *2012 IEEE International Geoscience and Remote Sensing Symposium* (2012), pp. 79–82. https://doi.org/10.1109/IGARSS.2012.6351632
66. M. Grabisch, The application of fuzzy integrals in multicriteria decision making. Eur. J. Oper. Res. **89**(3), 445–456 (1996)
67. C. Labreuche, Construction of a Choquet integral and the value functions without any commensurateness assumption in multi-criteria decision making, in *EUSFLAT Conference* (2011), pp. 90–97
68. D.T. Anderson, P. Elmore, F. Petry, T.C. Havens, Fuzzy Choquet integration of homogeneous possibility and probability distributions. Inf. Sci. **363**, 24–39, (2016). https://doi.org/10.1016/j.ins.2016.04.043. http://www.sciencedirect.com/science/article/pii/S0020025516302961. ISSN: 0020-0255

69. D.T. Anderson, T.C. Havens, C. Wagner, J.M. Keller, M.F. Anderson, D.J. Wescott, Extension of the fuzzy integral for general fuzzy set-valued information **22**(6), 1625–1639, (2014). https://doi.org/10.1109/TFUZZ.2014.2302479. ISSN: 1063-6706
70. M. Anderson, D.T. Anderson, D.J. Wescott, Estimation of adult skeletal age-at-death using the sugeno fuzzy integral. Am. J. Phys. Anthropol. **142**(1), 30–41 (2010)
71. L. Tomlin, D.T. Anderson, C. Wagner, T.C. Havens, J.M. Keller, *Fuzzy integral for rule aggregation in fuzzy inference systems* (Springer International Publishing, Berlin, 2016), pp. 78–90. https://doi.org/10.1007/978-3-319-40596-4_8
72. A.J. Pinar, J. Rice, L. Hu, D.T. Anderson, T.C. Havens, Efficient multiple kernel classification using feature and decision level fusion. PP(99), 1 (2016). ISSN: 1063-6706. https://doi.org/10.1109/TFUZZ.2016.2633372
73. A. Pinar, T.C. Havens, D.T. Anderson, L. Hu, Feature and decision level fusion using multiple kernel learning and fuzzy integrals, in *2015 IEEE International Conference on Fuzzy Systems (FUZZIEEE)* (2015), pp. 1–7. https://doi.org/10.1109/FUZZ-IEEE.2015.7337934
74. L. Hu, D.T. Anderson, T.C. Havens, J.M. Keller, Efficient and scalable nonlinear multiple kernel aggregation using the choquet integral, in *Information Processing and Management of Uncertainty in Knowledge-Based Systems: 15th International Conference, IPMU, Montpellier, France, July 15–19, 2014, Proceedings. Part I* (Springer International Publishing, Berlin, 2014), pp. 206–215
75. L. Hu, D.T. Anderson, T.C. Havens, Multiple kernel aggregation using fuzzy integrals, in *2013 IEEE International Conference on Fuzzy Systems (FUZZ-IEEE)* (2013), pp. 1–7. https://doi.org/10.1109/FUZZ-IEEE.2013.6622312
76. X. Du, A. Zare, J.M. Keller, D.T. Anderson, Multiple instance Choquet integral for classifier fusion, in *2016 IEEE Congress on Evolutionary Computation (CEC)* (2016), pp. 1054–1061. https://doi.org/10.1109/CEC.2016.7743905
77. M. Al Boni, D.T. Anderson, R.L. King, Hybrid measure of agreement and expertise for ontology matching in lieu of a reference ontology. Int. J. Intell. Syst. **31**(5), 502–525 (2016). https://doi.org/10.1002/int.21792. ISSN: 1098-111X
78. M.A. Islam, D.T. Anderson, F. Petry, D. Smith, P. Elmore, The fuzzy integral for missing data, in *2017 IEEE International Conference on Fuzzy Systems (FUZZ-IEEE)* (2017), pp. 1–8. https://doi.org/10.1109/FUZZ-IEEE.2017.8015475
79. M. Sugeno, Theory of fuzzy integrals and its applications. Ph.D. thesis, Tokyo Institute of Technology, (1974)
80. M.A. Islam, D.T. Anderson, A.J. Pinar, T.C. Havens, Data-driven compression and efficient learning of the Choquet Integral. IEEE Trans. Fuzzy Syst. PP(99), 1 (2017). https://doi.org/10.1109/TFUZZ.2017.2755002. ISSN: 1063-6706
81. J.M. Keller, J. Osborn, Training the fuzzy integral. Int. J. Approx. Reas. **15**(1), 1–24 (1996)
82. D.T. Anderson, S.R. Price, T.C. Havens, Regularization-based learning of the Choquet integral, in *2014 IEEE International Conference on Fuzzy Systems (FUZZ-IEEE)* (2014), pp. 2519–2526. https://doi.org/10.1109/FUZZ-IEEE.2014.6891630
83. A.J. Pinar, D.T. Anderson, T.C. Havens, A. Zare, T. Adeyeba, Measures of the shapley index for learning lower complexity fuzzy integrals. Granul. Comput. 1–17 (2017)
84. T. Murofushi, S. Soneda, Techniques for reading fuzzy measures (iii): interaction index, in *9th Fuzzy System Symposium* (Sapporo, Japan, 1993)
85. M. Grabisch, M. Roubens, An axiomatic approach to the concept of interaction among players in cooperative games. Int. J. Game Theory **28**(4), 547–565 (1999)
86. M. Grabisch, An axiomatization of the shapley value and interaction index for games on lattices, in *SCIS-ISIS* (2004)
87. S.R. Price, D.T. Anderson, C. Wagner, T.C. Havens, J.M. Keller, *Indices for introspection on the Choquet integral, in Advance Trends in Soft Computing* (Springer, Berlin, 2014), pp. 261–271
88. K. He, X. Zhang, S. Ren, J. Sun, *Deep Residual Learning for Image Recognition* (2015). arXiv preprint arXiv:1512.03385
89. G.J. Scott, M.R. England, W.A. Starms, R.A. Marcum, C.H. Davis, Training deep convolutional neural networks for land-cover classification of high-resolution imagery. IEEE Geosci. Remote Sens. Lett. **14**(4), 549–553 (2017)

90. S.D. Newsam, *UC Merced Land Use Dataset* (2010), http://vision.ucmerced.edu/datasets/landuse.html
91. Y. Yang, S. Newsam, Bag-of-visual-words and spatial extensions for land-use classification, in *ACM SIGSPATIAL International Conference on Advances in Geographic Information Systems (ACM GIS)* (2010), 666 p
92. C. Chen, B. Zhang, H. Su, W. Li, L. Wang, Land-use scene classification using multi-scale completed local binary patterns. Signal Image Video Proc. **10**(4), 745–752 (2016)
93. D. Dai, W. Yang, Satellite image classification via two-layer sparse coding with biased image representation. IEEE Geosci. Remote Sens. Lett. **8**(1), 173–176 (2011)
94. D.T. Anderson, M. Islam, R. King, N.H. Younan, J.R. Fairley, S. Howington, F. Petry, P. Elmore, A. Zare, Binary fuzzy measures and Choquet integration for multi-source fusion, in *6th International Conference on Military Technologies* (2017)

Deep Neural Networks for Structured Data

Monica Bianchini, Giovanna Maria Dimitri, Marco Maggini and Franco Scarselli

Abstract Learning machines for pattern recognition, such as neural networks or support vector machines, are usually conceived to process real–valued vectors with predefined dimensionality even if, in many real–world applications, relevant information is inherently organized into entities and relationships between them. Instead, Graph Neural Networks (GNNs) can directly process structured data, guaranteeing universal approximation of many practically useful functions on graphs. GNNs, that do not strictly meet the definition of deep architectures, are based on the unfolding mechanism during learning, that, in practice, yields networks that have the same depth of the data structures they process. However, GNNs may be hindered by the long–term dependency problem, i.e. the difficulty in taking into account information coming from peripheral nodes within graphs — due to the local nature of the procedures for updating the state and the weights. To overcome this limitation, GNNs may be cascaded to form layered architectures, called Layered GNNs (LGNNs). Each GNN in the cascade is trained based on the original graph "enriched" with the information computed by the previous layer, to implement a sort of incremental learning framework, able to take into account progressively further information. The applicability of LGNNs will be illustrated both with respect to a classical problem in graph–theory and to pattern recognition problems in bioinformatics.

Keywords Graph neural networks · Deep neural networks · Protein structure prediction

M. Bianchini (✉) · M. Maggini · F. Scarselli
Department of Information Engineering and Mathematics,
University of Siena, Siena, Italy
e-mail: monica@diism.unisi.it

M. Maggini
e-mail: maggini@diism.unisi.it

F. Scarselli
e-mail: franco@diism.unisi.it

G. M. Dimitri
Department of Computer Science, University of Cambridge, Cambridge, UK
e-mail: gmd43@cam.ac.uk

W. Pedrycz and S.-M. Chen (eds.), *Computational Intelligence for Pattern Recognition*, Studies in Computational Intelligence 777,
https://doi.org/10.1007/978-3-319-89629-8_2

1 Introduction

The formal representation of objects is a key issue in pattern recognition problems. Actually there are two fundamental ways for implementing a pattern recognition system, namely using statistical or structural approaches. In the statistical framework, objects are described by feature vectors, whereas structural approaches exploit symbolic data structures, such as strings, trees, or graphs. Both approaches show pros and cons. If, in fact, evaluating the similarity between two entities is easily defined in vectorial spaces, and can be efficiently obtained, nevertheless this representation is not suited to explicitly describe relationships between the different subparts of objects. On the other hand, graphs can easily describe relations among subparts of complex data, but with a significant increase in computational complexity. For instance, the problem of evaluating the identity of two feature vectors has a linear complexity (w.r.t. the vector dimension), whereas testing the isomorphism between graphs is just an exponential problem [1].

Anyway, in several pattern recognition applications, data can actually be suitably represented in form of structures, f.i. in image processing [2] and bioinformatics [3]. Indeed, for these applications, information is inherently represented by atomic entities, sharing some common properties and dependent on each other, as described by relationships that encode their mutual influence and interactions. Contextual, hierarchical or causal connections between parts of a given pattern provide crucial information. The intrinsic subsymbolic nature of these tasks prevents a natural representation of data by vectors, since the feature extraction procedure is problem–dependent, heuristic, computationally expensive, and may also produce loss of information. The way in which structured data should be processed can neither be simply related to the "symbolic" information (normally collected in labels, describing the atomic entities) nor to the "subsymbolic" organization of entities. Hence, the processing scheme for these pattern representations should be designed in order to take into account both the entity labels and their relationships. In particular, when dealing with patterns encoded as general graphs, we may consider *graph–focused* and *node–focused* applications. In the graph–focused framework, the decision to be taken is related to the entire structure, while for node–focused problems, an output is expected for each node in the graph. For instance, a chemical compound can be modeled by a graph G, where nodes represent atoms or small molecules, whereas edges describe covalent bonds. The problem of estimating the probability that a molecule is mutagenic, i.e. may induce a particular disease [4], is graph–focused, since it is a property of the compound as a whole, not of a specific atom. Instead, the prediction of secondary structure elements and disordered regions (or loops) in proteins [5] is configured as a node–focused task. In computer vision, images can be modeled by Region Adjacency Graphs (RAGs) [6], with labeled nodes denoting homogeneous regions (from the visual point of view) and arcs defining their adjacency relationship. In this context, the localization and detection of an object is a node–focused task [7], while image classification is graph–focused [8].

Since the late '90s, different neural network models for processing graphs have been proposed, both for supervised and unsupervised tasks [9]. In Recursive Neural Networks (RNNs) [10], subsymbolic and symbolic information collected within graphs, and respectively related to the topological organization of nodes and to their labels, is encoded into a set of state variables, associated with each node. The states are updated dynamically following the graph topology, and used to calculate the outputs. RNN models are only able to process directed positional acyclic graphs (DPAGs) and are normally used to address graph–focused problems. Instead, Graph Neural Networks (GNNs) [11] process input data encoded as general labeled graphs. GNNs are trained based on a supervised learning algorithm that incorporates, into the error function, a criterion aimed at enforcing a contractive dynamics, to stabilize the learning procedure also for cyclic input graphs. Moreover, GNNs are able to naturally model both node–focused and graph–focused functions. Therefore, GNNs have recently been applied to many forefront applications. In particular, in bioinformatics, GNNs were employed for modelling quantitative structure–activity relationship problems, i.e. for the prediction of the mutagenicity and of the biodegradability of molecules [12]. Besides, GNNs were successfully exploited for object localization in images [13] and, with respect to web applications, for page ranking [14], sentence extraction [15], and document clustering and classification [16].

Unfortunately, all recursive models are plagued by the long–term dependency pathology, i.e. they struggle in processing deep structures, because of the local nature of the learning procedure. Actually, the error contribution, which vanishes during backpropagation, prevents a sensible update of the weights and, therefore, of the states, related to far nodes. In fact, practical issues are detected when dynamic neural network models are expected to learn tasks in which the relevant events in the input/output sequence span long intervals or, equivalently in the case of graphs, the dependencies involve nodes connected by a "long" path in the graph structure. This concept has been theoretically formalized in [17] in the case of recurrent networks, proving that the system is unable to latch temporal information robustly, since the gradient contribution due to information t time steps away vanishes as t increases. It is worth noting that the long–term dependency problem conflicts with the idea that deep networks, composed of a large number of layers, are necessary to cope with complex applications, being able to implement, with a reduced number of neurons, functions that cannot be realized with shallow architectures [18]. In fact, controlling the number of resources (in terms of neurons) needed to solve a given problem means also stemming the exponential growth in computational complexity (both in space and in time) of the learning procedure. In this perspective, many recent studies have been focused on how to train networks layer by layer, mixing unsupervised and supervised algorithms [19].

In this chapter, we describe a stacked architecture, referred to as Layered GNN (LGNN), that is composed of a cascade of GNNs, each of which takes in input the original graph and the information computed by the previous GNN in the cascade. LGNNs, that can be properly defined as deep architectures, are trained layer by layer, exploiting the provided targets and training each network in the cascade to correct the solution computed by the previous one. Basically, each GNN is expected to focus only on those patterns that were misclassified by the previous GNNs in the hierarchy, thus implementing a kind of incremental learning, that is progressively enriched by taking into account information from further nodes.

An experimental evaluation is reported, based on synthetic and real–world datasets. Synthetic data are used as a benchmark for a central task in graph theory, i.e. finding cliques — fully connected subgraphs — inside graphs. Instead, publicly available datasets come from the bioinformatic field and are related to the prediction of the secondary structure of proteins, and to the classification of chemical compounds with respect to mutagenicity. The reported results are promising and guarantee a significantly improved accuracy with respect to standard GNNs. Finally, as a pattern recognition application of the clique search problem, we describe how to model the protein surface in order to identify the onset of transient pockets — typically involved in protein–protein and protein–drug interactions — for drug design. Indeed, transient pockets are cavities that can appear and disappear on a protein surface, often containing active sites where pharmaceutical agents can be anchored. Standard methods for protein 3D structure prediction are unable to identify and localize transient pockets, due to their short persistence and complex pattern, whereas connectionist models seem to capture sufficient information in the vicinity of the putative pocket to predict its appearance (dimension and duration). In particular, in drug discovery, it often happens that several molecules of apparently unrelated structures are active on the same drug target. In fact, if the interatomic distances among a subset of atoms in a drug match the distances among a similar–sized subset in a second drug, a putative shared pharmacophore exists, which guarantees the interaction of both drugs with the target. Constructing a complete edge–weighted graph representing each drug (with atoms as vertices and edge weights that stand for interatomic distances), the problem of interest is reformulated as that of finding the Maximum Common Subgraph (MCS) between them, where the MCS is defined as the largest subset of atom pairs that have matching distances. To solve the MCS problem, couples of protein graphs are first converted to *correspondence* graphs, identifying all graph–to–graph pairs of elements with matching distances, on which the application of a clique–detection algorithm efficiently identifies the MCS. Recently, the described approach has been also extended to the identification of complementary surfaces on proteins, a core task within the problem of computational docking of biomolecules.

The chapter is organized as follows. Section 2 presents the GNN model with a detailed explanation of the related learning procedure, while Sect. 3 describes the Layered GNNs, showing some possible design choices for this architecture. Some applications of LGNNs to pattern recognition problems are illustrated in Sect. 4, together with some experimental results. Finally, conclusions and future works are reported in Sect. 5.

2 Graph Neural Networks

Complex patterns can be described by a structured representation encoded as a graph. In fact, graphs allow us to represent data as a set of atomic parts, the *nodes*, and binary relationships among them, the *arcs* or *edges*. For instance, an image can be represented as a set of regions. Each region corresponds to a node of the graph and the edge connecting two nodes may encode that the two corresponding regions are adjacent. In this case, the relationship between the two linked entities is symmetric and the corresponding link will not be oriented (usually in this case the term *edge* is used). Other type of relationships (for instance inclusion between two regions) may be asymmetrical and the corresponding link will be oriented (the term *arc* is preferable in this case).

Graph Neural Networks (GNNs) provide a computational scheme able to compute a *node–focused* function on a given graph. In particular, the output of the computation is, generally, a set of real valued vectors attached to the nodes of the input graph. For pattern recognition applications, like for instance classification, where the graph represents the object to be processed, a single output can be assigned to the graph by selecting a node, in order to implement a function $\tau(G, v)$ from the graph G to a vector in $\mathbb{R}^r$, given the node v in G. The function implemented by the GNN depends on parameters, the *weights*, that can be adapted by a supervised learning procedure.

2.1 Graphs

A graph is defined as a pair $G = (V, E)$, being V the set of nodes and $E \subseteq V \times V$ the set of edges. An edge is denoted by the pair of connected nodes as (u, v), where $u, v \in V$. If the graph is *undirected* the order of the two nodes in the pair is not considered, whereas for *directed* graphs the pair defines an arc where the first node is the source and the second node the destination of the link.

The pair (V, E) describes the pattern by representing its parts and its structure through the connection topology. The representation can be completed by adding attributes to each part (node) and, eventually, to the relationships (arcs or edges). For instance, when a node stands for a region in an image, perceptual or geometrical features of the region, such as its average color, extension, perimeter, etc., can be stored as attributes of the node. If an edge stands for the adjacency relationship, its attributes may correspond to the relative distance of the barycenters of the two regions, the relative orientation of the two shapes and so on [7]. The attributes attached to each node or edge/arc are referred to as a *label*. We can assume that the same label space L is used for all the nodes. In the following, we will consider only the case of graphs without labeled edges/arcs, but the definitions and computational models can be easily extended by introducing the edge/arc label space [20]. Since we examine neural network models to process graphs, we assume that the attributes are encoded as real values such that the label is a real vector, i.e. $\mathcal{L}(v) \in L \subset \mathbb{R}^q$. A labeled graph is defined as a triple $G_L = (V, E, \mathcal{L})$, being $\mathcal{L} : V \rightarrow L$ the node labeling function that allows us to retrieve the label $\mathcal{L}(v) \in L$ of the node $v \in G_L$. To simplify the notation, the label of node v will be directly referred to as $\mathbf{l}_v \in \mathbb{R}^q$.

Given a graph, we can define several properties that depend on the topology of its connections. In particular, in the following, we will consider the *neighbourhood* of a given node v as the set of nodes directly connected to it. Formally it can be defined as $\text{ne}[v] = \{u \in V | (u, v) \in E \vee (v, u) \in E\}$. For undirected graphs there is no need to distinguish between the pairs (u, v) and (v, u), whereas for directed graphs the concept of neighbourhood considers both incoming and outgoing arcs. In fact, in this latter case, the *parents* of a node v are defined as $\text{pa}[v] = \{u \in V | (u, v) \in E\}$ and the *children* of v as $\text{ch}[v] = \{u \in V | (v, u) \in E\}$. In this case, $\text{ne}[v] = \text{ch}[v] \bigcup \text{pa}[v]$ holds. In the definition of the GNN processing scheme we will refer to the general concept of neighbours of a node as described before, without making any distinction between directed or undirected graphs, to simplify the presentation. In practice, the GNN architecture can be designed to implement a different processing to incoming arcs with respect to the outgoing ones [20]. Finally, we define the *node degree* as the cardinality of its neighbourhood $|\text{ne}[v]|$.

2.2 Graph Neural Networks

The computation of the GNN is performed locally at each node by means of a *state vector* $\mathbf{x}_v \in \mathbb{R}^s$, that stores a hidden representation which depends on the node and its context in the graph. In fact, the state is computed for each node through a diffusion mechanism based on the graph topology, that allows us to locally encode the information concerning both the graph structure and the node labels. The encoding is performed by a learnable function that can be tuned to extract the relevant features to obtain the desired processing. The state variables $\mathbf{x}_v$, $v \in V$, are additional labels stored into the graph nodes that represent the state of the computation. Thus, the GNN model can be thought as a set of identical computational units that calculate a local state for each node, based on the states of its neighbours and on its label and links (see Fig. 1). The computational scheme is obtained by exploiting a *state transition function* f that models how $\mathbf{x}_v$ is obtained given the context of v. Its arguments are the label $\mathbf{l}_v$ of node v, the states and the labels of the neighbours, $\mathbf{x}_{\text{ne}[v]}$ and $\mathbf{l}_{\text{ne}[v]}$ respectively. The transition function will also depend on a vector of parameters $\theta_f \in \mathbb{R}^p$, that are adapted during the learning process. Hence, the state update equations can be formally expressed as

$$\mathbf{x}_v = f(\mathbf{l}_v, \mathbf{x}_{\text{ne}[v]}, \mathbf{l}_{\text{ne}[v]} | \theta_f) \ . \tag{1}$$

The state transition function f can be realized by an Artificial Neural Network (ANN). The network will have an appropriate architecture to deal with a variable number of arguments (i.e. inputs), since the degree of each node v can be different. Apart from this requirement, that will be clarified in the following showing a possible

Fig. 1 Local state computation with the GNN state transition function

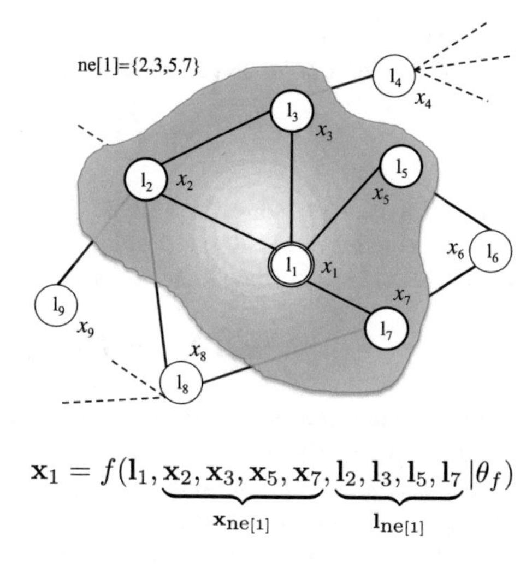

implementation, the number of outputs will be equal to the state space dimension s and the network architecture will be defined by the choice of the type and number of neurons, the number of layers, etc., as for the classical ANNs. The parameter vector θ_f will collect all the neural network connection weights. In this processing scheme, the same function (i.e. the same weight vector) is exploited for all the nodes of the input graph.

Given the state transition function and an input graph G, the state computation yields a vector assigned to each node v in G. By ordering the nodes, we can stack all the node variables into a single global vector $\mathbf{x} \in \mathbb{R}^{s|V|}$. Similarly, all the node labels define the vector $\mathbf{l} \in \mathbb{R}^{q|V|}$. Given this notation, the global computation carried out on the input graph is described by the equation

$$\mathbf{x} = F(\mathbf{x}, \mathbf{l}|\theta_f) \ , \tag{2}$$

being F the *global transition function*, whose entries result from stacking the transition functions f applied to each node v, with an appropriate projection of the two vectors $\mathbf{x}$ and $\mathbf{l}$ to yield the variables related to the neighbourhood of v. The vector of the learnable parameters is still θ_f, since all the local instances of the transition function share the same weights.

The result of the state computation is the vector $\mathbf{x}$ that satisfies Eq. (2), that is a system of non–linear equations in the variable $\mathbf{x}$. This equation can have multiple solutions, but we are interested only in those cases when the solution is unique. This requirement can be satisfied if the function F is a *contraction map* with respect to the variable $\mathbf{x}$. In this case, Eq. (2) has a unique solution by the Banach Fixed Point Theorem [21]. The condition, that guarantees the function F to be a contraction map, requires the existence of a value μ, $\mu \in [0, 1)$, for which $\| F(\mathbf{x}, \mathbf{l}|\theta_f) - F(\mathbf{y}, \mathbf{l}|\theta_f)\| \leq \mu \|\mathbf{x} - \mathbf{y}\|$, for any $\mathbf{x}$ and $\mathbf{y}$. This condition can eventually be forced in the learning process.

The Banach Theorem defines also the procedure to compute the solution of Eq. (2). In fact, the fixed point can be obtained by an iterative process of *state update* with the equation

$$\mathbf{x}(t+1) = F(\mathbf{x}(t), \mathbf{l}|\theta_f) \ , \tag{3}$$

being $\mathbf{x}(t)$ the state value at iteration t. The sequence defined by the recurrent equation (3) converges to the fixed point when F is a contraction map. The iterative computation is performed locally at each node by applying the state function as

$$\mathbf{x}_v(t+1) = f(\mathbf{l}_v, \mathbf{x}_{\text{ne}[v]}(t), \mathbf{l}_{\text{ne}[v]}|\theta_f) \ , \tag{4}$$

until the states converge to the solution. Notice that the final state is dependent on the input graph. The node labels are explicitly considered as an input of the computation, whereas the graph topology is implicitly exploited by the diffusion process among neighbour nodes in the graph. In fact, the computation is performed by a set of computational units, connected with the same topology of the edges in the input graph. When processing different graphs, the same units are arranged in a different structure, yielding different encodings. This computational structure constitutes the *encoding network* (see Fig. 2). This network is obtained by replicating the computational unit f for each node and by connecting the units using the same topology of the edges. The same transition function yields different encoding networks when applied to different input graphs.

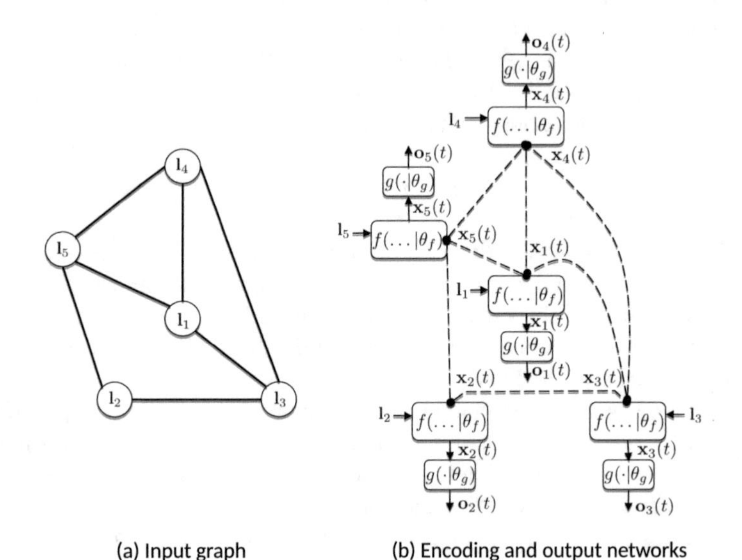

Fig. 2 The *encoding* and the *output* networks (**b**) when processing an input graph (**a**). The *graph neural network*, realizing $f(\dots, \theta_f)$ and $g(\cdot, \theta_g)$, is replicated on the input graph nodes and its weights, θ_f and θ_g, are shared among all the replicas

The output vector $\mathbf{o}_v \in \mathbb{R}^m$ is computed at each node v, depending on its label and state. The *local output function* g is exploited at each node after the computation of the fixed point for the state vector. g is parametrized by a vector of adaptable parameters θ_g, that can be adjusted during the learning process. For each node v the output is calculated as

$$\mathbf{o}_v = g(\mathbf{x}_v, \mathbf{l}_v | \theta_g) \ . \tag{5}$$

Figure 2 depicts the output computation as obtained after the state encoding step. To implement a graph–focused function, $\tau(G)$, the output is considered only for a predefined node having specific properties [20]. As shown in Fig. 2, the *Graph Neural Network* is defined by the functions f and g.

The function g can be realized by a feedforward neural network with $s + q$ inputs and m outputs, without any other restriction on the network architecture. The parameter vector θ_g collects the neural network connection weights.

The model for the state transition function f depends on the properties of the considered graphs. For *positional graphs* with bounded degree, i.e. graphs in which the edges connected to a node have a predefined order, the function f has a predefined maximum number of arguments and there is a given correspondence between each argument and an edge attached to the node. Hence, f can be realized by a feedforward neural network with s outputs and a fixed number of inputs. If d_M is the maximum node degree, the number of inputs will be $q + d_M(q + s)$. For nodes having a degree $d < d_M$ some arguments will not be available and a specific `nil` vector can be exploited to encode the label or the state corresponding to the missing edges (f.i. a vector with all 0s can be used as `nil`). This approach is feasible when d_M is sufficiently small and the variability of the node degree in the input graphs is not too high. If the graphs are *non–positional*, f can be realized with a model that makes it independent of the neighbourhood size and of the edge order. In this case, the state update function can be implemented as

$$\mathbf{x}_v(t+1) = \sum_{u \in \mathrm{ne}[v]} h(\mathbf{l}_v, \mathbf{x}_u(t), \mathbf{l}_u | \theta_h) \ . \tag{6}$$

The function h computes the contribution of each node u in the neighbourhood of v and all the contributions are summed up yielding the independence from the edge order and number. Two different implementations (*linear* and *neural*) for the function h in Eq. (6) have been proposed in [11, 20]. For neural GNNs, the function h is implemented by a feedforward neural network with $2q + s$ inputs and s outputs. However, this solution may not produce a contraction map for any value of θ_h. In this case, the learning objective can be defined as to include a cost term that penalizes mappings that are not contractions.

2.3 The GNN Learning Procedure

We consider a supervised learning scheme in which a supervisor provides a set of examples with a target value for the network output. In this case, each example consists of a graph and a target value for one or more of its nodes. Formally, the learning set is defined as

$$\mathcal{L}_e = \big\{(G_p, v_{pj}, \mathbf{t}_{pj})|\ G_p \in \mathcal{G}, p = 1, \ldots, P, \\ v_{pj} \in V_{G_p}, \mathbf{t}_{pj} \in \mathbb{R}^m, j = 1, \ldots, n_p\big\}\ , \tag{7}$$

where $\mathcal{G}$ is the set of graphs provided as examples. For the input graph G_p, the supervisor may provide a set of target vectors $\mathbf{t}_{pj}$, $j = 1, \ldots, n_p$, for a subset $S_{G_p} = \big\{v_{pj} \in V_{G_p}, j = 1, \ldots, n_p\big\}$. For graph–focused tasks, the supervision is given only for one node and a single target is provided for each graph in $\mathcal{G}$. The learning process corresponds to the minimization of an objective function that penalizes the error of the GNN response with respect to the given targets. Usually, the quadratic loss is exploited, defined as

$$E(\theta|\mathcal{L}_e) = \frac{1}{P}\sum_{p=1}^{P} e(\theta|G_p) = \frac{1}{2P}\sum_{p=1}^{P}\sum_{v_{pj}\in S_{G_p}} ||\mathbf{t}_{pj} - \mathbf{o}_{v_{pj}}(\theta|G_p)||_2^2\ , \tag{8}$$

where $\theta = [\theta'_f \theta'_g]'$ collects the parameters of the encoding and the output functions. When the transition and output functions, f and g, are differentiable with respect to θ, the loss $E(\theta|\mathcal{L}_e)$ is a differentiable function and it can be minimized by gradient descent. In fact, it can be proven (see Theorem 1 in [11]) that, when $F(\mathbf{x}, \mathbf{l}|\theta_f)$ is a contraction mapping, the loss $E(\theta|\mathcal{L}_e)$ depends on its unique fixed point that is a continuous and continuously differentiable function of the parameters θ_f.

If the model, selected for the state transition function f, is not guaranteed to implement a contraction for any value of the parameters θ_f, the learning objective must include a term to penalize non–contractive behaviours. The penalty term is obtained from the gradient of the function F as

$$p(\theta_f) = \beta L\left(\left\|\frac{\partial F(\mathbf{x}, \mathbf{l}|\theta_f)}{\partial \mathbf{x}}\right\|\right)\ , \tag{9}$$

where β is the weight of the penalty term, $L(y)$ is usually the hinge $\max(0, y - \mu)$, able to penalize values of y greater than the target contraction constant $\mu \in [0, 1)$ (see [11] for more details). Given the objective function of Eq. (8), with eventually the addition of the penalty in Eq. (9), for each graph $G_p \in \mathcal{G}$, the gradient of the loss with respect to the parameters θ is computed by the following procedure.

1. *State computation.* Starting from the initial state $\mathbf{x}(0|G_p)$, the state transition function

$$\mathbf{x}(t+1|G_p) = F(\mathbf{x}(t|G_p), \mathbf{l}_{G_p}|\theta_f)$$

is iterated until the condition $\|\mathbf{x}(T|G_p) - \mathbf{x}(T-1|G_p)\| < \epsilon$ is verified at the iteration T, where $\epsilon > 0$.

2. *Error evaluation.* The output network $g(\mathbf{x}_{v_{pj}}(T|G_p), \mathbf{l}_{v_{pj}}|\theta_g)$ is used to compute the outputs at the supervised nodes $v_{pj} \in S_{G_p}$ and the partial loss $e(\theta|G_p)$ of Eq. (8) is accumulated.
3. *Error backpropagation and gradient computation.* The value of the partial gradient $\nabla_\theta e(\theta|G_p)$ is computed by error backprogation on the output and encoding networks, and the computed value is accumulated to yield the total gradient, as $\nabla_\theta E(\theta|\mathcal{L}_e) = \frac{1}{P}\sum_{p=1}^{P} \nabla_\theta e(\theta|G_p)$.

After processing all the graphs in $\mathcal{L}_e$, the total gradient $\nabla_\theta E(\theta|\mathcal{L}_e)$ can be exploited to update the GNN weights using a gradient descent rule as, for instance,

$$\theta_{k+1} = \theta_k - \eta_{k+1}\nabla_\theta E(\theta|\mathcal{L}_e)|_{\theta=\theta_k}\ , \tag{10}$$

where k is the learning epoch, the gradient of $E(\theta|\mathcal{L}_e)$ is computed for $\theta = \theta_k$ and η_{k+1} is the *learning rate*. An adaptive learning rate scheme can be possibly employed in the learning process. The initial weights θ_0 are usually set to small random values (f.i. sampled using a uniform distribution with mean 0). The learning process can be stopped after a given maximum number of epochs, when the loss is below a predefined value, or when the gradient norm is below a threshold.

The error backprogation on the encoding network follows a scheme that is obtained by combining Backpropagation Through Structure and Backpropagation Through Time [22]. In fact, the relaxation process, applied for T steps to compute the state fixed point, corresponds to a layered unfolding network having T layers, where each layer is a replica of the function $F(\mathbf{x}(t|G_p), \mathbf{l}_{G_p}|\theta_f)$ for a given time step $t = 1, \ldots, T$. The output of layer t is the input for layer $t+1$, such that their units are interconnected by the state variables $\mathbf{x}(t)$. The resulting unfolding network of Fig. 3 has a multilayer structure, where the connection topology between adjacent layers is determined by the edges in the input graph. The replicas of the function F in the unfolding network share the same parameters θ_f. Finally, the state units of layer T are connected to the output functions to compute the GNN outputs.

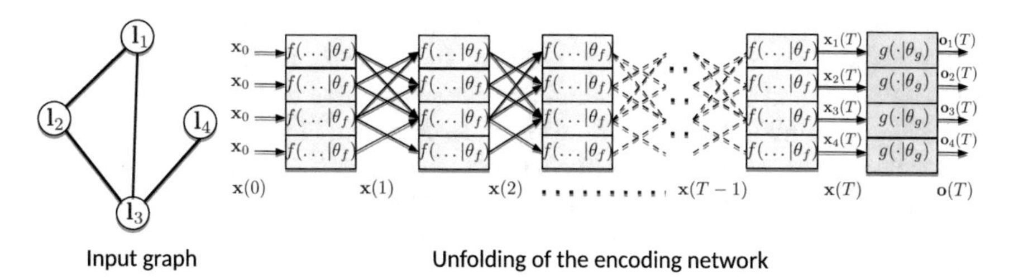

Fig. 3 The unfolding network obtained when computing the state on an input graph G_p

The unfolding network represents an efficient schema for the computation of the gradient of the loss function with respect to the GNN parameters θ. In fact, the original Backpropagation algorithm for feed–forward neural networks can be directly exploited on the unfolding network (see [20] for more details).

3 Layered Graph Neural Networks

The computation performed by GNNs can be layered to yield a deep processing architecture [23]. Given an input graph G in the learning set, we can add GNN layers such that each new layer exploits the original graph and the computation performed by the GNN of the previous layer. Formally, the GNN of layer k, $\mathcal{N}^k$, will process a graph G^{k-1} having the same topology of $G = G^0$, i.e. $V^{k-1} = V$ and $E^{k-1} = E$, but enriched labels assigned to the nodes. In particular, three different cases can be considered for layers $k = 2, \ldots, K$, being K the number of layers:

1. the node labels in G^{k-1} are obtained by concatenating the original labels in G with the outputs computed by the previous GNN layer, i.e., $\mathbf{l}_v^{k-1} = [\mathbf{l}_v, \mathbf{o}_v^{k-1}] \in \mathbb{R}^{q+m}$, where $\mathbf{l}_v \in \mathbb{R}^q$ is the label of node v in G and $\mathbf{o}_v^{k-1} \in \mathbb{R}^m$ is the output of the GNN $\mathcal{N}^{k-1}$ on node v;
2. the node labels in G^{k-1} are the concatenation of the original labels with the node states computed by the GNN in the previous layer, i.e., $\mathbf{l}_v^{k-1} = [\mathbf{l}_v, \mathbf{x}_v^{k-1}] \in \mathbb{R}^{q+s}$, where $\mathbf{x}_v^{k-1} \in \mathbb{R}^s$ is the state for node v computed by $\mathcal{N}^{k-1}$;
3. the node labels in G^{k-1} are obtained by adding both the outputs and the states computed by the GNN in the previous layer to the original labels, i.e., $\mathbf{l}_v^{k-1} = [\mathbf{l}_v, \mathbf{x}_v^{k-1}, \mathbf{o}_v^{k-1}]$.

The first layer $k = 1$ will process the original graph. The resulting computational scheme is depicted in Fig. 4. The outputs of the Layered GNN (LGNN) are the output values computed by the last GNN layer $\mathcal{N}^K$.

For instance, let us assume that the input G represents the region adjacency graph (RAG) of an image, i.e. a graph where nodes stand for homogeneous regions of the image and edges denote their adjacency relationships. Moreover, let us assume that the goal of the application is that of classifying the regions, i.e. the nodes, according to the class of the object they represent. In such a case, the first GNN takes in input a RAG G and produces a preliminary classification of the nodes. The resulting classification (and/or the states of the GNN, as described in points 1–3) is added to the labels of the RAG G in order to construct the graph G^1. Then, the second GNN is fed by G^1 and the procedure is repeated for all the layers of the LGNN.

The LGNN is trained by adding one layer after the other, until the last one. When training layer k, the weights of the GNNs in the previous layers can be kept fixed or can be fine tuned, as it happens in deep learning frameworks. Each network is trained using the original targets and the graphs constructed by the previous layer, as described before. The targets of each supervised node v in G^{k-1} are equal to original targets, i.e. $\mathbf{t}_v^{k-1} = \mathbf{t}_v$.

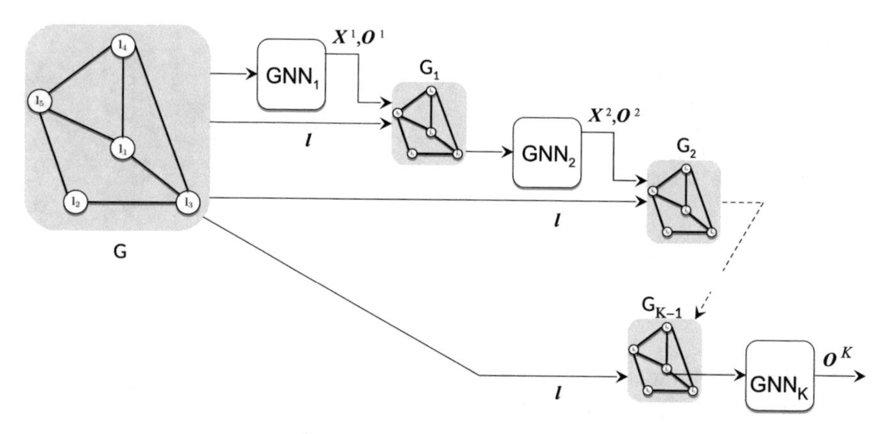

Fig. 4 The Layered GNN processing scheme

The layer–by–layer training procedure makes the learning feasible, reducing problems caused by the gradient vanishing effect due both to the network depth and to the long–term dependency impact on the graph. Each layer develops its expertise exploiting the supervisions provided in the learning set. The stacking of the layers allows the upper levels to take advantage of the knowledge already extracted in the previous layers, as it happens in other cascade neural architectures, following in some sense a boosting scheme. Basically, the added layers can focus their processing only to correct the errors made by previous layers. This processing scheme can reduce the long–term dependency problem, i.e. the difficulty in learning dependencies that involve nodes far from each other in the graph (distance is in terms of the shortest path through the graph edges). In fact, for a given layer k, the state $\mathbf{x}_v^k$ and the output $\mathbf{o}_v^k$ are able to summarize information up to a certain distance in the neighbourhood of node v. The GNN in the following layer will be actually able to exploit the information related to a larger neighbourhood since the nodes in $\text{ne}[v]$ contain additional information on their context in the added input components.

4 Applications to Pattern Recognition

We used an artificial dataset and two real–life benchmarks to test the LGNN model. In particular we decided to test on: clique localization, classification of mutagenic molecules and protein secondary structure prediction.

4.1 Dataset Descriptions

The clique localization problem Given an undirected graph $G = (V, E)$, a clique of size k can be defined as a complete subgraph of k nodes (see Fig. 5). The clique

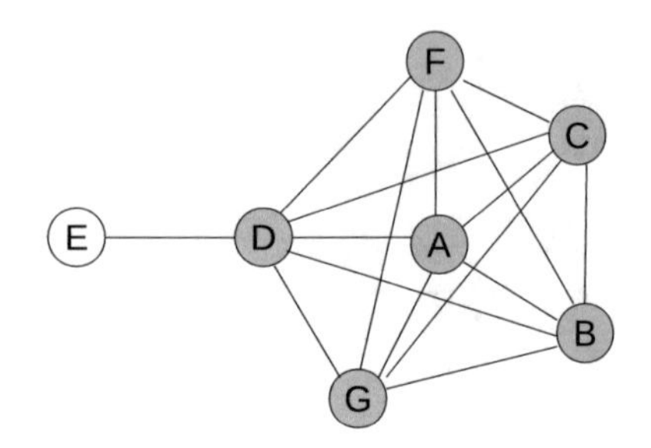

Fig. 5 An example of a clique in a graph: the gray nodes (A, B, C, D, F, G) are part of the maximum clique, while the white node (E) is out of the identified subgraph

localization can be identified as a particular instance of the subgraph localization task. Maximum clique identification in a graph is NP hard [24] and it has been widely addressed in the computer science literature. This is due to the fact that the applications of maximum clique algorithms are widely spread among various disciplines since cliques are associated with important network properties. For example, in chemistry, bioinformatics and computational biology, clique detection is exploited to identify similarities among molecules, and, in particular, for drug database screening [25]. Another important application of clique searching is for protein structure comparison. The understanding of such structure is in fact vital for the identification of protein functionalities and for the study of protein interactions [26].

Clique–detection methods are also used for clustering gene expression data [27]. With most of the recent DNA sequencing technologies, the task of finding interesting patterns in long DNA strings is one of the most challenging. In [27], this problem was approached by implementing a parallel algorithm for Maximum Clique detection. Other interesting applications of maximum clique localization are related to other fields. For instance, in social network analysis, cliques correspond to communities, while in the World Wide Web they represent web clusters that share common topics [28]. In [29], a new method for studying and defining overlapping communities was proposed, by defining universal network features and showing their application to the case of protein–protein interaction and web community detection.

The dataset used in our experiments is made of $1,000$ random graphs having 20 nodes, containing a clique of size 5. The GNN was trained to approximate the function defined by $\tau(G, v) = 1$ if v belongs to a clique, and $\tau(G, v) = -1$, otherwise. This problem setting was inspired by similar experiments described in [30].

The Mutagenesis problem The study of mutagenic compounds is a well known chemistry and biological challenge. Such compounds may be the cause of cancers and damages to DNA strand molecules [4] as well as they can be responsible for ageing, evolution and human genetic diseases [31]. Therefore the study of mutagenic compounds is of particular interest for pharmaceutical industries, to identify molecules that potentially have a mutagenic activity.

The dataset [4] collects data for 230 nitroaromatic compounds. Such compounds are commonly used as intermediate subproducts of industrial chemical reactions. In this case, the objective is to learn how to predict mutagenic molecules. Many different types of information are included in the dataset, for example the structure of the atom

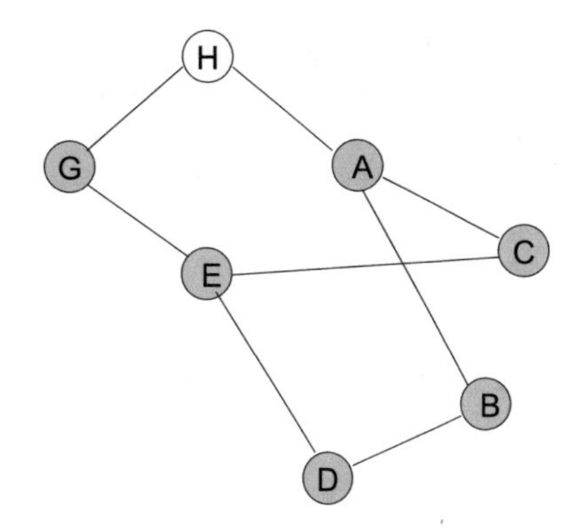

Fig. 6 Mutagenicity classification. The computation is graph–focused: the mutagenicity of the molecule is computed at node H

bonds and the chemical properties of the molecules. The graphical representation of the compound was designed to encode the bond structure. In particular, the nodes represent the atoms and the edges represent the bonds between them.

Each node is labeled using the information on the chemical properties. The label is a 13 dimensional vector, and it is constituted by features regarding the single atom and also by 4 global features [4]. The single atom features represent the charge of the atom and a one–hot representation of the atom type. The global features, on the other hand, represent the lowest unoccupied molecule orbital, the water/octanol partition coefficient and two precoded structural attributes. Moreover, we decided to adopt the following convention: in each graph, a single node is supervised. This has been chosen to be the first atom of the original dataset (see Fig. 6). The output is 1 in the case of a mutagenic molecule and -1 in the opposite case.

Secondary protein structure prediction Predicting the protein structure from the amino acid sequence is considered a fundamental task in molecular biology. The high throughput of large scale genome sequencing makes available a huge amount of sequence data, that can be exploited by the most recent machine learning and statistical tools to obtain reliable prediction models for protein secondary structures. Helixes, strands and coils are considered the most typical secondary structure regions. The knowledge of the secondary structure is particularly important, as it can be used for the tertiary structure prediction. This last one is of particular interest, as it defines the 3D amino acid positioning and, therefore, determines the functionality of the protein itself.

Secondary structure prediction can be easily modeled as a classification problem by using a graphical structure. Amino acids can be represented as nodes, whereas edges stand for peptide bonds. In this way, we can encode the protein primary structure. The final objective becomes therefore to predict one of the three categories mentioned before, which the amino acid belongs to.

This task can be faced with GNNs learning a function $\tau(G, v)$ that associates the class of the region containing the amino acid corresponding to the node v in the protein represented by the graph G.

The dataset is composed by 2171 proteins and $344,653$ amino acids [32]. As before, we store the amino acid available features as a 20 dimensional vector, that represents the one–hot encoding of the type of amino acid. A three–dimensional vector is the target associated to each node, and this contains information regarding the one–hot encoding of the three classes, i.e. *helix*, *coil* or *strand*.

4.2 Experimental Settings

The base GNN model used in the experiments is realized as follows. We implemented the transition function f and the output function g as a static network with three layers (where one is a hidden layer). In particular, we used a linear activation function in the output layer and a hyperbolic tangent function as activation for the hidden layer. The number of hidden units for f and g networks are 5, with a 2–dimensional state. We initialized the GNN weights randomly, in the range $[-0.1, 0.1]$. For each experiment, we split the dataset into three parts: training, test, and validation. The clique localization training set has 600 graphs, while the validation and the test contain 200 examples each. We used a 10–fold cross validation strategy for splitting the mutagenesis dataset and a 5–fold cross validation procedure for the secondary protein structure prediction. This was done accordingly to the papers in which the datasets were originally used.

By a trial–and–error procedure, we chose 2000, 500 and 1000 as the number of training epochs for the clique localization, the mutagenesis and the secondary protein structure prediction, respectively. The network generalization performance was calculated every 10 epochs on the validation set. Then, the network yielding the best validation error was evaluated on the test set.

We exploited the Resilient BackPropagation (RPROP) algorithm for the learning procedure, setting its parameters to standard values [33]. We report the standard deviation and the average accuracy on five repetitions of each experiment. For the case of the clique localization task, we rebuilt the dataset and reinitialized the weights at each repetition. For real–world benchmarks, instead, weights were recomputed for every fold. More precisely, the reported results consist in the macro–average and the macro–standard deviation.

4.3 Experimental Results

Figure 7 shows the results achieved on the clique localization problem. The plots correspond to the three LGNN architectures: propagation of the previous layer output (first column), of the previous layer state (second column), and of the previous layer state and output (third column). In the plots, points represent the average accuracies, while vertical bars represent their standard deviations. These results have been

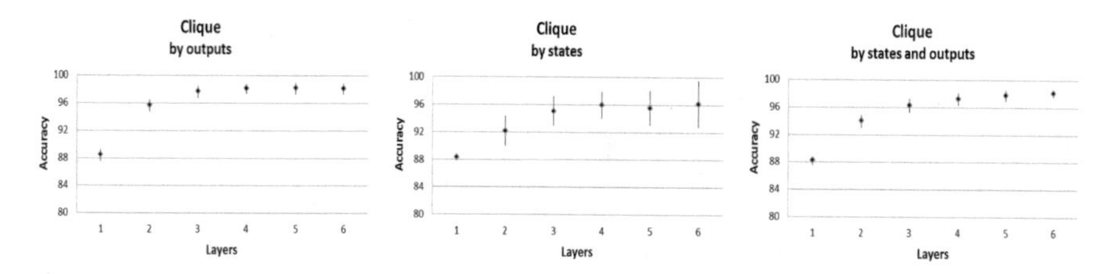

Fig. 7 Results for the clique localization problem. The plots report the average accuracy and the standard deviation (vertical bars). Each column shows the results for the three different LGNN configurations, based on the addition to the labels of the previous layer outputs, of the previous layer state and of both of them

obtained by varying the number of layers of the LGNN. The result for a standard GNN architecture with only one layer is reported as the first value.

The accuracy increases with the number of layers, as Fig. 7 shows. This is particularly clear when the first layer is added, but it appears of less impact when more layers are added to the architecture. In particular, in this case, we can actually observe an oscillation of the results. Moreover, the effect is more evident when propagating the GNN outputs rather than the states. An interesting fact is that using states seems to have an effect on yielding larger variances. Actually, even if, in theory, the state contains the information to reproduce the outputs, however the richness of the state configurations may confuse the learning algorithm, preventing it to extract the information relevant for the specific task. However, overall, we can say that the proposed multilayer architecture always performs better than the standard GNN for all the considered cases.

The average accuracy and the standard deviation for the LGNN in the case of the mutagenesis and the secondary protein structure benchmarks are shown in Fig. 8, in the first and in the second row, respectively. We only report the results in the case in which the node label is enriched by the previous layer output, since this configuration was shown to yield the best results in the case of the artificial dataset. For the secondary protein structure benchmark, the trend of the accuracy with respect to the number of layers has a similar behaviour to that observed for the clique dataset. The results for the mutagenesis dataset, instead, show a larger variance for the test set. Even though the average accuracy improves, the variance is too large to consider such an improvement statistically significant. Overfitting issues may cause the larger variance, as a comparison of test and training sets suggests. In particular, the dimension of the mutagenesis dataset is small and this can generate problems in the generalization. This is a well known issue in machine learning: when the number of layers increases, the architecture is able to approximate a larger set of functions, but if the training set is not statistically significant with respect to the real data distribution, the generalization capabilities of the model may be hindered. We also performed experiments with a single layer GNN and an increased number of states and hidden neurons in order to understand if the performance improvements for LGNNs were actually due to the fact that we are using a larger set of parameters. We evaluated

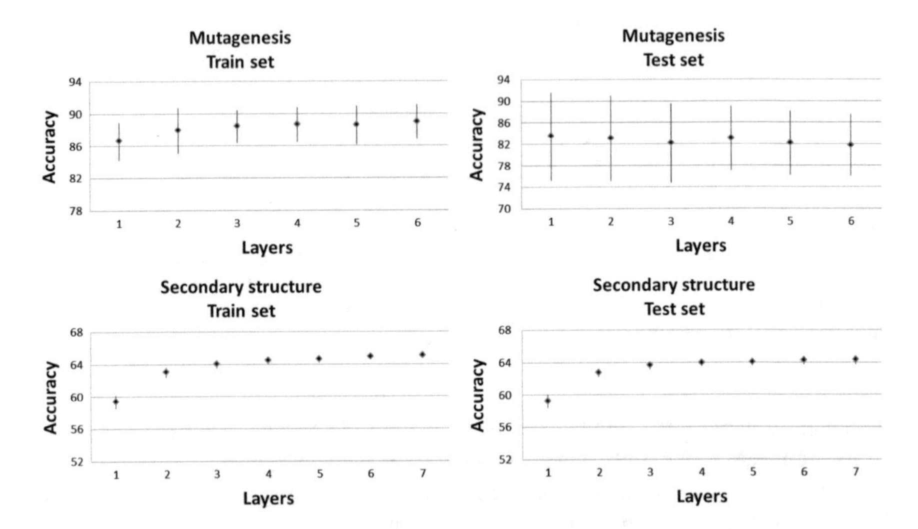

Fig. 8 Average accuracy and standard deviation (vertical bars) on the mutagenesis (first row) and on the secondary protein structure (second row) benchmarks. The first and the second column show the results on the training and the test set, respectively

GNNS with 5, 15, 25, 35 hidden nodes, in the output and in the transition networks, and with state dimensions varying in {2, 5, 10, 15}. In Table 1, we show the best performance achieved together with the best configuration. These results show that the same performance of an LGNN architecture cannot be obtained only by increasing the number of parameters in a shallow GNN. Moreover, in the case of an LGNN architecture, the amount of computational resources needed by a k–layered architecture is smaller than those required in the case of a GNN, if we consider the case of a GNN with k times more parameters or state variables. In fact, we can prove that the computational time in GNNs is quadratic w.r.t. the state dimension [11].

A comparison with the state–of–the–art on the considered benchmarks was not a main goal of this evaluation. Nevertheless, it is worth mentioning that the mean accuracy obtained by LGNNs on the mutagenesis dataset (83.47%) is not far from the best accuracy reported in literature (88%) [34].

For what concerns the secondary protein structure prediction, the LGNN performance is 10% smaller than the one achieved by Porter [32]. However these two results are not directly comparable. In fact, Porter exploits also a pre– and a post–processing methodology, that was not applied in the reported experiments, due to time and resource constraints. The classifier used in Porter is based on recursive neural networks (RNNs). GNNs and LGNNs are an extension of RNNs and, therefore, we expect that, by applying the Porter pre– and post–processing procedures, similar or better results can be achieved.

Table 1 Accuracy for the best configuration of a single layer GNN when varying the state dimension and the number of hidden nodes, both for transition and output networks.

	State dim.	Num. hiddens	Accuracy
Clique Loc.	10	10	97.75
Mutagenesis	2	5	83.47
Secondary Struct.	15	5	60.53

4.4 Transient Pocket Prediction

In the last few years, there has been an increasing interest in developing algorithms for finding transient pockets on the protein 3D surface. Transient pockets are cavities that can appear and disappear on a protein surface due to the protein dynamics. As common pockets, transient pockets can contain active sites, useful for chemical interactions with *ad hoc* ligands. Protein–protein and protein–ligand interfaces are in fact of particular interest for pharmaceutical companies, since the identification of new disease pathways leads also to the development and identification of new drug compounds [35]. The geometrical similarity between the protein surface in the vicinity of the active site and the target molecule gives an important insight to establish how good is a putative pocket to act as an anchorage for a specific drug.

Several methodologies and algorithms have been applied to this task. For example, the Q–site finder [36] is a method used to identify pockets which are unable to accomodate a particular compound, due to potentially conflicting geometries or because of their small size. Other interesting and widely used softwares are LIGSITECSC and LIGSITE, that take into account the information regarding the protein surface in the vicinity of the pocket to establish its usability [37]. On the other hand, PocketPicker [35] is a grid–based approach used for the prediction of the presence of a binding site inside the pocket. PocketPicker has been successfully applied to a set of protein–ligand complexes and an extensive evaluation was carried out to establish the quality of the protein binding site prediction.

Nevertheless, standard methods for protein structure prediction do not allow to efficiently and affordably localize transient pockets due to their short persistence and complex pattern, whereas connectionist models for structures seem to capture sufficient information in the vicinity of the putative pocket to predict its appearance, dimension and duration. In particular, in drug discovery, it often happens that many molecules of apparently unrelated structures are active on the same drug target. In this case, constructing a complete edge–weighted graph for describing drugs — with atoms as vertices and edge weights that stand for interatomic distances — means finding the Maximum Common Subgraph between them, on which a clique–detection algorithm can be efficiently applied [38]. Recently, the described approach has been also extended to the case of the identification of complementary protein surfaces. This is particularly important for biomolecules docking. In [39], a graph kernel method is presented for the prediction of DNA–binding sites with proteins. Labeled graphs are used to model the local structure of the two macromolecules, searching for possible similarities. Finally, in [40], surface local motifs, that are frequently observed in a specific protein group, are extracted, to establish — via common subgraph matching — if they are able to accommodate similar molecules.

Also GNNs and, more specifically, LGNNs, can be used to predict both the appearance and the dimension of pockets and, thanks to the incremental learning procedure, implemented by the cascade of networks, they are also able to take into account information not strictly related to the neighbourhood of the pocket, allowing to predict in advance, with respect to molecular dynamics models, the formation of a new pocket. Preliminary experimental results actually show both a good accuracy and a big time gain compared with classical simulations of molecular dynamics.

5 Conclusions

Graphs are commonly adopted as abstract representations for complex data, including DNA sequences, protein structures, documents, text, and images. In pattern recognition problems, graphs have been used since the early '70s, when they were employed for classifying visual patterns, especially in structural methods. In these approaches, objects are considered as constituted by "basic entities" (i.e. parts of the object homogeneous with respect to some visual features) related to each other. Under this assumption, node labels are descriptors of the parts, whereas edges encode relationships among them. The use of a pattern representation based on graphs implies reformulating all the standard tasks of a recognition system in terms of these data structures, such as classification, that in many approaches requires the comparison between an input and a set of prototypes, and learning, that consists in the process of creating a model for a set of classes given a set of known samples.

To this aim, in the present chapter, we have presented Graph Neural Networks (GNNs) and, in particular, a composite, deep architecture, referred to as Layered GNN (LGNN). In this model, each layer is an expert, implemented by a GNN, that solves the given task exploiting the original data and the computation provided by the previous layers. Intuitively, training the cascaded architecture layer by layer can help in summarizing, within the state of each node, the information related to gradually increasing neighbourhoods. Such an intuition, i.e. that the LGNNs can deal with and reduce the long–term dependency problem, when processing large structures, was experimentally investigated on both artificial and real–world data.

Nevertheless, the strategy, proposed to extend the basic GNN model, is just one of a set of potential alternatives, aimed at recursively learning deep structures. In fact, when the problem to be solved can be naturally partitioned into subtasks, each of them can be addressed separately, gaining partial solutions that can be subsequently used to move towards the overall solution. An example of such a situation is represented by the clique localization problem: a node v, not belonging to an N–clique, is not certainly part of a clique of dimension $N-1$; instead, a node that belongs to an N–clique can possibly belong also to an $(N+1)$–clique. Of course, this decomposition strategy is a less general approach to efficient learning in GNNs, but a preliminary experimental evaluation reports significant performance improvements whenever it is applicable. Another possibility for learning large structured data consists in decomposing the procedure into two steps, first solving a classification task given the graph labels — based on standard classifiers —, and then exploiting this information to enrich the original labels, keeping intact the graph topology. Even in this case, preliminary results show improved performances w.r.t. the original GNN model. Finally, different combinations of these methods may be considered as the object of future research.

References

1. H. Bunke, K. Riesen, Recent advances in graph-based pattern recognition with applications in document analysis. Pattern Recognit. **44**(5), 1057–1067 (2011)
2. M. Bianchini, F. Scarselli, Artificial neural networks for processing graphs with applications to image understanding: a survey, in *Multimedia Techniques for Device and Ambient Intelligence*, ed. by E. Damiani, J. Jeong (Springer, Berlin, 2009), pp. 179–199
3. C. Mooney, G. Pollastri, Beyond the twilight zone: automated prediction of structural properties of proteins by recursive neural networks and remote homology information. Proteins **77**(1), 181–190 (2009)
4. A. Srinivasan, S. Muggleton, R.D. King, M.J.E. Sternberg, Mutagenesis: ILP experiments in a non–determinate biological domain, In *Proceedings of the 4th International Workshop on Inductive Logic Programming*, Gesellschaft für Mathematik und Datenverarbeitung MBH, pp. 217–232, 1994
5. J. Cheng, M.J. Sweredoski, P. Baldi, Accurate prediction of protein disordered regions by mining protein structure data. Data Min. Knowl. Discov. **11**(3), 213–222 (2005)
6. M. Bianchini, M. Maggini, L. Sarti, Recursive neural networks and their applications to image processing, in *Advances in Imaging and Electron Physics*, vol. 140, ed. by P.W. Hawkes, (Elsevier – Academic Press, 2006), pp. 1–60
7. M. Bianchini, M. Maggini, L. Sarti, F. Scarselli, Recursive neural networks for processing graphs with labelled edges: theory and applications. Neural Netw. **18**, 1040–1050 (2005)
8. V. Di Massa, G. Monfardini, L. Sarti, F. Scarselli, M. Maggini, M. Gori, A comparison between recursive neural networks and graph neural networks, in *Proceedings of the IEEE International Joint Conference on Neural Networks*, pp. 778–785, 2006
9. A.-C. Tsoi, M. Hagenbuchner, R. Chau, V. Lee, Unsupervised and supervised learning of graph domains, *Studies in Computational Intelligence - Innovations in Neural Information Paradigms and Applications*, vol. 247 (Springer, Berlin, 2009), pp. 43–65
10. P. Frasconi, M. Gori, A. Sperduti, A general framework for adaptive processing of data structures. IEEE Trans. Neural Netw. **9**(5), 768–786 (1998)
11. F. Scarselli, M. Gori, A.-C. Tsoi, M. Hagenbuchner, G. Monfardini, The graph neural network model. IEEE Trans. Neural Netw. **20**(1), 61–80 (2009)
12. W. Uwents, G. Monfardini, H. Blockeel, M. Gori, F. Scarselli, Neural networks for relational learning: an experimental comparison. Mach. Learn. **82**(3), 315–349 (2011)
13. G. Monfardini, V. Di Massa, F. Scarselli, M. Gori, Graph neural networks for object localization, in *Proceedings of ECAI 2006* (IOS Press, 2006), pp. 665–669
14. L. Di Noi, M. Hagenbuchner, F. Scarselli, A.-H. Tsoi, Web spam detection by probability mapping graph–SOMs and graph neural networks, in *International Conference on Artificial Neural Networks* (Springer, Berlin, 2010), pp. 372–381

15. D. Muratore, M. Hagenbuchner, F. Scarselli, A.-H. Tsoi, Sentence extraction by graph neural networks, in *International Conference on Artificial Neural Networks* (Springer, 2010), pp. 237–246
16. R. Chau, A.-H. Tsoi, M. Hagenbuchner, V. Lee, A conceptlink graph for text structure mining, in *Proceedings of the Thirty–Second Australasian Conference on Computer Science*, vol. 91, (Australian Computer Society, Inc., 2009), pp. 141–150
17. Y. Bengio, P. Simard, P. Frasconi, Learning long-term dependencies with gradient descent is difficult. IEEE Trans. Neural Netw. **5**(2), 157–166 (1994)
18. M. Bianchini, F. Scarselli, On the complexity of neural network classifiers: a comparison between shallow and deep architectures. IEEE Trans. Neural Netw. Learn. Syst. **25**(8), 1553–1565 (2014)
19. G.E. Hinton, S. Osindero, Y. Teh, A fast learning algorithm for deep belief nets. Neural Comput. **2**, 1527–1554 (2006)
20. M. Bianchini, M. Maggini, Supervised neural network models for processing graphs, in *Handbook on Neural Information Processing* (Springer, Berlin, 2013), pp. 67–96
21. M.A. Khamsi, *An Introduction to Metric Spaces and Fixed Point Theory* (Wiley, New York, 2001)
22. F.J. Pineda, Recurrent back-propagation and the dynamical approach to adaptive neural computation. Neural Comput. **1**, 161–172 (1989)
23. N. Bandinelli, M. Bianchini, F. Scarselli, Learning long–term dependencies using layered graph neural networks, in *The 2010 International Joint Conference on Neural Networks (IJCNN)*, pp. 1–8, July 2010
24. D.R. Wood, An algorithm for finding a maximum clique in a graph. Oper. Res. Lett. **21**(5), 211–217 (1997)
25. C. Hofbauer, H. Lohninger, A. Aszó, SURFCOMP: a novel graph-based approach to molecular surface comparison. J. Chem. Inf. Comput. Sci. **44**(3), 837–847 (2004)
26. J. Konc, D. Janežič, A branch and bound algorithm for matching protein structures, in *Adaptive and Natural Computing Algorithms*, pp. 399–406, 2007
27. Q. Ouyang, P.D. Kaplan, S. Liu, A. Libchaber, DNA solution of the maximal clique problem. Science **278**(5337), 446–449 (1997)
28. A. Broder, R. Kumar, F. Maghoul, P. Raghavan, S. Rajagopalan, R. Stata, A. Tomkins, J. Wiener, Graph structure in the web. Comput. Netw. **33**(1), 309–320 (2000)
29. G. Palla, I. Derényi, I. Farkas, T. Vicsek, Uncovering the overlapping community structure of complex networks in nature and society. Nature **435**, 814–818 (2005)
30. F. Scarselli, M. Gori, A.-C. Tsoi, M. Hagenbuchner, G. Monfardini, Computational capabilities of graph neural networks. IEEE Trans. Neural Netw. **20**(1), 81–102 (2009)
31. B.A. Kunz, K. Ramachandran, E.J. Vonarx, DNA sequence analysis of spontaneous mutagenesis in saccharomyces cerevisiae. Genetics **148**(4), 1491–1505 (1998)
32. G. Pollastri, A. Mclysaght, Porter: a new, accurate server for protein secondary structure prediction. Bioinformatics **21**(8), 1719–1720 (2005)
33. M. Riedmiller, H. Braun, A direct algorithm method for faster backpropagation learning: the RPROP algorithm, in *Proceedings of the International Conference on Neural Networks*, vol. 1, (Portland (USA), 1993), pp. 586–591
34. W. Uwents, G. Monfardini, H. Blockeel, F. Scarselli, M Gori, Two connectionist models for graph processing: an experimental comparison on relational data, In *European Conference on Machine Learning*, pp. 211–220, 2006
35. M. Weisel, E. Proschak, G. Schneider, Pocket picker: analysis of ligand binding-sites with shape descriptors. Chem. Cent. J. **1**, 1–7 (2007)
36. A.T.R. Laurie, R.M. Jackson, Q-SiteFinder: an energy-based method for the prediction of protein-ligand binding sites. Bioinformatics **21**(9), 1908–1916 (2005)
37. B. Huang, M.M. Schroeder, LIGSITE csc: predicting ligand binding sites using the Connolly surface and degree of conservation. BMC Struct. Biol. **6**(1), 19 (2006)
38. S. Butenko, W.E. Wilhelm, Clique-detection models in computational biochemistry and genomics. Eur. J. Oper. Res. **173**(1), 1–17 (2006)

39. C. Yan, Y. Wang, A graph kernel method for DNA-binding site prediction. BMC Syst. Biol. **8**(4), S10 (2014)
40. N. Kurumatani, H. Monji, T. Ohkawa, Binding site extraction by similar subgraphs mining from protein molecular surfaces, in *IEEE 12th International Conference on Bioinformatics & Bioengineering (BIBE), 2012* (IEEE, New York, 2012), pp. 255–259

Granular Computing Techniques for Bioinformatics Pattern Recognition Problems in Non-metric Spaces

Alessio Martino, Alessandro Giuliani and Antonello Rizzi

Abstract Computational intelligence and pattern recognition techniques are gaining more and more attention as the main computing tools in bioinformatics applications. This is due to the fact that biology by definition, deals with complex systems and that computational intelligence can be considered as an effective approach when facing the general problem of complex systems modelling. Moreover, most data available on shared databases are represented by sequences and graphs, thus demanding the definition of meaningful dissimilarity measures between patterns, which are often non-metric in nature. Especially in such cases, evolutive and fully automatic machine learning systems are mandatory for dealing with parametric dissimilarity measures and/or for performing suitable feature selection. Besides other approaches, such as kernel methods and embedding in dissimilarity spaces, granular computing is a very promising framework not only for designing effective data-driven modelling systems able to determine automatically the correct representation (abstraction) level, but also for giving to field-experts (biologists) the possibility to investigate information granules (frequent substructures) that have been discovered by the machine learning system as the most relevant for the problem at hand. We expect that many important discoveries in biology and medicine in the next future will be determined by an increasingly stronger integration between the ongoing research efforts of natural sciences and modern inductive modelling tools based on computational intelligence, pattern recognition and granular computing techniques.

A. Martino (✉) · A. Rizzi
Department of Information Engineering, Electronics and Telecommunications, University of Rome "La Sapienza", Via Eudossiana 18, 00184 Rome, Italy
e-mail: alessio.martino@uniroma1.it

A. Rizzi
e-mail: antonello.rizzi@uniroma1.it

A. Giuliani
Department of Environment and Health, Istituto Superiore di Sanità, Via Regina Elena 299, 00161 Rome, Italy
e-mail: alessandro.giuliani@iss.it

W. Pedrycz and S.-M. Chen (eds.), *Computational Intelligence for Pattern Recognition*, Studies in Computational Intelligence 777,
https://doi.org/10.1007/978-3-319-89629-8_3

Keywords Computational intelligence · Pattern recognition · Machine learning
Granular computing · Bioinformatics · Computational biology · Systems biology
Non-metric spaces analysis

1 Introduction

1.1 Bioinformatics, Computational Intelligence and Pattern Recognition

The word 'bioinformatics' took different meanings since its introduction around forty years ago [17]. The definition of an autonomous 'bioinformatics' field started with the need to efficiently analyse and store increasing amounts of sequence data. Consequently, in the first years of the application of computational science in biology, bioinformatics was mainly devoted to technical and instrumental problems with no relation at all with the core of biological sciences. Computational scientists were hired to give a service to biologists because 'they were able to play with computers' in a way not too dissimilar of any laboratory technician taking care of a spectrophotometer properly working.

It is worth noting that the relation between biology and statistical methodology (the first root of pattern recognition approaches in life sciences) started with completely different premises. From the beginning of their relation, in the first years of the last century, biology and statistics interacted on a peer-to-peer basis and many statistical tools were developed in the core of biological community (e.g. Ronald Fisher, one of the fathers of modern statistics, was a geneticist and he developed linear regression in the frame of human genetics and evolution studies [65, 84, 95]).

During the years, the relation of biology with bioinformatics became something more than a purely occasional affair and approachedv the 'true-love wedding' level of the one-hundred years lasting relation between biology and statistics. Notwithstanding that, the term 'bioinformatics' is still largely prevalent with respect to other terms lexically more suited for describing the growing maturity of Biology and Computational Intelligence relation, such as 'computational biology' and 'systems biology'.

Besides the terminology, pattern recognition and computational intelligence techniques are nowadays gaining attention from the bioinformatics community [43, 71]. Many machine learning problems that can be instantiated in both biology and medicine are defined on domains in which each entry of the database at hand is a data structure far more complex than a plain real-valued feature vector, such as sequences, graphs, images or often even more complex structures arising from the concatenation of different data types (unconventional, structured data). Dealing with such structured domains usually demands to be able to define custom and meaningful (dis)similarity measures between elements in such unconventional domains, relying on sequence and graph matching techniques. Specifically, networks (graphs) are nowadays the most powerful approaches to describe the complexity behind biological systems.

In fact, the application of computationally intensive methods to biological problems became strictly intermingled with the actual frontiers of biomedicine and went well beyond the biological polymers sequence analysis, directly tackling the archetypal form of biological objects from protein science to ecology: complex networks, interpreted as simplified, yet powerful, representations of complex systems.

Complex systems are everywhere in nature, as well as in most artificial systems designed and built by mankind (telecommunications and energy distribution systems, as instances). Complex systems are by far more frequent than 'simple' ones, which are the true outliers in our world. However, a precise definition of what should be a 'complex system' is still a disputable issue. This challenge is due to the fact that complex systems are nowadays a research topic faced by many different scientific areas, such as mathematics, biology, physics, chemistry and engineering, each one bringing its own point of view, concepts and terms into the discussion. Since 1995, when John Horgan published his famous paper entitled "From Complexity to Perplexity" [36] evidencing the lack of a shared and precise definition about complex systems, the debate is still well alive. However, most authors agree in considering the following characteristics as necessary conditions to consider a given system as 'complex':

- The system is composed by many mutually interactive elements
- Elements behaviour is characterised by nonlinear dynamics
- The graph representing the causal relationships between elements contains loops

Elements are usually defined as atomic entities at the semantic level chosen for system description. For example, proteins can be considered as atomic entities in the network of chemical reactions in a biological cell; neurons are the basic constituents of the brain, when focusing on purely computational issues; each individual can be considered as an atomic entity in an ecosystem or in a social network. These examples of complex systems underline a property frequently found in such systems, concerning the fact the usually complexity arises in the form of a hierarchical organisation, as nested Systems of Systems. From this last point of view, it is possible to consider causal relations between elements belonging to different levels in the hierarchical organisation. When the network of these relations contains a loop, sometimes it is referred to as 'strange loop', i.e. a causal loop between different levels of the hierarchy [35]. This property is strongly related with the emergence of the most interesting behaviours of a given Systems of Systems, when considered as a whole.

In a fundamental paper appeared in 1948 entitled "Science and Complexity" [92], Warren Weaver, one of the fathers of modern information science together with Claude Shannon, proposed a tri-partition of science styles.

Scientific themes can be sub-divided into:

1. Problems of simplicity
2. Problems of disorganised complexity
3. Problems of organised complexity

The first class (simplicity) roughly corresponds to problems that can be solved in terms of differential equations. These 'simple problems' are the ones allowing for a high degree of abstraction (e.g. a planet could be considered an abstract dimensionless 'material point' for sketching general gravitational laws on the pure basis of its mass and distance from the sun).

Problems belonging to the second class (disorganised complexity) allow for a higher degree of generalisation than first class problems without losing in precision. These problems imply a somewhat opposite style of reasoning: the efficiency does not stem from the possibility to get an efficient abstract description of the involved players, but from totally discarding such 'atomic' knowledge in favour of very coarse grain macroscopic descriptors corresponding to gross averages on a transfinite number of atomic elements. This is the case of thermodynamic parameters (e.g. pressure, volume, temperature, etc.). The two above mentioned approaches have drastic limitations of their applicability range: class 1 needs the presence of very few involved players interacting in a stable way with a practically null boundary conditions effect, whereas class 2 needs very large numbers of particles with only negligible interactions among them.

Problems of organised complexity (class 3) arise in all those situations in which many (even if not-so-many as in class 2) elements are involved with non-negligible interactions among them. This is the 'middle kingdom' of complexity, where biological systems live and where computational intelligence and pattern recognition can 'make the difference'.

Network (or graph) is the archetype of organised complexity: a set of nodes (e.g. genes, brain areas, animal species) are each other connected by mutual correlations (edges). The wiring architecture of these graphs can vary in both space and time and it is of utmost importance to get quantitative similarities and differences among them. When graphs are adopted to represent only topological information concerning a set of objects and their relations, the network approach can roughly be described as the answer to the question "what can we derive from the sole knowledge of the wiring diagram of a system?" [28, 58].

The most crucial questions at the frontiers of biomedical sciences demands a reliable answer to the above question. Fields (just to name a few) that are increasing their formalisation in terms of network representations are: neuroscience at both clinical and basic research level [11, 68], biochemistry [5], cancer research [94], structural biology [46], ecology [27].

Moreover, when dealing with fully labelled graphs (where both nodes and edges are associated with possibly structured data), a fundamental topic is how to define proper dissimilarity measure between pairs of such patterns (the graph matching problem [47]).

Modelling a complex system is a matter of identifying the correct level of abstraction, which usually means to extract a hierarchy of information granules, searching for the level of the hierarchy better related to the semantic of the problem at hand. At any level, information granules are nodes of a network, so that the granulation process must deal with the problem of searching for frequent substructures in labelled graphs which, in turn, means to define algorithms able to automatically identify suitable dissimilarity measures in graph spaces. To this aim granular computing techniques are nowadays a promising tool.

Keeping this general frame in mind, in order to fix clear boundaries to this Chapter, a general definition of computational intelligence and pattern recognition is sketched in the following.

1.2 Theoretical Background and Definitions

Computational Intelligence, formerly known as *Soft Computing* thanks to the seminal work [96], is a set of data processing techniques tolerant to imprecisions, uncertainty, partial truth and approximation (in the data and/or models), aimed to provide robust and low-cost solutions and to achieve tractability when dealing with complexity. Such toolbox includes mostly biologically-inspired algorithms, usually exploiting inductive reasoning (i.e. based on generative logic inferences, such as analogy and induction) [13]. Basically, in this toolbox it is possible to find:

- Artificial Neural Networks
- Fuzzy Logic and Neuro-Fuzzy Systems
- Evolutionary Computation and derivative-free optimisation metaheuristics, such as genetic algorithms and swarm intelligence

Such a (heterogeneous) set of computational tools are usually combined to design powerful data-driven modelling systems. Being able to synthesise a (predictive) model of a given (physical or even abstract) process P is a fundamental topic in all natural sciences, as well as in engineering.

Before the pervasive widespread of digital computing devices, modelling was performed 'by hand', mostly relying on field-experts (*analytical modelling*), consisting in identifying meaningful quantities and relations among them and finally writing a system of integro-differential equations as the final output. This implies a clear understanding of the process at hand to be modelled.

However, when a meaningful sampling S of the process P to be modelled is available, a second approach (*data-driven modelling*) consists in writing an algorithm (often suitable to be run on a Von Neumann computing architecture) able to automatically synthesise a model M of P according to some predefined optimality criteria. This modelling approach is nowadays usually referred to as *Machine Learning*. The design and development of such learning systems is basically an engineering problem.

A formal machine learning definition has been given in [59], where the author considers machine learning as the following, well-posed problem:

> A computer program is said to learn from experience E with respect to some class of tasks T and performance measure P, if its performance at tasks in T, as measured by P, improves with experience E.

More broadly, machine learning can be defined as a (set of) complex intelligent processing system(s), usually defined by means of adaptive learning algorithms, able to act without being explicitly programmed or, in other words, able to learn from data and experience.

Pattern recognition techniques fall under the machine learning umbrella, focusing on classification of objects in a given number of categories (classes). Indeed, pattern recognition includes a wide range of techniques employed to solve (properly said) classification problems and clustering problems. Broadly, pattern recognition techniques can generally be divided into two main families: *supervised* and *unsupervised learning*, both of which fall under the aforementioned data-driven modelling paradigm.

For a more formal definition, let us consider an orientated process $P : \mathcal{X} \rightarrow \mathcal{Y}$ where $\mathcal{X}$ is the input space (domain) and $\mathcal{Y}$ is the output space (codomain). Moreover, let $\langle x; y \rangle$ be a generic input-output sample drawn from P, i.e. $y = P(x)$. In supervised learning, a finite set S of input-output observations drawn from P are supposed to be known. Common supervised learning tasks can be divided into two families, depending on the output space nature: *classification* and *function approximation*. In classification, outputs take values from a set of categorical labels, each of which correspond to a given problem-related class (e.g. "sick" or "healthy" in a predictive diagnosis/medicine problem). Conversely, in function approximation (such as regression, interpolation, extrapolation, fitting) outputs take values usually in the real field. Formally, in the former case, $\mathcal{Y}$ is a discrete label set where it is not possible to establish any total ordering between its elements, whereas, in the latter case, $\mathcal{Y}$ can be considered as a normed space.

In unsupervised learning there are no output classes or labels and regularities have to be discovered by considering mutual relations between elements drawn from the input space only. One of the mostly acclaimed unsupervised learning approaches relies on data *clustering* [37]. Aim of a clustering algorithm is to discover groups (clusters) of patterns in such a way that similar pattern will fall into the same cluster, whereas dissimilar pattern will fall into different clusters. Formally, let S be a sampling of a non-orientated process P and let c be the number of clusters, constrained to $2 \leq c \leq |S|$. Aim of a clustering algorithm is to assign to every $x \in \mathcal{X}$ an integer $h \in [1, c]$ starting from the set of c clusters induced over S.

In both of these cases, the goal of a learning machine is to build a predictive model from observations, aiming to discover the underlying model structure. Moreover, learning machines must be able to generalise their discrimination capabilities to previously unseen patterns or, in plain terms, they must be able to assign a label (either a class label or a cluster label) to patterns not belonging to S.

For the sake of completeness, it is worth stressing that clustering and classification algorithms might as well co-operate and shall not be considered as two diametrically opposed techniques. For classification purposes, a rather common approach relies on clustering labelled data without considering their respective labels, then assigning a label to each cluster by considering, for example, the most frequent label amongst the patterns belonging to the cluster itself. Finally, each new pattern is classified according to the nearest cluster's label. An example of such workflow can be found in [21, 22].

1.3 Chapter Scope

Aim of this Chapter is to review and discuss major issues when dealing with pattern recognition problems in non-metric spaces, namely input spaces for which a (dis)similarity measure might not be metric. As a case study, bioinformatics and computational biology-related problems will be investigated, since in these fields not only pattern recognition has emerged as a breakthrough discipline, but it is also very common to find structured data such as graphs or sequences which lie in non-metric spaces (see Sect. 1). Moreover, biological processes are excellent examples of complex systems, strongly suggesting the use of granular computing techniques for facing the challenging problem of (data-driven) model synthesis.

In Sect. 2 the data-driven modelling steps at the basis of pattern recognition problems will be described in detail, with particular emphasis on classification and clustering, underlying the role of computational intelligence techniques in designing pattern recognition systems.

Section 3 will regard non-metric spaces, remarking some examples of bioinformatics and computational biology-related problems in which structured data are commonly used. Moreover, some important issues when dealing with pattern recognition in non-metric spaces and possible solutions, including information granulation-based techniques, will be discussed.

In Sect. 4 some real case studies of bioinformatics/computational biology problems faced by means of pattern recognition techniques design to work in structured and non-metric domains will be summarised.

Finally, Sect. 5 will draw some conclusions, stressing major advantages of granular computing-based techniques over more 'traditional' approaches.

2 Machine Learning Systems Design

In conventional machine learning, a *pattern* is defined by a set of measures related to the original object to be represented, arranged in an array. Each entry (*feature*) is usually a real-valued variable. When a metric dissimilarity measure is implicitly or explicitly fixed in order to compare a pair of such simple data structures, usually it is referred to as a *feature vector*. The multi-dimensional space spanned by feature vectors forms the *feature space*. A well-defined feature space is able to facilitate the modelling process. For example, in the classification (supervised) case a well-designed feature space yields simpler decision surfaces in terms of structural complexity (smooth and regular).

Let us consider a plain supervised pattern recognition (classification) problem, as an instance of the more general machine data-driven modelling paradigm. Recalling Sect. 1.2, aim of a classification system is to assign an input pattern (represented by its feature vector) to one amongst the class labels defining the problem at hand.

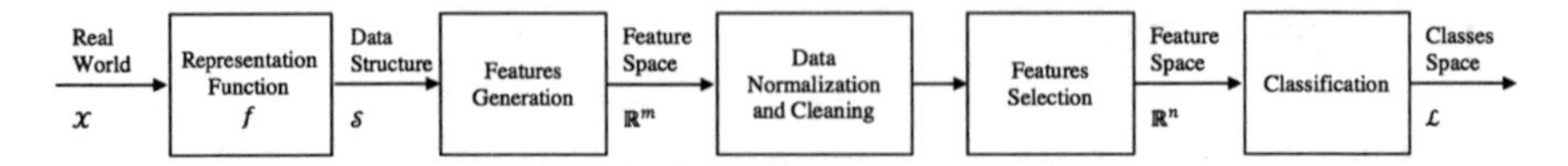

Fig. 1 A simplified pattern recognition system workflow

In Fig. 1, the main steps in order to build a classification system are summarised. First, real-world data, belonging to a generic (and possibly abstract) space $\mathcal{X}$ are casted into a proper data structure $\mathcal{S}$, processable by a computational device, by means of a representation function f which must ad-hoc be chosen for the problem at hand.

From structured data $\mathcal{S}$, a given number m of (usually numerical) features is extracted, thus casting data in $\mathcal{S}$ towards $\mathbb{R}^m$ (the aforementioned feature space).

The two following blocks are not mandatory, but they have been added for the sake of completeness and in order to take into account inevitable uncertainties in data collection and processing. The first block is in charge of data normalisation and cleaning: the former task is sometimes crucial in order to facilitate the classification algorithm under particular circumstances[1]; the latter deals with missing and noisy data. An intuitive data cleaning task is, for example, outliers' removal.[2] Conversely, the Feature Selection block allows to select a significant subset of the previously generated features; indeed, as a general rule, the feature vector should be small, yet informative,[3] in order to avoid undesired phenomena such as overfitting and/or the so-called *curse of dimensionality*. Further, it is recommended to get rid of unreliable features and correlations with existing features. At the end of this selection stage, feature vectors will lie in a (possibly) reduced features space $\mathbb{R}^n$, where $n \leq m$. Finally, the set of feature vectors will be used in order to train the classification system, with the final goal of estimating the correct label (identified, for the sake of ease, as an instance of a nominal value set $\mathcal{L}$ in Fig. 1) for any input vector.

For a better understanding of Fig. 1 and all of its steps, let us consider a real-world, Bioinformatics-related scenario, where $\mathcal{X}$ corresponds to the protein space (i.e. the set of real macromolecules). Let us suppose to represent proteins as graphs (cf. Sect. 3.1), then f is an (hypothetical) function which must convert macromolecules into graphs ($\mathcal{S}$). Fortunately, at least from a machine learning point of view, molecular

[1]For example, let us consider a classification/clustering algorithm driven by the Euclidean distance. A common problem with the Euclidean distance is that features spanning a wider range of values have more influence in the resulting distance measure, therefore normalising all attributes in the same range (usually [0, 1] or $[-1, +1]$) ensures fair contribution from all attributes, regardless of their original range.

[2]In Statistics, *outliers* are "anomalous data" that for a given dissimilarity measure lie far away from most observations.

[3]*Non sunt multiplicanda entia sine necessitate (Entities are not to be multiplied without necessity)*, commonly known as "The Ockham's Razor" Criterion (William of Ockham, circa 1287–1347). This criterion states that among a set of predicting models sharing the same performances, the simplest one (i.e. the one with the simplest decision surfaces) should be preferred. It is for sure one of the fundamental axioms for thoughtful and practical data-driven modelling.

biology helps: 3-dimensional protein structures, mainly gathered by crystallography, are available in online databases (e.g. Protein Data Bank [7]), therefore it is rather easy to build graph-based protein representations, either labelled or unlabelled on nodes and/or edges. The Features Generation block is in charge of extracting numerical features from graphs in $\mathcal{S}$ (cf. Sect. 3.2.1) which, after possible further processing, will be directly fed into the classification/clustering system.

The training phase for a classification system is a rather delicate task and it needs a separate discussion. Indeed, thanks to the training phase, the classification system learns how to map and discriminate input patterns according to their class labels. In other words, it learns the decision surfaces (decision regions boundaries) which separates patterns corresponding to different classes.

A usual procedure for measuring in a fair way the generalisation capability of a classification model consists in splitting the entire available dataset into two non-overlapping subsets, namely the Training Set and the Test Set. Specifically, as far as classification tasks are concerned, one shall figure both Training and Test Sets as composed by $\langle x; y \rangle$ pairs (see Sect. 1.2). The classification system, driven by a training algorithm which strictly depends on the chosen model (e.g. Support Vector Machine, Artificial Neural Network, K-Nearest Neighbours), will use the Training Set in order to learn the input-output mapping. The Test Set will then be used on such trained model, without further adaptive changes, in order to compute its performances (e.g. percentage of correctly classified patterns). For a thoughtful modelling, the two sets (albeit distinct) should satisfactorily represent the same statistical properties of the process to be modelled.

This double-split procedure, however, is not effective since every training algorithm depends on a set of parameters,[4] which must be tuned with the ultimate goal of maximising the generalisation capability of the synthesised model. In order to find the optimal set of hyperparameters (i.e. model selection) a three-split procedure is usually employed: the whole dataset is split into three non-overlapping parts, namely Training Set, Validation Set and Test Set. The training algorithm, driven by the set of hyperparameters Γ, will again exploit the Training Set and its performances will be evaluated on the Validation Set. The parameters Γ will be tuned in order to maximise the performances on the Validation Set and once the optimal $\Gamma^\star$ has been found, the final performances will be evaluated on the Test Set.

In literature, several ways to perform the aforementioned search for $\Gamma^\star$ have been proposed, amongst which grid search, random search [6] and evolutionary optimisation-based techniques emerge (see Sect. 2.1).

When dealing with unsupervised learning, the scheme reported in Fig. 1 does not change significantly, apart from the rightmost block. Indeed, rather than feature a Classification algorithm, a Clustering algorithm must be placed instead. A clustering algorithm is in charge of returning groups of data (clusters) according to a given (dis)similarity measure and to a predefined objective function.

In literature, three main families of clustering algorithms can be found, which mainly differ for their objective function (i.e. according to which criterion clusters

[4] Also known as *hyperparameters* in the Machine Learning terminology.

should be discovered): *partitional clustering* (e.g. k-means [51, 52], k-medians [10], k-medoids [39]), which split the dataset into k non-overlapping partitions; *hierarchical clustering* (e.g. BIRCH [97], CURE [32]), where clusters are found by building a dendrogram in either top-down or bottom-up approach; *density-based clustering* (e.g. DBSCAN [24], OPTICS [3]), which detect clusters as the most dense regions of the dataset.

Clustering algorithms do need some parameters tuning as well. Selecting their respective optimal value(s) can be done according to some internal validation measures, such as the Silhouette [74] or the Davies-Bouldin Index [19]. Both manual or fully automatic tuning by means of evolutionary optimisation techniques can be employed in unsupervised learning as well.

2.1 Evolutive and Fully Automatic Approaches

Evolutionary optimisation metaheuristics such as genetic algorithms [30], particle swarm optimisation [40], ant colony optimisation [16] and simulated annealing [41], are one of the main topics under the Computational Intelligence umbrella (Sect. 1.2). Such metaheuristics are well suited when the objective function to be optimised is not known in closed-form and gradient-based methods turn to be unfeasible.[5] Indeed, the decision boundary which separates two or more classes in a classification problem is determined thanks to a sampling of the boundary itself, namely the set of patterns which compose the dataset at hand, along with their respective class labels. As introduced in Sect. 2, they are often used in order to automatise the hyperparameters' tuning for classification and/or clustering algorithms. Further, they can help in conducting the feature selection phase (see Fig. 1). Indeed, one might ask which is the most relevant set of features in order to maximise the classification and/or clustering performances. To this end, evolutionary optimisation metaheuristics play a huge role.

Let us consider a genetic algorithm as an example. One can consider the genetic code to have the form $[\Gamma, \mathbf{w}]$ where Γ, as in Sect. 2, is a set of hyperparameters for the clustering/classification algorithm at hand, whereas $\mathbf{w}$ is an m-length real valued vector which tunes the (dis)similarity measure, core of the algorithm itself.

As far as classification tasks are concerned, a typical workflow might consist in letting each individual in the evolving population to be considered for training the classification model on the Training Set using both the hyperparameters and the (dis)similarity weights specified by its genetic code. The classification model's performance will later be evaluated on the Validation Set and such performance will serve as (part of) the fitness function.[6] Trivially, at the end of the evolutionary

[5]That is why evolutionary optimisation metaheuristics fall within the *derivative-free* methods.

[6]A common choice for a genetic algorithm fitness function takes into account both the model performance and its structural complexity. Specifically, whilst the former should be maximised, the latter should be minimised in order to avoid overfitting (cf. the Ockham's Razor Criterion).

stage, the best individual will be the one which maximises the performances on the Validation Set and its final performances will be evaluated on the Test Set. An example of such workflow can be found in [55] for classification algorithms and in [21, 22] for re-adaptation of clustering algorithms for classification purposes.

When dealing with clustering algorithms, the overall workflow does not change significantly. However, each individual will process the entire dataset according to the parameters stored in its genetic code and, similarly, since the performances cannot rely on any ground-truth labels, other internal validation measures should be used as the fitness function. An overview of clustering with evolutionary-driven feature selection can be found in [1].

It is worth stressing that, in both cases, the resulting best individual's genetic code contains the set of hyperparameters $\Gamma^\star$ which, along with the weights vector, maximise the algorithm's performances. Specifically, the latter deserves some further notes: if one considers $\mathbf{w} \in [0, 1]^m$, such vector acts as a feature selector, where 0's correspond to features which will not be considered in the (dis)similarity measure, and 1's correspond to features which, conversely, will be considered. The subset of n elements for which $\mathbf{w}$ is not-null can be seen as the reduced features space.

3 Dealing with Non-metric Spaces

So far, the design of a pattern recognition system has been described in its standard and most common form, where patterns are represented by means of real-valued vectors. In these cases, any Minkowski-based (e.g. Euclidean) distances can be good and straightforward candidates. Moreover, such (dis)similarity measures are metric.

Formally, a dissimilarity measure d defined on a generic space $\mathcal{S}$ is a function $d : \mathcal{S} \times \mathcal{S} \rightarrow \mathbb{R}$ satisfying the following properties:

1.
$$\exists d_0 \in \mathbb{R} \text{ such that } -\infty < d_0 \leq d(x, y) < \infty \tag{1}$$
2.
$$d(x, x) = d_0 \tag{2}$$
3.
$$d(x, y) = d(y, x) \tag{3}$$

for any two objects $x, y \in \mathcal{S}$. If, alongside Eqs. (1)–(3), d satisfies the following two properties

1.
$$d(x, y) = d_0 \text{ if and only if } x = y \tag{4}$$
2.
$$d(x, z) \leq d(x, y) + d(y, z) \tag{5}$$

for any three objects $x, y, z \in \mathcal{S}$, then d is said to be *metric*.

Similarly, in $\mathcal{S}$ it is possible to define a similarity measure $s : \mathcal{S} \times \mathcal{S} \rightarrow \mathbb{R}$ whether it satisfies the following properties:

1.
$$\exists s_0 \in \mathbb{R} \text{ such that } -\infty < s(x, y) \leq s_0 < \infty \tag{6}$$
2.
$$s(x, x) = s_0 \tag{7}$$
3.
$$s(x, y) = s(y, x) \tag{8}$$

for any two objects $x, y \in \mathcal{S}$. If, alongside Eqs. (6)–(8), s satisfies the following two properties

1.
$$s(x, y) = s_0 \text{ if and only if } x = y \tag{9}$$
2.
$$s(x, y) \cdot s(y, z) \leq (s(x, y) + s(y, z)) \cdot s(x, z) \tag{10}$$

for any three objects $x, y, z \in \mathcal{S}$, then s is said to be *metric*.

Moreover, it is possible to prove that:

Theorem 1 *If d is a metric dissimilarity measure with $d(x, y) > 0, \forall x, y \in \mathcal{S}$, then $s = a/d$ is a metric similarity measure for $a > 0$.*

Theorem 2 *If d is a metric dissimilarity measure, let d_{max} be the maximum pairwise distance between elements in $\mathcal{S}$, then $s = d_{max} - d$ is a metric similarity measure.*

The above two theorems demonstrate that, under particular circumstances, one can easily 'switch' between (metric) similarity and dissimilarity measures in a given input space. Indeed, dissimilarity measures quantify the degree of separation, whereas similarity measures estimate the complementary notion of closeness.[7]

3.1 Examples of Structured Data in Bioinformatics and Computational Biology

Dealing with non-metric spaces is a common issue when unconventional (structured) data, such as graphs or sequences, are considered as the input domain.

As introduced in Sect. 1, especially in bioinformatics and computational biology, patterns are usually described by means of data structures more complex than plain real-valued feature vectors: some common examples include proteins, DNA and RNA, metabolic pathways and brain connectivity networks.

[7]That is why in most of the Chapter, unless explicitly specified, the generic term *(dis)similarity* will be used.

Indeed, DNA and RNA transcripts are usually described as sequences of 4 possible nucleotides: adenine (A), cytosine (C), thymine (T), guanine (G) for DNA and adenine (A), cytosine (C), uracil (U), guanine (G) for RNA.

Proteins can be described by both sequences and graphs. The former representation is more straightforward: a protein is encoded in genes (DNA sequence) which is transcribed into pre-messenger RNA (RNA sequence). The RNA transcript is loaded into the ribosome which reads three nucleotides at the time (codons) and converts each triplet into one of the 20 amino-acids. It is clear that there exist up to three sequence-based protein representations, which mainly differ from their alphabet (4 nucleotides vs. 20 amino-acids) and their length (nucleotide-based sequences are three times longer than amino-acid-based ones). The protein representation as a sequence of amino-acids is also known as *primary structure*.

Graph representations result from a biological step forward in protein biosynthesis. Indeed, when the protein leaves the ribosome, a process called *protein folding* starts, during which the protein folds on itself, leading to a unique three-dimensional structure (also known as the *tertiary structure*). Protein Contact Networks [23] are an example of graph-based protein representation [29], where nodes correspond to amino-acids and edges between any two nodes exist whether their Euclidean distance falls within a given range, typically [4, 8]Å (e.g. [44–46, 54, 55]). The lower bound is usually considered in order to discard trivial backbone first-order neighbour contacts (i.e. sequence proximity), whereas the upper bound is usually defined by taking into account the peptide bonds geometry; indeed, 8Å roughly correspond to two peptide bond lengths or, equivalently, to two Van der Waals radii between residues' alpha-carbon atoms. In their original formulation, Protein Contact Networks are undirected graphs with no labels on nodes and edges: information regarding the type of amino-acid and their respective proximities are deliberately discarded in order to focus on proteins' topological structure and their complex nature.

Metabolic pathways are mainly described by graphs as they can be seen as protein networks and chemical networks. In the former, nodes correspond to proteins, whereas links correspond to physical (protein-protein interaction) and/or functional relations between them. In the latter, links correspond to chemical reactions (catalysed by specific enzymes) transforming the nodes (organic molecules produced – or used – in the metabolic processes) at their extremities into one another.

To our knowledge, the brain is probably the most complex circuit in the Universe, a complex system of nested subsystems, usually modelled as a network, since its functions strictly depend on the anatomical and functional wiring of billions of neurons [11, 31, 75, 88].

While in the case of brain networks based on the anatomical links between parts of the brains (macroscopic scale) or between single neurons in a small brain portion (microscopic scale) it is possible to rely on the assumption of a certain degree of invariance in time,[8] this is not the case as for functional brain networks (e.g. related to areas metabolic activity correlations observed by Nuclear Magnetic Resonance

[8]Indeed, the anatomical structure changes in the order of months/years depending on the age of subjects.

(NMR) or Positron Emission Tomography (PET)) that modify their wiring patterns on very short time scales [25, 81, 83].

Spontaneous neuronal activity in resting state depends on dynamic communication between brain regions allowing both local segregation and long-distance integration of neuronal processes. Several functional networks in which temporally or spatially coherent connections exist [18]. These networks have been identified in healthy subjects by functional Magnetic Resonance Imaging (fMRI) and by PET, respectively. Both these techniques deal with the quantification of metabolic rate correlation across different brain areas. Specifically, fMRI measures as 'marker' the variation of the amount of blood flowing across brain areas (coupled with metabolism by the dynamics between oxidised and reduced haemoglobin) [26], whereas PET focuses on the different metabolic rate of glucose (the most important energy source for brain cells) across different brain areas [79].

Both fMRI and PET techniques define a brain connectivity network in correlative terms: two nodes i, j of the network are linked by an edge if the metabolic rates of nodes i and j are each other correlated (given the quantitative character of the measures used by Pearson correlation coefficient metrics).

3.2 *Pattern Recognition in Non-metric Spaces*

When dealing with complex data structures such as graphs or sequences, the scheme from Fig. 1 should be revisited since patterns cannot be directly described by means of real-valued vectors.

In literature, three major approaches can be found [49, 50]:

1. directly working in the input data structure space, by defining ad-hoc (dis) similarity measures
2. by means of kernel transformations and kernel machines
3. by defining an embedding function from the input space to real-valued vectors

These approaches are summarised in Fig. 2 and, along with the 'classical' Feature Generation procedure, will be discussed separately.

3.2.1 Classical Processing Chain by Feature Generation

Recalling Sect. 2, given a generic and possibly non-metric input space $\mathcal{S}$, the most straightforward approach consists in defining a mapping function $\phi : \mathcal{S} \rightarrow \mathbb{R}^n$ specifically designed for the input space at hand. In this section, three examples of mapping function suitable for dealing with graphs will be described. Moreover, the additional challenge of dealing with patterns of different size in $\mathcal{S}$ will be discussed. For the sake of argument, let us consider graphs representing proteins since, notably, proteins have different sizes both in terms of primary and tertiary structures, meaning that

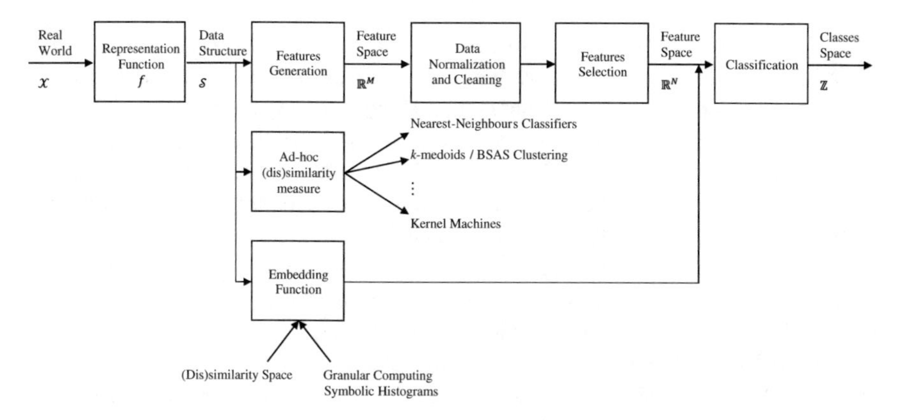

Fig. 2 Overview of possible approaches for pattern recognition in non-metric spaces

their amino-acids sequences and, by extension, their folded 3-dimensional structures have different size.

In [46, 54, 55] a mapping function based on graphs spectra has been proposed. Specifically, each graph has been described by means of its normalised spectrum, i.e. the set of eigenvalues evaluated from its corresponding normalised Laplacian matrix [38]. Such eigenvalues lie in range [0, 2], making such approach suitable for comparing graphs with different sizes. However, the number of eigenvalues composing the spectrum equals the number of nodes in the graph and, in order to overcome this problem, it is possible to estimate the spectral density by means of a kernel density estimator [63]. In this way, the distance between any two graphs can be evaluated by integrating the squared difference between their respective spectral densities all over the [0, 2] range. This evaluation can be performed also in the discrete domain by sampling a finite number (n) of points from such spectral densities (being the support domain equal for all graphs, regardless of their respective sizes). In such finite domain, the distance between two graphs can be evaluated as the considered distance (e.g. Euclidean) between their respective sets of samples.

In [45] a feature-engineering based approach has been employed in order to predict proteins' solubility starting from their topological structures. Several features have been manually selected such as the number of nodes and edges, the number of protein chains, some centrality measures (e.g. closeness and degree) and some physical characteristics (e.g. heat trace, energy). The union of these features forms the feature vector for a given graph.

Other feature extraction procedure(s) can rely on a rather novel field known as Topological Data Analysis [12, 91]. Topological Data Analysis consists in a set of techniques in order to extract information from data starting from topological information by means of dimensionality reduction, manifold estimation and persistent homology in order to study how components lying in a given multidimensional space are connected (e.g. in terms of loops and multidimensional surfaces). This can be done by starting either by so-called *point clouds*[9] or by explicitly providing a similarity matrix (cf. the *Kernel Methods* paragraph). Albeit this field has very solid and rigorous foundations (from algebraic topology to pure mathematics), there are very few 'numerical' features which can be extracted, mainly the sequence of Betti numbers. Formally, the ith Betti number corresponds to the rank of the ith homology group. In plain terms, the ith Betti number corresponds to the number of i-dimensional 'holes' in a topological surface. For example, let us consider a three-dimensional graph, its first three Betti numbers have the following interpretations: the 0-th Betti number corresponds to the number of connected components in the graph; the 1-st Betti number corresponds to the number of 1-dimensional holes (e.g. circular holes); the 2-nd Betti number corresponds to the number of 2-dimensional holes (e.g. cavities). If the multidimensional space under analysis has a finite dimension, the Betti numbers vanish after the spatial dimension (e.g. the number of 4-dimensional holes in a 3-dimensional space is always equal to zero). Whether the Betti numbers can be an effective mapping function for pattern recognition purposes it still an open question.

3.2.2 Ad-Hoc Dissimilarities in the Input Space

One of the mostly acclaimed ad-hoc (dis)similarity measures for structured data are the so-called *edit distances*, according to which the distance between two objects is given by the minimum number of atomic edit operations (usually insertions, deletions and substitutions of elements in the sequence) needed to transform the first object into the second object. As regards strings, the Levenshtein distance [42] is the seminal example of an edit distance, which can be seen as a generalised Hamming distance[10] [33].

The same approach can be used to define dissimilarity measures between graphs as well, leading to the Graph Edit Distances [47, 60], which inherit the idea at the basis of the Levenshtein distance, defining atomic edit operations in both the sets of nodes and edges. In many pattern recognition applications defined in sequences domains the Dynamic Time Warping [76] can be adopted, where the sequence support is explicitly related with time. Specifically, by applying a non-linear distortion on the support independent variable (time), it returns the optimal correspondence (i.e. similarity) between two sequences.

Amongst these methods, the Levenshtein/Hamming distances are well-known to be metric; the same might not be true for Graph Edit Distances (as they might violate the symmetry property – cf. Eqs. (3) and (8)) and Dynamic Time Warping (as it

[9] A finite set of points equipped with a notion of distance in a finite multidimensional space.

[10] According to which the distance between two strings of equal length is given by the number of mismatches.

might violate the triangle inequality – cf. Eqs. (5) and (10)). Also, edit distances are not recommended if patterns have a high dimension variability as deletion/insertion costs can easily prevail over substitution costs.

On the plus side, however, methods based on ad-hoc (dis)similarity measures notably work in cases where the pattern recognition system does not need to define an algebraic structure on the input space. For example, let us consider a clustering task to be performed directly into a non-metric space with an a-priori chosen (dis)similarity measure. Algorithms such as k-means or k-medians cannot be considered as good candidates since the former needs to evaluate the component-wise mean amongst the pattern in a given cluster in order to evaluate its representative, whereas the latter needs to evaluate the component-wise median. Therefore, the need to define a meaningful algebraic structure emerges which, however, turns into a non-sense as concerns non-metric input spaces. Suitable clustering algorithm candidates for dealing with non-metric spaces are k-medoids, as discussed in [56], and BSAS [85], since they do rely on (dis)similarity measures only in order to form clusters and to update their representatives. Similarly, as far as classification algorithms are concerned, a good candidate is the K-Nearest Neighbour, since it classifies patterns according to their respective distances rather than defining operators such as the inner product, mandatory in Artificial Neural Networks, or Support Vector Machines, whether equipped with an ad-hoc kernel transformation (see the *Kernel Methods* paragraph).

Kernel Methods

Typically, kernel methods can safely be employed whether the input space has an underlying Euclidean geometry, since they are based on inner products. Given a pair of patterns $\mathbf{x}, \mathbf{y} \in \mathbb{R}^n$, their inner product is given by:

$$\langle \mathbf{x}, \mathbf{y} \rangle = \mathbf{x} \cdot \mathbf{y} = \sum_{i=1}^{n} x_i \cdot y_i \tag{11}$$

Further, let us consider the instances matrix for the dataset at hand, $\mathbf{X} \in \mathbb{R}^{N_P \times n}$, namely a matrix where each row corresponds to a given pattern. Let N_P indicate the number of patterns. It is possible to define the *kernel matrix*[11] as

$$\mathbf{K}_{i,j} = \langle \mathbf{x}_i, \mathbf{x}_j \rangle \tag{12}$$

or, in batch fashion

$$\mathbf{K} = \mathbf{X} \cdot \mathbf{X}^T \tag{13}$$

[11] Also known as the *Gram Matrix*, after Danish mathematician Jørgen Pedersen Gram.

More in general, let k be a symmetric and positive semi-definite kernel function from the input space at hand $\mathcal{S}$ towards $\mathbb{R}$, i.e. $k : \mathcal{S} \times \mathcal{S} \rightarrow \mathbb{R}$ such that:

$$k(\mathbf{x}_i, \mathbf{x}_j) = k(\mathbf{x}_j, \mathbf{x}_i) \qquad \forall \mathbf{x}_i, \mathbf{x}_j \in \mathbf{X} \tag{14}$$

$$\sum_{i=1}^{N_P} \sum_{j=1}^{N_P} c_i c_j k(\mathbf{x}_i, \mathbf{x}_j) \qquad \forall c_i, c_j \in \mathbb{R}, \forall \mathbf{x}_i, \mathbf{x}_j \in \mathbf{X} \tag{15}$$

As in the inner product case, starting from $k(\mathbf{x}_i, \mathbf{x}_j)$ one can easily evaluate the Kernel Matrix as

$$\mathbf{K}_{i,j} = k(\mathbf{x}_i, \mathbf{x}_j) \tag{16}$$

and if $\mathbf{K}$ is a positive semi-definite kernel matrix, then k is a positive semi-definite kernel function. One of the most intriguing property of kernel methods relies in the so-called *kernel trick* [77]: kernel of the form (14)–(15) are also known as *Mercer's kernels* since they satisfy the Mercer's theorem [57]; they can be seen as the inner product evaluation on a (possibly) infinite-dimensional and usually unknown Hilbert space $\mathcal{H}$. The kernel trick is usually defined by means of the following, seminal equation:

$$k(\mathbf{x}_i, \mathbf{x}_j) = \langle \psi(\mathbf{x}_i), \psi(\mathbf{x}_j) \rangle_{\mathcal{H}} \tag{17}$$

where $\psi : \mathcal{S} \rightarrow \mathcal{H}$ is the implicit and usually unknown mapping function.

Several positive semi-definite functions commonly used as kernels include the linear, exponential, radial basis function and polynomial [77, 78], which are usually employed in kernel machines, such as (non-linear) Support Vector Machines.

However, in many cases, defining the kernel function might not be easy, especially when dealing with non-metric spaces. Regardless of the nature of the input space, it is possible to evaluate the similarity matrix (cf. Sect. 3) $\mathbf{S} \in \mathbb{R}^{N_P \times N_P}$ where

$$\mathbf{S}_{i,j} = s(\mathbf{x}_i, \mathbf{x}_j) \tag{18}$$

If s is a metric similarity measure, it is possible to directly use $\mathbf{S}$ as the kernel matrix, as suggested in [15], or include similarities in widely-known kernel functions (e.g. radial basis function), as suggested in [77].

Conversely, if the (dis)similarity measure is not metric, two mainstream approaches can be followed. The former relies on moving the pattern recognition problem towards a dissimilarity space (as explained in the next paragraph), the latter relies on 'modifying' the similarity matrix in order to be a valid kernel matrix (i.e. satisfying Mercer's theorem) [14, 15, 89].

Embedding Functions and Information Granulation

Embedding functions can be seen as particular cases of mapping functions as defined in the *Classical processing chain by Feature Generation* paragraph. Indeed, while both of them aim at moving the problem from a generic input space $\mathcal{S}$ towards

$\mathbb{R}^n$, embedding functions, at least in this context, do rely on other patterns or on substructures extracted from the dataset at hand in order to build such mapping.

A first example of embedding consists in moving the pattern recognition problem into a *dissimilarity space* [66].

In turn, dissimilarity representations can follow two further approaches:

1. Each pattern is described by its own row[12] from the similarity matrix **S** (cf. the *Kernel Methods* paragraph); that is, each pattern is described by the distance(s) vector with respect to other patterns (including self-distance)
2. Each pattern is described by the distance(s) vector with respect to a given number of representatives drawn from the input space at hand. Certainly, the selection of such representatives is a crucial task since a) they must well-characterise the decision boundary between patterns in the input space and b) there should be few of those since the number of representatives has a major impact on the model complexity. Representatives selection heuristics range from class-aware random selections to clustering procedures directly in the input space [48] (cf. the *Ad-hoc Dissimilarities in the Input Space* paragraph).

Regardless of which of the two methods is employed, a dissimilarity space can be equipped with algebraic structures and operators, such as the inner product, in order to be suitable with traditional kernel methods [48]. But, more in general, since patterns are now casted in $\mathbb{R}^{N_P}$ (former case) or $\mathbb{R}^R$ (latter case – where R indicates the number of representatives), any "standard" pattern recognition algorithm can be used.

In order to introduce the embedding by means of substructures, let us introduce widely known embedding functions for sequences. Since sequences are finite collections of objects drawn from a finite alphabet (cf. RNA/DNA sequences or proteins' primary structure, Sect. 3.1) one of the most intuitive approaches relies on histograms. Indeed, a sequence can effectively be described as the number of occurrences of any alphabet symbol within the sequence itself. For 'simple' sequences such as nucleotides or amino-acids sequences, histograms defined as above suffice. For example, in [90] a double experiment has been proposed in order to classify proteins starting from their primary structure according to their physiological role; in a first experiment, each protein is described by the number of occurrences of each amino-acid within the primary structure and, in a second experiment, such histogram-based representation has been extended to triplets of amino-acids in order to take into account also information about proximity and ordering. Further, in [53], the histogram-based representation considers pairs of amino-acids whose distance along the protein backbone is within a minimum and maximum value, a-priori defined.

For more complex sequences such as sentences or entire text documents, bag-of-words and word-count models have been proposed, where the alphabet is composed by the set of unique words in the sentence or document. 'Complex sequences' such

[12]If the similarity measure at hand is not symmetric, patterns' distance vectors as taken by rows or columns will be different. In order to overcome this problem, one can 'force' a similarity measure to be symmetric by considering $\mathbf{S} := (\mathbf{S} + \mathbf{S}^T)/2$ (e.g. [14]).

as sentences or entire text documents are rather rare (if non-existent altogether) in bioinformatics as such, but bag-of-words models, along with statistical and/or machine learning techniques, have been successfully employed for health analysis and forecasting (e.g. [82] for anastomosis leakage detection, [93] for diabetes-related notes in electronic health records).

In the last years, granular computing [4] emerged as a novel and promising information processing paradigm. In granular computing, atomic quantities known as *information granules* have to be extracted in order to be further studied and analysed, for gathering useful knowledge and insights from data, but finding the adequate level of abstraction for the problem at hand might be a challenging task. Along with symbolic histograms, granular computing can play the role of a promising data-driven framework which can simultaneously deal with embedding functions in non-metric spaces and knowledge discovery. In [8, 9, 69, 72] have been proposed fully automated data-driven and granular computing-based classification systems both for graphs and sequences. These systems are composed by four main macroblocks: motifs extractor, granulator, embedder and classifier.

The motifs extractor is in charge of extracting, according to some heuristics (possibly exhaustively), substructures (i.e. subgraphs/subsequences) from the dataset at hand.

The set of motifs is then forwarded to the granulator which runs a clustering algorithm on it, relying on a suitable inexact matching procedure (i.e. on a given dissimilarity measure in the substructures space), yielding a set of frequent sub-structures (clusters), whose representatives can be considered as candidate information granules (symbols). It is worth stressing that the clustering algorithm works in the input space since motifs are frequent substructures, and that free-clustering algorithms such as BSAS should be preferred, in order to automatically return a suitable number of clusters, avoiding to set it is advance. Further, since the input space might not be metric, a suitable cluster's representative is the medoid (or MinSoD) [20, 56].

The set of information granules are the main input for the embedder block which, according to the *symbolic histograms* approach, maps each pattern into an integer-valued vector. Specifically, each pattern is represented as the number of occurrences of each information granule within the pattern itself. The embedder, therefore, returns a set of vectors which can feed any standard pattern recognition algorithm for classification or clustering purposes.

The whole cascade is driven by a genetic algorithm, following the workflow as described in Sect. 2.1, in order to maximise the classifier's performances. The genetic algorithm acts as an orchestrator, and is in charge of optimising the final classifier synthesis, accomplishing two tasks: under an algorithmic point of view, it automatically tunes the clustering algorithm and possible (dis)similarity measure parameters, maximising the classifier's performances and selecting the subset of information granules better related with the classification task at hand (cf. Feature Selection block in Fig. 1); under a knowledge discovery point of view, since it returns the (sub)optimal set of information granules for the problem at hand.

The latter deserves some further observations. It is clear that embracing a granular computing/symbolic histograms approach is more computationally expensive than any other technique discussed so far. Indeed, the embedding procedure requires a clustering phase, searching for candidate granules. Even for small datasets, an exhaustive approach for the extractor might be unfeasible (since its complexity is combinatorial with respect to the pattern size and the substructure order), and it must be replaced by a stochastic approach (random subsampling). Moreover, when dealing with sequences or graphs, the (dis)similarity measure adopted by the core clustering procedure is by far more computationally demanding with respect to Euclidian distance performed on plain real-valued vectors. Furthermore, the selection of the most informative information granules, as well as of the best (dis)similarity measure parameters, demands additional computational burden by the evolutionary optimisation, since for every candidate solution it is needed to launch a full classification model synthesis procedure (for example a Support Vector Machine) in order to evaluate its fitness, computed as the performance of the classifier on the Validation Set (cf. Sect. 2.1). For these reasons, the symbolic histograms approach is practically feasible only when relying on parallel/distributed computing software/hardware environments.

But, on the plus side, granular computing-based techniques unleash an invaluable potential thanks to information granules. Indeed, if the training procedure yields a classification model with satisfying performances, able to correctly discriminate the input patterns for the problem at hand, the resulting information granules subset brings useful knowledge on the problem at hand, since information granules are at the basis of the embedding feature space. Information granules selected by the evolutionary optimisation are therefore responsible for the final definition of decision surfaces in that space and, consequently, they can show useful information that can be exploited by field-experts. This is the main advantage of granular computing techniques with respect to competitive approaches: extracting automatically meaningful information granules is useful both under an algorithmic point of view and under the application field point of view (biology, in this case).

As a more concrete example, let us consider a metabolic pathways problem, where metabolic pathways are described by graphs as in Sect. 3.1. One of the information granules might be the citric acid cycle.[13] The Krebs cycle (in network terms, a motif with a set of nodes lined to form a closed loop) is driven by oxygen and therefore it might be a key granule in order to discriminate between aerobic and anaerobic organisms. For this example a well-known chemical reaction has been considered, but the opposite might also happen: indeed, information granules can *pose* questions other than *confirm* statements: why these information granules are considered as significant for the discrimination/classification problem at hand?

[13]Also known as the *Krebs cycle*.

4 Case Studies and Applications

In most of the cases introduced in Sect. 3.1, it is almost impossible to project the analysed objects into a proper metric space spanned by a shared set of descriptors without considering some global features (e.g. classical network invariants like degree, characteristic length, closeness centrality, etc.) and thus losing a considerable part of information linked to 'who-is-connected-with-whom'. On the contrary, such information can be easily recovered projecting the objects into a non-metric space defined by motifs and/or frequent substructures (Sect. 3.2.2).

The need of a non-metric approach is evident in many biologically relevant cases. This need not necessarily derive by the lack of a common feature space, but it is motivated by the importance to individuate particular motifs endowed with a meaningful semantics.

In the field of protein sequence analysis this is the case of the identification of 'natively unfolded' tracts. This is a particularly intriguing problem in structural biology [87]. Until the end of last century, the general view of structure/function relation in protein molecules was apparently straightforward (cf. Sect. 3.1): protein primary structures correspond to the amino-acid residues linear ordering along the sequence. The primary structure determines both the mutual position of nearby (secondary structure) and distant along the sequence amino-acid residues (tertiary structure). The specific 3-dimensional arrangement of the protein molecule in turn determines its physiological role [70]. This view was questioned some years ago [87] by the discovery of 'natively unfolded' proteins that are molecules that do not have a definite 3-dimensional structure but that, on the contrary, remain in a random coil state until they interact with some partners (e.g. other proteins) and, after the binding, assume a specific 3-dimensional configuration. The same natively unfolded protein (and thus with only one specific sequence) can assume completely different 3-dimensional structures (and functions) depending on the different partners it interacts with. All the vital functions of a cell are managed by the creation of aggregates of different proteins generating a sort of nano-machine performing a specialised task (e.g. energy production, biosynthesis, immune response, DNA repair and duplication, etc.), where natively unfolded proteins are the 'hubs' of such protein-protein interaction networks, given their ability to change structure 'on demand' and thus to participate to different nano-machines (protein aggregations) [80]. Besides proteins that are natively unfolded in their entirety, all the proteins do have (smaller or longer) tracts that are natively unfolded corresponding to their interaction sites. If the goal is to modify the behaviour of a protein aggregate for a therapeutic intervention (e.g. by a drug binding to the protein molecule) it is of utmost importance to recognise such natively unfolded parts of the molecule from their sequence.

This is a very challenging task for classical machine learning approaches, due to the following reasons:

1. The context dependence of the problem: the same subsequence can be natively unfolded in protein *A* and perfectly folded in protein *B* due the general properties of the entire protein molecule [67]

2. The ambiguous character of the definition of 'unfolding': many of the so-called unfolded proteins (or tracts) could be only highly flexible systems that have only one preferred fold without structuring on-demand [34]
3. The dependence on the chemico-physical micro-environment the protein experiences (i.e. pH, molecular crowding, etc.) deciding the disordered/ordered condition [34]
4. The highly variable length of the disordered patterns [34]

This is why (even if never defining explicitly in these terms) all the tentative solutions of the problem used non-metrics approaches that in turn allowed to both select some 'relatively context independent unfolded motifs' and individuating some regularities in these motifs [73].

A somewhat related problem is to predict the relative solubility in water of protein molecules. Again, there exist a similar context dependences of the disordered/ordered case and, in [44], the problem was approached by considering several different representations. The protein folding problem has interested biologists for many years: if the native protein structure is 'encoded' in its primary structure, is it possible to predict its folded state? Relative solubility in water is the major feature for proteins' folding propensity. However, some proteins spontaneously fold, whereas other proteins need so-called *chaperones*[14] in order to fold correctly.

Recall from Sect. 3.1 that a protein can be described in different ways by either taking into account its primary or tertiary structure; therefore in [44] a subset of the *Escherichia Coli* proteome has been considered in three different representations: the plain primary structure; an 'extended' Protein Contact Network representation (cf. Sect. 3.1) where labels exist on both nodes and edges (nodes labels correspond to one of the 20 amino-acids, edges labels correspond to the Euclidean distance between the two vertices at their extremities); a serialised version of the graph-representation, where each vertex is associate with a 3-dimensional real-valued vector derived from the graph transition matrix. The goal was to predict the relative water solubility of each protein *in vitro* (i.e. without the help of chaperones). Given that water solubility encompasses the ability to reach of a correctly folded structure, this prediction task can be considered as an explorative study in the chemico-physical drivers of folding process.

The different representations allowed us to grasp different aspects of 'relative folding propensity' of proteins, being the extended Protein Contact Network the most promising representation.

The impossibility to design a data set on a shared feature space (and consequently the need of non-metric approaches) is evident in neuroscience in the case of comparing different brain connectivity networks [88]. Recall from Sect. 3.1, both fMRI and PET outputs are images of the brain: a quantitative value is attached to each voxel corresponding to the entity of metabolism activity in that location. The voxels are in the order of tens of thousands and their actual quantitative value is not rele-

[14]Protein molecules driving the folding of other protein systems.

vant per-se[15]: what differentiates healthy and pathological subjects is the degree of organisation (correlation among areas) of the system. The selective breakdown of intrinsic brain networks during the progression from the healthy state to mild cognitive impairment to Alzheimer's disease has been observed using both fMRI and PET. Using the single voxels as nodes can be highly misleading in the comparison of images across patients: not only their high number produces networks very difficult to analyse but the pairing of the voxels across different subjects (i.e. to recognise that the j-th voxel of patient A corresponds to the j-th voxel of patient B) is virtually impossible. To solve the problem anatomical knowledge is considered: the physician segments the brain image into ROIs (Region of Interest) correspondent to the well-known anatomical areas of the brain (e.g. hippocampus, amygdala, cerebellum, etc.) that all patients do have, so ROIs become the nodes and edges correspond to the scoring of a strong correlation between pairs of ROIs.

Alzheimer's disease risk scales with the progressive disruption of 'long range' correlations in favour of 'small scale' correlation between nearby areas [62]. This implies that for discriminating different risk levels it is not possible to rely on shared 'global correlation measures' on the brain, nor on the focusing on 'specific relations' between key areas because they can be very different across different patients, while maintaining the above described pattern of 'decrease in long range and increase in small range correlations'. This situation is solved by non-metric approaches, in which different brain connectivity networks are compared on the basis of the dynamics of 'attachment'-'detachment' from the giant component of the network (the bulk of connected ROIs) on a subject by subject basis [61].

In the case of brain connectivity studies, computational intelligence is having a great expansion and the search for suitable context-dependent metrics for comparing different conditions is highly debated in both clinical and basic research communities.

5 Conclusions and Future Directions

In synthesis, we can surely affirm that non-metric approaches rely on a sufficiently stable and reliable theoretical basis implemented on very efficient algorithms. On the other hand, the 'biological side' generates an ever-increasing amount of data amenable to be faced by computational intelligence approaches. The crucial point (deciding for the success/failure of the particular application) is the choice of a representation located at the most 'fruitful' level of biological organisation. The search for the scale maximising 'non-trivial determinism' is a crucial issue in applied statistics [64] and roughly corresponds to the search of the level where the number (and strength) of correlations between the different pieces of information (e.g. different descriptors) reaches a maximum.

[15] Indeed, the absolute entity of metabolic rate can vary for a lot of reasons going from anatomical differences among patients to their actual nutrition state.

Convesely to the classical reductionist tenet, this level (in the case of organised complexity [92]) is seldom located at the most detailed scale of analysis (e.g. single patients in epidemiological studies, single genes, primary structures of proteins, single pixels of an NMR or PET image of the brain, etc.) that in the great majority of cases are dominated by noise [86].

The search for the optimal representation (see the protein solubility case described in Sect. 4) asks for a conscious (and knowledge oriented) decision about the representation level to adopt (e.g. sequence-graph-labelled graph). This choice can only be a mixture of theory and data-driven choices, and thus asks for a real interaction of data scientists and biologists. For this interaction to be fruitful, both the communities must develop a similar language and share at least the basic principles of both the fields.

We think that, beside some 'bombastic exaggerations' on the 'death of science' to be substituted by a purely data-driven theoretically blind approach [2], the future will be characterised by an increasingly stronger integration between computational intelligence and pattern recognition techniques, and the different application fields. Indeed, computational intelligence techniques rely on data-driven modelling (see Sect. 2), which particularly suits problems where the process to be modelled – or at the heart of the problem itself – is unknown or hard to determine in closed-form (e.g. by analytical modelling).

As far as biology (and related fields) are concerned, computational intelligence and pattern recognition can be seen as useful methodological tools in order to perform "in-vitro experiments" and formulate hypotheses to be, if needed, further investigated by means of proper laboratory equipment by field-experts.

In this Chapter, we reviewed and discussed the major challenges and related modus-operandi when dealing with non-metric input spaces in computational intelligence and pattern recognition. By considering bioinformatics and computational biology as application fields, we explored several case studies in which data are conveniently represented by means of complex structures.

We stress that, amongst the three main macro-techniques for solving pattern recognition in non-metric spaces (Sect. 3.2), granular computing seems to be the most appealing in terms of results interpretability and knowledge discovery. Indeed, the automatically extracted information granules are the ones which maximize the classification performances, therefore the most informative and significant for the problem at hand. The set of information granules which, recall, is a set of motifs (i.e. recurrent substructures) can be analysed by field-experts in order to check whether they have some biological soundness and, possibly, boost further research, not only in granular computing and computational intelligence as such, but also in the proper application field in which such techniques have been employed.

References

1. S. Alelyani, J. Tang, H. Liu, Feature selection for clustering: a review. Data Clust. Algorithms Appl. **29**, 110–121 (2013)
2. C. Anderson, The end of theory: the data deluge makes the scientific method obsolete. Wired mag. **16**(7), 16–07 (2008)
3. M. Ankerst, M.M. Breunig, H.P. Kriegel, J. Sander, Optics: ordering points to identify the clustering structure. ACM Sigmod Rec. **28**, 49–60 (1999)
4. A. Bargiela, W. Pedrycz, *Granular Computing: An Introduction* (Kluwer Academic Publishers, Boston, 2003)
5. V. Beckers, L.M. Dersch, K. Lotz, G. Melzer, O.E. Bläsing, R. Fuchs, T. Ehrhardt, C. Wittmann, In silico metabolic network analysis of arabidopsis leaves. BMC Syst. Biol. **10**(1), 102 (2016)
6. J. Bergstra, Y. Bengio, Random search for hyper-parameter optimization. J. Mach. Learn. Res. **13**, 281–305 (2012)
7. H.M. Berman, J. Westbrook, Z. Feng, G. Gilliland, T. Bhat, H. Weissig, I.N. Shindyalov, P.E. Bourne, The protein data bank. Nucleic Acids Res. **28**(1), 235–242 (2000)
8. F.M. Bianchi, L. Livi, A. Rizzi, A. Sadeghian, A granular computing approach to the design of optimized graph classification systems. Soft Comput. **18**(2), 393–412 (2014)
9. F.M. Bianchi, S. Scardapane, A. Rizzi, A. Uncini, A. Sadeghian, Granular computing techniques for classification and semantic characterization of structured data. Cogn. Comput. **8**(3), 442–461 (2016)
10. P.S. Bradley, O.L. Mangasarian, W.N. Street, Clustering via concave minimization, in *Advances in Neural Information Processing Systems* (1997), pp. 368–374
11. E. Bullmore, O. Sporns, Complex brain networks: graph theoretical analysis of structural and functional systems. Nat. Rev. Neurosci. **10**(3), 186–198 (2009)
12. G. Carlsson, Topology and data. Bull. Am. Math. Soc. **46**(2), 255–308 (2009)
13. C. Cellucci, *Rethinking Logic: Logic in Relation to Mathematics, Evolution, and Method* (Springer Science & Business Media, 2013)
14. Y. Chen, E.K. Garcia, M.R. Gupta, A. Rahimi, L. Cazzanti, Similarity-based classification: concepts and algorithms. J. Mach. Learn. Res. **10**, 747–776 (2009)
15. Y. Chen, M.R. Gupta, B. Recht, Learning kernels from indefinite similarities, in *Proceedings of the 26th Annual International Conference on Machine Learning* (ACM, 2009), pp. 145–152
16. A. Colorni, M. Dorigo, V. Maniezzo, Distributed optimization by ant colonies, in *Toward a Practice of Autonomous Systems: Proceedings of the First European Conference on Artificial Life* (Mit Press, 1992), p. 134
17. D. Counsell, A review of bioinformatics education in the uk. Brief. Bioinform. **4**(1), 7–21 (2003)
18. J. Damoiseaux, S. Rombouts, F. Barkhof, P. Scheltens, C. Stam, S.M. Smith, C. Beckmann, Consistent resting-state networks across healthy subjects. Proc. Natl. Acad. Sci. **103**(37), 13848–13853 (2006)
19. D.L. Davies, D.W. Bouldin, A cluster separation measure. IEEE Trans. Pattern Anal. Mach. Intell. **2**, 224–227 (1979)
20. G. Del Vescovo, L. Livi, F.M. Frattale Mascioli, A. Rizzi, On the problem of modeling structured data with the minsod representative. Int. J. Comput. Theory Eng. **6**(1), 9 (2014)
21. A. Di Noia, P. Montanari, A. Rizzi, Occupational diseases risk prediction by cluster analysis and genetic optimization, in *Proceedings of the International Joint Conference on Computational Intelligence* (SCITEPRESS-Science and Technology Publications, Lda, 2014), pp. 68–75
22. A. Di Noia, P. Montanari, A. Rizzi, Occupational diseases risk prediction by genetic optimization: Towards a non-exclusive classification approach, in *Computational Intelligence* (Springer, Berlin, 2016), pp. 63–77
23. L. Di Paola, M. De Ruvo, P. Paci, D. Santoni, A. Giuliani, Protein contact networks: an emerging paradigm in chemistry. Chem. Rev. **113**(3), 1598–1613 (2012)
24. M. Ester, H.P. Kriegel, J. Sander, X. Xu et al., A density-based algorithm for discovering clusters in large spatial databases with noise. Kdd **96**, 226–231 (1996)

25. M.D. Fox, M.E. Raichle, Spontaneous fluctuations in brain activity observed with functional magnetic resonance imaging. Nat. Rev. Neurosci. **8**(9), 700–711 (2007)
26. K.J. Friston, C.D. Frith, R.S. Frackowiak, R. Turner, Characterizing dynamic brain responses with fmri: a multivariate approach. Neuroimage **2**(2), 166–172 (1995)
27. J. Gao, B. Barzel, A.L. Barabási, Universal resilience patterns in complex networks. Nature **530**(7590), 307–312 (2016)
28. A. Giuliani, S. Filippi, M. Bertolaso, Why network approach can promote a new way of thinking in biology. Front. Genet. **5** (2014)
29. A. Giuliani, A. Krishnan, J.P. Zbilut, M. Tomita, Proteins as networks: usefulness of graph theory in protein science. Curr. Protein Peptide Sci. **9**(1), 28–38 (2008)
30. D.E. Goldberg, *Genetic Algorithms in Search, Optimization and Machine Learning* (Addison-Wesley, USA, 1989)
31. M.D. Greicius, B. Krasnow, A.L. Reiss, V. Menon, Functional connectivity in the resting brain: a network analysis of the default mode hypothesis. Proc. Natl. Acad. Sci. **100**(1), 253–258 (2003)
32. S. Guha, R. Rastogi, K. Shim, Cure: an efficient clustering algorithm for large databases. ACM Sigmod Rec. **27**, 73–84 (1998)
33. R.W. Hamming, Error detecting and error correcting codes. Bell Labs Tech. J. **29**(2), 147–160 (1950)
34. B. He, K. Wang, Y. Liu, B. Xue, V.N. Uversky, A.K. Dunker, Predicting intrinsic disorder in proteins: an overview. Cell Res. **19**(8), 929–949 (2009)
35. D.R. Hofstadter, *I Am a Strange Loop*, Basic Books (2007)
36. J. Horgan, From complexity to perplexity. Sci. Am. **272**(6), 104–109 (1995)
37. A.K. Jain, M.N. Murty, P.J. Flynn, Data clustering: a review. ACM Comput. Surv. (CSUR) **31**(3), 264–323 (1999)
38. G. Jurman, R. Visintainer, C. Furlanello, An introduction to spectral distances in networks. Front. Artif. Intell. Appl. **226**, 227–234 (2011)
39. L. Kaufman, P. Rousseeuw, Clustering by means of medoids. Stat. Data Anal. Based L1-Norm Relat. Methods, 405–416 (1987)
40. J. Kennedy, R. Eberhart, Particle swarm optimization, in *Proceedings of the IEEE International Conference on Neural Networks*, vol. 4 (IEEE, 1995), pp. 1942–1948
41. S. Kirkpatrick, C.D. Gelatt, M.P. Vecchi, Optimization by simulated annealing. Science **220**(4598), 671–680 (1983)
42. V.I. Levenshtein, Binary codes capable of correcting deletions, insertions, and reversals. Soviet physics doklady. **10**, 707–710 (1966)
43. A.W.C. Liew, H. Yan, M. Yang, Pattern recognition techniques for the emerging field of bioinformatics: A review. Pattern Recognition **38**(11), 2055–2073 (2005)
44. L. Livi, A. Giuliani, A. Rizzi, Toward a multilevel representation of protein molecules: comparative approaches to the aggregation/folding propensity problem. Inf. Sci. **326**, 134–145 (2016)
45. L. Livi, A. Giuliani, A. Sadeghian, Characterization of graphs for protein structure modeling and recognition of solubility. Curr. Bioinform. **11**(1), 106–114 (2016)
46. L. Livi, E. Maiorino, A. Giuliani, A. Rizzi, A. Sadeghian, A generative model for protein contact networks. J. Biomol. Struct. Dyn. **34**(7), 1441–1454 (2016)
47. L. Livi, A. Rizzi, The graph matching problem. Pattern Anal. Appl. **16**(3), 253–283 (2013)
48. L. Livi, A. Rizzi, A. Sadeghian, Optimized dissimilarity space embedding for labeled graphs. Inf. Sci. **266**, 47–64 (2014)
49. L. Livi, A. Rizzi, A. Sadeghian, Granular modeling and computing approaches for intelligent analysis of non-geometric data. Appl. Soft Comput. **27**, 567–574 (2015)
50. L. Livi, A. Sadeghian, Granular computing, computational intelligence, and the analysis of non-geometric input spaces. Granul. Comput. **1**(1), 13–20 (2016)
51. S. Lloyd, Least squares quantization in pcm. IEEE Trans. Inf. Theory **28**(2), 129–137 (1982)
52. L. MacQueen, Some methods for classification and analysis of multivariate observations, in *Proceedings of the Fifth Berkeley Symposium on Mathematical Statistics and Probability*, vol. 1 (Oakland, USA, 1967), pp. 281–297

53. H.A. Maghawry, M.C. Mostafa, M.H. Abdul-Aziz, T.E. Gharib, A modified cutoff scanning matrix protein representation for enhancing protein function prediction, in *9th International Conference on Informatics and Systems (INFOS)* (IEEE, 2014), pp. DEKM–40
54. E. Maiorino, A. Rizzi, A. Sadeghian, A. Giuliani, Spectral reconstruction of protein contact networks. Phys. A: Stat. Mech. Appl. **471**, 804–817 (2017)
55. A. Martino, E. Maiorino, A. Giuliani, M. Giampieri, A. Rizzi, Supervised approaches for function prediction of proteins contact networks from topological structure information, in *Scandinavian Conference on Image Analysis* (Springer, Berlin, 2017), pp. 285–296
56. A. Martino, A. Rizzi, F.M. Frattale Mascioli, Efficient approaches for solving the large-scale k-medoids problem, in *Proceedings of the 9th International Joint Conference on Computational Intelligence. IJCCI*, vol. 1 (INSTICC, 2017), pp. 338–347
57. J. Mercer, Functions of positive and negative type, and their connection with the theory of integral equations, in *Philosophical Transactions of the Royal Society of London. Series A, Containing Papers of a Mathematical or Physical Character*, vol. 209 (1909), pp. 415–446
58. D.C. Mikulecky, Network thermodynamics and complexity: a transition to relational systems theory. Comput. Chem. **25**(4), 369–391 (2001)
59. T.M. Mitchell, *Machine Learning* (McGraw-Hill Boston, MA, 1997)
60. M. Neuhaus, H. Bunke, *Bridging the Gap Between Graph Edit Distance and Kernel Machines*, vol. 68 (World Scientific, 2007)
61. M. Pagani, A. Giuliani, J. Öberg, A. Chincarini, S. Morbelli, A. Brugnolo, D. Arnaldi, A. Picco, M. Bauckneht, A. Buschiazzo et al., Predicting the transition from normal aging to alzheimer's disease: a statistical mechanistic evaluation of fdg-pet data. NeuroImage **141**, 282–290 (2016)
62. M. Pagani, A. Giuliani, J. Öberg, F. De Carli, S. Morbelli, N. Girtler, F. Bongioanni, D. Arnaldi, J. Accardo, M. Bauckneht et al., Progressive disgregation of brain networking from normal aging to alzheimer's disease. independent component analysis on fdg-pet data. J. Nucl. Med. jnumed–116 (2017)
63. E. Parzen, On estimation of a probability density function and mode. Ann. Math. Stat. **33**(3), 1065–1076 (1962)
64. M. Pascual, S.A. Levin, From individuals to population densities: searching for the intermediate scale of nontrivial determinism. Ecology **80**(7), 2225–2236 (1999)
65. K. Pearson, Mathematical contributions to the theory of evolution. iii. regression, heredity, and panmixia, in *Philosophical Transactions of the Royal Society of London. Series A, Containing Papers of a Mathematical or Physical Character*, vol. 187 (1896), pp. 253–318
66. E. Pękalska, R.P. Duin, *The Dissimilarity Representation for Pattern Recognition: Foundations and Applications* (World Scientific, 2005)
67. K. Peng, P. Radivojac, S. Vucetic, A.K. Dunker, Z. Obradovic, Length-dependent prediction of protein intrinsic disorder. BMC Bioinform. **7**(1), 208 (2006)
68. J.B. Pereira, M. Mijalkov, E. Kakaei, P. Mecocci, B. Vellas, M. Tsolaki, I. Kłoszewska, H. Soininen, C. Spenger, S. Lovestone et al., Disrupted network topology in patients with stable and progressive mild cognitive impairment and alzheimer's disease. Cereb. Cortex **26**(8), 3476–3493 (2016)
69. F. Possemato, A. Rizzi, Automatic text categorization by a granular computing approach: facing unbalanced data sets, in *The International Joint Conference on Neural Networks (IJCNN)* (IEEE, 2013), pp. 1–8
70. J.S. Richardson, The anatomy and taxonomy of protein structure. Adv. Protein Chem. **34**, 167–339 (1981)
71. D. de Ridder, J. de Ridder, M.J. Reinders, Pattern recognition in bioinformatics. Brief. Bioinform. **14**(5), 633–647 (2013)
72. A. Rizzi, F. Possemato, L. Livi, A. Sebastiani, A. Giuliani, F.M. Frattale Mascioli, A dissimilarity-based classifier for generalized sequences by a granular computing approach, in *The International Joint Conference on Neural Networks (IJCNN)* (IEEE, 2013), pp. 1–8
73. P. Romero, Z. Obradovic, X. Li, E.C. Garner, C.J. Brown, A.K. Dunker, Sequence complexity of disordered protein. Proteins Struct. Funct. Bioinform. **42**(1), 38–48 (2001)

74. P.J. Rousseeuw, Silhouettes: a graphical aid to the interpretation and validation of cluster analysis. J. Comput. Appl. Math. **20**, 53–65 (1987)
75. M. Rubinov, O. Sporns, Complex network measures of brain connectivity: uses and interpretations. Neuroimage **52**(3), 1059–1069 (2010)
76. H. Sakoe, S. Chiba, Dynamic programming algorithm optimization for spoken word recognition. IEEE Trans. Acoust. Speech Signal Process. **26**(1), 43–49 (1978)
77. B. Schölkopf, A.J. Smola, *Learning with Kernels: Support Vector Machines, Regularization, Optimization, and Beyond* (MIT press, 2002)
78. J. Shawe-Taylor, N. Cristianini, *Kernel Methods for Pattern Analysis* (Cambridge university press, Cambridge, 2004)
79. D.H. Silverman, G.W. Small, C.Y. Chang, C.S. Lu, M.A.K. de Aburto, W. Chen, J. Czernin, S.I. Rapoport, P. Pietrini, G.E. Alexander et al., Positron emission tomography in evaluation of dementia: regional brain metabolism and long-term outcome. Jama **286**(17), 2120–2127 (2001)
80. G.P. Singh, M. Ganapathi, D. Dash, Role of intrinsic disorder in transient interactions of hub proteins. Proteins Struct. Funct. Bioinform. **66**(4), 761–765 (2007)
81. J. Smucny, K.P. Wylie, J.R. Tregellas, Functional magnetic resonance imaging of intrinsic brain networks for translational drug discovery. Trends Pharmacol. Sci. **35**(8), 397–403 (2014)
82. C. Soguero-Ruiz, K. Hindberg, J.L. Rojo-Álvarez, S.O. Skrøvseth, F. Godtliebsen, K. Mortensen, A. Revhaug, R.O. Lindsetmo, K.M. Augestad, R. Jenssen, Support vector feature selection for early detection of anastomosis leakage from bag-of-words in electronic health records. IEEE J. Biomed. Health Inf. **20**(5), 1404–1415 (2016)
83. P.G. Spetsieris, J.H. Ko, C.C. Tang, A. Nazem, W. Sako, S. Peng, Y. Ma, V. Dhawan, D. Eidelberg, Metabolic resting-state brain networks in health and disease. Proc. Natl. Acad. Sci. **112**(8), 2563–2568 (2015)
84. J.M. Stanton, Galton, pearson, and the peas: A brief history of linear regression for statistics instructors. J. Stat. Education **9**(3), 1–16 (2001)
85. S. Theodoridis, K. Koutroumbas, *Pattern Recognition*, 4th edn. (Academic Press, 2008)
86. M.K. Transtrum, B.B. Machta, K.S. Brown, B.C. Daniels, C.R. Myers, J.P. Sethna, Perspective: sloppiness and emergent theories in physics, biology, and beyond. J. Chem. Phys. **143**(1), 07B201_1 (2015)
87. V.N. Uversky, Natively unfolded proteins: a point where biology waits for physics. Protein Sci. **11**(4), 739–756 (2002)
88. B.C. Van Wijk, C.J. Stam, A. Daffertshofer, Comparing brain networks of different size and connectivity density using graph theory. PloS one **5**(10), e13701 (2010)
89. J.P. Vert, K. Tsuda, B. Schölkopf, *Kernel Methods in Computational Biology*, A primer on kernel methods (2004), pp. 35–70
90. Y.C. Wang, Y. Wang, Z.X. Yang, N.Y. Deng, Support vector machine prediction of enzyme function with conjoint triad feature and hierarchical context. BMC Syst. Biol. **5**(1), S6 (2011)
91. L. Wasserman, Topological data analysis. Ann. Rev. Stat. Appl. **5**(1) (2018)
92. W. Weaver, Science and complexity. Am. Sci. **36**(4), 536 (1948)
93. A. Wright, A.B. McCoy, S. Henkin, A. Kale, D.F. Sittig, Use of a support vector machine for categorizing free-text notes: assessment of accuracy across two institutions. J. Am. Med. Inf. Assoc. **20**(5), 887–890 (2013)
94. Y. Yang, L. Han, Y. Yuan, J. Li, N. Hei, H. Liang, Gene co-expression network analysis reveals common system-level properties of prognostic genes across cancer types. Nat. Commun. **5**, 3231 (2014)
95. F. Yates, K. Mather, Ronald aylmer fisher, 1890–1962. Biogr. Mem. Fellows R. Soc. **9**, 91–129 (1963)
96. L.A. Zadeh, Soft computing and fuzzy logic. IEEE Softw. **11**(6), 48–56 (1994)
97. T. Zhang, R. Ramakrishnan, M. Livny, Birch: an efficient data clustering method for very large databases. ACM Sigmod Rec. **25**, 103–114 (1996)

Multi-classifier-Systems: Architectures, Algorithms and Applications

Peter Bellmann, Patrick Thiam and Friedhelm Schwenker

Abstract In this work multi-classifier-systems (MCS) are discussed. Several fixed and trainable aggregation rules are presented. The most famous examples of MCS, namely bagging and boosting, are explained. Diversity between the base classifiers is a crucial point in order to build accurate MCS. Several criteria to measure diversity in MCS are defined and a motivation for diversity measures, based on the base classifiers' outputs is given. A case study on pain intensity estimation, based on physiological data streams, is conducted. Within the framework of the case study, different MCS and fusion approaches are evaluated. The case study is conducted on two different data sets, with four and five pain levels respectively, which were induced to the test persons under strictly controlled conditions. The aim of the case study is to implement an automatic pain intensity application system and analyse its effectiveness.

Keywords Pain intensity estimation · Decision fusion · Bagging · Boosting
Random forests · Diversity

1 Introduction

Classification is a huge part of pattern recognition and constitutes an every day phenomenon. Vehicle owners, for example, have to identify (classify) their parking cars before using them. Usually they don't need the licence number for that. It is enough to take a look at other characteristics (features), like the car's brand, colour

P. Bellmann (✉) · P. Thiam · F. Schwenker
Institute of Neural Information Processing, Ulm University,
James-Franck-Ring, 89081 Ulm, Germany
e-mail: peter.bellmann@uni-ulm.de

P. Thiam
e-mail: patrick.thiam@uni-ulm.de

F. Schwenker
e-mail: friedhelm.schwenker@uni-ulm.de

W. Pedrycz and S.-M. Chen (eds.), *Computational Intelligence for Pattern Recognition*, Studies in Computational Intelligence 777,
https://doi.org/10.1007/978-3-319-89629-8_4

and/or the shape, or more specific things, like rims or tinted windows. Identifying one's own car is one of the easiest imaginable daily issues. But there are several tasks, which are too time-consuming or too complex to solve without computational intelligence. Examples for these kinds of tasks include fingerprint- or DNA-matching. Intelligent software can also be applied as an additional supporter for human beings in classification tasks, such as in lie detection.

Classification per se is the last step of a processing chain in supervised pattern recognition. The first step is the data acquisition, i.e. the reading and recording of data from one or several sensors. Step number two would be some pre-processing, like removing noise or outliers, for example. The third step is usually the feature extraction, which can be done manually based on experts' knowledge or automatically with deep learning architectures. Feature extraction is then usually followed by step four, the post-processing. Post-processing could consist of standardisation, feature reduction, etc. The final step of the processing chain is then the classification.

The first sub-step of classification is to choose an adequate classifier for the given task. There exists a huge variety of different classifiers, for example linear discriminant classifiers [1], k-nearest-neighbour classifiers [2], decision trees [3] and artificial neural networks [4]. Even if one finds the best type of classifiers for the given task, one would not build just one single classifier, since a single classifier might not be able to learn an appropriate decision boundary for the given classification task, or perform unexpectedly bad on unseen data. Thus, it is common to use multi-classifier-systems (MCS) [5–10], i.e. a set of classifiers. Once one has fixed the type(s) of the classifiers to use, the size of the MCS, i.e. the number of classifiers, has to be chosen. After fixing the type(s) and amount of classifiers there are still two crucial design steps to be undertaken. First, one has to choose an aggregation rule for the set of classifiers, which depends on the classifiers' output types. Second, one has to determine the fusion architecture of the MCS. As the data is usually recorded by several sensors, one has to specify at which level and in what way the extracted features from the different data streams have to be merged.

The design of an MCS itself constitutes an essential step. In the ideal case one would like to get a set of classifiers, which are always correct, and take one of it for the whole task. But this is an unrealistic scenario, especially in real world applications. Therefore, the aim is to build an MCS with classifiers which complete each other as good as possible. This means one would like to have classifiers which, first, are as accurate as possible (in regard to their generalisation ability), and second, make different errors. More precisely, one would like to have a set of classifiers, which have a certain level of diversity. Diversity in MCS has been studied over the last years. There is a plethora of approaches to define diversity [11]. The explicit use of diversity measures in MCS can help to improve the performance in classification and related fields [12].

The latter three examples of the here aforementioned classification tasks (fingerprint-, DNA-matching and lie detection) have one common characteristic, which is the fact that the signals used for the classification task are biometric. One challenging field of application, that deals with biometric signals, is affective computing [13–18] (e.g. emotion or pain recognition). Pain recognition is an important field in e-health, since there are people who are not able to express their health condition in the correct way, or not at all. This applies, for example, to newborn babies, infants, comatose patients, or people with Alzheimer's or other mental diseases. The implementation of an automatic pain recognition system, based on multiple input channels, would help to improve the condition of people, affected by the aforementioned problem of not being able to express their state of health.

The remainder of this work is structured as follows. In Sect. 2, we will list and explain the main reasons for the use of an MCS instead of a single classifier. Section 3 provides common approaches for the combination of classifiers. In Sect. 4, we explain the main methods for creating MCS. Section 5 summarises the different fusion architectures of MCS. In Sect. 6, we discuss diversity in MCS. Different diversity measures are presented and adequate optimisation methods are introduced. Section 7 presents our case study, which deals with pain intensity estimation. Finally, in Sect. 8 we conclude the paper.

2 Motivation for Multi-classifier-Systems

Using an MCS instead of a single classifier increases the complexity of a classification task. Not only one, but a set of classifiers has to be build, trained and stored. And an additional aggregation rule has to be applied. This all is linked to a higher memory usage and a longer running time during the training and operational phase in comparison to a single classifier. Despite the arising difficulty, it is still common to use MCS in classification tasks, since the resulting advantages of an MCS outweigh the aforementioned drawbacks. Dietterich [19] and Polikar [20] suggest different reasons for the use of an MCS instead of a single classifier. In the following, we point out two of the most important ones.

2.1 Computational Motivation

The main goal of each classification task is to learn an underlying decision boundary of the given data which separate points from different classes. The general idea of the divide-and-conquer approach is to divide one complex problem into several easier tasks.

Regarding classification tasks, two problems can arise. First, the underlying decision boundaries might be quite difficult to learn by the given training algorithm. One might get classifiers whose decision boundaries represent local optima. Second, the underlying decision boundaries of the given data might lie outside the given classifier space.

For example, a single linear classifier can't solve a linearly non-separable problem, but by using a set of linear classifiers with an appropriate aggregation rule, we might get a possible solution. Figure 1 shows a basic example for this issue.

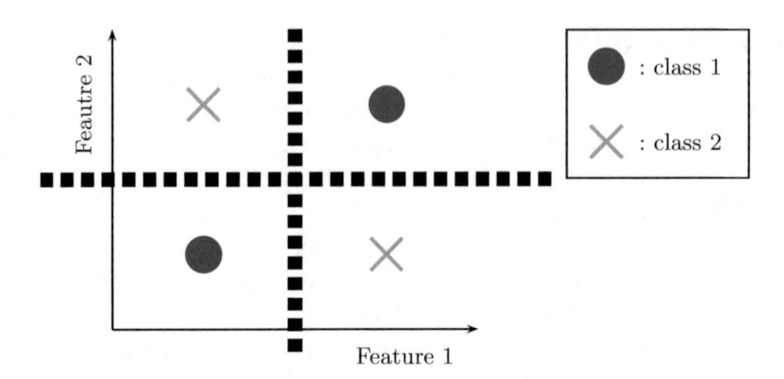

Fig. 1 Linearly non-separable two-dimensional two-class problem that can not be solved by a linear classifier, but it can be solved by multiplying the outputs of two linear classifiers (*parallel to the axis*)

2.2 Statistical Motivation

Given a data set one can divide the set in a training and a validation set. We can build several classifiers and compute their training and validation errors. But the generalization ability can only be estimated, for example by the hold-out or cross-validation methods. It is not guaranteed that classifiers with similar generalization approximations will perform similarly in the test case. By using an MCS of classifiers, we avoid the risk of taking one single classifier which might perform poorly on unseen data.

2.3 Basic Results for Single Classifier Versus MCS

In the following we see the results for a single classifier against an MCS of seven classifiers for the most common combination approaches, namely bagging, boosting and random forests, which will be explained in Sect. 4.1. The data sets used in Table 1 constitute pain intensity estimation tasks, whereby the Pain1 data set represents a 5-class problem, and both Pain2L and Pain2R data sets represent a 4-class problem each. More information will be provided in Sect. 7.1.

Table 1 shows that using an MCS of already a small amount of classifiers can improve the results significantly. However, it is important to mention that the use of an MCS does not guarantee an improvement over one single classifier or its best member. But empirical studies on this topic prove the effectiveness of MCS.

Table 1 Comparison of one single decision tree classifier against an MCS of 7 decision tree classifiers. Every MCS is significantly better than the single classifier and its best member in regard to the Wilcoxon signed rank test with a significance level of $p < 0.05$. Exceptions are: the bagging MCS is not significantly better than the single classifier on Pain2-Left, but still significantly better than its best member; the boosting MCS is obviously worse than its best member on Pain2-Left. The table shows the mean classification accuracies and standard deviations in % in a leave-one-subject-out setting for the recorded biopotentials. The random performance accuracy for Pain1 is 20%, whereas the random performance accuracy for Pain2-Left and Pain2-Right is 25% each, since Pain1 represents a 5-class classification task while Pain2 is a 4-class classification task

Dataset	Single	Bagging		Boosting		Random Forests	
		MCS	Best	MCS	Best	MCS	Best
Pain1	31.17 ± 8.18	34.32 ± 8.98	29.79 ± 7.48	33.30 ± 5.96	33.08 ± 5.80	33.70 ± 8.46	29.44 ± 7.70
Pain2L	35.35 ± 5.04	37.42 ± 7.74	34.28 ± 5.88	39.35 ± 6.11	39.39 ± 5.90	38.65 ± 7.09	34.87 ± 6.33
Pain2R	34.83 ± 6.78	37.95 ± 8.29	34.65 ± 6.29	40.39 ± 6.83	40.19 ± 6.87	37.78 ± 7.90	34.35 ± 5.78

3 Aggregating Classifiers

Once one has built a set of classifiers, it is indispensable to choose an aggregation rule to form the final MCS decision from the individual classifiers' outputs. There exists a variety of different combination rules for classifiers, which can be grouped in trainable and fixed groups or in regard to the used types of the base classifier' outputs. We stick to the first separation. Xu et al. [21] suggest to differentiate between the following types of classifier outputs:

- **Label Outputs**: Given an input x to the classifier, the classifier computes a class label as the output.
- **Continuous Outputs**: Here it is assumed that the classifiers are able to produce a vector of class memberships for each class, for instance with values between 0 and 1. One could also call this kind of outputs *fuzzy outputs*.

Xu et al. [21] also list an additional type of outputs, the so-called *rank outputs*, where the classifier produces an ordered (sub)set of the label set Ω. Hereby, the first element corresponds to the class with the highest probability to be the true class. This kind of outputs can directly be gained from the continuous outputs by ordering the class with the highest support in the first place, the class with the second highest support in the second place, and so on.

Based on the notations defined in [22], we denote the size of an MCS by L and the base classifiers by D_i. Thus, for a given input $x \in X \subset \mathbb{R}^d$, $d \in \mathbb{N}$, the label output of the base classifier i is denoted by $D_i(x)$. Whereas the continuous output of the base classifier i is denoted by $d_i(x) = (d_{i,1}(x), \ldots, d_{i,c}(x))$, where c is the number of class labels from the label set $\Omega = \{\omega_1, \ldots, \omega_c\}$. The MCS support for the class

ω_j is denoted by μ_j. In each case the MCS chooses the label ω_k as the final decision if

$$k = \underset{j=1}{\overset{c}{argmax}}\, \mu_j(x) \ .$$

3.1 Fixed Combination Rules

Fixed combination rules don't need extra training. Once a set of classifiers has been trained, the classifiers' outputs can be directly combined.

(Weighted) Majority Vote

Probably one of the most intuitive combination methods is the majority vote. Let $x \in \mathbb{R}^d$ be a data point and $v(x) \in \Omega^L$ be the vector with the label outputs from each classifier, i.e. $v(x) := \{D_1(x), \ldots, D_L(x)\}$. The majority vote is defined as

$$D(x) = mode\ v(x) \ .$$

In case of a draw, one can choose a class randomly among the winners, thus we may assume that the *mode* operator yields a unique result.

We can also weight the votes of each classifier. Let $b \in \mathbb{R}^L$ be the weight vector for the MCS. The weighted majority vote is then given as

$$D(x) = \underset{j=1}{\overset{c}{argmax}} \sum_{\substack{i=1 \\ D_i(x)=\omega_j}}^{L} b_i \ .$$

The weights of the classifiers can be determined by computing the training or validation error. Usually one would take the validation error, since the classifiers' training accuracies tend to be too optimistic and could lead to overfitting. However, in cases, when there is only a small set of training data available, one might take the training error into consideration. When each classifier is trained on a different subset of the training set, one might also take the training error, or the error on the points, which were unseen by the individual classifiers during the training phase (this kind of error is called *out-of-bag error* in the bagging approach).

Algebraic Combiners

One easy way to merge classifiers is to take their outputs and simply combine them by using some algebraic function,

$$\mu_j(x) = \mathcal{F}(d_{1,j}, \ldots, d_{L,j}) \ .$$

In the following, some existing examples for $\mathcal{F}$ are listed.

Product Rule:

$$\mu_j(x) = \prod_{i=1}^{L} d_{i,j}(x) \ .$$

Generalized Mean:

$$\mu_j(x, \alpha) = \left(\frac{1}{L} \sum_{i=1}^{L} d_{i,j}(x)^{\alpha} \right)^{1/\alpha} ,$$

with the special cases minimum and maximum for $\alpha \to \mp\infty$, the geometric mean and the harmonic mean for $\alpha \to 0$ and $\alpha \to -1$ respectively, and the arithmetic mean for $\alpha = 1$.

The most common rules are the maximum rule and the arithmetic mean rule. One could also take measures derived from order statistics, like the median, a trimmed mean, or any other algebraic function.

The mean rule is also known as the sum rule, since the maximal sum corresponds to the maximal mean value. In the case of label outputs one can regard the continuous output of each classifier as a binary vector $d_i(x) \subset \{0, 1\}^c \, \forall 1 \leq i \leq L, \forall x \in X$, where just one entry is equal to one. Taking the mean rule for these kind of vectors leads to the majority vote.

Kittler et al. [23] and Tax et al. [24] conclude that the sum rule is robust and the most resilient one to estimation errors compared to the min, max, product, median and the majority vote rule.

Borda Count

Concerning the original Borda count [25], each classifier creates a rank order of the classes. Let c be the number of classes, then the first ranked one gets $c - 1$ votes, the second placed one gets $c - 2$ votes and the ith placed one $c - i$ votes. The last ranked one gets 0 votes. The votes of each classifier are added for each class and the class with the most votes is defined as the winner.

The Borda count method can be varied by changing the amount of votes given to each candidate or by changing the number of candidates that will receive votes from each classifier. Different variations of Borda count are used all over the world, for example in some political elections, for granting sports awards or in other competitions, like the Eurovision Song Contest.

3.2 Trainable Combination Rules

In contrast to the fixed aggregation rules, trainable combination methods require an extra training before the MCS can be used for the classification task.

Behaviour Knowledge Space Method

The behaviour knowledge space (BKS) method [26] creates a look-up table during the training phase. For each training data point the MCS produces a combination of the classifiers' label outputs. For each combination one has to note the true class of the data point. The look-up table consists of the occurred combinations, each with the class label with the highest appearance.

For example, let's assume we use three classifiers for a two-class problem and the combination $(D_1(x), D_2(x), D_3(x)) = (\omega_1, \omega_2, \omega_1)$ occurred for 10 data points during the training phase. If the majority of these 10 data points had the true class label ω_1, then every point which leads to the same output combination will be labelled as ω_1, otherwise as ω_2.

If there is a draw, then a random choice has to be taken. In the case that one combination does not appear during the training phase, one can stick to the simple majority vote.

Decision Templates

The idea of the decision templates method is based on the so-called decision profiles, introduced by Kuncheva [27]. The decision profile is a matrix with all the vector outputs of the whole MCS:

$$DP(x) = \begin{bmatrix} d_{1,1}(x) & \dots & d_{1,j}(x) & \dots & d_{1,c}(x) \\ \vdots & \vdots & \vdots & \vdots & \vdots \\ d_{L,1}(x) & \dots & d_{L,j}(x) & \dots & d_{L,c}(x) \end{bmatrix} \in \mathbb{R}^{L \times c} \ .$$

The idea of the decision templates method is to save the typical decision profile for each class ω_j. Such a profile is called decision template DT_j and is simply the average among all data points from one class,

$$DT_j = \frac{1}{N_j} \sum_{\substack{x \in X \\ l(x)=\omega_j}} DP(x) \ ,$$

whereby $l(x)$ denotes the true label of x and N_j the number of objects with true class label ω_j. To label an input data point x, one has to compare $DP(x)$ with all DT_j by using a similarity measure S,

$$\mu_j(x) = S(DP(x), DT_j) \ .$$

Sample x is then labelled as the class with the highest similarity. Thus, this method represents a 1-*NN* classifier, where the prototypes are the decision templates and each data point first has to be transferred to its decision profile before finding its most similar (nearest) neighbour. Extensions of the decision templates approach have been introduced in [28].

Pseudo-Inverse

The Pseudo-Inverse [29] aggregation rule consists of a least-squares linear mapping obtained by computing the Pseudo-Inverse of the base classifiers horizontally concatenated outputs

$$D = [D_1 \dots D_L], \text{ with } D_i = \begin{bmatrix} d_i(x_1) \\ \vdots \\ d_i(x_m) \end{bmatrix} \forall i \in \{1, \dots, L\} ,$$

where $m \in \mathbb{N}$ is the size of the training set and multiplying it with the corresponding class labels Z:

$$M = \lim_{\alpha \to \infty} D^T \left(DD^T + \alpha I\right)^{-1} Z .$$

The mapping is subsequently applied to the base classifiers' outputs of an unseen sample and the assigned label corresponds to the class with the highest estimated score. Relations between the Pseudo-Inverse solution, sum rule and decision templates are discussed in [30].

3.3 Classifier Selection

Regarding classifier selection [31] the combination method has a selecting rule that can be fixed or learned during the training process. By applying the rule one chooses one or more classifiers for making the final decision for a given input. Classifier selection approaches can be grouped in two categories.

Dynamic Selection:

Given an input sample x. Consider the training or validation set. For each classifier determine

- the k nearest neighbours to x; or
- the k nearest neighbours to x, which are labelled to the same class as x by the classifier.

Then compute the accuracy among the k neighbours to choose which classifiers have to make the final decision. As in the usual k-*NN* classification, one can also take the distances to the neighbours into account.

Predefined Selection Regions:

Given a set of classifiers. One can cluster the training or validation set and compute the accuracy of each classifier for every cluster or train the classifiers separately on the clusters. Given an input sample x, one has first to find the cluster of x. Then, one can choose the best classifier(s) assigned to that cluster to label x.

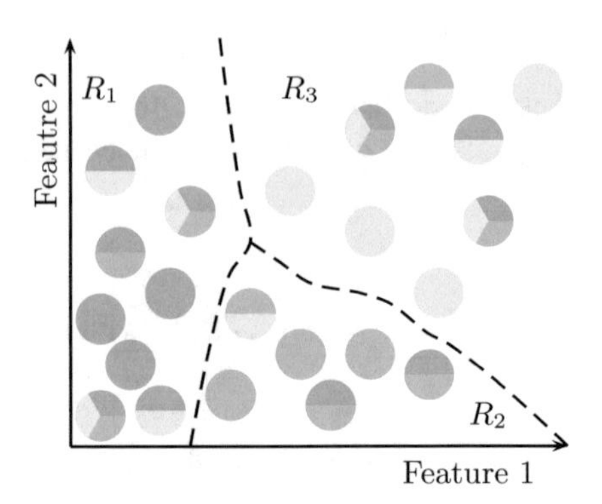

Fig. 2 A two-dimensional classification task where three classifiers (*represented by specific colours*) are used. The colours of each data point show which classifier labels the sample correctly. The dashed curves separate the feature space into three regions

Regarding Fig. 2, a possible selector could choose the classifier which is represented by the blue colour to classify samples from region R_1, the one represented by the green colour for samples from region R_2 and data points which lie in region R_3 by the classifier represented by yellow. One famous classifier selection method is the mixture of experts [32] which was originally constructed for neural networks.

4 Designing a Multi-classifier-System

Building an MCS is not a straight forward procedure. By training each classifier on the whole data set, one might get a set of classifiers which are quite similar, or in the worst case, all identical. This contradicts the main idea of the MCS approach. One needs to apply intelligent techniques to build an appropriate MCS. Two of the most popular methods for creating MCS are boosting and bagging with its variants. The main ideas of these methods will be explained in the following subsections.

4.1 Bagging, Random Subspace Method and Random Forests

Bagging: Breiman introduced the bagging method [33] as an acronym for **b**ootstrap **agg**regat**ing**. The idea of bagging is the building of the MCS by using different training sets. For each classifier one draws a certain amount of training data with replacement.

Random Subspace Method: The random subspace method [34] is equivalent to bagging. But instead of drawing a random subspace of the training set one draws a random subspace of the feature space for each classifier.

Bagging/Random Subspace Method/Random Forests (Training)
Input: Training set $X \subset \mathbb{R}^d$, number of classifiers L.

1. Initialization: $\mathcal{D} = \emptyset$.
2. FOR $k = 1, \ldots, L$
   - Determine training data S_k for classifier D_k:
     - Draw a subset $S_k \subset X$ (**Bagging**); OR
     - Set $S_k = X$ and draw a subset of the feature space s.t. $S_k \subset \mathbb{R}^p$ with $p \leq d$ (**Random Subspace Method**); OR
     - Draw a subset $S_k \subset X$ and take a subset of the feature space s.t. $S_k \subset \mathbb{R}^p$ with $p \leq d$ (**Random Forests**).
   - Train classifier D_k on S_k and add it to the MCS, i.e. $\mathcal{D} = \mathcal{D} \cup \{D_k\}$.

Output: MCS $\mathcal{D}$.

Fig. 3 Training of bagging, the random subspace method and random forests

Random Forests: Random forests [35] is the combination of bagging and the random subspace method. It is constructed for decision trees. Additionally, one can also vary the parameters of each tree, i.e. minimum leaf size, impurity measure, and so on.

The training phase for each of these methods is shown in Fig. 3 with the notations from [22]. It is proposed to use decision trees and take the majority vote for the final decision. But one could also take a different combination method and choose different classifiers to build the MCS. One of the main advantages of these methods is that each classifier can be created independently, thus the building of the MCS can be parallelised.

4.2 Boosting

The idea of boosting is to add the classifiers sequentially. Like in bagging, one uses a subset of the training data for each classifier. At the beginning, the samples of the individual training set are selected according to uniform distribution. At the end of each step, after a classifier is build, the distribution is updated. The probability to draw data points, that have been misclassified so far, increases. Boosting was created for two-class problems. In the original version one trained three classifiers, and built the final decision via majority voting.

AdaBoost [36], acronym for **ada**ptive **boost**ing, is a generalisation of the boosting method. It was also created for two-class problems. The generalisation was made for the number of classifiers. One of the most famous versions for multi-class problems is AdaBoost.M1. Its training phase is explained in Fig. 4, also with the notations from [22].

AdaBoost.M1 (Training)

Input: Training set $X \subset \mathbb{R}^d$, number of classifiers L.

1. Initialization: $\mathcal{D} = \emptyset$, $\boldsymbol{\omega}^1 \in [0,1]^N$, $\sum_{j=1}^{N} \omega_j^1 = 1$.
2. FOR $k = 1, \ldots, L$
   - Draw a subset $S_k \subset X$ in regard to the distribution of $\boldsymbol{\omega}^k$ and train D_k.
   - Compute the weighted MCS error

$$\epsilon_k = \sum_{j=1}^{N} \omega_j^k l_k^j,$$

   whereby, $l_k^j = 1$, if D_k misclassifies x^j and 0 otherwise.
   - If $\epsilon_k = 0$ or $\epsilon_k \geq 0.5$, discard D_k, set ω_j^k to $\frac{1}{N}$ and continue.
   - Else compute

$$\beta_k = \frac{\epsilon_k}{1 - \epsilon_k}, \text{ with } \epsilon_k \in (0, 0.5).$$

   - Update the weights

$$\omega_j^{k+1} = \frac{\omega_j^k \beta^{1-l_k^j}}{\sum_{i=1}^{N} \omega_j^k \beta^{1-l_i^j}}, \quad j = 1, \ldots, N.$$

Output: MCS $\mathcal{D}$ and the weights $\beta_1, \ldots \beta_L$.

Fig. 4 Training of the AdaBoost.M1 method

Once one trained the MCS $\mathcal{D} = \{D_1, \ldots, D_L\}$ with the corresponding weights $\beta_1, \ldots \beta_L$, one can compute the support for class ω_i by

$$\mu_i(x) = \sum_{D_k(x)=\omega_i} ln\left(\frac{1}{\beta_k}\right).$$

The class with the highest support defines the label for x.

Further variants of boosting are AdaBoost.M2 [36], LogitBoost [37], MultiBoosting [38], AveBoost [39] and AdaBoost-VC [40].

5 Fusion Architectures

Once an MCS is build and an aggregation rule determined, one has to design the MCS architecture. Usually, classification tasks make use of data recorded from different sources, especially in real world applications. After the features are extracted from the different data streams, one has to merge them. There exist different fusion architectures. The two main approaches are the early and late fusion. But one can also combine both approaches to get the hybrid fusion. In the following, the fusion architectures are discussed.

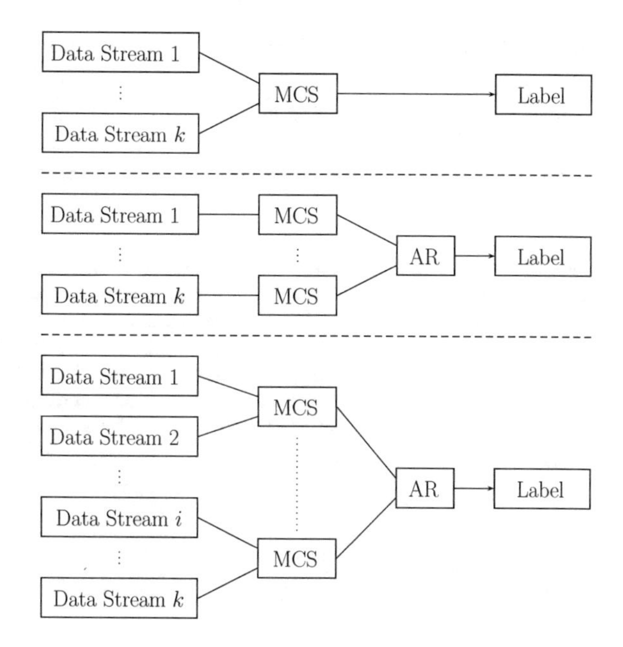

Fig. 5 Fusion Architectures. **Top**: Early Fusion. **Middle**: Late Fusion. **Bottom**: Hybrid Fusion. AR stands for aggregation rule

- **Early Fusion**: Early fusion is the most straightforward approach. One takes the features and merges them to one vector for each data point.
- **Late Fusion**: In the late fusion approach the feature space is divided naturally (based on the sources of the data streams, e.g. audio, video, etc.) or artificially into several subspaces. For each subspace, one trains a classifier or an MCS and aggregates their outputs for the final decision.
- **Hybrid Fusion**: The hybrid fusion approach is a combination of the early and late fusion.

Figure 5 shows the differences between the three fusion methods.

Both main fusion architectures have their benefits and disadvantages. Regarding early fusion, one does not need an extra layer for the aggregation of the classifier systems, which is needed in the late fusion approach.

In case that some data from one or more of the data streams are missing, one has to compensate this problem in the early fusion approach by taking the average, median or taking the values from the nearest neighbours of the given data. In the case of the late fusion approach, one can just leave the MCS out, which are responsible for the missing data streams. This can be a comfortable solution for on-line classification tasks. Table 2 summarises the main benefits and disadvantages of both fusion approaches.

As in Sect. 2, some basic results for the data sets described in Sect. 7.1, are given in Table 3 to show the effectiveness of the fusion approaches.

Table 2 Main benefits and disadvantages of the early and late fusion architectures

Approach	+	−
Early	No additional aggregation rule	Complexity due to high dimensionality dealing with missing data
Late	Missing data can be left out	Additional aggregation rule

Table 3 **Comparison of one single decision tree classifier against bagged late fusion MCS of 7 and 7 × 7 decision tree classifiers.** Every 7 × 7 MCS is significantly better than the single classifier in regard to the Wilcoxon signed rank test with a significance level of $p < 0.05$. To mention is, the MCS with 7 classifiers is significantly worse than the single classifier on the Pain1 data set. The table shows the mean accuracies and standard deviations in % for the leave-one-subject-out setting for the biopotentials of the given data sets. The random performance accuracy for Pain1 is 20%, whereas the random performance accuracy for Pain2-Left and Pain2-Right is 25% each

Data Set	Single	Late fusion 7 classifiers	Late fusion 7 × 7 classifiers
Pain1	31.17 ± 8.18	28.89 ± 7.93	33.61 ± 9.44
Pain2-Left	35.35 ± 5.04	35.54 ± 5.46	38.91 ± 7.44
Pain2-Right	34.83 ± 6.78	34.70 ± 6.35	39.45 ± 7.68

6 Diversity

Methods like bagging and boosting were constructed to create MCS that are somehow diverse. But what does diversity among classifiers mean? One could say that classifiers with different accuracies on the same sets are diverse. However, two classifiers with a similar or even the same accuracy might make different errors. This banal observation is the motivation to apply so-called pruning methods, whereby a *large* MCS is decreased or a *small* MCS is increased until a certain size or an *optimal* MCS, in relation to the classification performance, is found. The removing/adding step in the pruning process is done by using an evaluation measure, which consists of a specific diversity measure or a combination of diversity and accuracy measures. There are many approaches to measure the diversity of an MCS explicitly. Usually the measures are separated into two groups, pairwise and non-pairwise measures. In the following we will list some of the most intuitive measures of each group and add an additional category of diversity measures, which are defined for label outputs of classifiers.

6.1 *Pairwise Measures*

Let $\mathcal{M}$ be a diversity measure. With $\mathcal{M}_{i,j}$ we denote the diversity value for the pair of classifiers (D_i, D_j). The diversity value of the whole MCS is then computed for all pairwise measures as

Table 4 Table with the variables used for defining pairwise diversity measures

$a+b+c+d=1$	D_j correct	D_j wrong
D_i correct	a	b
D_i wrong	c	d

$$\mathcal{M} = \frac{2}{L(L-1)} \sum_{i=1}^{L-1} \sum_{j=i+1}^{L} \mathcal{M}_{i,j} \ .$$

Most of the pairwise measures are based on Table 4.

Two simple and quite intuitive pairwise measures are defined as follows:

- **Disagreement Measure**. The disagreement measure is the probability that two classifiers disagree:
$$\mathcal{M}_{i,j} = b + c \ . \tag{1}$$
- **Double-Fault Measure**. The double-fault measure is the probability that both classifiers make an incorrect decision:
$$\mathcal{M}_{i,j} = d \ . \tag{2}$$

A higher value in (1) means higher diversity, while a higher value in (2) means lower diversity. Further pairwise measures are, for example, the Q-statistic [41], correlation and product-moment correlation [42].

6.2 *Non-pairwise Measures*

The non-pairwise measures compute the diversity value by using the whole MCS at once. Let $Y(x)$ be the number of classifiers that classify instance x correctly, i.e.

$$Y(x) := \sum_{i=1}^{L} \mathbb{1}_{\{D_i(x)=l(x)\}}(x) \ .$$

The Entropy Measure:

The entropy measure is based on the thought that for an object $x \in X$ the MCS has the highest diversity if $\lfloor L/2 \rfloor$ votes are correct and the other $L - \lfloor L/2 \rfloor$ votes are wrong. Based on this, the entropy measure [43] is defined as

$$E = \frac{1}{N} \frac{1}{L - \lceil L/2 \rceil} \sum_{x \in X} \min\{Y(x), L - Y(x)\} \ .$$

Measurement of Interrater Agreement:

Let $\bar{p}$ denote the mean accuracy of the classifiers, then the interrater agreement is defined as

$$\kappa = 1 - \frac{\frac{1}{L}\sum_{x \in X} Y(x)(L - Y(x))}{N(L-1)\bar{p}(1-\bar{p})} .$$

Kuncheva and Whitaker [11] have shown that the interrater agreement is a multiple of the disagreement measure, i.e.

$$\kappa = 1 - \left(\frac{1}{2\bar{p}(1-\bar{p})} \times \text{disagreement value of the MCS} \right) .$$

This is an interesting observation, since the interrater agreement is a non-pairwise measure. Further non-pairwise measures are, for example, the measure of difficulty [44], the Kohavi-Wolpert Variance [45], the generalised diversity and the coincident failure diversity [46]. Kuncheva [22] calls the variety of measures the *diversity of diversity*.

6.3 Measures for Label Outputs

Most of the diversity measures are defined for the artificial correct-false-output of the classifiers, based on Table 4. Kuncheva calls this kind of outputs the *oracle output* [22]. In the following, we introduce an example to justify why it might be better to use the classifiers' label outputs to compute diversity, in regard to multi-class problems.

Motivation

Let's consider a three class problem ($c = 3$) with the class labels $\Omega = \{1, 2, 3\}$ and a set of five elements $X = \{x^1, \ldots, x^5\}$. Furthermore, let $D_1, \ldots, D_4$ be four classifiers which are combined to two different MCS, $\mathcal{D}_1 = \{D_1, D_2, D_3\}$ and $\mathcal{D}_2 = \{D_1, D_2, D_4\}$. Let the true and the guessed labels be given as follows, in Table 5.

Considering all diversity measures based on the correct-false-output (see Table 4) the MCS $\mathcal{D}_1$ and $\mathcal{D}_2$ will be regarded as equally diverse. But intuitively one would say that MCS $\mathcal{D}_2$ is more diverse than MCS $\mathcal{D}_1$, since classifiers D_2 and D_3 produce exactly the same outputs on the given data set X, whereas the classifiers of $\mathcal{D}_2$ produce all different outputs on four out of five samples.

Disagreement and Double-Fault

According to [47], the disagreement measure can be naturally extended for the label outputs of classifiers. The disagreement measure for two classifiers is then defined as the probability that the classifiers provide different label outputs. In this sense, we naturally extend the double-fault measure for the classifiers' label outputs. The double-fault measure of two classifiers is then the probability that the two classifiers make the *same* mistakes.

Table 5 Motivational example for the use of diversity measures, based on the classifiers' label outputs. The MCS $\mathcal{D}_1$ (rows 1 to 8) contains two classifiers which provide exactly the same outputs on all of the given samples. The MCS $\mathcal{D}_2$ (rows 9 to 16) consists of three classifiers which provide all different outputs on all of the given samples, except for x^2. However, in regard to the correct-false-outputs, both MCS are considered the same

MCS $\mathcal{D}_1$	x^1	x^2	x^3	x^4	x^5
$l(x^j)$	1	1	2	2	3
$D_1(x^j)$	1	2	2	2	3
$D_2(x^j)$	2	2	1	1	2
$D_3(x^j)$	2	2	1	1	2
$\mathbb{1}_{\{D_1(x^j)=l(x^j)\}}$	1	0	1	1	1
$\mathbb{1}_{\{D_2(x^j)=l(x^j)\}}$	0	0	0	0	0
$\mathbb{1}_{\{D_3(x^j)=l(x^j)\}}$	0	0	0	0	0
MCS $\mathcal{D}_2$	x^1	x^2	x^3	x^4	x^5
$l(x^j)$	1	1	2	2	3
$D_1(x^j)$	1	2	2	2	3
$D_2(x^j)$	2	2	1	1	2
$D_4(x^j)$	3	2	3	3	1
$\mathbb{1}_{\{D_1(x^j)=l(x^j)\}}$	1	0	1	1	1
$\mathbb{1}_{\{D_2(x^j)=l(x^j)\}}$	0	0	0	0	0
$\mathbb{1}_{\{D_4(x^j)=l(x^j)\}}$	0	0	0	0	0

- **Disagreement measure for classifiers' label outputs**:

$$\mathcal{M}_{i,j}^{\text{lab}} = \frac{1}{|X|} \sum_{x \in X} \mathbb{1}_{\{D_i(x) \neq D_j(x)\}}(x) \ .$$

- **Double-Fault measure for classifiers' label outputs**:

$$\mathcal{M}_{i,j}^{\text{lab}} = \frac{|\{x \in X : D_i(x) = D_j(x),\ D_i(x) \neq l(x)\}|}{|X|} \ .$$

Further measures for label outputs are, for example, the Rényi and Shannon entropy [47]. Considering the MCS $\mathcal{D}_1$ and $\mathcal{D}_2$ we can compute the proposed measure values. If we had computed the disagreement and double-fault values using the correct-false-outputs of the classifiers than the values would be the same for both MCS, $\mathcal{D}_1$ and $\mathcal{D}_2$. But using the measures for the label outputs we get the result that MCS $\mathcal{D}_2$ is more diverse than MCS $\mathcal{D}_1$ and does less similar mistakes (see Table 6). This observation is obviously true.

Therefore, the benefits of using diversity measures for label outputs are:

- We get a deeper differentiation by using label outputs instead of the correct-false-outputs for multi-class problems.

Table 6 Diversity values for label outputs for the MCS $\mathcal{D}_1$ and $\mathcal{D}_2$

	$\mathcal{D}_1$	$\mathcal{D}_2$
Disagreement Measure	$\frac{8}{15}$	$\frac{12}{15}$
Double-Fault Measure	$\frac{7}{15}$	$\frac{3}{15}$

- Usually, we don't need a labelled set for computing the diversity, which means that we can even use artificial data to compute the diversity among a set of classifiers.

Exception for the second bullet point is the double-fault measure.

6.4 Pruning

One way to make use of the diversity measures are the so-called MCS pruning methods [48]. There are two main groups of pruning methods.

- **Backward Elimination**: One creates an MCS and gradually removes classifiers by using a given measure.
- **Forward Selection**: One starts with an empty set and gradually adds classifiers by using a given measure.

The measure can be an accuracy measure, any diversity measure or a combination of both. Applying the backward elimination, one can also first build the MCS, then create subsets of the MCS with similar classifiers in regard to a diversity measure, and finally perform the pruning algorithm for each subset separately. Figure 6 shows one step of the backward elimination pruning.

By using a pairwise diversity measure $\mathcal{M}$ we propose to save computational effort by calculating the diversity matrix $DM(\mathcal{M})$,

One Step Backward Elimination Pruning
Input: MCS $\mathcal{D} = \{D_1, \ldots, D_L\}$, evaluation measure $\mathcal{E}$.

1. Define
   - $\mathcal{D}_j := \mathcal{D} \setminus \{D_j\}$
   - $\mathcal{E}_j$ evaluation value for $\mathcal{D}_j$
2. FOR $k = 1, \ldots, L$
   - Remove D_k from $\mathcal{D}$ just for step k.
   - Compute $\mathcal{E}_k$.

Output: $\mathcal{D}_{j^*}$ s.t. $\mathcal{E}_{j^*}$ is the optimum value among all $\mathcal{E}_j$.

Fig. 6 One step of the backward elimination pruning in the general form

$$DM(\mathcal{M}) = \begin{bmatrix} 0 & \mathcal{M}_{1,2} & \cdots & \cdots & \mathcal{M}_{1,L} \\ 0 & 0 & \mathcal{M}_{2,3} & \cdots & \mathcal{M}_{2,L} \\ \vdots & \vdots & \vdots & \ddots & \vdots \\ 0 & 0 & \cdots & 0 & 0 \end{bmatrix} \in \mathbb{R}^{L \times L} .$$

To compute the diversity value for the whole MCS one has to sum up each element of $DM(\mathcal{M})$ and divide it by the number of pairs. To compute the diversity value after removing classifier D_k from the MCS $\mathcal{D}$ one can just cancel row k and column k in $DM(\mathcal{M})$ and do the same calculation for the corresponding number of classifier pairs. The main reasons for applying pruning methods are

1. A smaller MCS needs less memory and computational time for classification.
2. By taking a subset of the original MCS one tries to increase the accuracy.

To increase accuracy one should not try to build the MCS as diverse as possible. Figure 7 shows the relationship of accuracy versus both disagreement measures. Obviously the most diverse MCS are not the most accurate ones. To compute the results in Fig. 7 we used the Pain1 data set (see Sect. 7.1). Furthermore, to induce a range, in which the accuracy values vary significantly, we applied the following steps. First, we partitioned the data set into a relatively *huge* training set and a relatively *small* test set. The training set was used to design the MCS, whereby the test set was used to calculate the accuracy, and the diversity values as well. Then, we gradually

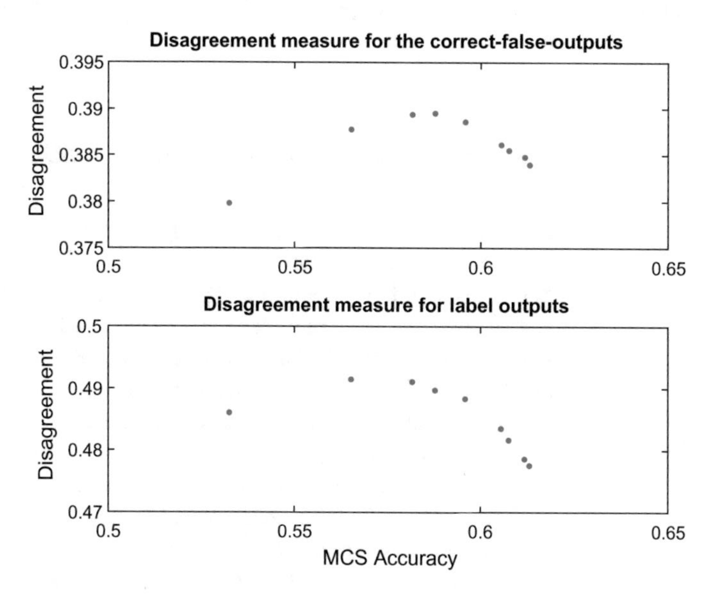

Fig. 7 MCS accuracy versus both disagreement measures on one of the skin conductance level channels of the Pain1 data set with the classes $\Omega = \{T_0, T_2, T_T\}$. The MCS with the highest accuracies are not the most diverse ones. The points form a parabolic shape

changed the ratio of both sets and repeated our experiments, until the training set became relatively *small* and the test set relatively *huge*.

Brown and Kuncheva [49] have shown that by using the disagreement measure - based on the correct-false-outputs - for two-class problems, the MCS majority vote error can be decomposed in three parts, the average individual loss of the classifiers and two diversity terms. One of the diversity terms decreases the MCS error ("good" diversity) and the other diversity term increases the MCS error ("bad" diversity).

7 Case Study: Pain Intensity Estimation

As we mentioned in the introduction, there are several reasons why an automatic pain recognition system would be useful. To take this thought one step further, it might be necessary to be able to estimate different levels of pain, to determine the right therapy for the affected patients, for example. Since pain is a *subjective* experience in the first place, we will ignore data, which is extracted from video and audio signals. Some people may scream when they experience pain, others stay quiet - some people may heavily shake, others remain mainly motionless. Also the recordings of video and audio data can be well-interpreted by human experts. Therefore, the subject of our case study is pain intensity estimation based on biophysiological data. In the first place, we will give a short description of the used data sets, then explain the feature extraction, list some related work on the proposed data sets and finally present the results, followed by a short discussion.

7.1 BioVid Heat Pain Database and SenseEmotion Database

In our case study we use the BioVid Heat Pain Database [50] and the SenseEmotion Database [51] that have been collected at Ulm University. These data sets were specifically recorded for the research field of automatic pain intensity estimation.

The data sets consist of 87 (44 male, 43 female) and 40 (20 male, 20 female) participants, respectively. A thermode was attached to one of the participants' forearms. In a controlled environment, the participants were subjected to several levels of pain stimuli elicited through the thermode. The pain free baseline temperature was $32\,^{\circ}C$ (T_0) for all participants. The different pain levels were calibrated for each participant individually. Therefore, one raised the temperature slowly to define two thresholds. First, one determined the temperature T_P (pain threshold), when the person felt a change from heat to pain. Then, the temperature T_T (tolerance threshold) was determined, when the person felt that the pain was barely manageable (The maximum allowed temperature was set to $50.5\,^{\circ}C$). Then the interval $[T_P, T_T]$ was partitioned in three and two sections of equal size for the BioVid Heat Pain Database and the SenseEmotion Database respectively.

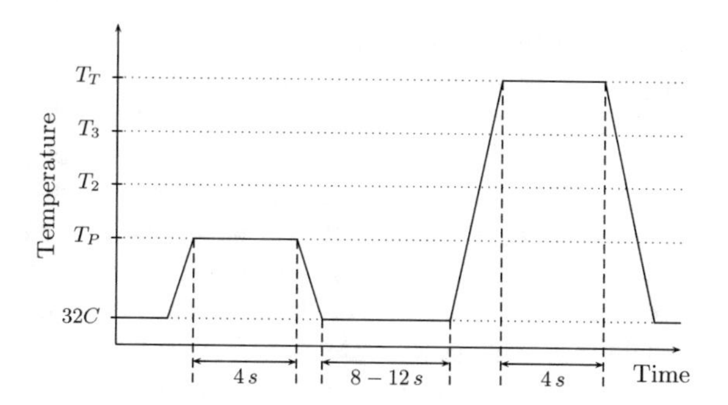

Fig. 8 An example for the stimulation and recovering phase for one participant for the BioVid Heat Pain Database. In the SenseEmotion Database there is only one intermediate temperature between T_P and T_T

Table 7 Characteristics of both data sets, namely BioVid Heat Pain Database (Pain1) and SenseEmotion Database (Pain2)

Differences between the both data sets	Pain1	Pain2
Number of participants	87 (44 m, 43 f)	40 (20 m, 20 f)
Number of classes	5 ($T_0, \ldots, T_4$)	4 ($T_0, \ldots, T_3$)
Number of samples per class for each person	20	30
ECG/EMG/SCL/Video	✓	✓
Audio/RSP	–	✓

Each participant was stimulated with all the pain levels 20 and 30 times in a randomized order for the BioVid Heat Pain Database and the SenseEmotion Database respectively. The pain level was held for 4 s each time. Between two pain stimuli there was a break for a random time of 8–12 s (see Fig. 8). During the experimental sessions different video and audio signals were recorded. The physiological data streams consisted of the skin conductance level (SCL), electrocardiography (ECG), electromyography (EMG) and respiration (RSP).

The SenseEmotion Database was also recorded for an emotion recognition task. But since our case study deals with automatic pain intensity recognition we will focus on the experimental part regarding pain stimulation. The reader is referred to [51] to get the full data set description. Furthermore, the experiments on the SenseEmotion Database were carried out twice for each participant. Once for the right forearm and once for the left forearm. We will denote the BioVid Heat Pain Database by Pain1 and the two parts of the SenseEmotion Database by Pain2-Left and Pain2-Right. Table 7 summarizes the main differences between the data sets.

7.2 *Related Work on both BioVid Heat Pain Database and SenseEmotion Database*

Based on the given data sets, several studies have been conducted. Kächele et al. [52] studied different fusion approaches for the video and biopotential features of the BioVid Heat Pain Database, for the following person independent (leave-one-subject-out cross-validation setting), binary classification tasks: (T_0, T_1), (T_0, T_4), (T_1, T_2), (T_2, T_3), (T_3, T_4) and (T_1, T_4). Furthermore, Kächele et al. [53] also examined different experiments, based uniquely on the biopotentials of the BioVid Heat Pain Database. Thereby, the best performance for the person independent task was reached for (T_0, T_4), i.e. the task T_0 versus T_4, with an average accuracy of 85.7%.

Kessler et al. [54] proposed to include remote Photoplethysmography features from the video channels for the pain intensity estimation on the SenseEmotion Database. This approach was used in [55] by Kessler et al. Thereby, for the person independent task T_0 versus T_3, the fusion of video and biopotential features led to an average accuracy of 71.10% and 71.85%, on the left and right forearm respectively.

Thiam et al. [56] proposed a personalised pain level recognition on the SenseEmotion Database for the task T_0 versus T_3, based uniquely on video features. In [57] Thiam et al. combined audio and video features of the SenseEmotion Database in both person dependent and person independent scenarios. The following binary classification tasks have been conducted: (T_0, T_1), (T_0, T_2) and (T_0, T_3). For the person independent setting, the best performance was reached for the task T_0 versus T_3, with an average accuracy of 65.89 and 66.76%, on the left and right forearm respectively. By combining all available modalities (audio, video and biopotentials) Thiam et al. [58] reached an average accuracy of 82 and 83%, on the left and right forearm respectively for the task T_0 versus T_3, in a person independent setting.

7.3 Feature Extraction of the Biopotentials

Prior to the classification experiments, a finite set of descriptors specific to each modality and characterising typical responses of the autonomic nervous system to external stimuli was extracted. The autonomic nervous system regulates visceral activities such as heart rate, respiratory rate, perspiration and muscle activities during periods of stress and emergency [59]. These descriptors are subsequently used as features for each specific modality during the classification experiments. The recorded modalities were first individually pre-processed using a wide range of signal filtering and smoothing techniques (low- and bandpass filtering, Gaussian filtering, signal detrending). This pre-processing step is necessary in order to substantially reduce the amount of noise and artefacts within the signals resulting from the sensibility of the sensors used to perform the recordings, combined with several sources of noise such as unconstrained body motion or progressive detachment of electrodes from the skin during data acquisition.

Subsequently, statistical descriptors from both the temporal domain (*mean signal value, standard deviation, maxima, minima, skewness, kurtosis*) and the frequency domain (*Fourier coefficients, bandwidth, central frequency, mean frequency*) were extracted. Additional descriptors based on entropy (*shannon entropy, fuzzy entropy, sample entropy*) and the signal amplitude (*waveform length, modified mean value of the absolute value of the signal*) were also extracted. Furthermore, typical descriptors from the ECG signal were extracted based on the differences between consecutive heart beats and the detected PQRST waves (*root mean sum of squared differences, amplitudes of P, Q, R, S and T wavelets*) [53, 60]. Finally, additional statistical descriptors were extracted from both tonic and phasic components of the SCL signal. A total of 194 features were extracted from the BioVid Heat Pain Database, while a total of 335 features were extracted from the SenseEmotion Database. After removing unusable features, i.e. ones that contained NaN (Not-a-Number) values or that were constant for all data points, the bio-physiological feature set specific to the SenseEmotion Database was reduced to 307 features.

7.4 Results

In our case study we did several experiments on person independent pain intensity estimation. For each of the proposed data sets from Sect. 7.1 we did a cross-validation, whereby in each step the data of one person was used as the test set and the remaining data as the training set (leave-one-subject-out setting). We repeated the cross-validation for the methods bagging, boosting and random forests. For each method we computed the accuracies for each single modality and the three different fusion approaches, explained in Sect. 5. We used decision trees as base classifiers. Furthermore, we used the mean as the aggregation rule for both, the late and hybrid fusion approach. In the late fusion approach we trained one MCS for each channel, in the hybrid fusion approach one MCS for each modality. Each MCS consisted of 200 and 300 decision trees for the Pain1 and Pain2 data sets respectively. As shown in Table 7, Pain1 is a 5-class classification task, whereas the Pain2 data sets represent a 4-class classification task each. All of the experiments were conducted with the built-in functions in MATLAB. The parameter setting was left to default. Tables 8, 9 and 10 show the mean accuracies and standard deviations of the evaluated cross-validations on the three proposed data sets respectively.

The boosting results on the Pain2-Right data set (see Table 10) and the basic boosting results from Table 1 for the early fusion lead to one interesting observation regarding the number of classifiers, which are summarized in Table 11. From Table 11 we can see that the increase of the MCS size from 7 to 300 base classifiers could only slightly improve the mean accuracy. In contrast, the median accuracy even decreased, and the standard deviation increased.

Regarding Table 11, it might lead to better results by creating MCS of smaller size, when using the boosting method. Since the boosting method is choosing the training samples selectively in each step, it might lead to overfitting by taking too many samples from one single or a specific group of test persons. Furthermore, the late and hybrid fusion approaches might perform better by using a different aggregation rule, for example a trainable one, like the Pseudo-Inverse proposed in Sect. 3.2. From the Tables 8, 9 and 10 we get the following observations:

Table 8 Comparison between single modalities and fusion approaches on Pain1. Bagging and random forests are both significantly better compared to the boosting method for each fusion approach in regard to the Wilcoxon signed rank test with a significance level of $p < 0.05$. For bagging and random forests the late fusion approach is significantly worse than the other two approaches. The table shows the mean accuracies and standard deviations in % for the leave-one-subject-out setting for the Pain1 data set. The best performing method is represented in bold numbers. The figures in brackets denote the number of channels per data stream. The random performance accuracy for Pain1 is 20%, since it represents a 5-class classification task

Pain1	Single Modalities			Fusion Approaches		
Method	EMG (1)	ECG (3)	SCL (3)	Late	Hybrid	Early
Bagging	23.91 ± 6.82	26.30 ± 8.11	38.46 ± 11.21	37.08 ± 10.77	38.69 ± 10.83	**39.34 ± 10.21**
Boosting	24.17 ± 6.12	24.16 ± 7.51	35.90 ± 7.43	35.63 ± 7.28	**35.94 ± 7.78**	35.74 ± 7.19
RF	23.85 ± 7.15	26.52 ± 8.57	38.40 ± 10.90	37.51 ± 11.06	38.25 ± 11.11	**39.09 ± 10.85**

Table 9 Comparison between single modalities and fusion approaches on Pain2-Left. Only for the boosting method one of the fusion approaches outperforms the SCL modality. But it's still worse than the mean accuracy of the bagged SCL single modality. The table shows the mean accuracies and standard deviations in % for the leave-one-subject-out setting for the Pain2-Left data set. The best performing method is represented in bold numbers. The figures in brackets denote the number of channels per data stream. The random performance accuracy for Pain2-Left is 25%, since it represents a 4-class classification task

Pain2-Left	Single Modalities			Fusion Approaches		
Method	RSP (1)	ECG (2)	SCL (3)	Late	Hybrid	Early
Bagging	33.16 ± 7.09	33.34 ± 7.11	**43.50 ± 8.65**	41.96 ± 8.85	42.27 ± 8.68	43.03 ± 8.51
Boosting	33.03 ± 7.52	33.25 ± 6.24	42.17 ± 7.93	41.88 ± 8.16	**42.75 ± 7.85**	42.31 ± 8.08
RF	33.81 ± 7.08	32.56 ± 8.27	**42.72 ± 8.64**	42.08 ± 8.95	41.31 ± 7.57	42.09 ± 8.79

- The SCL is the best single modality. The accuracy results are significantly better, compared to the other single modalities.
- The SCL modality leads to the best results twice, both times on the Pain2-Left data set.
- The hybrid fusion approach led to the best results on all of the three data sets when using the boosting method.
- The bagging method led to the best overall results on all of the three data sets.

Figure 9 shows some of the presented results in detail.

Tables 8, 9 and 10 show the mean accuracies, which are all below 50% and exhibit relatively high standard deviations. Figure 10 presents the individual accuracies for some of the proposed methods. From Fig. 10 we can observe:

Table 10 Comparison between single modalities and fusion approaches on Pain2-Right. For each method all fusion approaches outperform the best single modality (SCL). The boosting hybrid approach is significantly better than the boosting SCL single modality in regard to the Wilcoxon signed rank test with a significance level of $p < 0.05$, but it's not significantly better than the bagged SCL single modality. The table shows the mean accuracies and standard deviations in % for the leave-one-subject-out setting for the Pain2-Right data set. The best performing method is represented in bold numbers. The figures in brackets denote the number of channels per data stream. The random performance accuracy for Pain2-Right is 25%, since it represents a 4-class classification task

Pain2-Right	Single Modalities			Fusion Approaches		
Method	RSP (1)	ECG (2)	SCL (3)	Late	Hybrid	Early
Bagging	34.23 ± 6.79	33.82 ± 8.71	42.26 ± 9.85	**43.16 ± 8.16**	42.15 ± 7.96	42.34 ± 8.83
Boosting	33.13 ± 7.20	33.49 ± 7.65	40.68 ± 8.14	41.90 ± 8.08	**42.74 ± 8.88**	40.81 ± 8.14
RF	33.07 ± 7.73	33.78 ± 8.23	41.79 ± 9.82	**42.51 ± 8.10**	42.24 ± 8.51	41.89 ± 8.65

Table 11 Comparison of the number of classifiers for the boosting method in the early fusion approach evaluated on the Pain2-Right data set. The table shows the accuracies and standard deviations in % for the leave-one-subject-out setting

Number of classifiers	Mean	Median
$L = 7$	40.39 ± 6.83	41.84
$L = 300$	40.81 ± 8.14	41.20

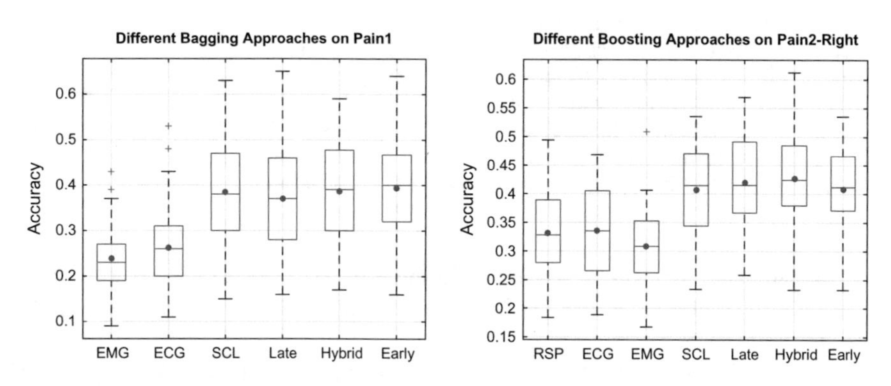

Fig. 9 Boxplots of accuracies of different methods. The dots denote the mean values. **Left**: The early fusion is the best approach regarding the median and mean value. **Right**: The Hybrid fusion approach is significantly better than the best single modality (SCL) in regard to the Wilcoxon signed rank test with a significance level of $p < 0.05$

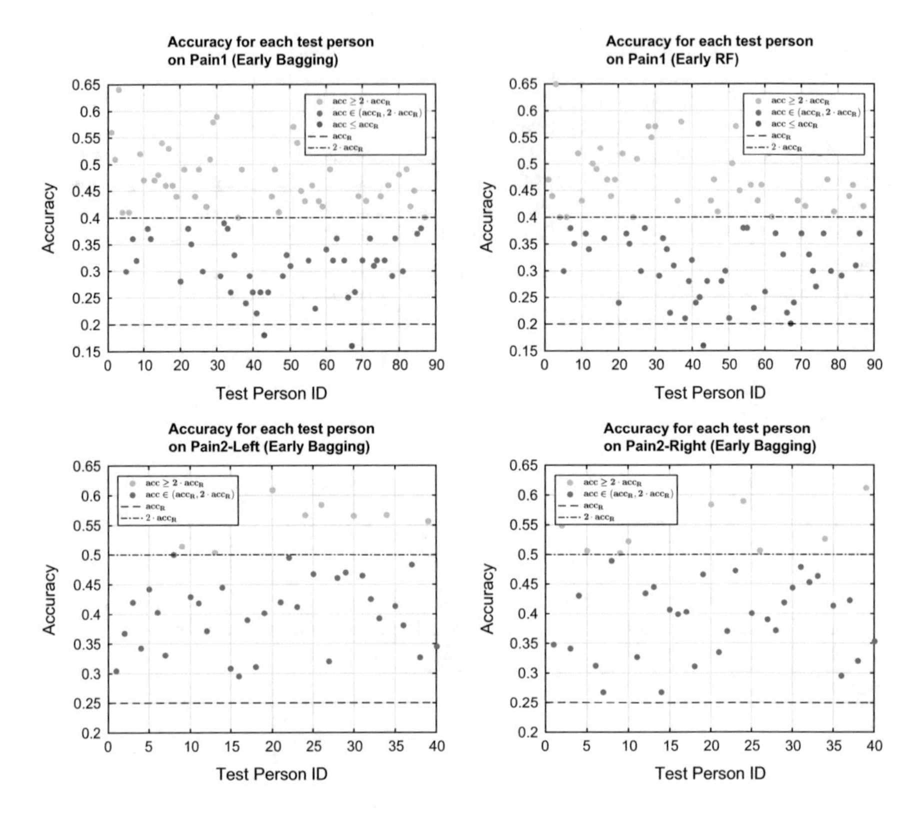

Fig. 10 Individual accuracies for the leave-one-subject-out setting with the early fusion approach. Accuracy is denoted by acc, whereas acc_R indicates the chance level of random guessing. **Top** (Pain1): In the bagging approach there are 44 out of 87 test persons with an accuracy above $2 \cdot acc_R$. In the random forests approach there are 43 out of 87 test persons with an accuracy above $2 \cdot acc_R$. The test persons with the IDs 43 and 67 are in both approaches below the chance level acc_R. **Bottom** (Pain2): There are 8 and 9 out of 40 test persons with accuracies above $2 \cdot acc_R$, on Pain2-Left and Pain2-Right respectively. There is also not even one test person below the chance level acc_R

- The automatic person independent pain intensity estimation (leave-one-subject-out setting) seems to work better for a subset of the test persons, but unfortunately not for all.
- Especially in the Pain1 data set there seem to be at least two test persons, which don't fit to the rest of the set at all.

7.5 Discussion

In our case study we dealt with person independent pain intensity estimation. Thereby, we focused on the physiological data streams of the BioVid Heat Pain Database and the SenseEmotion Database. The bagging method performed best on all of the three data sets.

The EMG is the worst performing modality. The ECG and RSP modalities lead to similar accuracy results, which lie between the results obtained from the EMG and SCL channels. The SCL is a very strong modality to measure pain. From the experiments we can conclude that the skin conductance is correlated to the temperature stimuli experienced by the test persons. However the information gained from the SCL modality might be only relevant for the pain scenario, in which the pain is induced by heat. Anyway, the results seem not to be reliable enough for a real world application. At this stage, it seems to be reasonable, to simplify the task of the estimation of different pain intensity levels to the distinction between no pain and (strongly experienced) pain, at least for the person independent scenario. An other approach is to train an additional selecting rule, which finds *similar* test persons and restricts the training data in regard to each test person, to improve accuracy, as proposed in [53].

To show the effectiveness of the different fusion approaches one could repeat the experiments with all available modalities, including audio and video features. Furthermore, one could compare the accuracies obtained by using single modalities against all fusion approaches without taking the SCL channels into account.

It also might be useful to repeat the experiments with fixed temperature levels for all test persons. The individual calibration of the stimulation temperatures might have possibly led to a multi-class problem, where each temperature level represents an own class.

8 Conclusion

In this paper we summarised the main aspects of MCS. First, we reflected the main reasons for the use of an MCS and added some basic results to substantiate the effectiveness of the MCS approach. Therefore, we showed that already a *small* amount of classifiers significantly increases the classification performance in regard to the accuracy reached by a single classifier. Then, we described some existing combination methods for classifiers and MCS building approaches to show the variety of MCS designs. We continued by summarising the main fusion architectures, namely early, late and hybrid fusion and pointed out their essential differences, advantages and drawbacks. Every combination of all the three fusion architectures was each combined with the bagging, boosting and random forests approach in our experiments. The bagging method led to the best overall classification results on our proposed data sets in regard to accuracy.

In the second part of the paper we explained the meaning of diversity in the context of MCS. We proposed a short list of existing diversity measures and introduced an example why it might be useful to stick to the classifier's label outputs when dealing with diversity. In this sense we naturally extended the existing double-fault measure to the classifier's label outputs. Then, we discussed the main application of diversity measures, the MCS pruning. We added another basic result, which underpins that diversity can also be detrimental to the performance of an MCS.

The last part of this paper includes a case study on pain intensity estimation, which is important in the field of e-health. First, we described the BioVid Heat Pain Database and the SenseEmotion Database. We continued with an explanation of the implemented feature extraction of the biopotentials of the proposed data sets. Our experiments show that the implementation of different fusion architectures is useful and is able to lead to an overall accuracy improvement most of the times. Unfortunately, it is not possible to undertake one of the presented pruning methods to achieve higher classification accuracies. The *complexity* of the proposed data sets is obviously quite high, especially for the leave-one-subject-out setting. To undertake effective pruning steps one needs a suitable pruning set (validation set), which is *similar* to the test set. This constitutes the main problem, since the test persons seem to differ significantly, concerning the recorded data. Therefore, to improve the automatic pain intensity estimation for a person independent setting it is necessary to propose new concepts. One possible approach, which was mentioned in Sect. 7.5, is to train the MCS on an appropriate subset of the training data for each test set. Therefore, one has to find test persons which are similar to the test person represented by the test data. The similarity can be calculated in regard to some meta information (test persons' gender, age, etc.) or machine learning based techniques (Euclidean distance, Hausdorff distance, etc.).

Acknowledgements Peter Bellmann is supported by a scholarship of the Landesgraduierten-förderung Baden-Württemberg at Ulm University. Patrick Thiam is supported by the Federal Ministry of Education and Research (BMBF) within the project *SenseEmotion*. The work of Friedhelm Schwenker is partially supported by the Transregional Collaborative Research Centre SFB/TRR 62 *Companion-Technology for Cognitive Technical Systems*, funded by the German Research Foundation (DFG). We gratefully acknowledge the support of NVIDIA Corporation with the donation of the Tesla K40 GPU used for this research.

References

1. R.A. Fisher, The use of multiple measurements in taxonomic problems. Ann. Eugenics **7**(2), 179–188 (1936)
2. B.V. Dasarathy, Nearest neighbor (nn) norms: Nn pattern classification techniques. In: IEEE Computer Society Press, Los Alamitos, California (1990)
3. L. Breiman, J.H. Friedman, R.A. Olshen, C.J. Stone, Classification and Regression Trees. (Wadsworth, 1984)
4. J.A. Anderson, E. Rosenfeld (eds.), *Neurocomputing: Foundations of Research* (MIT Press, Cambridge, 1988)
5. Kittler, J., Roli, F. (eds.): Multiple Classifier Systems, First International Workshop, MCS 2000, Cagliari, Italy, June 21-23, 2000, Proceedings, Lecture Notes in Computer Science, vol. 1857 (Springer, 2000)
6. Schwenker, F., Roli, F., Kittler, J. (eds.): Multiple Classifier Systems - 12th International Workshop, MCS 2015, Günzburg, Germany, June 29 - July 1, 2015, Proceedings, Lecture Notes in Computer Science, vol. 9132. Springer (2015)

7. C. Dietrich, F. Schwenker, G. Palm, Classification of time series utilizing temporal and decision fusion, in *Multiple Classifier Systems, Second International Workshop, MCS 2001 Cambridge, UK, 2–4 July 2001, Proceedings* ed. by J. Kittler, F. Roli. Lecture Notes in Computer Science, vol. 2096 (Springer, 2001), pp. 378–387
8. C. Dietrich, F. Schwenker, G. Palm, Multiple classifier systems for the recognition of orthoptera songs, in *Pattern Recognition, 25th DAGM Symposium, Magdeburg, Germany, 10–12 Sept 2003, Proceedings* ed. by B. Michaelis, G. Krell. Lecture Notes in Computer Science, vol. 2781 (Springer, 2003), pp. 474–481
9. C. Thiel, F. Schwenker, G. Palm, Using dempster-shafer theory in MCF systems to reject samples, in *Multiple Classifier Systems, 6th International Workshop, MCS 2005, Seaside, CA, USA, 13–15 June 2005, Proceedings* ed. by N.C. Oza, R. Polikar, J. Kittler, F. Roli. Lecture Notes in Computer Science, vol. 3541 (Springer, 2005), pp. 118–127
10. M.F.A. Hady, F. Schwenker, Decision templates based RBF network for tree-structured multiple classifier fusion, in *Multiple Classifier Systems, 8th International Workshop, MCS 2009, Reykjavik, Iceland, 10–12 June 2009, Proceedings* ed. by J.A. Benediktsson, J. Kittler, F. Roli. Lecture Notes in Computer Science, vol. 5519 (Springer, 2009), pp. 92–101
11. L.I. Kuncheva, C.J. Whitaker, Measures of diversity in classifier ensembles and their relationship with the ensemble accuracy. Mach. Learn. **51**(2), 181–207 (2003)
12. S. Faußer, F. Schwenker, Neural network ensembles in reinforcement learning. Neural Process. Lett. **41**(1), 55–69 (2015)
13. F. Schwenker, S. Scherer, M. Schmidt, M. Schels, M. Glodek, Multiple classifier systems for the recognition of human emotions, in *Multiple Classifier Systems, 9th International Workshop, MCS 2010, Cairo, Egypt, 7–9 Apr 2010, Proceedings* ed. by N.E. Gayar, J. Kittler, F. Roli. Lecture Notes in Computer Science, vol. 5997 (Springer, 2010), pp. 315–324
14. M. Schels, F. Schwenker, A multiple classifier system approach for facial expressions in image sequences utilizing GMM supervectors, in *20th International Conference on Pattern Recognition, ICPR 2010* (IEEE Computer Society, Istanbul, Turkey, 2010), pp. 4251–4254
15. M. Schels, P. Schillinger, F. Schwenker, Training of multiple classifier systems utilizing partially labeled sequential data sets, in *ESANN 2011, Proceedings of 19th European Symposium on Artificial Neural Networks*, Bruges, Belgium, 27–29 Apr 2011
16. M. Glodek, S. Tschechne, G. Layher, M. Schels, T. Brosch, S. Scherer, M. Kächele, M. Schmidt, H. Neumann, G. Palm, F. Schwenker, Multiple classifier systems for the classification of audio-visual emotional states, in *Affective Computing and Intelligent Interaction - Fourth International Conference, ACII 2011, Memphis, TN, USA, 9–12 Oct 2011, Proceedings* ed. by S.K. D'Mello, A.C. Graesser, B.W. Schuller, Martin, J. (eds.), Part II. Lecture Notes in Computer Science, vol. 6975 (Springer, 2011), pp. 359–368
17. M. Glodek, F. Honold, T. Geier, G. Krell, F. Nothdurft, S. Reuter, F. Schüssel, T. Hörnle, K.C.J. Dietmayer, W. Minker, S. Biundo, M. Weber, G. Palm, F. Schwenker, Fusion paradigms in cognitive technical systems for human-computer interaction. Neurocomputing **161**, 17–37 (2015)
18. S. Meudt, D. Zharkov, M. Kächele, F. Schwenker, Multi classifier systems and forward backward feature selection algorithms to classify emotional coloured speech, in *2013 International Conference on Multimodal Interaction, ICMI '13*, ed. by J. Epps, F. Chen, S. Oviatt, K. Mase, A. Sears, K. Jokinen, B.W. Schuller, 9–13 Dec 2013. (ACM, Sydney, NSW, Australia, 2013), pp. 551–556
19. T.G. Dietterich, Ensemble methods in machine learning, in *Multiple Classifier Systems, First International Workshop, MCS 2000, Cagliari, Italy, 21–23 June 2000, Proceedings* ed. by J. Kittler, F. Roli. Lecture Notes in Computer Science, vol. 1857 (Springer, 2000). pp. 1–15
20. R. Polikar, Ensemble based systems in decision making. IEEE Circuits Syst. Mag. **6**(3), 21–45 (2006)
21. L. Xu, A. Krzyzak, C.Y. Suen, Methods of combining multiple classifiers and their applications to handwriting recognition. IEEE Trans. Syst. Man Cybern. **22**(3), 418–435 (1992)
22. L.I. Kuncheva, *Combining Pattern Classifiers: Methods and Algorithms* (Wiley, New Jerssey, 2004)

23. J. Kittler, M. Hatef, R.P.W. Duin, J. Matas, On combining classifiers. IEEE Trans. Pattern Anal. Mach. Intell. **20**(3), 226–239 (1998)
24. D.M.J. Tax, M. van Breukelen, R.P.W. Duin, J. Kittler, Combining multiple classifiers by averaging or by multiplying? Pattern Recognit. **33**(9), 1475–1485 (2000)
25. D.G. Saari, Mathematical structure of voting paradoxes. Econ. Theory **15**(1), 55–102 (2000)
26. Y.S. Huang, C.Y. Suen, A method of combining multiple experts for the recognition of unconstrained handwritten numerals. IEEE Trans. Pattern Anal. Mach. Intell. **17**(1), 90–94 (1995)
27. L. Kuncheva, J.C. Bezdek, R.P.W. Duin, Decision templates for multiple classifier fusion: an experimental comparison. Pattern Recognit. **34**(2), 299–314 (2001)
28. C. Dietrich, G. Palm, F. Schwenker, Decision templates for the classification of bioacoustic time series. Inf. Fusion **4**(2), 101–109 (2003)
29. R. Penrose, A generalized inverse for matrices. Proc. Camb. Philos. Soc. **51**, 406–413 (1955)
30. F. Schwenker, C.R. Dietrich, C. Thiel, G. Palm, Learning of decision fusion mappings for pattern recognition. Int. J. Artif. Intell. Mach. Learn. (AIML) **6**, 17–21 (2006)
31. G. Giacinto, F. Roli, Design of effective neural network ensembles for image classification purposes. Image Vision Comput. **19**(9–10), 699–707 (2001)
32. R.A. Jacobs, M.I. Jordan, S.J. Nowlan, G.E. Hinton, Adaptive mixtures of local experts. Neural Comput. **3**(1), 79–87 (1991)
33. L. Breiman, Bagging predictors. Mach. Learn. **24**(2), 123–140 (1996)
34. T.K. Ho, The random subspace method for constructing decision forests. IEEE Trans. Pattern Anal. Mach. Intell. **20**(8), 832–844 (1998)
35. L. Breiman, Random forests. Mach. Learn. **45**(1), 5–32 (2001)
36. Y. Freund, R.E. Schapire, A decision-theoretic generalization of on-line learning and an application to boosting. J. Comput. Syst. Sci. **55**(1), 119–139 (1997)
37. J. Friedman, T. Hastie, R. Tibshirani, Additive logistic regression: a statistical view of boosting (with discussion and a rejoinder by the authors). Ann. Stat. **28**(2), 337–407 (2000)
38. G.I. Webb, Multiboosting: a technique for combining boosting and wagging. Mach. Learn. **40**(2), 159–196 (2000)
39. N.C. Oza, Boosting with averaged weight vectors, in *Multiple Classifier Systems, 4th International Workshop, MCS 2003, Guilford, UK, 11–13 June 2003, Proceedings* ed. by T. Windeatt, F. Roli. Lecture Notes in Computer Science, vol. 2709 (Springer, 2003), pp. 15–24
40. P.M. Long, SN, V.B.V.: Boosting and microarray data. Mach. Learn. **52**(1–2), 31–44 (2003)
41. G.U. Yule, On the association of attributes in statistics. Phil. Trans. **A**, 257–319 (1900)
42. D. Ruta, B. Gabrys, Analysis of the correlation between majority voting error and the diversity measures in multiple classifier systems, in *Proceedings of the SOCO/ISFI'2001 Conference*, No. 1824-0, ICSC-NAISO (Academic Press, 2001), p. 50
43. P. Cunningham, J. Carney, Diversity versus quality in classification ensembles based on feature selection, in *Machine Learning: ECML 2000, 11th European Conference on Machine Learning, Barcelona, Catalonia, Spain, May 31 - June 2, 2000, Proceedings*, ed. by R.L. de Mántaras, E. Plaza, vol. 1810. Lecture Notes in Computer Science (Springer, 2000), pp. 109–116
44. L.K. Hansen, P. Salamon, Neural network ensembles. IEEE Trans. Pattern Anal. Mach. Intell. **12**(10), 993–1001 (1990)
45. R. Kohavi, D. Wolpert, Bias plus variance decomposition for zero-one loss functions, in *Machine Learning, Proceedings of the Thirteenth International Conference (ICML '96), Bari, Italy, 3–6 July 1996* ed. by L. Saitta (Morgan Kaufmann, 1996), pp. 275–283
46. D. Partridge, W.J. Krzanowski, Software diversity: practical statistics for its measurement and exploitation. Inf. Softw. Technol. **39**(10), 707–717 (1997)
47. A. Mikami, M. Kudo, A. Nakamura, Diversity measures and margin criteria in multi-class majority vote ensemble, in *Multiple Classifier Systems - 12th International Workshop, MCS 2015, Günzburg, Germany, June 29–July 1, 2015, Proceedings* ed. by F. Schwenker, F. Roli, J. Kittler. Lecture Notes in Computer Science, vol. 9132 (Springer, 2015), pp. 27–37
48. G. Tsoumakas, I. Partalas, I.P. Vlahavas, An ensemble pruning primer, in *Applications of Supervised and Unsupervised Ensemble Methods, Studies in Computational Intelligence*, ed. by O. Okun, G. Valentini, vol. 245 (Springer, 2009), pp. 1–13

49. G. Brown, L.I. Kuncheva, "good" and "bad" diversity in majority vote ensembles, *Multiple Classifier Systems, 9th International Workshop, MCS 2010, Cairo, Egypt, 7–9 Apr 2010, Proceedings* ed. by N.E. Gayar, J. Kittler. Lecture Notes in Computer Science, vol. 5997 (Springer, 2010), pp. 124–133
50. S. Walter, S. Gruss, H. Ehleiter, J. Tan, H.C. Traue, S.C. Crawcour, P. Werner, A. Al-Hamadi, A.O. Andrade, The biovid heat pain database data for the advancement and systematic validation of an automated pain recognition system, in *2013 IEEE International Conference on Cybernetics, CYBCO 2013, 13–15 June 2013* (IEEE, Lausanne, Switzerland, 2013), pp. 128–131
51. M. Velana, S. Gruss, G. Layher, P. Thiam, Y. Zhang, D. Schork, V. Kessler, S. Meudt, H. Neumann, J. Kim, F. Schwenker, E. André, H.C. Traue, S. Walter, The senseemotion database: A multimodal database for the development and systematic validation of an automatic pain- and emotion-recognition system, in *Multimodal Pattern Recognition of Social Signals in Human-Computer-Interaction - 4th IAPR TC 9 Workshop, MPRSS 2016, Cancun, Mexico, December 4, 2016, Revised Selected Papers* ed. by F. Schwenker, S. Scherer. Lecture Notes in Computer Science, vol. 10183 (Springer, 2016), pp. 127–139
52. M. Kächele, P. Werner, A. Al-Hamadi, G. Palm, S. Walter, F. Schwenker, Bio-visual fusion for person-independent recognition of pain intensity, in *Multiple Classifier Systems - 12th International Workshop, MCS 2015, Günzburg, Germany, June 29–July 1, 2015, Proceedings* ed. by F. Schwenker, F. Roli, J. Kittler. Lecture Notes in Computer Science, vol. 9132 (Springer, 2015), pp. 220–230
53. M. Kächele, P. Thiam, M. Amirian, F. Schwenker, G. Palm, Methods for person-centered continuous pain intensity assessment from bio-physiological channels. J. Sel. Topics Signal Process. **10**(5), 854–864 (2016)
54. V. Kessler, P. Thiam, A. Mohammadreza, F. Schwenker, Pain recognition with camera photoplethysmography, in *International Conference on Image Processing Theory, Tools and Applications* (2017)
55. V. Kessler, P. Thiam, M. Amirian, F. Schwenker, Multimodal fusion including camera photoplethysmography for pain recognition, in *International Conference on Companion Technology. IEEE International Conference on Companion Technology* (2017)
56. P. Thiam, V. Kessler, F. Schwenker, Hierarchical combination of video features for personalised pain level recognition, in *25th European Symposium on Artificial Neural Networks, Computational Intelligence and Machine Learning* (2017), pp. 465–470
57. P. Thiam, V. Kessler, S. Walter, G. Palm, F. Schwenker, Audio-visual recognition of pain intensity. *Multimodal Pattern Recognition of Social Signals in Human-Computer-Interaction - 4th IAPR TC 9 Workshop, MPRSS 2016, Cancun, Mexico, December 4, 2016, Revised Selected Papers* ed. by F. Schwenker, S. Scherer. Lecture Notes in Computer Science, vol. 10183. (Springer, 2016), pp. 110–126
58. P. Thiam, F. Schwenker, Multi-modal data fusion for pain intensity assessment and classification, in *International Conference on Image Processing Theory, Tools and Applications* (2017)
59. S. Balters, M. Steinert, Capturing emotion reactivity through physiology measurement as a foundation for affective engineering in engineering design science and engineering practices. J. Intell. Manuf. **28**(7), 1585–1607 (2017)
60. M. Kächele, M. Amirian, P. Thiam, P. Werner, S. Walter, G. Palm, F. Schwenker, Adaptive confidence learning for the personalization of pain intensity estimation systems. Evol. Syst. **8**(1), 71–83 (2017)

Learning Label Dependency and Label Preference Relations in Graded Multi-label Classification

Khalil Laghmari, Christophe Marsala and Mohammed Ramdani

Abstract Graded multi-label classification (GMLC) is a supervised machine learning task where the association between each data and a label has a membership degree from an ordinal scale of membership degrees: for example, an odorous molecule can be associated to the graded subset of odors {strong musc, moderate animal} based on the ordinal scale of odor intensity: {very weak, weak, moderate, strong, very strong}, and a movie can be associated to the graded subset of labels {action ⋆ ⋆ ⋆⋆, suspense ⋆⋆, humour ⋆⋆} based on the ordinal scale of one-to-five star rating. The aim in GMLC is to build a predictive model called classifier, in order to predict the graded set of labels based on descriptive attributes of data. For example, predicting the graded set of molecule odors based on molecular properties such as the molecular structure and weight. Or predicting the graded set of genres for a movie based on the synopsis and the main actors. An interesting challenge in GMLC is learning label relations and exploiting them to enhance the prediction performance of classifiers. A label relation can be a dependency relation: for example, movies containing a lot of 'action' often contains also some 'suspense'. Another type of label relations is preference relations: for example, it is preferred to associate a movie containing a lot of movements to the label 'action' than to the label 'humour'. The limitation of existing approaches is that they can either learn dependency relations or preference relations. This work reviews state of the art GMLC approaches, and introduces a new GMLC approach that can learn both dependency and preference label relations. Experiments on real datasets show that the new approach outperforms baseline approaches according the used prediction evaluation measures.

K. Laghmari (✉) · M. Ramdani
Laboratoire Informatique de Mohammedia, FSTM Hassan II University of Casablanca, BP 146, 20650 Mohammedia, Maroc
e-mail: laghmari.khalil@gmail.com

M. Ramdani
e-mail: christophe.marsala@lip6.fr

K. Laghmari · C. Marsala
Sorbonne Universités, CNRS, Laboratoire Informatique de Paris 6, 75005 Paris, France
e-mail: ramdani@fstm.ac.ma

W. Pedrycz and S.-M. Chen (eds.), *Computational Intelligence for Pattern Recognition*, Studies in Computational Intelligence 777,
https://doi.org/10.1007/978-3-319-89629-8_5

Keywords Graded multi-label classification · Label dependency relations · Label preference relations

1 Introduction

One of the most fundamental human cognitive skills is pattern recognition. For example, given a set of categories such as 'happy', 'sad', and 'angry', a human is able to label music records with the given categories depending on the emotion expressed by each music record. Even without a given set of categories, a human is still able to regroup the given set of recorded musics depending on his perception. For example, long music records can be grouped and separated from the group of short music records. Music records using violin as the main instrument can be grouped and separated from music records using guitar or piano as the main instrument.

The challenge of transferring the ability of pattern recognition to machines belongs to an area of artificial intelligence called machine learning. Machine learning can be unsupervised in case a set of unlabelled data described by features is given, and the machine learns to regroup data into clusters so that feature values of data in the same cluster are similar, and feature values of data from different clusters are dissimilar. This task of unsupervised machine learning is called clustering. Machine learning can be supervised in case a set of labelled data described with a set of features is given, and the machine learns to label unlabelled data by extracting from labelled data the features discriminating labels. This supervised machine learning task is called classification.

Single label classification is the case where each instance is associated to exactly one label from a set of available labels. Binary classification is the case where there are only two available labels. Multi-class classification is the case where there are more than two available labels. Multi-class ordinal classification is the case where the set of available labels is ordered [1]. Multi-label classification (MLC) is the case where each instance can be associated to a subset of labels instead of a single label [2]. The task of MLC is handled either by extending single label classification approaches, or by exploiting existing single label classifiers after transforming multi-label instances to single label instances [3]. Fuzzy classification is a MLC task where the association between an instance and a label has a membership degree ranging between 0 and 1. Graded multi-label classification (GMLC) is different from fuzzy classification by the fact that membership degrees are from an ordinal scale of membership grades instead of being necessarily in the range [0, 1] [4, 5].

The GMLC task is handled either by transforming it into a set of MLC subtasks, then labels associated to each membership degree are predicted, or by transforming it into a set of ordinal classification subtasks, then the membership degree corresponding to each label is predicted. The MLC and the ordinal classification subtasks can be decomposed to a set of binary classification subtasks. The advantage of transforming the GMLC task is to benefit from existing binary classifiers instead of modifying their algorithms to handle graded multi-label instances.

Learning label relations and exploiting them to improve the prediction quality is an interesting task that has been addressed in MLC [6, 7]. GMLC can benefit from MLC strategies of learning label relations since GMLC can be transformed into a set of MLC subtasks. There are two main types of label relations in MLC:

- preference relations where the decision is the preferred label to be associated to an instance among a pair of labels. For example, in case of a movie containing a lot of movements, the label 'action' is preferred over the label 'humour'.
- dependency relations where the decision to associate a label to an instance depends on whether the instance is associated to another label or not. For example the decision to associate a movie to the label 'suspense' can be supported by the fact that the movie is strongly associated to the label 'action'.

The problem of label dependencies is that a prediction error for a label may be propagated to another label depending on it: for example, a classifier using the rule: an image associated to the label 'beach' should be also associated to the label 'ocean', may predict the label 'beach' for an image that is not really associated to the label 'beach', and then the prediction of the label 'ocean' may also be wrong.

Label dependencies can help to output coherent prediction, however, they may also cause prediction error propagation. An interesting challenge in learning label dependencies is to learn an optimal set of label dependencies minimizing the risk of error propagation.

Another challenging task in learning label dependencies is handling the problem of cyclic dependencies. Indeed, a multi-label classifier cannot output a set of associated labels in case there are for example two labels where the prediction of each label depends on the outputted prediction of the other label. Most of existing approaches allow learning only a subset of label dependencies to avoid cyclic dependencies. The drawback of this type of approaches is that a restriction is forced on allowed label relations before the learning stage. Therefore, some label dependencies cannot be learned, and there may be no label dependency to discover in the subset of allowed relations.

Approaches that can learn label dependencies are called binary relevance based approaches, and approaches that can learn label preferences are called pairwise based approaches. State of the art approaches can either learn label preference or label dependency relations [8]. This work introduces a new approach to overcome the above limitations based on two main ideas:

- the first idea is to allow learning label relations without any restriction, then some dependencies may be removed or replaced in order to reduce the risk of error propagation or to eliminate a cyclic dependency.
- the second idea is to combine a binary relevance based approach and a pairwise based approach in order to benefit from both label dependency and label preference relations.

Experiments are conducted on real multi-label (ML) and graded multi-label (GML) datasets. The aim is to confirm that allowing learning different types of label relations without a restriction before the learning phase can improve the predictive performance of classifiers.

In this work, state of the art classification approaches are reviewed in Sect. 2. Transformation approaches of ordinal classification approaches to a set of binary classification subtasks are reviewed in Sect. 2.1. Transformation approaches of MLC to a set of binary classification subtasks are reviewed in Sect. 2.2. GMLC transformation approaches to a set of MLC subtasks or to a set of ordinal classification subtasks are reviewed in Sect. 2.3. The GMLC task can be transformed to a set of binary classification subtasks by combining transformation approaches in Sect. 2. The new proposed approach that allows learning both label dependency and label preference relations is presented in Sect. 3. Experiments on real datasets comparing our new proposed approach to state of the art MLC approaches are presented in Sect. 4. Experiments on real GML datasets comparing some extensions of our new proposed approach to existing GMLC approaches are presented in Sect. 5.

2 State of the Art

2.1 *Single Label Classification*

Let $A = \{a_j\}_{1 \leq j \leq p}$ be a set of descriptive attributes. Let $C = \{c_l\}_{1 \leq l \leq k}$ be a set of labels. Let $X = \{x_i\}_{1 \leq i \leq n}$ be a set of instances. Each instance $x_i \in a_1 \times \cdots \times a_p$ is described by a vector of attribute values $x_i = (x_{i,1}, \ldots, x_{i,p})$, and it is associated to a label $y_i \in C$. Let $\lambda : X \rightarrow C$ be a function that outputs for each instance $x_i \in X$ the true associated label $\lambda(x_i) = y_i$. The function λ is called a training function. The single label classification task is to learn from the training set X and the training function λ a classifier $H : a_1 \times \cdots \times a_p \rightarrow C$ that predicts for any given instance $x \in a_1 \times \cdots \times a_p$ the associated label $H(x)$.

Some classifiers such as decision trees [9] are adapted to multi-class classification ($k > 2$), however, there are classifiers such as support vector machines [10] that are adapted only to binary classification ($k = 2$). The advantage of transforming multi-class classification to binary classification is to benefit from all existing binary classifiers.

The three main transformation approaches of multi-class classification to binary classification are presented in the following.

2.1.1 One Vs All Approach

The One Vs All approach (OVA) builds a multi-class classifier H based on k binary classifiers $\{H_l\}_{1 \leq l \leq k}$ [11]. The training set of a classifier H_l is the set X, however, the training function λ is replaced by the function $\lambda_l : X \rightarrow \{0, 1\}$ that outputs 1 for an instance associated to the label c_l, and outputs 0 otherwise. Each classifier $H_l : a_1 \times \cdots \times a_p \rightarrow \{0, 1\}$ predicts whether a given instance $x \in a_1 \times \cdots \times a_p$ should be associated to the label c_l ($H_l(x) = 1$) or not ($H_l(x) = 0$). The predicted label with the highest confidence is outputted by the multi-class classifier: $H(x) = c_l$ where $\forall l' \neq l$: $Conf(H_l(x) = 1) \geq Conf(H_{l'}(x) = 1)$.

2.1.2 One Vs One Approach

The One Vs One approach (OVO) builds a multi-class classifier H based on $\frac{k(k-1)}{2}$ binary classifiers $\{H_{l,l'}\}_{1 \leq l < l' \leq k}$ [12]. The training function λ does not change for a classifier $H_{l,l'}$: $\lambda_{l,l'} = \lambda$, however, the training set of a classifier $H_{l,l'}$ is a subset of X that contains only instances associated either to the label c_l or to the label $c_{l'}$: $X_{l,l'} = \{x_i \in X,\ y_i = c_l\ or\ y_i = c_{l'}\}$. Each classifier $H_{l,l'} : a_1 \times \cdots \times a_p \rightarrow \{c_l, c_{l'}\}$ predicts for a given instance $x \in a_1 \times \cdots \times a_p$ whether it is preferred to associate it to the label c_l or to the label $c_{l'}$. The most preferred label by the classifiers $\{H_{l,l'}\}_{1 \leq l < l' \leq k}$ is predicted by the multi-class classifier: $H(x) = c_l$ where $\forall l' \neq l$:
$|\{H_{e,f}, H_{e,f}(x) = c_l\}_{1 \leq e < f \leq k}| \geq |\{H_{e,f}, H_{e,f}(x) = c_{l'}\}_{1 \leq e < f \leq k}|$.

2.1.3 Ordinal Classification Transformation Approach

Ordinal classification is the case where labels are ordered $c_1 < \cdots < c_k$. A simple approach to handle this case is to build an ordinal classifier H based on $k - 1$ binary classifiers $\{H_l\}_{2 \leq l \leq k}$ [1]. The training set of a classifier H_l does not change $X_l = X$, however, the training function λ is replaced by the function $\lambda_l : X \rightarrow \{0, 1\}$ that outputs for an instance x_i whether it is associated to a label $c_{l'} \geq c_l$ ($\lambda_l(x_i) = 1$) or not ($\lambda_l(x_i) = 0$). Each classifier H_l should output the corresponding probability to its prediction. The probability that a given instance $x \in a_1 \times \cdots \times a_p$ is associated to a label c_l ($y = c_l$) is given by the following:

- $Pr(y = c_1) = 1 - Pr(y \geq c_2) = 1 - Pr(H_2(x) = 1)$
- $Pr(y = c_k) = Pr(y \geq c_k) = Pr(H_k(x) = 1)$
- $\forall l \in [\![2, k-1]\!]$:
  $Pr(y = c_l) = Pr(y \geq c_l) - Pr(y \geq c_{l+1}) = Pr(H_l(x) = 1) - Pr(H_{l+1}(x) = 1)$

The ordinal classifier H predicts the label having the highest corresponding probability: $H(x) = c_l$ where $\forall l' \neq l$: $Pr(y = c_l) \geq Pr(y = c_{l'})$.

2.2 *Multi-label Classification*

The difference between single label classification and MLC is that each instance $x_i \in X$ is associated to a subset of labels from C instead of a single label: $y_i \in \mathscr{P}(C)$

($\mathscr{P}(C)$ is the set of all subsets of labels in C). The training function $\lambda : X \rightarrow \mathscr{P}(C)$ outputs the subset of associated labels to a given instance x_i: $\lambda(x_i) = y_i$. The aim in MLC is to learn from the training set X and the training function λ a classifier $H : a_1 \times \cdots \times a_p \rightarrow \mathscr{P}(C)$ that predicts for any given instance $x \in a_1 \times \cdots \times a_p$ the associated subset of labels $H(x)$. In this work, we are interested also to learning and exploiting label relations in order to output coherent predictions [13, 14].

A strategy of handling the MLC task is to adapt single label classifiers to the case of multi-label data [15–17]. The drawback of this strategy is that to modify the approach of learning label relations the classifier itself should be modified.

Another strategy of handling the MLC task is to transform multi-label data to single label data in order to exploit existing single label classifiers. The three main categories of transformation based approaches are presented in the following.

2.2.1 Label Set Category

The key idea of label set category of approaches is to treat each different subset of labels $y_i \in \mathscr{P}(C)$ as a single new label [18]. The set of new labels $\mathscr{C}$ can contain at most n distinct labels $\mathscr{C} = \{\mathscr{C}_l\}_{1 \leq l \leq n}$ in case all subsets of labels y_i, $i \in [\![1, n]\!]$ are different.

Let $LS : X \rightarrow \mathscr{C}$ be the label mapping function that outputs for each instance $x_i \in X$ the corresponding new label $LS(x_i) \in \mathscr{C}$. Let $LS^{-1} : \mathscr{C} \rightarrow X$ be a function that outputs for a given new label $\mathscr{C}_l \in \mathscr{C}$ one instance $x_i \in X$ so that: $LS^{-1}(\mathscr{C}_l) = x_i$ and $LS(x_i) = \mathscr{C}_l$.

A single label classifier $h : a_1 \times \cdots \times a_p \rightarrow \mathscr{C}$ is trained using the training set X and the training function LS. The multi-label classifier H predicts a subset of labels for a given instance x by converting back the outputted new label $h(x)$ to a subset of original labels $H(x) = \lambda(LS^{-1}(h(x)))$.

The drawback of the label set category of transformation approaches is that two different subsets of labels with some common labels are considered as different labels in the transformed problem. Therefore, label relations between labels from different subsets cannot be learned.

2.2.2 Binary Relevance Category

Binary relevance category of approaches is based on the idea of the OVA transformation approach (Sect. 2.1.1). The Binary Relevance approach (BR), which is the baseline approach in this category, builds a multi-label classifier H based on a set of k binary classifiers $\mathscr{H} = \{H_l\}_{1 \leq l \leq k}$ [8]. Each classifier $H_l : a_1 \times \cdots \times a_p \rightarrow \{0, 1\}$ predicts for a given instance x the relevance of the label c_l (whether x should be associated to c_l: $H_l(x) = 1$, or not: $H_l(x) = 0$). The multi-label classifier H outputs for a given instance x all labels predicted as relevant for x:

$H(x) = \{c_l \in C, H_l(x) = 1\}$. The drawback of the BR approach is that label relations cannot be learned.

The Classifier Chains approach (CC) [19] allows learning label dependencies by considering relevancies of labels as additional descriptive attributes. Let $r_l : X \rightarrow \{0, 1\}$ be the relevance function of the label c_l given by:

$r_l(x_i) = 1$ if $c_l \in y_i$, and $r_l(x_i) = 0$ otherwise.

The training set X is extended for each classifier $H_l, l \geq 2$ as follows:

$X_l = \{(x_{i,1}, \ldots, x_{i,p}, r_1(x_i), \ldots, r_{l-1}(x_i)\}_{1 \leq i \leq n}$. The training set for the classifier H_1 is not extended: $X_1 = X$. To output a prediction for a given instance x, the binary classifier H_1 predicts the relevance of c_1 to x. Then the instance x is extended by $H_1(x)$: $(x_{i,1}, \ldots, x_{i,p}, H_1(x))$. The classifier H_2 predicts the relevance of c_2 to x considering the predicted relevance by H_1. Indeed, each classifier $H_l, l \geq 2$ outputs a prediction considering the extended instance x by predictions of all previous classifiers: $x = (x_{i,1}, \ldots, x_{i,p}, H_1(x), \ldots, H_{l-1}(x))$. The drawback of the CC approach is that label dependencies that can be learned depend on the initial order of labels. Indeed, a classifier H_l is not allowed to depend on the outputted prediction of a classifier $H_{l'}, l' > l$.

The CC approach can be extended by allowing each classifier $H_l, l \in [\![1, k]\!]$ to depend on all other classifiers. This is done by extending the training set for each binary classifier H_l as follow:

- $X_1 = \{(x_{i,1}, \ldots, x_{i,p}, r_2(x_i), \ldots, r_k(x_i)\}_{1 \leq i \leq n}$
- $X_k = \{(x_{i,1}, \ldots, x_{i,p}, r_1(x_i), \ldots, r_{k-1}(x_i)\}_{1 \leq i \leq n}$
- $\forall l \in [\![2, k-1]\!]$: $X_l = \{(x_{i,1}, \ldots, x_{i,p}, r_1(x_i), \ldots, r_{l-1}(x_i), r_{l+1}(x_i), \ldots, r_k(x_i)\}_{1 \leq i \leq n}$

This approach is referred in the following as the Full Binary Relevance approach (FBR). Note that the fact of allowing label dependencies for a classifier H_l does not mean necessarily that H_l should depend on any other classifier. Indeed, the learned classifier H_l can be eventually independent.

Let $D^{\rightarrow} : \mathcal{H} \rightarrow \mathcal{P}(\mathcal{H})$ be the function that outputs for a classifier $H_l \in \mathcal{H}$ the set of classifiers on which it depends.

Let $D^{\leftarrow} : \mathcal{H} \rightarrow \mathcal{P}(\mathcal{H})$ be the function that outputs for a classifier $H_l \in \mathcal{H}$ the set of classifiers depending on it.

At the prediction phase for a given instance x, each classifier H_l should output its prediction only after predictions of all classifiers $D^{\rightarrow}(H_l)$ on which H_l depends are collected. The problem is that in CC approach the prediction order of classifiers is known before, while in the FBR approach the prediction order can not be known before building the classifiers. The drawback of the FBR approach is that the prediction order cannot be selected in case of a cyclic dependency:

For example a cyclic dependency may be caused by the fact that two classifiers depend on each other: $\exists\, l, l' \in [\![1, k]\!] : H_{l'} \in D^{\rightarrow}(H_l)\ \&\ H_l \in D^{\rightarrow}(H_{l'})$. Therefore, it is not possible for any classifier of H_l and $H_{l'}$ to output its prediction because the prediction of the other classifier is needed. The FBR approach is not a complete MLC approach because it does not give a solution for cyclic dependencies, however it is the base approach of other MLC approaches handling the problem of cyclic dependencies.

The approach of Aggregating Independent and Dependent classifiers (AID) [20] combines the FBR and the BR approaches to overcome the problem of the prediction order in the FBR approach. The idea of the AID approach is to build a set of k binary classifiers $\{h_l\}_{1 \leq l \leq k}$ as in the BR approach, and another set of k binary classifiers $\{H_l\}_{1 \leq l \leq k}$ as in the FBR approach. At the prediction phase for a given instance x, classifiers of the BR approach output their predictions to extend the instance x: $(x_1, \ldots, x_p, h_1(x), \ldots, h_k(x))$. Then each classifier H_l of the FBR approach outputs its prediction based on the extended instance x instead of the outputted predictions of other classifiers $\{H_{l'}\}_{l' \neq l}$. The multi-label classifier H outputs the predicted set of labels as follows: $H(x) = \{c_l, H_l(x) = 1\}$. The drawback of this approach is that the prediction $H(x)$ is not necessarily coherent with learned label dependencies by classifiers of the FBR approach. The following example illustrates this remark:

- the classifier H_2 outputs 1 only if the label c_1 is not associated to the given instance x
- the instance x is not associated to c_1 according to the classifier h_1, therefore c_2 is predicted for the instance x
- the instance x is associated to c_1 according to the classifier H_1

In this example, both labels c_1 and c_2 are predicted by the AID approach which is not a coherent prediction considering the dependency learned by the classifier H_2.

An approach has been recently proposed to ensure coherent predictions with learned label dependencies without forcing a prediction order before the learning phase [21]. The idea is to learn a set of k initial classifiers $\mathscr{H} = \{h_l\}_{1 \leq l \leq k}$ as in the FBR approach. Then cyclic dependencies are eliminated by replacing some classifiers based on three measures called Pre-selection, Selection, and Interest of chaining (PSI measures). The obtained set of final classifiers $\mathbb{H} = \{H_l\}_{1 \leq l \leq k}$ does not contain any cyclic dependency.

Let $MaxD^{\rightarrow} : \mathscr{P}(\mathscr{H}) \rightarrow \mathscr{P}(\mathscr{H})$ be the function that outputs for a given set of classifiers $\mathscr{H}' \subseteq \mathscr{H}$ the subset of classifiers $\mathscr{H}'' \subseteq \mathscr{H}'$ depending on the largest number of classifiers:

- $\forall h_l \in \mathscr{H}'', \forall h_{l'} \in \mathscr{H}' - \mathscr{H}''$: $|D^{\rightarrow}(h_l)| > |D^{\rightarrow}(h_{l'})|$
- $\forall h_l, h_{l'} \in \mathscr{H}''$: $|D^{\rightarrow}(h_l)| = |D^{\rightarrow}(h_{l'})|$

Let $MaxD^{\leftarrow} : \mathscr{P}(\mathscr{H}) \rightarrow \mathscr{P}(\mathscr{H})$ be the function that outputs for a given set of classifiers $\mathscr{H}' \subseteq \mathscr{H}$ the subset of classifiers $\mathscr{H}'' \subseteq \mathscr{H}'$ on which depend the largest number of classifiers:

- $\forall h_l \in \mathscr{H}'', \forall h_{l'} \in \mathscr{H}' - \mathscr{H}''$: $|D^{\leftarrow}(h_l)| > |D^{\leftarrow}(h_{l'})|$
- $\forall h_l, h_{l'} \in \mathscr{H}''$: $|D^{\leftarrow}(h_l)| = |D^{\leftarrow}(h_{l'})|$

The PSI measures work as follows:

- the pre-selection measure $\mathbb{P} : \mathscr{H} \rightarrow \{0, 1\}$ splits the set of initial classifiers to a set of candidate classifiers to be replaced $\{h_l \in \mathscr{H}, \mathbb{P}(h_l) = 1\}$, and a set of classifiers that can be moved to the set of final classifiers $\{h_l \in \mathscr{H}, \mathbb{P}(h_l) = 0\}$. The pre-selection measure that pre-selects dependent classifiers that may be involved in cyclic dependencies can be given as: $\mathbb{P}(h_l) = 1$ if $D^{\rightarrow}(h_l) \neq \emptyset$, and $\mathbb{P}(h_l) = 0$ otherwise. All non pre-selected classifiers are removed from $\mathscr{H}$ and added to $\mathbb{H}$. A classifier that remains in $\mathscr{H}$ and does not depend on another classifier that remains in $\mathscr{H}$ is considered independent using the pre-selection measure $\mathbb{P}$. Therefore, the pre-selection measure is applied iteratively to move classifiers from $\mathscr{H}$ until the remaining classifiers in $\mathscr{H}$ are all pre-selected. The remaining classifiers in $\mathscr{H}$ using this strategy are involved in cyclic dependencies.

- the selection measure $\mathbb{S} : \mathscr{P}(\mathscr{H}) \rightarrow \mathscr{H}$ selects one initial classifier h_l to be replaced by a final classifier H_l from a set of candidate classifiers $\{h_l \in \mathscr{H}, \mathbb{P}(h_l) = 1\}$. A classifier h_l on which depend the largest number of classifiers $h_l \in MaxD^{\leftarrow}(\mathscr{H})$ can be selected to be replaced and removed from $\mathscr{H}$ in order to help making independent the largest number of classifiers. In case there are more than one classifier $h_l \in MaxD^{\leftarrow}(\mathscr{H})$, the classifier h_l depending on the largest number of classifiers $h_l \in MaxD^{\rightarrow}(MaxD^{\leftarrow}(\mathscr{H}))$ can be selected to reduce the risk of prediction error propagation by replacing the classifier having the largest number of dependencies. A selection measure that outputs exactly one classifier can be given as follows: $\mathbb{S}(\mathscr{H}) = h_l$ where:

  - $h_l \in MaxD^{\rightarrow}(MaxD^{\leftarrow}(\mathscr{H}))$
  - $\forall h_{l'} \in MaxD^{\rightarrow}(MaxD^{\leftarrow}(\mathscr{H})) - \{h_l\}$: $l' > l$

- the interest of chaining measure $\mathbb{I} : \mathbb{H} \rightarrow \{0, 1\}$ selects the final classifiers $\{H \in \mathbb{H}, \mathbb{I}(H_l) = 1\}$ on which the classifier H_l can depend to replace the selected classifier h_l. The interest of chaining measure can be given as: $\forall H_l \in \mathbb{H}$: $\mathbb{I}(H_l) = 1$ in order to allow learning the maximum of label dependencies that are not cyclic. Indeed, the order of adding classifiers to the set of final classifiers $\mathbb{H}$ is the prediction order of classifiers $\{H_l\}_{1 \leq l \leq k}$. The new classifier H_l replacing h_l is then added to $\mathbb{H}$, and the initial classifier h_l is removed from $\mathscr{H}$. Some remaining classifiers in $\mathscr{H}$ may appear independent after removing h_l according to the pre-selection measure $\mathbb{P}$. Hence, the pre-selection measure is applied iteratively to move all independent classifiers from $\mathscr{H}$ to $\mathbb{H}$.

The pre-selection, the selection, and the interest of chaining measures are applied iteratively until the set of final classifiers is completed $|\mathbb{H}| = k$ and the set of initial classifiers becomes empty $\mathscr{H} = \emptyset$. This approach based on the three PSI measures is called PSI. It allows learning different set of label dependencies according to the given PSI measures [21].

2.2.3 Pairwise Comparisons Category

Pairwise comparisons category of MLC transformation approaches is based on the OVO transformation approach (Sect. 2.1.2). The Ranking by Pairwise Comparisons approach (RPC) is the baseline approach in this category. It builds a multi-label classifier H based on a set of $\frac{k(k-1)}{2}$ binary classifiers $\mathscr{H} = \{H_{l,l'}\}_{1\leq l<l'\leq k}$ [22]. Each classifier $H_{l,l'} : a_1 \times \cdots \times a_p \rightarrow \{c_l, c_{l'}\}$ predicts for a given instance x the preferred label among c_l and $c_{l'}$ to be associated to the instance x. The training set for the classifier $H_{l,l'}$ is a subset of X containing only instances associated to exactly one of the labels c_l or $c_{l'}$:

$X_{l,l'} = \{x_i \in X, (c_l \in y_i \;\&\; c_{l'} \notin y_i) or (c_l \notin y_i \;\&\; c_{l'} \in y_i)\}.$

Let $\mathbb{V}_{\mathscr{H},c_l} : a_1 \times \cdots \times a_p \rightarrow [\![0, |\mathscr{H}| - 1]\!]$ be the function that outputs for a given instance x the number of times the label c_l was preferred by the binary classifiers in $\mathscr{H}$: $\mathbb{V}_{\mathscr{H},c_l}(x) = |\{H_{l',l''} \in \mathscr{H}, H_{l',l''}(x) = c_l\}_{l'=l or l''=l}|$.

In order output a multi-label prediction for a given instance x, a threshold $v \in [\![0, |\mathscr{H}| - 1]\!]$ can be used to output labels preferred more times than the threshold value: $H(x) = \{c_l \in C, \mathbb{V}_{\mathscr{H},c_l}(x) \geq v\}$.

The Calibrated Label Ranking approach (CLR) is an extension of the RPC approach that allows to output a multi-label prediction using a virtual label instead of a threshold value [23]. The CLR approach learns k more binary classifiers $\{H_{l,0}\}_{1\leq l\leq k}$ than the RPC approach. Each classifier $H_{l,0}$ predicts the relevance of the label c_l to a given instance x. The training set is transformed for the classifier $H_{l,0}$ so that instances not associated to c_l are considered as associated to a virtual label c_0. In order to output a multi-label prediction for a given instance x, the set of labels is ordered according to the number of times each label is preferred. Then all labels that are preferred more times than the virtual label are predicted: $H(x) = \{c_l \in C, \mathbb{V}_{\mathscr{H},c_l}(x) \geq \mathbb{V}_{\mathscr{H},c_0}(x)\}$.

2.3 Graded Multi-label Classification

The difference between MLC and GMLC is that in GMLC an ordinal scale of membership degrees is given: $M = \{m_1 < \cdots < m_s\}$. For each membership degree m_g the function $\lambda_{m_g} : X \rightarrow \mathscr{P}(C)$ outputs for an instance $x_i \in X$ the set of labels associated to x_i with the membership degree m_g. For each label $c_l \in C$ the function $\mu_{c_l} : X \rightarrow M$ outputs for an instance $x_i \in X$ the corresponding membership degree to the label c_l. Each instance $x_i \in X$ is associated to a vector $y_i \in M^k$ of k membership degrees: $y_i = (\mu_{c_1}(x_i), \ldots, \mu_{c_k}(x_i)) = (\mu_{c_l}(x_i))_{1\leq l\leq k}$. The training function $\mu : X \rightarrow M^k$ is given by: $\mu(x_i) = (\mu_{c_l}(x_i))_{1\leq l\leq k}$. The task of GMLC is to learn from the training set X and the training function μ a classifier $H : a_1 \times \cdots a_p \rightarrow M^k$ that outputs for a given instance x the corresponding vector of membership degrees $H(x) \in M^k$.

The task of GMLC can be decomposed either into a set of ordinal classification subtasks, or into a set of MLC subtasks, or into a set of binary classification subtasks. The three main categories of decomposition approaches of the GMLC task are detailed in the following.

2.3.1 Vertical Decomposition Category of GMLC Approaches

The idea of the vertical decomposition (VD) is to build a GML classifier H based on a set of k classifiers $\{H_{c_l}\}_{1 \leq l \leq k}$ [4]. Each classifier $H_{c_l} : a_1 \times \cdots \times a_p \rightarrow M$ predicts the membership grade of the label c_l. The VD category is similar to the binary relevance category of MLC transformation based approaches (Sect. 2.2.2). The difference is that each classifier H_{c_l} in MLC is a binary classifier $H_{c_l}(x) \in \{0, 1\}$, while in the VD task each classifier H_{c_l} is an ordinal classifier $H_{c_l}(x) \in \{m_1 < \cdots < m_s\}$ (Sect. 2.1.3). The GML classifier H outputs a prediction for a given instance $x \in a_1 \times \ldots \times a_p$ as follows: $H(x) = (H_{c_l}(x))_{1 \leq l \leq k}$.

The advantage of the VD is that the generated ordinal classification subtasks can be handled by any ordinal classifier, and the strategy of generating ordinal classification subtasks can be combined with any transformation approach in the binary relevance category (Sect. 2.2.2):

- the approach referred as Vertical_BR combines the vertical decomposition with the Binary Relevance approach. Therefore each learned classifier H_{c_l} is independent.
- the approach referred as Vertical_CC combines the vertical decomposition with the Classifier Chains approach. Therefore each learned classifier H_{c_l} is allowed to depend on classifiers $\{H_{c_{l'}}\}_{1 \leq l' < l \leq k}$.
- the approach referred as Vertical_PSI combines the vertical decomposition with the PSI approach. Therefore each learned classifier H_{c_l} is allowed initially to depend on all other classifiers $\{H_{c_{l'}}\}_{l' \neq l}$. Then in case of a cyclic dependency some classifiers may be rebuilt considering a restricted set of allowed dependencies.

2.3.2 Horizontal Decomposition Category of GMLC Approaches

The idea of the horizontal decomposition (HD) is to build a GML classifier H based on a set of $s - 1$ classifiers $\{h_{\geq m_g}\}_{2 \leq g \leq s}$ [4]. Each classifier

$h_{\geq m_g} : a_1 \times \cdots \times a_p \rightarrow \mathscr{P}(C)$ predicts for a given instance x the set of labels that are associated to the instance x with a membership degree $m_{g'}$ at least equals to m_g: $m_{g'} \geq m_g$. The classifier $h_{\geq m_1}$ is ignored because m_1 is the lowest available membership grade, therefore $h_{\geq m_1}(x) = \{c_l\}_{1 \leq l \leq k}$. The idea of building a classifier $h_{\geq m_g}$ for each membership degree $m_g \geq m_2$ is taken from the idea of decomposing an ordinal classification task into a set of binary classification subtasks (Sect. 2.1.3). The difference is that each classifier $h_{\geq m_g}$ in the ordinal classification is a binary classifier $h_{\geq m_g}(x) \in \{0, 1\}$, while in the HD task each classifier $h_{\geq m_g}$ is a multi-label classifier $h_{\geq m_g}(x) \subseteq C$. For the HD approach, each label c_l may be predicted for a given instance x by multiple classifiers $h_{\geq m_g}$, $m_g \geq m_2$. Therefore, an aggregation function H_{c_l} that outputs a final membership degree for the label c_l by combining predictions of classifiers $\{h_{\geq m_g}\}_{2 \leq g \leq s}$ is needed. H_{c_l} can be given as the function that outputs the highest membership grade corresponding to a classifier $h_{\geq m_g}$ that predicts the label c_l: $H_{c_l}(x) = m_g$ where:

- $c_l \in h_{\geq m_g}(x)$

- $\forall m_{g'} > m_g$: $c_l \notin h_{\geq m_{g'}}(x)$

The GML classifier H outputs a prediction for a given instance x as follows:

$H(x) = (H_{c_l}(x))_{1 \leq l \leq k}$. The advantage of the HD is that the strategy of generating MLC subtasks can be combined with any ordinal classification transformation approach, and the generated MLC subtasks can be handled by any existing multi-label classifier. The drawback of the HD is that a classifier $h_{\geq m_g}$ may ignore the output of all other classifiers $\{h_{\geq m_g}\}_{2 \leq g \leq s}$:

For example, let $M = \{m_1, m_2, m_3, m_4\}$ be the set of membership grades. The fact that for a given instance x the label c_l is not predicted by the classifiers $h_{\geq m_3}$ and $h_{\geq m_2}$ means that the membership degree of c_l is not $\geq m_2$ and it is not $\geq m_3$. However, if the label c_l is predicted by the classifier $h_{\geq m_4}$ the outputted membership degree for c_l is m_4. The predictions of classifiers $h_{\geq m_3}$ and $h_{\geq m_2}$ are not considered in this case.

The Horizontal Calibrated Label Ranking approach (Horizontal_CLR) [24] solves each generated MLC task by the horizontal decomposition using the CLR approach (Sect. 2.2.3). The Horizontal_CLR approach introduces a set of virtual labels

$W = \{w_g\}_{2 \leq g \leq s}$, and a set of virtual membership degrees $V = \{v_g\}_{2 \leq g \leq s}$. Each virtual label w_g has a fixed membership degree v_g for any given instance

$x \in a_1 \times \cdots \times a_p$. The position of virtual membership grades in regards to the original membership grades is given as follows:

$M \cup V = \{m_1 < v_2 < m_2 < v_3 \cdots < m_{s-1} < v_s < m_s\}$. Each multi-label classifier $h_{\geq m_g}$ is built as in the CLR approach using a set of $\frac{k(k+1)}{2}$ binary classifiers $\mathscr{H}_{\geq m_g} = \{h_{\geq m_g, c_l, c_{l'}}\}_{1 \leq l < l' \leq k} \cup \{h_{\geq m_g, c_l, w_g}\}_{1 \leq l \leq k}$.

The training set of a classifier $h_{\geq m_g, c_l, c_{l'}}$ contains only instances that are associated with a membership degree $m_{g'} \geq m_g$ either to the label c_l or to the label $c_{l'}$:

$X_{\geq m_g, c_l, c_{l'}} =$

$\{x_i \in X, (\mu_{c_l}(x_i) \geq m_g)\ \&\ (\mu_{c_{l'}}(x_i) < m_g)\} \cup$

$\{x_i \in X, (\mu_{c_l}(x_i) < m_g)\ \&\ (\mu_{c_{l'}}(x_i) \geq m_g)\}$.

Let $\mu_{\geq m_g, c_l, c_{l'}} : X_{\geq m_g, c_l, c_{l'}} \to \{c_l, c_{l'}\}$ be the training function given by

$\mu_{\geq m_g, c_l, c_{l'}}(x_i) = c_l$ if $\mu_{c_l}(x_i) > \mu_{c_{l'}}(x_i)$, and $\mu_{\geq m_g, c_l, c_{l'}}(x_i) = c_{l'}$ otherwise.

Each classifier $h_{\geq m_g, c_l, c_{l'}}$ learns from the training set $X_{\geq m_g, c_l, c_{l'}}$ and the training function $\mu_{\geq m_g, c_l, c_{l'}}$. The classifier $h_{\geq m_g, c_l, c_{l'}}$ predicts for any given instance x the preferred label among the labels c_l and $c_{l'}$ to be associated to x with a membership degree at least equals to m_g.

The training set of each classifier $h_{\geq m_g, c_l, w_g}$ is not modified: $X_{\geq m_g, c_l, w_g} = X$. The corresponding training function $\mu_{\geq m_g, c_l, w_g} : X \to \{c_l, w_g\}$ is given by: $\mu_{\geq m_g, c_l, w_g} = c_l$ if $\mu_{c_l}(x_i) > v_g$, and $\mu_{\geq m_g, c_l, w_g}(x_i) = w_g$ otherwise.

For a given instance x, the multi-label classifier $h_{\geq m_g}$ collects predictions of all binary classifiers in $\mathscr{H}_{\geq m_g}$. Then the classifier $h_{\geq m_g}$ outputs the set of labels that are preferred more times than the virtual label w_g.

The Horizontal_CLR approach adapts the CLR transformation approach to the GMLC task using multiple calibration labels, however it does not provide an alternative to the used strategy by the horizontal decomposition to aggregate predictions of classifiers $\{h_{\geq m_g}\}_{2 \leq g \leq s}$.

The Full Calibrated Label Ranking approach (Full_CLR) [24] extends the Horizontal_CLR approach in order to overcome the problem of classifiers $\{h_{\geq m_g}\}_{2\leq g\leq s}$ that may ignore the outputted predictions of each other. The Full_CLR approach builds $\frac{(k+s-1)(k+s-2)}{2}$ preference classifiers $\mathscr{H} = \{h_{c_l,c_{l'}}\}_{1\leq l<l'\leq k} \cup \{h_{c_l,w_g}\}_{\substack{1\leq l\leq k\\ 2\leq g\leq s}}$ $\cup \{h_{w_g,w_{g'}}\}_{2\leq g<g'\leq s}$ to predict the label having the highest membership degree among a pair of labels

$(\alpha, \beta) \in C \cup W \times C \cup W$. The training set for a preference classifier $h_{\alpha,\beta}$ where $\alpha, \beta \in C \cup W$ is given by: $X_{\alpha,\beta} = \{x_i \in X, \mu_\alpha(x_i) \neq \mu_\beta(x_i)\}$. The corresponding training function $\mu_{\alpha,\beta} : X_{\alpha,\beta} \rightarrow \{\alpha, \beta\}$ is given by: $\mu_{\alpha,\beta}(x_i) = \alpha$ if $\mu_\alpha(x_i) > \mu_\beta(x_i)$, and $\mu_{\alpha,\beta}(x_i) = \beta$ otherwise.

To output a prediction for a given instance x, labels $C \cup W$ are ordered according to the number of times each label is preferred. The aggregation function H_{c_l} that outputs for a given instance x the membership degree corresponding to the label c_l is given by: $H_{c_l}(x) = m_g$ where

- $\mathbb{V}_{\mathscr{H},c_l}(x) \geq \mathbb{V}_{\mathscr{H},w_g}(x)$
- $\forall g' \in [\![g+1, s]\!]$: $\mathbb{V}_{\mathscr{H},c_l}(x) < \mathbb{V}_{\mathscr{H},w_{g'}}(x)$

The GML classifier H outputs a prediction for the instance x as follows: $H(x) = (H_{c_l}(x))_{1\leq l\leq k}$. The advantage of the Full_CLR approach is that predictions of all preference classifiers are aggregated to output a total order for labels. The drawback of the Full_CLR approach is that preference classifiers ignore the difference between membership degrees of labels:

For example, a label c_l with a membership degree m_4 is preferred over a label $c_{l'}$ with a membership degree m_3 and over a label $c_{l''}$ with a membership degree m_2. Therefore there is no difference between the preference of c_l over $c_{l'}$ and the preference of c_l over $c_{l''}$.

The Joined Calibrated Label Ranking approach (Joined_CLR) [24] combines the Horizontal_CLR and the Full_CLR to overcome the limitations of both approaches. The main ideas off the Joined_CLR are as follows:

- build a set of preference classifiers $\mathscr{H} = \bigcup_{2\leq g\leq s} \mathscr{H}_{\geq m_g}$ as in the Horizontal_CLR approach so that the difference between membership degrees is considered
- output predictions as in the Full_CLR approach so that predictions of classifiers in $\mathscr{H}$ are not ignored.

The problem of combining the previous two ideas is that there are $k(s-1)$ preference classifiers for each label $c_l \in C$, while there are only k preference classifiers for each label $w_g \in W$. The joined_CLR solves this problem by adding a set of preference classifiers $\{h_{\geq m_g,c_l,w_{g'}}\}_{\substack{1\leq l\leq k\\ g'\in[\![2,s]\!]-\{g\}}} \cup \{h_{\geq m_g,w_{g'},w_{g''}}\}_{2\leq g'<g''\leq s}$ for all membership grades m_g, $g \in [\![2, s]\!]$. The classifiers $\{h_{\geq m_g,c_l,w_{g'}}\}_{\substack{1\leq l\leq k\\ g'\in[\![2,s]\!]-\{g\}}}$ are learned as the classifiers $\{h_{\geq m_g,c_l,w_g}\}_{1\leq l\leq k}$ in the Horizontal_CLR approach. Classifiers $\{h_{\geq m_g,w_{g'},w_{g''}}\}_{2\leq g'<g''\leq s}$ do not need to be trained. Indeed, each classifier $h_{\geq m_g,w_{g'},w_{g''}}$ outputs the label $w_{g''}$ as the preferred label because it has the highest membership grade. Considering the

added classifiers, each label c_l and each label w_g can have at most $(k+s-2)(s-1)$ votes.

2.3.3 Complete Decomposition Category of GMLC Approaches

The idea of the complete decomposition (CD) is to build a GML classifier H based on a set of $k\ (s\ -\ 1)$ binary classifiers $\{h_{\geq m_g,c_l}\}_{\substack{2\leq g\leq s\\1\leq l\leq k}}$ [4]. Each classifier $h_{\geq m_g,c_l} : a_1 \times \cdots \times a_p \rightarrow \{0,1\}$ predicts for a given instance x whether the label c_l has a membership grade $m_{g'} \geq m_g$ or not. The complete reduction is equivalent to a vertical decomposition where each ordinal classifier H_{c_l} is built based on a set of binary classifiers $\{h_{\geq m_g,c_l}\}_{2\leq g\leq s}$ (Sect. 2.1.3). The complete decomposition is equivalent also to an horizontal decomposition where each multi-label classifier $h_{\geq m_g}$ is solved using the binary relevance approach: $\{h_{\geq m_g,c_l}\}_{1\leq l\leq k}$. The Complete decomposition can be combined with other learning strategies to allow learning label relations

- the approach referred as Complete_BR combines the complete decomposition with the Binary Relevance approach. Therefore each learned classifier $h_{\geq m_g,c_l}$ is independent.
- the approach referred as Complete_PSI combines the complete decomposition with the PSI approach. Therefore each learned classifier $h_{\geq m_g,c_l}$ is allowed to depend initially on all other classifiers, then in case of a cyclic dependency some classifiers may be rebuilt.

The drawback of the vertical decomposition and the complete decomposition is that preferences between labels cannot be learned. The advantage of the horizontal decomposition is that any MLC approach can be used to solve the generated MLC subtasks, however, existing MLC approaches can learn either label dependency or label preference relations.

3 A New Approach to Learning both Label Dependency and Label Preference Relations

Our idea to allow learning both label dependencies and label preferences in GMLC is to use an horizontal decomposition, and solve each generated MLC subtask using a new MLC approach combining a MLC approach from the binary relevance category to learn label dependencies (Sect. 2.2.2), and a MLC approach from the pairwise comparisons category to learn label preferences (Sect. 2.2.3).

In this work, two combining strategies are introduced: one using the CLR approach and the other using the RPC approach as the base pairwise comparisons approach. Both strategies are combined with the PSI approach because it has the advantage of learning different label dependencies according to the given PSI measures (Sect. 2.2.2) [25].

The first combining strategy called CLR_PSI builds $\frac{k(k+1)}{2}$ binary classifiers $\{H_{l,l'}\}_{1\leq l<l'\leq k} \cup \{H_{l,0}\}_{1\leq l\leq k}$ and outputs the predicted set of labels as in the CLR approach (Sect. 2.2.3). The difference is that classifiers $\{H_{l,0}\}_{1\leq l\leq k}$ that predict the preference between labels $c_l \in C$ and the virtual label c_0 are built using additional attributes as in the PSI approach (Sect. 2.2.2).

In order to learn dependencies between classifiers, there should be enough common instances between classifiers at the training step. The problem of preference classifiers $\{H_{l,l'}\}_{1\leq l<l'\leq k}$ is that each classifier $H_{l,l'}$ learns from its one subset of instances. Therefore there may be only few or no common instances at all between the preference classifiers. This explains why only classifiers $\{H_{l,0}\}_{1\leq l\leq k}$ are built using the PSI approach.

The second combining strategy called Stacked_RPC_PSI answers the challenge of learning label dependencies based on preference classifiers $\{H_{l,l'}\}_{1\leq l<l'\leq k}$.

The Stacked_RPC_PSI approach builds $\frac{k(k-1)}{2}$ binary classifiers as in the RPC approach. Then each preference classifier $H_{l,l'}$ predicts the preferred label for all training instances $x_i \in X$. The Stacked_RPC_PSI adds a set of k classifiers $\{H_l\}_{1\leq l\leq k}$ as in the PSI approach. The difference is that each classifier H_l is allowed to depend on preference classifiers $\{H_{l,l'}\}_{1\leq l<l'\leq k}$ by extending the training set X using the outputted predictions of the preference classifiers:

$X' = \{(x_{i,1}, \ldots, x_{i,p}, H_{1,2}(x_i), H_{1,3}(x_i), \ldots, H_{k-1,k}(x_i))\}_{1\leq i\leq n}$. The Stacked_RPC_PSI approach outputs a prediction for a given instance x by collecting predictions from classifiers of the RPC approach, then from classifiers of the PSI approach.

The Stacked_RPC_PSI approach outputs the set of labels outputted by the PSI approach.

3.1 The CLR_PSI Approach on a Multi-label Dataset Example

Let $X = \{x_i\}_{1\leq i\leq 10}$ be the training set in Table 1. Each instance x_i is described using two attributes a_1 and a_2, and it can be associated to one or more labels from a set of 3 available labels $C = \{c_1, c_2, c_3\}$.

The dataset in Table 1 is extended according to the CLR_PSI approach as shown in Table 2. Each preference classifier $H_{l,l'}$, $1 \leq l < l' \leq 3$ learns to predict the preferred label between c_l and $c_{l'}$ (the added attribute $c_l\#c_{l'}$ in Table 2) based on the descriptive attributes $\{a_1, a_2\}$. The attribute $c_l\#c_{l'}$ is called a target attribute for the classifier $H_{l,l'}$. Instances that are not considered by the classifier $H_{l,l'}$ are marked with an empty-set symbol $\emptyset$ for the target attribute $c_l\#c_{l'}$ (Table 2).

Each preference classifier $H_{l,0}$, $l \in [\![1, 3]\!]$ learns to predict the preferred label between c_l and the virtual label c_0 based on an extended descriptive attribute set

Table 1 Multi-label dataset example

Instances	a_1	a_2	Labels
x_1	20	30	$\{c_2, c_3\}$
x_2	35	35	$\{c_2, c_3\}$
x_3	15	40	$\{c_2, \}$
x_4	20	50	$\{c_2\}$
x_5	30	45	$\{c_1, c_2\}$
x_6	35	30	$\{c_1, c_2\}$
x_7	10	40	$\{c_1, c_2\}$
x_8	15	45	$\{c_1, c_3\}$
x_9	25	55	$\{c_1, c_3\}$
x_{10}	30	60	$\{c_1, c_3\}$

Table 2 Dataset extension for the CLR_PSI approach

Instances	a_1	a_2	$c_1\#c_2$	$c_1\#c_3$	$c_2\#c_3$	$c_1\#c_0$	$c_2\#c_0$	$c_3\#c_0$
x_1	20	30	c_2	c_3	$\emptyset$	c_0	c_2	c_3
x_2	35	35	c_2	c_3	$\emptyset$	c_0	c_2	c_3
x_3	15	40	c_2	$\emptyset$	c_2	c_0	c_2	c_0
x_4	20	50	c_2	$\emptyset$	c_2	c_0	c_2	c_0
x_5	30	45	$\emptyset$	c_1	c_2	c_1	c_2	c_0
x_6	35	30	$\emptyset$	c_1	c_2	c_1	c_2	c_0
x_7	10	40	$\emptyset$	c_1	c_2	c_1	c_2	c_0
x_8	15	45	c_1	$\emptyset$	c_3	c_1	c_0	c_3
x_9	25	55	c_1	$\emptyset$	c_3	c_1	c_0	c_3
x_{10}	30	60	c_1	$\emptyset$	c_3	c_1	c_0	c_3

$\{a_1, a_2\} \cup \{c_{l'}\}_{l' \neq l}$. Then in case of a cyclic dependency, some classifiers may be rebuilt depending on the given PSI measures.

In the following, decision trees [9] without pruning are used as the base binary classifiers. Classifiers $H_{1,2}$, $H_{1,3}$, and $H_{2,3}$ are learned as in the CLR approach (Figs. 1, 2 and 3).

The root node in the decision tree $H_{1,2}$ contains 3 instances associated to the label c_1, and 4 instances associated to the label c_2 (Fig. 1). For instances with a value $x_{i,2} \leq 40$ the label c_2 is predicted (Fig. 1).

The initial set of classifiers $\mathscr{H} = \{h_{1,0}, h_{2,0}, h_{3,0}\}$ is built as in the PSI approach (Figs. 4, 5 and 6). The decision tree $h_{1,0}$ contains a node $c_2\#c_0$ and a node $c_3\#c_0$ (Fig. 4), therefore $h_{1,0}$ depends on the output of decision trees $h_{2,0}$ and $h_{3,0}$. The decision tree $h_{2,0}$ contains only descriptive attribute nodes (Fig. 5), hence it's independent. The decision tree $h_{3,0}$ contains a node $c_1\#c_0$ and a node $c_2\#c_0$ (Fig. 6), therefore $h_{3,0}$ depends on the output of decision trees $h_{1,0}$ and $h_{2,0}$. There is a cyclic dependency between decision trees $h_{1,0}$ and $h_{3,0}$ because they depend on each other. Figure 7

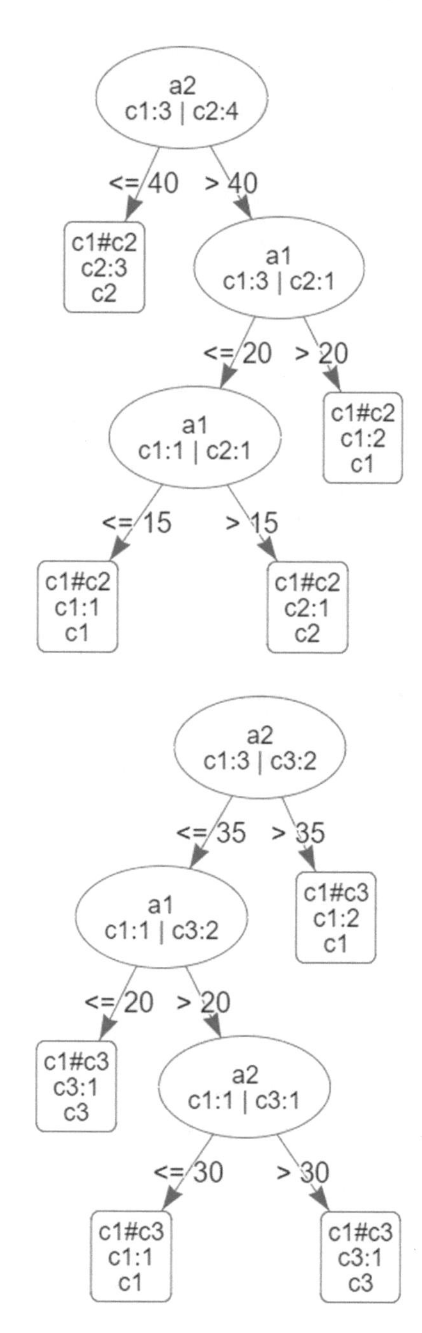

Fig. 1 The binary classifier $H_{1,2}$ learned by the CLR_PSI approach

Fig. 2 The binary classifier $H_{1,3}$ learned by the CLR_PSI approach

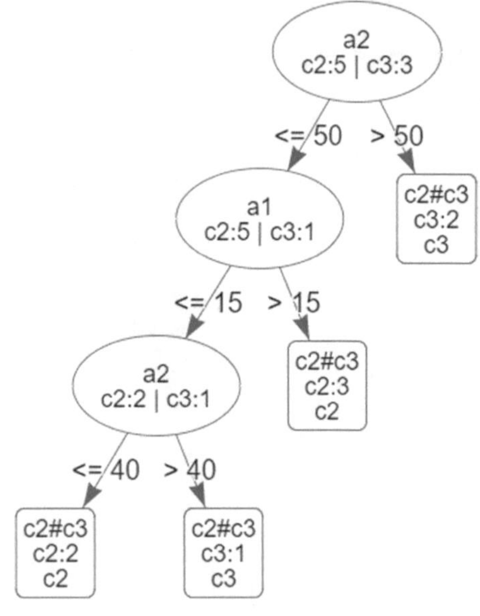

Fig. 3 The binary classifier $H_{2,3}$ learned by the CLR_PSI approach

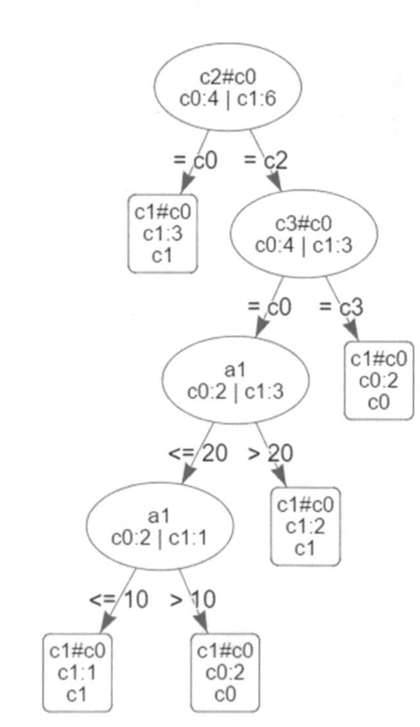

Fig. 4 The binary classifier $h_{1,0}$ learned by the CLR_PSI approach

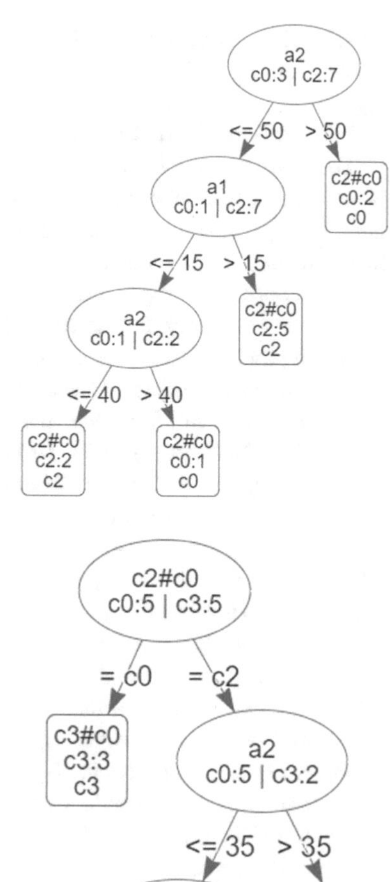

Fig. 5 The binary classifier $h_{2,0}$ learned by the CLR_PSI approach

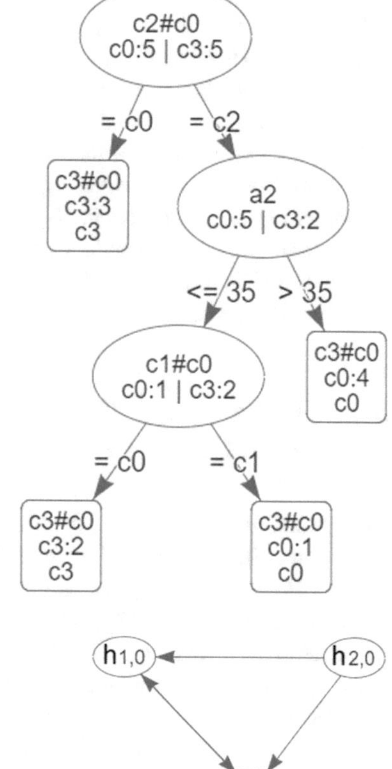

Fig. 6 The binary classifier $h_{3,0}$ learned by the CLR_PSI approach

Fig. 7 Initial dependency structure of classifiers learned by the CLR_PSI approach

presents the dependency structure of the initial set of classifiers $\{h_{1,0}, h_{2,0}, h_{3,0}\}$. An arrow from a classifier $h_{l,0}$ to another classifier $h_{l',0}$ indicates that $h_{l',0}$ depends on $h_{l,0}$.

In the following, the PSI measures presented in Sect. 2.2.2 are considered.

The classifier $h_{2,0}$ is independent, hence it is not pre-selected ($\mathbb{P}(h_{2,0}) = 0$). $h_{2,0}$ is moved from the set of initial classifiers: $\mathscr{H} = \{h_{1,0}, h_{3,0}\}$ to the set of final classifiers: $\mathbb{H} = \{h_{2,0}\}$ ($H_{2,0} = h_{2,0}$). The remaining classifiers $h_{1,0}$ and $h_{3,0}$ are dependent, therefore they are pre-selected ($\mathbb{P}(h_{1,0}) = \mathbb{P}(h_{3,0}) = 1$). $h_{1,0}$ and $h_{3,0}$ depend only on each other after moving $h_{2,0}$ from $\mathscr{H}$. The selection measure $\mathbb{S}$ selects the classifier with the lowest index in this case because $h_{1,0}$ and $h_{3,0}$ have the same number of depending and dependent classifiers: $\mathbb{S}(\mathscr{H}) = h_{1,0}$.

The interest of chaining measure allows the classifier $H_{1,0}$ replacing $h_{1,0}$ to depend on all classifiers in $\mathbb{H}$. Hence, the new classifier $H_{1,0}$ is learned considering the extended attribute set $\{a_1, a_2\} \cup \{c_2\#c_0\}$ (Fig. 8). The classifier $h_{1,0}$ is removed from $\mathscr{H}$: $\mathscr{H} = \{h_{3,0}\}$, and the classifier $H_{1,0}$ is added to $\mathbb{H}$: $\mathbb{H} = \{H_{2,0}, H_{1,0}\}$.

The classifier $h_{3,0}$ is considered independent after moving $h_{2,0}$ and $h_{1,0}$ from $\mathscr{H}$. Hence $h_{3,0}$ is not pre-selected: $\mathbb{P}(h_{3,0}) = 0$. $h_{3,0}$ is then moved from $\mathscr{H}$ to $\mathbb{H}$: $\mathscr{H} = \emptyset$, and $\mathbb{H} = \{H_{2,0}, H_{1,0}, h_{3,0}\}$ ($H_{3,0} = h_{3,0}$). The obtained final dependency structure does not contain any cyclic dependency (Fig. 9).

Note that attributes for the training set of the classifier $H_{2,0}$ are extended to the set $\{a_1, a_2\} \cup \{c_1\#c_0, c_3\#c_0\}$. However, the learned classifier $H_{2,0}$ is independent. The drawback of most of existing approaches compared to the PSI approach is that an assumption is made on the dependency structure before the learning phase: For example, in the CC approach, $H_{1,0}$ cannot depend on any other classifier. However, in the PSI approach a dependency of $H_{1,0}$ on $H_{2,0}$ is discovered (Fig. 8). The classifier $H_{2,0}$ in the CC approach is allowed to depend only on the classifier $H_{1,0}$, however, no dependency is discovered after building the classifier $H_{2,0}$ (Fig. 5).

3.2 *The Stacked_RPC_PSI Approach on a Multi-label Dataset Example*

The initial training set X and the PSI measures described in Sect. 3.1 are considered to build the Stacked_RPC_PSI multi-label classifier.

Classifiers $\{H_{l,l'}\}_{1 \le l < l' \le 3}$ in the Stacked_RPC_PSI approach are built as in the CLR_PSI approach (Figs. 1, 2 and 3). Then they are used to output predictions and extend all instances $x_i \in X$ (Table 3). Values for each added attribute $H_{l,l'}$, $1 \le l < l' \le 3$ in Table 3 are the predictions outputted by the classifier $H_{l,l'}$.

Classifiers $\{H_l\}_{1 \le l \le 3}$ in the Stacked_RPC_PSI approach are built using the PSI approach and considering the extended training set in Table 3. The initial classifiers $\{h_l\}_{1 \le l \le 3}$ learned by the PSI approach are independent, hence they become the final classifiers (Figs. 10, 11 and 12).

The classifier H_3 learned by the Stacked_RPC_PSI approach (Fig. 12) depends on two label preferences. Indeed, the classifier H_3 predicts the relevancy of the label

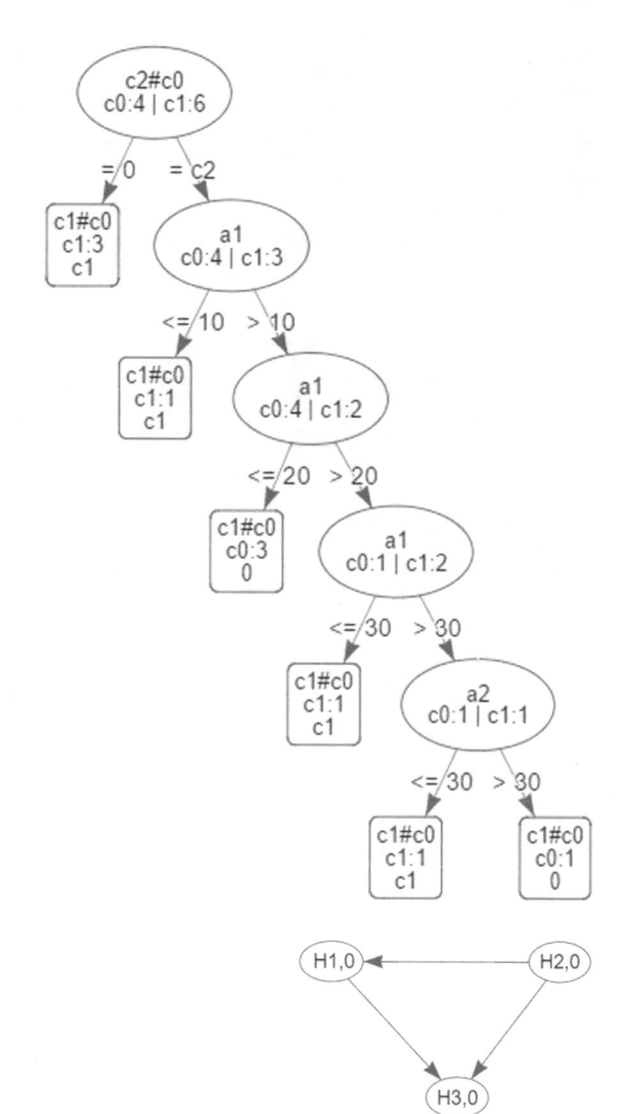

Fig. 8 The new binary classifier $H_{1,0}$ learned by the CLR_PSI approach

Fig. 9 Final dependency structure of classifiers learned by the CLR_PSI approach

c_3 to a given instance x based on the predicted preferences between labels c_1 and c_2 by the classifier $H_{1,2}$, and between labels c_2 and c_3 by the classifier $H_{2,3}$.

The learned dependency structure by the Stacked_RPC_PSI approach considering all classifiers $\{H_{l,l'}\}_{1\leq l<l'\leq 3} \cup \{H_l\}_{1\leq l\leq 3}$ is illustrated in Fig. 13.

Table 3 Extended dataset using predictions of classifiers $\{H_{1,2}, H_{1,3}, H_{2,3}\}$

Instances	a_1	a_2	$H_{1,2}$	$H_{1,3}$	$H_{2,3}$	c_1	c_2	c_3
x_1	20	30	c_2	c_3	c_2	0	1	1
x_2	35	35	c_2	c_3	c_2	0	1	1
x_3	15	40	c_2	c_1	c_2	0	1	0
x_4	20	50	c_2	c_1	c_2	0	1	0
x_5	30	45	c_1	c_1	c_2	1	1	0
x_6	35	30	c_2	c_1	c_2	1	1	0
x_7	10	40	c_2	c_1	c_2	1	1	0
x_8	15	45	c_1	c_1	c_3	1	0	1
x_9	25	55	c_1	c_1	c_3	1	0	1
x_{10}	30	60	c_1	c_1	c_3	1	0	1

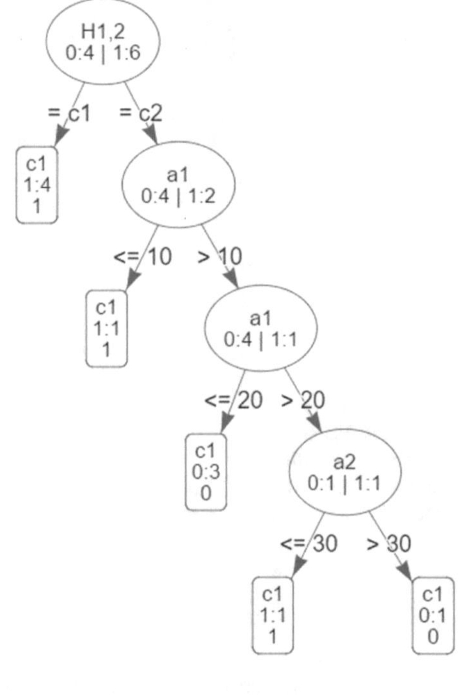

Fig. 10 The binary classifier h_1 learned by the Stacked_RPC_PSI approach

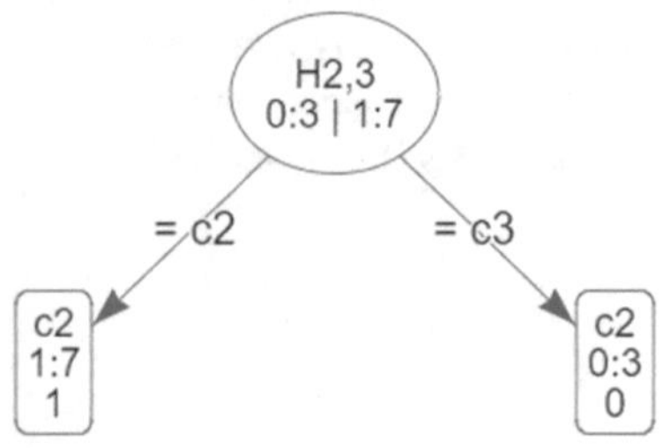

Fig. 11 The binary classifier h_2 learned by the Stacked_RPC_PSI approach

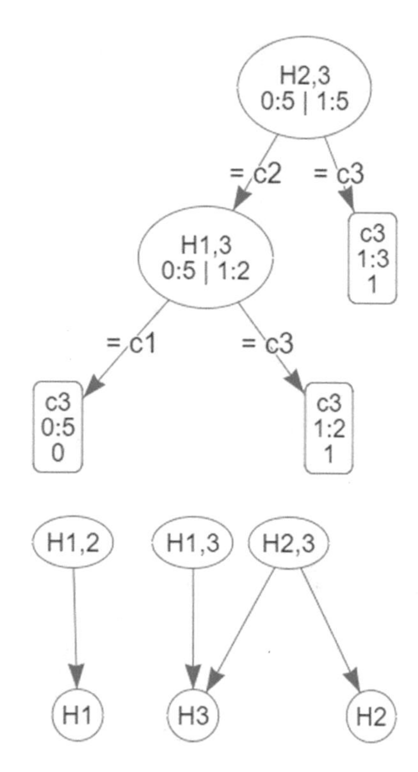

Fig. 12 The binary classifier h_3 learned by the Stacked_RPC_PSI approach

Fig. 13 The learned dependency structure by the Stacked_RPC_PSI approach

3.3 GMLC Decompositions on a Dataset Example

Let $X = \{x_i\}_{1 \leq i \leq 10}$ be the set of training instances in Table 4. Each instance $x_i \in X$ is described by a set of descriptive attributes $\{a_1, a_2\}$, and it is associated to each label from $C = \{c_1, c_2, c_3\}$ with a membership grade from $M = \{m_1 < m_2 < m_3\} = \{1 < 2 < 3\}$. A set of calibration labels $W = \{w_2, w_3\}$ and a set of virtual membership grades $V = \{v_2 < v_3\}$ are introduced so that:

- w_2 has a fixed membership grade v_2, and w_3 had a fixed membership grade v_3
- $M \cup V = \{m_1 < v_2 < m_2 < v_3 < m_3\}$

In Table 4, v_2 corresponds to the value 1.5 and v_3 corresponds to the value 2.5. The attribute w_2 is denotes by '@1.5' and the attribute w_3 is denoted by '@2.5'.

The Horizontal_CLR approach builds a set of $(|M| - 1)(\frac{|C|(|C| + 1)}{2}) = 12$ preference classifiers to predict values for target attributes illustrated in Table 5:

For example, the attribute denoted by '$\geq 2|c_1\#c_2$' is the target attribute for the preference classifier that predicts which label of c_1 and c_2 is preferred to be associated to a given instance with a membership grade $m_g \geq 2$.

Table 4 Example of GML data with multiple calibration labels: 3 membership grades, 3 labels, 2 calibration labels

Instances	a_1	a_2	c_1	c_2	c_3	@1.5	@2.5
x_1	20	30	1	2	1	1.5	2.5
x_2	35	35	1	2	2	1.5	2.5
x_3	15	40	2	2	3	1.5	2.5
x_4	20	50	2	3	3	1.5	2.5
x_5	30	45	3	3	2	1.5	2.5
x_6	35	30	3	3	1	1.5	2.5
x_7	10	40	1	1	1	1.5	2.5
x_8	15	45	1	1	2	1.5	2.5
x_9	25	55	2	1	3	1.5	2.5
x_{10}	30	60	2	1	3	1.5	2.5

The Full_CLR approach builds a set of $\frac{(|C|+|M|-1)(|C|+|M|-2)}{2} = 10$ preference classifiers to predict values for target attributes illustrated in Table 6.

The Joined_CLR approach builds a set of $(|M|-1)\frac{(|C|+|M|-1)(|C|+|M|-2)}{2} = 20$ preference classifiers to predict values for target attributes illustrated in Table 5 and in Table 7.

The CLR_PSI approach can be used to solve the GMLC task as follows:

- a pairwise decomposition such as Horizontal_CLR, Full_CLR, or Joined_CLR is used to build the GML classifier based on a set of preference classifiers.
- classifiers that predict the preference between a label and a calibration label are built using all training instances. Hence, the PSI approach can be applied to learn dependency relations between those classifiers.
- the remaining classifiers and the prediction strategy of the base pairwise decomposition approach are not modified.

The corresponding combined approaches are called Horizontal_CLR_PSI, Full_CLR_PSI, and Joined_CLR_PSI.

The Stacked_RPC_PSI approach can be used to solve the GMLC task as follows:

- a pairwise decomposition such as Horizontal_CLR, Full_CLR, or Joined_CLR is used to build the GML classifier based on preference classifiers.
- classifiers that predict the preference between a label and a calibration label are removed: only classifiers generated by the RPC approach are kept.
- the remaining preference classifiers are built, then their predictions for the training set of instances are collected to extend training instances.

Table 5 Generated target attributes by the Horizontal_CLR approach

Instances	$\geq 2\vert c_1\#c_2$	$\geq 2\vert c_1\#c_3$	$\geq 2\vert c_1\#@1.5$	$\geq 2\vert c_2\#c_3$	$\geq 2\vert c_2\#@1.5$	$\geq 2\vert c_3\#@1.5$
x_1	c_2	$\emptyset$	@1.5	c_2	c_2	@1.5
x_2	c_2	c_3	@1.5	$\emptyset$	c_2	c_3
x_3	$\emptyset$	$\emptyset$	c_1	$\emptyset$	c_2	c_3
x_4	$\emptyset$	$\emptyset$	c_1	$\emptyset$	c_2	c_3
x_5	$\emptyset$	$\emptyset$	c_1	$\emptyset$	c_2	c_3
x_6	$\emptyset$	c_1	c_1	c_2	c_2	@1.5
x_7	$\emptyset$	$\emptyset$	@1.5	$\emptyset$	@1.5	@1.5
x_8	$\emptyset$	c_3	@1.5	c_3	@1.5	c_3
x_9	c_1	$\emptyset$	c_1	c_3	@1.5	c_3
x_{10}	c_1	$\emptyset$	c_1	c_3	@1.5	c_3
Instances	$\geq 3\vert c_1\#c_2$	$\geq 3\vert c_1\#c_3$	$\geq 3\vert c_1\#@2.5$	$\geq 3\vert c_2\#c_3$	$\geq 3\vert c_2\#@2.5$	$\geq 3\vert c_3\#@2.5$
x_1	$\emptyset$	$\emptyset$	@2.5	$\emptyset$	@2.5	@2.5
x_2	$\emptyset$	$\emptyset$	@2.5	$\emptyset$	@2.5	@2.5
x_3	$\emptyset$	c_3	@2.5	c_3	@2.5	c_3
x_4	c_2	c_3	@2.5	$\emptyset$	c_2	c_3
x_5	$\emptyset$	c_1	c_1	c_2	c_2	@2.5
x_6	$\emptyset$	c_1	c_1	c_2	c_2	@2.5
x_7	$\emptyset$	$\emptyset$	@2.5	$\emptyset$	@2.5	@2.5
x_8	$\emptyset$	$\emptyset$	@2.5	$\emptyset$	@2.5	@2.5
x_9	$\emptyset$	c_3	@2.5	c_3	@2.5	c_3
x_{10}	$\emptyset$	c_3	@2.5	c_3	@2.5	c_3

- The PSI approach is used to learn dependencies between a set of ordinal classifiers trained using the extended instances. Each ordinal classifier learns to predict the membership grade of a specific label based on the outputted predictions of preference classifiers.
- The original prediction strategy of the base GMLC pairwise decomposition is not applied. Indeed, the outputted corresponding membership degree to a label is the one predicted by the PSI approach.

The corresponding combined approaches are called Horizontal_Stacked_RPC_PSI, Full_Stacked_RPC_PSI, and Joined_Stacked_RPC_PSI.

Table 6 Generated target attributes by the Full_CLR approach

Instances	$c_1\#c_2$	$c_1\#c_3$	$c_1\#@1.5$	$c_1\#@2.5$	$c_2\#c_3$	$c_2\#@1.5$	$c_2\#@2.5$	$c_3\#@1.5$	$c_3\#@2.5$	@1.5#@2.5
x_1	c_2	∅	@1.5	@2.5	c_2	c_2	@2.5	@1.5	@2.5	@2.5
x_2	c_2	c_3	@1.5	@2.5	∅	c_2	@2.5	c_3	@2.5	@2.5
x_3	∅	c_3	c_1	@2.5	c_3	c_2	@2.5	c_3	c_3	@2.5
x_4	c_2	c_3	c_1	@2.5	∅	c_2	c_2	c_3	c_3	@2.5
x_5	∅	c_1	c_1	c_1	c_2	c_2	c_2	c_3	@2.5	@2.5
x_6	∅	c_1	c_1	c_1	c_2	c_2	c_2	@1.5	@2.5	@2.5
x_7	∅	∅	@1.5	@2.5	∅	@1.5	@2.5	@1.5	@2.5	@2.5
x_8	∅	c_3	@1.5	@2.5	c_3	@1.5	@2.5	c_3	@2.5	@2.5
x_9	c_1	c_3	c_1	@2.5	c_3	@1.5	@2.5	c_3	c_3	@2.5
x_{10}	c_1	c_3	c_1	@2.5	c_3	@1.5	@2.5	c_3	c_3	@2.5

Table 7 Generated target attributes by the Joined_CLR approach that are not generated by the Horizontal_CLR approach: (all Horizontal_CLR generated attributes are also generated by the Full_CLR)

Instances	$\geq 2\|c_1\#@2.5$	$\geq 2\|c_2\#@2.5$	$\geq 2\|c_3\#@2.5$	$\geq 2\|@1.5\#@2.5$	$\geq 3\|c_1\#@1.5$	$\geq 3\|c_2\#@1.5$	$\geq 3\|c_3\#@1.5$	$\geq 3\|@1.5\#@2.5$
x_1	@2.5	c_2	@2.5	@2.5	@1.5	@1.5	@1.5	@2.5
x_2	@2.5	c_2	c_3	@2.5	@1.5	@1.5	@1.5	@2.5
x_3	c_1	c_2	c_3	@2.5	@1.5	@1.5	c_3	@2.5
x_4	c_1	c_2	c_3	@2.5	@1.5	c_2	c_3	@2.5
x_5	c_1	c_2	c_3	@2.5	c_1	c_2	@1.5	@2.5
x_6	c_1	c_2	@2.5	@2.5	c_1	c_2	@1.5	@2.5
x_7	@2.5	@2.5	@2.5	@2.5	@1.5	@1.5	@1.5	@2.5
x_8	@2.5	@2.5	c_3	@2.5	@1.5	@1.5	@1.5	@2.5
x_9	c_1	@2.5	c_3	@2.5	@1.5	@1.5	c_3	@2.5
x_{10}	c_1	@2.5	c_3	@2.5	@1.5	@1.5	c_3	@2.5

4 Experiments on Multi-label Datasets

4.1 Measures to Describe Multi-label Datasets

The multi-label dataset complexity can be quantified using three main measures [8]:

- the label cardinality which quantifies the average number of labels over all instances:
  $\mathbb{LC} = \frac{1}{n}\sum_{i=1}^{n} |y_i|.$
- the label density which quantifies the ratio of label cardinality to the number of labels:
  $\mathbb{LD} = \frac{1}{n}\sum_{i=1}^{n} \frac{|y_i|}{k} = \frac{\mathbb{LC}}{k}.$
- the distinct label combinations which quantifies the number of distinct label sets associated to instances:
  $\mathbb{DLC} = |\{y_i\}_{1\leq i\leq n}|.$

4.2 Evaluation Measures of Predictions in MLC

The prediction performance can be evaluated using different measures [26]:

- the Hamming-loss measure [27] given by
  $\mathbb{HL} = \frac{|y_i \Delta H(x_i)|}{k}$, where $y_i \Delta H(x_i)$ is the symmetric difference between the set of true associated labels and the set of predicted labels given by:
  $y_i \Delta H(x_i) = \{c_l \in y_i - H(x_i)\}_{1\leq l\leq k} \bigcup \{c_l \in H(x_i) - y_i\}_{1\leq l\leq k}.$
  The drawback of Hamming-loss measure is that it is too optimistic for datasets with low label cardinality and density. Indeed, because of low label cardinality the values $\{|y_i|\}_{1\leq i\leq n}$ are small. The cardinality of predictions $\{|H(x_i)|\}_{1\leq i\leq n}$ are also small since the classifier H learned from the labelled set $\{(x_i, y_i)\}_{1\leq i\leq n}$. The symmetric difference $y_i \Delta H(x_i)$ in that case gives also small values.
  Hamming-loss measure in case of low label cardinality combined with low label density $k >> |y_i|$ gives always small error values because the cardinality of the symmetric difference is too small compared to the number of labels
  $y_i \Delta H(x_i) << k.$
- the closely related Hamming score measure [28] is not sensitive to label cardinality and density. It measures the number of correctly predicted labels divided by the number of truly associated and predicted labels: $\mathbb{CRHS} = \frac{|y_i \cap H(x_i)|}{|y_i \cup H(x_i)|}.$

- the precision quantifies the probability that a predicted label is truly associated to the instance:
$\mathbb{PRECISION} = \frac{|y_i \cap H(x_i)|}{|H(x_i)|}$.
- the recall quantifies the probability that a truly associated label is predicted by the classifier:
$\mathbb{RECALL} = \frac{|y_i \cap H(x_i)|}{|y_i|}$.
- the F_β measure [29] combines both precision and recall measures. It is given for any $\beta > 0$ by:
$\mathbb{F}_\beta = (1+\beta^2)\frac{\mathbb{PRECISION} \times \mathbb{RECALL}}{\beta^2 \times \mathbb{PRECISION} + \mathbb{RECALL}}$.

  More importance is given to precision for $\beta < 1$, and more importance is given to recall for $\beta > 1$. The precision and recall are given the same importance for $\beta = 1$.
- the $\mathbb{GMEAN}$ measure [30] combines the positive accuracy:
$\mathbb{ACC}^+ = \frac{|\{c_l, c_l \in y_i \ \& \ c_l \in H(x_i)\}_{1 \le l \le k}|}{|y_i|} = \mathbb{RECALL}$, and the negative accuracy:
$\mathbb{ACC}^- = \frac{|\{c_l, c_l \notin y_i \ \& \ c_l \notin H(x_i)\}_{1 \le l \le k}|}{|C - y_i|}$ using a geometric mean:
$\mathbb{GMEAN} = \sqrt{\mathbb{ACC}^+ \times \mathbb{ACC}^-}$ in order to obtain a more reliable measure.
- the exact match is the most strict evaluation measure. It determines whether the predicted label set is exactly the true label set or not: $\mathbb{EM} = 1$ if $y_i = H(x_i)$, 0 otherwise.

4.3 Dataset Descriptions

Experiments are conducted on three datasets from different domains described in Table 8. Emotion dataset [31] contains a set of 594 instances. Each instance represents a song described by 72 attributes and associated to one or more emotions out of 6 classes of emotions {amazed-suprised, happy-pleased, relaxing-calm, quiet-still, sad-lonely, angry-aggresive}. Scene dataset [32] contains a set of 2407 instances. Each instance represents an image described by 294 attributes and associated to a subset of classes from a set of 6 available classes {Beach, Sunset, FallFoliage, Field, Mountain, Urban}. Yeast dataset [33] contains a set of 2417 instances. Each instance is a protein described by 103 attributes and associated to one or more localization sites from a set of 14 localization sites. The aim is to predict the localizations (called cellular components) of proteins in a yeast cell.

Table 8 Datasets

Dataset	Domain	Instances	Attributes	Labels	$\mathbb{LC}$	$\mathbb{LD}$	$\mathbb{DLC}$
Emotions	Music	593	72	6	1.869	0.311	27
Scenes	Image	2407	294	6	1.074	0.179	15
Yeast	Biology	2417	103	14	4.237	0.303	198

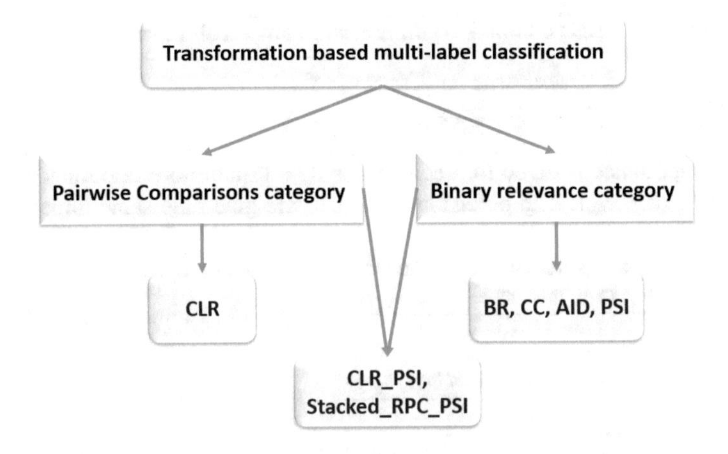

Fig. 14 Category of approaches compared in the experiments

4.4 Experimental Setup

The new introduced approaches CLR_PSI and Stacked_RPC_PSI are compared to five other MLC transformation based approaches (Fig. 14). All of them use decision trees [9] as the base single label classifier. In order to avoid over-fitting, a node in a decision tree becomes a leaf if any condition of the following is met:

- the number of instances is less than 10
- the proportion of instances associated to the majority class is greater than 0.9
- the depth of the node is 20

A 10-fold cross validation experimentation is conducted on each dataset. For each fold, prediction evaluation measures are computed for all approaches. Results corresponding to the 10 folds are then averaged for each approach, and the best results corresponding to each measure are marked with bold characters (Tables 9, 10 and 11).

Obtained values for each approach in Tables 9, 10, and 11 are converted to rankings (the ranking for $\mathbb{HL}$ is reversed because it is a measure that should be minimized). The mean ranking considering the three datasets emotions, scenes, and yeast is presented for each approach in Table 12.

Table 9 Prediction evaluation on emotion dataset

Approach	CRHS	FMEASURE	GMEAN	EM	HL	PRECISION	RECALL	ACC−
AID	0.46	0.55	0.60	0.19	0.24	0.58	0.58	0.84
BR	0.45	0.54	0.60	0.18	0.24	0.59	0.56	0.85
CC	0.46	0.55	0.60	0.21	0.25	0.59	0.56	0.85
CLR	0.45	0.54	0.59	0.17	0.24	0.55	0.60	0.83
CLR_PSI	**0.49**	**0.58**	**0.64**	0.20	0.24	0.59	**0.64**	0.83
PSI	0.48	0.56	0.61	**0.25**	0.24	0.59	0.58	0.84
Stacked_RPC_PSI	0.48	0.57	0.62	0.24	**0.23**	**0.60**	0.59	**0.86**

Table 10 Prediction evaluation on scene dataset

Approach	CRHS	FMEASURE	GMEAN	EM	HL	PRECISION	RECALL	ACC−
AID	0.55	0.58	0.62	0.46	0.15	0.57	0.63	0.91
BR	0.51	0.54	0.56	0.44	**0.12**	0.53	0.57	**0.95**
CC	0.57	0.59	0.60	0.54	0.13	0.60	0.59	0.93
CLR	0.50	0.53	0.58	0.40	0.13	0.51	0.59	0.94
CLR_PSI	**0.60**	**0.63**	**0.67**	0.50	**0.12**	0.62	**0.68**	0.93
PSI	0.57	0.59	0.60	0.54	0.13	0.59	0.59	0.93
Stacked_RPC_PSI	0.60	0.62	0.63	**0.56**	**0.12**	**0.63**	0.62	0.94

Table 11 Prediction evaluation on yeast dataset

Approach	CRHS	FMEASURE	GMEAN	EM	HL	PRECISION	RECALL	ACC–
AID	0.40	0.53	0.63	0.06	0.27	0.56	0.55	0.81
BR	0.42	0.54	0.63	0.06	0.25	0.60	0.55	0.85
CC	0.43	0.52	0.59	**0.16**	0.24	0.60	0.52	0.87
CLR	**0.47**	**0.59**	**0.66**	0.10	**0.21**	**0.67**	**0.59**	0.88
CLR_PSI	**0.47**	0.58	0.65	0.12	0.22	**0.67**	0.58	0.88
PSI	0.43	0.53	0.60	0.15	0.26	0.58	0.53	0.84
Stacked_RPC_PSI	0.44	0.54	0.61	0.13	0.23	0.62	0.53	0.88

Table 12 Ranking of multi-label classification approaches averaged considering emotion, scene, and yeast datasets

Approach	CRHS	FMEASURE	GMEAN	EM	HL	PRECISION	RECALL	ACC–
AID	5.67	5.33	3.67	5.67	5.67	6.00	3.33	6.00
BR	6.33	5.33	5.33	6.00	3.67	5.00	5.67	2.67
CC	4.00	5.00	5.67	2.00	5.33	3.67	6.67	3.67
CLR	4.67	5.00	4.67	6.33	3.00	5.00	2.33	3.33
CLR_PSI	**1.67**	**1.33**	**1.33**	4.00	**2.00**	2.67	**1.33**	5.33
PSI	3.67	3.67	4.33	**1.67**	6.00	4.00	4.67	5.33
Stacked_RPC_PSI	2.00	2.33	3.00	2.00	2.33	**1.67**	4.00	**1.67**

4.5 Discussion

All approaches exploiting label relations outperforms the BR approach that does not allow learning label relations (Tables 9, 10, and 11).

The CC approach allows learning label relations according to a predefined chained structure. The PSI approach learns label relations without any restriction. The AID approach allow learning label relations without any restriction but the prediction is not necessarily coherent with learned label relations. This explains the fact that CC and PSI approaches have better results for the $\mathbb{EM}$ measure then the AID approach in emotion, scene and yeast datasets (Tables 9, 10 and 11). Indeed, the $\mathbb{EM}$ measure rewards coherent predictions because a prediction is considered correct only if the predicted label set corresponds exactly to the associated label set.

The drawback of BR, CC, AID, and PSI approaches is that they learn a binary classifier for each label to predict whether it is associated to an instance or not. Therefore, the training set of a binary classifier H_l may contains very few instances associated to the label c_l compared to instances not associated to the label c_l (associated to other labels). Hence, the more available labels has a dataset, the more classifiers are affected by this class imbalance problem. A binary classifier H_l may always predicts for any given instance that c_l is not associated to that instance. The advantage of CLR approach is that it learns a binary classifier for each pair of labels to predict the preferred label between them. Only instances associated to one of the two labels are considered and the remaining instances are ignored. Hence, the problem of imbalanced classes is reduced for CLR approach compared to BR, CC, AID, and PSI approaches. This explains the fact that CLR gives better results for yeast dataset (Table 11) because it has 14 available labels compared to emotion and scene datasets that have only 6 available labels. The CLR_PSI approach gives good results for yeast dataset as the CLR approach because they use the same prediction strategy (Table 11). The obtained results for the Stacked_RPC_PSI approach on yeast dataset are not as good as for the CLR_PSI approach. This is because the Stacked_RPC_PSI approach uses the PSI approach prediction strategy instead of the CLR prediction strategy.

Table 12 shows that the best results considering the three datasets: emotions, scenes, and yeast are obtained by the CLR_PSI approach followed by the Stacked_CLR_PSI approach. This confirms that combining a pairwise classification approach with a binary relevance approach that allows label relations can improve the prediction performance in multi-label classification.

5 Experiments on Graded Multi-label Datasets

5.1 Measures to Describe GML Datasets

GML data can be decomposed horizontally to a set of ML data, hence, measures to describe ML data (Sect. 4.1) can be used to describe each sub MLC problem.

GML data can be described using specific measures to the GML case [24]:

- the average grade of labels given by $\mathbb{AG} = \frac{\sum_{1 \le l \le k} \mu_{c_l}(x_i)}{nk}$
- the distribution of grades: the average number of times each membership grade is used to associate a label to an instance over all labels and instances. It is given by: $\mathbb{DG}(m_g) = \frac{|\{(x_i, c_l) \in X \times C, \mu_{c_l}(x_i) = m_g\}|}{nk}$.

5.2 Evaluation Measures of Predictions in GMLC

Prediction evaluation measures in MLC (Sect. 4.2) can be used to evaluate each MLC task corresponding to a membership grade, then evaluations can be averaged over all membership grades. Predictions in GMLC can be evaluated using specific measures to the GML case such as:

- the extended Hamming-loss given for an instance x by: $\mathbb{HL}^* = \sum_{1 \le l \le k} \frac{|H_{c_l}(x) - \mu_{c_l}(x)|}{k|m_s - m_1|}$
- the Pairwise Ranking Error measure that outputs for an instance x the average number of label pairs incorrectly ranked. It is given by: $\mathbb{PRE}(x) = \frac{1}{k(k-1)/2} |\{(c_l, c_{l'}) \in C^2, (\mu_{c_l}(x) > \mu_{c_{l'}}(x)\ \&\ H_{c_l}(x) < H_{c_{l'}}(x))$ $or (\mu_{c_l}(x) < \mu_{c_{l'}}(x)\ \&\ H_{c_l}(x) > H_{c_{l'}}(x))\}_{1 \le l < l' \le k}|$

Note that Hamming-loss $\mathbb{HL}$, extended Hamming-loss $\mathbb{HL}^*$, and pairwise ranking errors $\mathbb{PRE}$ are all loss functions that should be minimized.

5.3 Dataset Descriptions

Experiments are conducted on two datasets described in Table 13:

BelaE dataset [34] contains a set of 1930 instances. Each instance represents a graduate student described by two attributes: the gender and the age. Each student is asked to assign an importance degree to 48 properties of its future job. Five importance degrees ranging from 'completely unimportant' to 'very important' are allowed. Due to the lack of descriptive attributes, only the last 10 job properties are considered as labels, and the remaining 28 properties are considered as attributes Table 13. The five membership grades are mapped to values in $\{0, 1, 2, 3, 4\}$ so that the grade 'completely unimportant' corresponds to the value 0. The average membership grade given to a label in BelaE dataset is $\mathbb{AG} = 2.66$. The distribution of grades over instances and labels is illustrated in Table 13.

The dataset of odors is extracted from a set containing about 3000 odorous molecules [35]. Each molecule can be associated up to 7 odors ordered by intensity.

Table 13 Description of graded multi-label datasets: BelaE and Odors datasets

Dataset	Domain	Instances	Attributes	Labels	Grades	$\mathbb{AG}$	$\mathbb{DG}(0)$	$\mathbb{DG}(1)$	$\mathbb{DG}(2)$	$\mathbb{DG}(3)$	$\mathbb{DG}(4)$	$\mathbb{DG}(5)$	$\mathbb{DG}(6)$	$\mathbb{DG}(7)$
BelaE	Psychology	1930	40	10	5	2.66	0.0379	0.1044	0.2490	0.3786	0.2302	-	-	-
Odors	Chemistry	1623	1838	30	8	0.48	0.918	0.0001	0.0007	0.0022	0.0069	0.0152	0.0259	0.0311

An ordinal scale of 8 membership grades $M = [\![0, 7]\!]$ is used to transform ordered sets to graded sets. Odors that are not associated to a molecule are given the membership grade 0, and odors that are strongly associated to a molecule are given the membership grade 7. The software 'Dragon' [36] is used to generate 3839 molecular descriptors by calculating values of physico-chemical properties of odorous molecules. Molecular descriptors having the same value for more than 0.99 of data are discarded. Odors that are associated to less than 30 molecules are discarded. The obtained dataset after this pre-processing and filtering step is illustrated in Table 13. The number of molecules associated to more than 4 odors is low. This explains the obtained low values $\{\mathbb{DG}(1), \mathbb{DG}(2), \mathbb{DG}(3)\}$ of the average occurrence of grades $\{1, 2, 3\}$ over labels and instances (Table 13).

5.4 Experimental Setup

Experiments on GML data are conducted using the same parameters as in experiments on ML data (Sect. 4.4) for the PSI measures and the base binary classifiers. The following approaches are evaluated:

- Vertical_BR and Vertical_PSI approaches as described in Sect. 2.3.1
- Complete_BR and Complete_PSI approaches as described in Sect. 2.3.3
- Horizontal_CLR, Full_CLR, and Joined_CLR approaches as described in Sect. 2.3.2
- Horizontal_CLR_PSI, Full_CLR_PSI, and Joined_CLR_PSI GMLC approaches that are based on the new introduced MLC approach CLR_PSI (Sect. 3.3)
- Horizontal_Stacked_CLR_PSI, Full_Stacked_CLR_PSI, and Joined_Stacked_CLR_PSI GMLC approaches that are based on the new introduced MLC approach Stacked_CLR_PSI (Sect. 3.3)

5.5 Discussion

MLC evaluation measures (Sect. 4.2) are used to evaluate the prediction performance on BelaE and on Odors datasets for each MLC subtask corresponding to a specific membership grade. Tables 14 and 15 show that all GMLC approaches can predict labels associated to low membership degrees better than labels associated to high membership degrees. Evaluations corresponding to grades ≥ 1, ≥ 2, and ≥ 3 are not illustrated for Odors dataset because they have almost the same values as evaluations corresponding to the grade ≥ 4. This is explained by the fact that molecules having more than 4 odors are very rare, therefore grades $\{7, 6, 5, 4\}$ are more frequent than grades $\{3, 2, 1\}$.

MLC evaluation measures are averaged over all membership grades as illustrated in Table 16 for BelaE dataset, and in Table 17 for Odors dataset.

The ordinal relation between membership degrees is ignored in approaches Vertical_BR and Vertical_PSI but not in approaches Complete_BR and Complete_PSI.

Table 14 Prediction evaluation on the BelaE dataset for all MLC subtasks

Grade	Approach	CRHS	FMEASURE	GMEAN	EM	HL	PRECISION	RECALL	ACC–
≥ 1	Vertical_BR	0.95	0.97	0.98	0.66	0.05	0.97	0.98	0.98
	Vertical_PSI	0.95	0.97	0.98	0.66	0.05	0.97	0.98	0.98
	Complete_BR	0.97	0.98	0.99	0.74	0.03	0.97	0.99	0.98
	Complete_PSI	0.96	0.98	0.98	0.70	0.04	0.97	0.98	0.98
	Full_CLR	0.94	0.96	0.96	0.53	0.06	0.98	0.96	0.97
	Full_CLR_PSI	0.94	0.96	0.96	0.54	0.06	0.98	0.96	0.97
	Full_Stacked_RPC_PSI	0.95	0.97	0.97	0.66	0.05	0.97	0.98	0.98
	Horizontal_CLR	0.96	0.98	0.99	0.74	0.03	0.97	0.99	0.98
	Horizontal_CLR_PSI	0.96	0.98	0.99	0.74	0.03	0.97	0.99	0.98
	Horizontal_Stacked_RPC_PSI	0.96	0.98	0.99	0.74	0.03	0.97	0.99	0.99
	Joined_CLR	0.89	0.94	0.94	0.29	0.10	0.98	0.91	0.97
	Joined_CLR_PSI	0.90	0.95	0.94	0.34	0.09	0.98	0.92	0.98
	Joined_Stacked_RPC_PSI	0.95	0.97	0.98	0.65	0.05	0.97	0.98	0.98
≥ 2	Vertical_BR	0.84	0.90	0.87	0.24	0.15	0.90	0.91	0.86
	Vertical_PSI	0.84	0.90	0.87	0.24	0.15	0.90	0.91	0.86
	Complete_BR	0.87	0.93	0.89	0.33	0.12	0.90	0.96	0.86
	Complete_PSI	0.85	0.91	0.89	0.28	0.14	0.91	0.93	0.87
	Full_CLR	0.75	0.84	0.82	0.13	0.22	0.94	0.80	0.88
	Full_CLR_PSI	0.75	0.84	0.82	0.14	0.22	0.94	0.80	0.88
	Full_Stacked_RPC_PSI	0.84	0.91	0.88	0.23	0.14	0.91	0.92	0.86
	Horizontal_CLR	0.87	0.93	0.90	0.33	0.12	0.91	0.96	0.86
	Horizontal_CLR_PSI	0.87	0.93	0.90	0.32	0.12	0.91	0.96	0.87
	Horizontal_Stacked_RPC_PSI	0.87	0.92	0.89	0.31	0.12	0.90	0.96	0.86
	Joined_CLR	0.84	0.91	0.87	0.21	0.14	0.93	0.89	0.87
	Joined_CLR_PSI	0.85	0.91	0.87	0.24	0.14	0.92	0.91	0.86

(continued)

Table 14 (continued)

Grade	Approach	CRHS	FMEASURE	GMEAN	EM	HL	PRECISION	RECALL	ACC−
	Joined_Stacked_RPC_PSI	0.84	0.91	0.88	0.23	0.15	0.91	0.92	0.86
≥ 3	Vertical_BR	0.61	0.74	0.70	0.04	0.29	0.75	0.76	0.68
	Vertical_PSI	0.61	0.73	0.70	0.04	0.29	0.75	0.76	0.69
	Complete_BR	0.65	0.77	0.72	0.05	0.26	0.76	0.80	0.68
	Complete_PSI	0.62	0.75	0.71	0.04	0.27	0.76	0.77	0.70
	Full_CLR	0.31	0.41	0.44	0.02	0.43	0.66	0.34	0.93
	Full_CLR_PSI	0.34	0.44	0.46	0.02	0.42	0.67	0.39	0.90
	Full_Stacked_RPC_PSI	0.61	0.73	0.70	0.04	0.29	0.74	0.76	0.68
	Horizontal_CLR	0.63	0.75	0.72	0.05	0.26	0.78	0.78	0.71
	Horizontal_CLR_PSI	0.63	0.75	0.72	0.06	0.26	0.78	0.78	0.71
	Horizontal_Stacked_RPC_PSI	0.64	0.76	0.72	0.05	0.27	0.77	0.79	0.69
	Joined_CLR	0.68	0.80	0.72	0.07	0.25	0.74	0.90	0.60
	Joined_CLR_PSI	0.68	0.80	0.71	0.07	0.25	0.72	0.92	0.59
	Joined_Stacked_RPC_PSI	0.61	0.74	0.70	0.03	0.29	0.75	0.77	0.68
≥ 4	Vertical_BR	0.26	0.34	0.40	0.12	0.23	0.37	0.37	0.84
	Vertical_PSI	0.26	0.33	0.39	0.11	0.23	0.37	0.36	0.84
	Complete_BR	0.27	0.34	0.39	0.15	0.21	0.39	0.35	0.87
	Complete_PSI	0.26	0.34	0.39	0.13	0.22	0.38	0.35	0.86
	Full_CLR	0.04	0.05	0.05	0.18	0.23	0.06	0.05	0.98
	Full_CLR_PSI	0.06	0.07	0.08	0.19	0.22	0.08	0.07	0.98
	Horizontal_CLR	0.24	0.31	0.35	0.18	0.19	0.39	0.30	0.92
	Horizontal_CLR_PSI	0.24	0.31	0.35	0.18	0.19	0.37	0.30	0.92
	Horizontal_Stacked_RPC_PSI	0.24	0.31	0.35	0.15	0.20	0.38	0.30	0.91
	Joined_CLR	0.30	0.41	0.50	0.00	0.45	0.31	0.73	0.46
	Joined_CLR_PSI	0.29	0.40	0.48	0.00	0.48	0.30	0.75	0.42
	Joined_Stacked_RPC_PSI	0.27	0.35	0.40	0.11	0.23	0.38	0.37	0.85

Table 15 Prediction evaluation on odors dataset for all MLC subtasks

Grade	Approach	CRHS	FMEASURE	GMEAN	EM	HL	PRECISION	RECALL	ACC−
≥ 4	Vertical_BR	0.23	0.30	0.37	0.05	0.09	0.39	0.28	0.97
	Vertical_PSI	0.23	0.30	0.37	0.06	0.09	0.39	0.27	0.97
	Complete_BR	0.35	0.46	0.61	0.05	0.09	0.49	0.51	0.95
	Complete_PSI	0.22	0.29	0.34	0.04	0.08	0.39	0.26	0.98
	Full_CLR	0.25	0.31	0.36	0.09	0.07	0.39	0.29	0.98
	Full_CLR_PSI	0.25	0.31	0.36	0.10	0.07	0.40	0.29	0.98
	Full_Stacked_CLR_PSI	0.27	0.35	0.43	0.08	0.08	0.43	0.34	0.97
	Horizontal_CLR	0.35	0.45	0.56	0.09	0.07	0.52	0.46	0.97
	Horizontal_CLR_PSI	0.33	0.44	0.57	0.09	0.08	0.49	0.47	0.96
	Horizontal_Stacked_CLR_PSI	0.34	0.45	0.58	0.07	0.08	0.49	0.48	0.96
	Joined_CLR	0.24	0.29	0.32	0.11	0.07	0.38	0.26	0.99
	Joined_CLR_PSI	0.25	0.32	0.38	0.09	0.08	0.40	0.31	0.98
	Joined_Stacked_CLR_PSI	0.25	0.32	0.41	0.06	0.08	0.39	0.32	0.97
≥ 5	Vertical_BR	0.23	0.31	0.37	0.06	0.08	0.38	0.29	0.97
	Vertical_PSI	0.24	0.31	0.37	0.07	0.08	0.38	0.28	0.97
	Complete_BR	0.32	0.41	0.53	0.06	0.08	0.44	0.44	0.96
	Complete_PSI	0.23	0.30	0.35	0.05	0.08	0.39	0.26	0.98
	Full_CLR	0.21	0.26	0.28	0.10	0.06	0.33	0.23	0.99
	Full_CLR_PSI	0.24	0.29	0.33	0.10	0.07	0.36	0.27	0.99
	Full_Stacked_CLR_PSI	0.27	0.35	0.43	0.09	0.08	0.42	0.35	0.97
	Horizontal_CLR	0.34	0.42	0.50	0.10	0.07	0.50	0.42	0.98
	Horizontal_CLR_PSI	0.35	0.43	0.52	0.15	0.07	0.50	0.43	0.97
	Horizontal_Stacked_CLR_PSI	0.33	0.43	0.54	0.09	0.07	0.49	0.45	0.96
	Joined_CLR	0.21	0.25	0.27	0.12	0.06	0.32	0.22	0.99
	Joined_CLR_PSI	0.25	0.31	0.35	0.11	0.07	0.39	0.29	0.98
	Joined_Stacked_CLR_PSI	0.25	0.33	0.41	0.07	0.08	0.38	0.34	0.97

(continued)

Table 15 (continued)

Grade	Approach	CRHS	FMEASURE	GMEAN	EM	HL	PRECISION	RECALL	ACC–
≥ 6	Vertical_BR	0.21	0.26	0.31	0.07	0.07	0.30	0.26	0.98
	Vertical_PSI	0.23	0.28	0.33	0.09	0.07	0.33	0.27	0.97
	Complete_BR	0.27	0.34	0.40	0.10	0.06	0.38	0.33	0.98
	Complete_PSI	0.22	0.27	0.31	0.08	0.06	0.33	0.25	0.98
	Full_CLR	0.17	0.20	0.21	0.10	0.05	0.24	0.18	0.99
	Full_CLR_PSI	0.20	0.23	0.26	0.10	0.06	0.28	0.22	0.99
	Full_Stacked_CLR_PSI	0.25	0.31	0.37	0.10	0.07	0.36	0.31	0.97
	Horizontal_CLR	0.30	0.35	0.40	0.15	0.05	0.40	0.35	0.98
	Horizontal_Stacked_CLR_PSI	0.29	0.35	0.40	0.14	0.06	0.40	0.35	0.98
	Joined_CLR_PSI	0.22	0.26	0.29	0.10	0.06	0.31	0.25	0.99
	Horizontal_CLR_PSI	0.30	0.35	0.40	0.15	0.06	0.40	0.35	0.98
	Joined_Stacked_CLR_PSI	0.26	0.31	0.37	0.10	0.06	0.35	0.32	0.97
	Joined_CLR	0.17	0.19	0.19	0.10	0.05	0.22	0.17	1.00
≥ 7	Vertical_BR	0.14	0.15	0.17	0.14	0.04	0.14	0.17	0.98
	Vertical_PSI	0.19	0.20	0.23	0.20	0.04	0.19	0.23	0.98
	Complete_BR	0.12	0.12	0.12	0.15	0.04	0.12	0.12	0.99
	Complete_PSI	0.19	0.20	0.20	0.20	0.04	0.19	0.20	0.98
	Full_CLR	0.09	0.09	0.09	0.12	0.03	0.09	0.09	1.00
	Full_CLR_PSI	0.15	0.15	0.15	0.19	0.03	0.15	0.15	0.99
	Full_Stacked_CLR_PSI	0.22	0.23	0.23	0.23	0.04	0.22	0.23	0.98
	Horizontal_CLR	0.17	0.17	0.17	0.20	0.03	0.17	0.17	1.00
	Horizontal_CLR_PSI	0.16	0.17	0.17	0.18	0.03	0.16	0.17	0.99
	Horizontal_Stacked_CLR_PSI	0.15	0.15	0.15	0.18	0.04	0.15	0.15	0.99
	Joined_CLR	0.12	0.12	0.12	0.14	0.03	0.12	0.12	1.00
	Joined_CLR_PSI	0.19	0.20	0.21	0.20	0.03	0.19	0.21	0.99
	Joined_Stacked_CLR_PSI	0.20	0.21	0.22	0.23	0.04	0.20	0.22	0.98

Table 16 Prediction evaluation on BelaE dataset by averaging evaluations of MLC subtasks

Approach	CRHS	FMEASURE	GMEAN	EM	HL	PRECISION	RECALL	ACC−
Vertical_BR	0.66	0.74	0.74	0.27	0.18	0.75	0.75	0.84
Vertical_PSI	0.66	0.74	0.73	0.26	0.18	0.75	0.75	0.84
Complete_BR	0.69	0.75	0.75	0.32	0.16	0.76	0.78	0.85
Complete_PSI	0.67	0.74	0.74	0.29	0.17	0.76	0.76	0.85
Full_CLR	0.51	0.57	0.57	0.22	0.23	0.66	0.54	0.94
Full_CLR_PSI	0.52	0.58	0.58	0.22	0.23	0.66	0.56	0.93
Full_Stacked_RPC_PSI	0.80	0.87	0.85	0.31	0.16	0.87	0.89	0.84
Horizontal_CLR	0.68	0.74	0.74	0.33	0.15	0.76	0.76	0.87
Horizontal_CLR_PSI	0.68	0.74	0.74	0.33	0.15	0.76	0.76	0.87
Horizontal_Stacked_RPC_PSI	0.68	0.74	0.74	0.31	0.16	0.76	0.76	0.86
Joined_CLR	0.68	0.76	0.76	0.14	0.24	0.74	0.86	0.73
Joined_CLR_PSI	0.68	0.76	0.75	0.16	0.24	0.73	0.88	0.71
Joined_Stacked_RPC_PSI	0.67	0.74	0.74	0.26	0.18	0.75	0.76	0.84

Table 17 Prediction evaluation on Odors dataset by averaging evaluations of MLC subtasks

Approach	CRHS	FMEASURE	GMEAN	EM	HL	PRECISION	RECALL	ACC–
Vertical_BR	0.21	0.26	0.31	0.07	0.07	0.30	0.26	0.98
Vertical_PSI	0.23	0.28	0.33	0.09	0.07	0.33	0.27	0.97
Complete_BR	0.27	0.34	0.40	0.10	0.06	0.38	0.33	0.98
Complete_PSI	0.22	0.27	0.31	0.08	0.06	0.33	0.25	0.98
Full_CLR	0.17	0.20	0.21	0.10	0.05	0.24	0.18	0.99
Full_CLR_PSI	0.20	0.23	0.26	0.10	0.06	0.28	0.22	0.99
Full_Stacked_CLR_PSI	0.25	0.31	0.37	0.10	0.07	0.36	0.31	0.97
Horizontal_CLR	0.30	0.35	0.40	0.15	0.05	0.40	0.35	0.98
Horizontal_CLR_PSI	0.30	0.35	0.40	0.15	0.06	0.40	0.35	0.98
Horizontal_Stacked_CLR_PSI	0.29	0.35	0.40	0.14	0.06	0.40	0.35	0.98
Joined_CLR	0.17	0.19	0.19	0.10	0.05	0.22	0.17	1.00
Joined_CLR_PSI	0.22	0.26	0.29	0.10	0.06	0.31	0.25	0.99
Joined_Stacked_CLR_PSI	0.26	0.31	0.37	0.10	0.06	0.35	0.32	0.97

This explains the fact that Complete_BR and Complete_PSI approaches gives lightly better results on BelaE dataset than Vertical_BR and Vertical_PSI approaches.

The difference between the Vertical and the Complete decompositions is clearer on Odors dataset because it has 8 available membership grades instead of just 5 as in BelaE dataset. An interesting remark in Odors dataset is that allowing label dependencies in the complete reduction using the PSI approach does not improve the prediction performance, but rather the opposite. This is explained by the fact that there is no separation between the generated binary classifiers as in the Horizontal decomposition. Therefore The PSI approach may learn some trivial dependencies such as:

'The membership grade for the label c_l is greater or equals to m_g only if the membership grade for the label c_l is greater or equals to m_{g-1}'.

Trivial dependencies are not helpful and they increase the risk of prediction error propagation.

Each MLC subtask is treated separately in the horizontal decomposition. The outputted membership degree for a label is the highest membership degree corresponding to a MLC subtask in which that label is predicted. Therefore, the outputted prediction is wrong if the label is not predicted in the correct MLC subtask. The Horizontal decomposition outputs good results on BelaE and Odors datasets for MLC subtasks corresponding to low membership degrees (Tables 14 and 15). This is explained by the fact that wrong predicted membership grades for labels are counted correct if they still be greater or equals to the membership degree cut of the MLC subtask. Labels corresponding to the highest membership grade are the most difficult to predict correctly because only correctly predicted labels are counted correct. The drawback of the Horizontal decomposition strategy of treating each horizontal problem separately is that a prediction error in a MLC subtask may not be compensated by correct predictions of other MLC subtasks.

The outputted prediction by the Full_CLR approach depends on the accumulated votes by all preference classifiers. The advantage of this prediction strategy is that a prediction error of one classifier can be compensated by a correct prediction of another classifier. The drawback of the Full_CLR is that the difference in membership grades is not considered when learning preferences. This explains the fact that the Full_CLR outputs the worst results for both BelaE and Odors datasets (Tables 14 and 15). This explains also the fact that the approach Full_CLR_PSI that allows learning label relations improves lightly the prediction performance. The Full_Stacked_RPC_PSI approach that allows learning label relations and predicts using the PSI approach instead of the voting strategy gives even better results that the Full_CLR_PSI.

The Joined decomposition includes all preference classifiers generated by the horizontal decomposition, however, the prediction strategy is the same as the Full_CLR approach. This explains the fact that the joined decomposition gives better results than the horizontal decomposition for highest membership degrees on both BelaE and Odors datasets (Tables 14 and 15).

Tables 16 and 17 show that combining GMLC decompositions with the PSI approach does not have a significant effect except for the Full_CLR decomposition. Indeed, the PSI approach allows learning label relations, however in case descriptive attributes are sufficient to discriminate data then learning label relations may not be needed. The drawback of the Full_CLR is that the prediction is based on the accumulated votes of preference classifiers that ignore the difference between membership degrees. The PSI approach learned labels relations in the Full_CLR decomposition that could not be learned in the Joined and the horizontal decompositions which do not ignore the difference between membership degrees. This explains the fact that the Full_Stacked_CLR_PSI approach outperforms all other approaches for BelaE dataset (Table 16). Combining the Stacked_CLR_PSI approach with the Horizontal decomposition for Odors dataset gave the second best results after the Horizontal_CLR approach (Table 17). That means that the the Full_Stacked_CLR_PSI could not learn enough good dependencies in Odors dataset as in BelaE dataset.

GMLC is a combination of MLC and ordinal classification. The extended Hamming loss $\mathbb{HL}^*$ and the pairwise error ranking $\mathbb{PER}$ measures are used to evaluate the ordinal aspect in GML predictions (Tables 18 and 19).

The difference between BelaE and Odors datasets is that in BelaE dataset labels can be associated to the same membership degree, and most of them have non-zero membership degree. This is because labels in BelaE dataset are job properties and membership grades represent the importance of properties. Therefore it is possible that multiple job properties have the same importance.

In Odors dataset labels are odors from a set of 30 available odors (Table 13), however, a molecule can be associated to at most 7 odors. Therefore, at least 23 labels have zero-membership degree for each molecule. Odors are ordered by intensity but without a given value for the intensity. Therefore membership grades in Odors dataset are just a mapping of label order and do not represent odor intensities.

Table 18 Prediction evaluation on BelaE dataset using adapted measures to GMLC

Approach	$\mathbb{HL}^*$	$\mathbb{PRE}$
Vertical_BR	0.18	0.50
Vertical_PSI	0.18	0.50
Complete_BR	0.16	0.48
Complete_PSI	0.17	0.49
Full_CLR	0.23	0.46
Full_CLR_PSI	0.23	0.46
Full_Stacked_RPC_PSI	0.18	0.50
Horizontal_CLR	0.15	0.47
Horizontal_CLR_PSI	0.15	0.47
Horizontal_Stacked_RPC_PSI	0.16	0.47
Joined_CLR	0.24	0.46
Joined_CLR_PSI	0.24	0.47
Joined_Stacked_RPC_PSI	0.18	0.49

Table 19 Prediction evaluation on odors dataset using adapted measures to GMLC

Approach	$\mathbb{HL}^*$	$\mathbb{PRE}$
Vertical_BR	0.08	0.16
Vertical_PSI	0.08	0.16
Complete_BR	0.08	0.19
Complete_PSI	0.07	0.16
Full_CLR	0.07	0.16
Full_CLR_PSI	0.07	0.15
Full_Stacked_CLR_PSI	0.08	0.17
Horizontal_CLR	0.07	0.15
Horizontal_CLR_PSI	0.07	0.17
Horizontal_Stacked_CLR_PSI	0.07	0.17
Joined_CLR	0.06	0.15
Joined_CLR_PSI	0.07	0.16
Joined_Stacked_CLR_PSI	0.07	0.16

The extended Hamming-loss and the pairwise ranking error for the odors dataset are low because of the number of labels associated to the zero-membership degree (Table 19).

The Vertical decomposition ignores label ordering, therefore it gives the worst results of pairwise label ordering for BelaE dataset (Table 18). The Complete decomposition gives lightly better results because it builds a binary classifier for each label and membership grade, hence label order is considered to some extent. The Full_CLR decomposition based approaches output the best pairwise ordering of labels, however, the difference in membership degrees for incorrectly ordered labels is high according to the extended Hamming loss measure (Table 18).

6 Conclusion

Learning label relations and exploiting them to output coherent predictions is an interesting challenge in multi-label classification. Transformation based multi-label classification approaches allow modifying the strategy of learning label relations without modifying the base classifier. The Calibrated Label Ranking approach (CLR) allows learning preference relations between labels by ranking them using pairwise comparisons (RPC). The PSI approach (PSI) allows learning label dependencies without any restriction. It is based on three measures called Pre-selection, Selection, and Interest of chaining. The role of the PSI measures is to eliminate cyclic dependencies by replacing some classifiers based on heuristics. This work introduces the CLR_PSI and the Stacked_RPC_PSI approaches that combine the CLR and the RPC approaches with the PSI approach in order to benefit from both label preference and label dependency relations. Experiments conducted on three datasets from different domains show that the best results according to the evaluated measures are obtained by the CLR_PSI approach followed by the Stacked_RPC_PSI approach. This confirms the ideas that combining label dependency and label preference relations can improve the predictive performance of multi-label classifiers. Graded multi-label classification is an extension of multi-label classification that allows labels to be associated to instances with membership degrees from an ordinal scale of membership grades. This work investigates the idea of decomposing the graded multi-label classification task into a set of multi-label classification subtasks in order to use the introduced approaches. Experiment on two different datasets show that the introduced approaches can have a significant effect on the prediction performance only in case label dependencies are learned. The introduced approaches may not learn label relations in case descriptive attributes are sufficient to discriminate labels. In the other case, the introduce approaches can improve significantly the predictive performance of graded multi-label classifiers, especially the Stacked_RPC_PSI approach because it forces a different and effective prediction strategy on the graded multi-label classifier. This work can be completed by investigating new learning and prediction strategies that can be combined to improve the prediction of classifiers. An extensive comparative study using more evaluation measures and considering other datasets and approaches is planed for future works.

References

1. E. Frank, M. Hall, A simple approach to ordinal classification, in *Proceedings of the 12th European Conference on Machine Learning*, ser. EMCL '01 (Springer, London, UK, 2001), pp. 145–156
2. Z.-H. Zhou, M.-L. Zhang, *Multi-label Learning* (Springer, Boston, MA, 2017), pp. 875–881
3. F. Herrera, F. Charte, A.J. Rivera, M.J. del Jesus, *Multilabel Classification Problem Analysis, Metrics and Techniques*. Multilabel Classification (2016), pp. 17–31
4. W. Cheng, K. Dembczynski, E. Hllermeier, Graded multilabel classification: the ordinal case, in *Proceedings of LWA2010 - Workshop-Woche: Lernen, Wissen & Adaptivitaet* ed. by M. Atzmller, D. Benz, A. Hotho, G. Stumme (Kassel, Germany, 2010)
5. C. Brinker, E.L. Menca, J. Frnkranz, Graded multilabel classification by pairwise comparisons, in *2014 IEEE International Conference on Data Mining* (2014), pp. 731–736
6. M.-L. Zhang and K. Zhang, "Multi-label learning by exploiting label dependency," in *Proceedings of the 16th ACM SIGKDD International Conference on Knowledge Discovery and Data Mining*, ser. KDD '10. New York, NY, USA: ACM, 2010, pp. 999–1008
7. R. Al-Otaibi, M. Kull, P. Flach, *Declaratively Capturing Local Label Correlations with Multi-Label Trees*. Frontiers in Artificial Intelligence and Applications (IOS Press, Netherlands, 2016), pp. 1467–1475
8. G. Tsoumakas, I. Katakis, Multi-label classification: an overview. Int. J. Data Wareh. Min. **2007**, 1–13 (2007)
9. J.R. Quinlan, *C4.5: Programs for Machine Learning* (Morgan Kaufmann Publishers Inc., San Francisco, CA, USA, 1993)
10. B.E. Boser, I.M. Guyon, V.N. Vapnik, A training algorithm for optimal margin classifiers, in *Proceedings of the Fifth Annual Workshop on Computational Learning Theory*, COLT '92 (ACM, New York, NY, USA, 1992), pp. 144–152

11. J. Friedman, Another approach to polychotomous classification, Department of Statistics, Stanford University, Technical Report (1996)
12. T. Hastie, R. Tibshirani, Classification by pairwise coupling, in *Proceedings of the 1997 Conference on Advances in Neural Information Processing Systems 10*, NIPS '97 (MIT Press, Cambridge, MA, USA, 1998), pp. 507–513
13. E. Loza Mencía, F. Janssen, Stacking label features for learning multilabel rules, in *Discovery Science - 17th International Conference on DS 2014, Bled, Slovenia, 8-10 Oct 2014, Proceedings* ed. by S. Deroski, P. Panov, D. Kocev, and L. Todorovski. Lecture Notes in Computer Science, vol. 8777 (Springer, 2014), pp. 192–203
14. E. Loza Mencía, F. Janssen, Learning rules for multi-label classification: a stacking and a separate-and-conquer approach. Mach. Learn. **105**(1), 77–126 (2016)
15. Z. Sun, Z. Guo, M. Jiang, X. Wang, C. Liu, *Research and Application of Fast Multi-label SVM Classification Algorithm Using Approximate Extreme Points* (Springer International Publishing, Cham, 2016), pp. 39–52
16. S. Agrawal, J. Agrawal, S. Kaur, S. Sharma, A comparative study of fuzzy pso and fuzzy svd-based rbf neural network for multi-label classification. *Neural Computing and Applications*, pp. 1–12, 2016
17. X. Wang, S. An, H. Shi, Q. Hu, *Fuzzy Rough Decision Trees for Multi-label Classification* (Springer International Publishing, Cham, 2015), pp. 207–217
18. J. Read, A Pruned Problem Transformation Method for Multi-label classification, in *Proceedings of 2008 New Zealand Computer Science Research Student Conference (NZCSRS 2008)* (2008), pp. 143–150
19. J. Read, B. Pfahringer, G. Holmes, E. Frank, Classifier chains for multi-label classification. Mach. Learn. **85**(3), 333–359 (2011)
20. E. Montas, J.R. Quevedo, J.J. del Coz, Aggregating independent and dependent models to learn multi-label classifiers. in *ECML/PKDD (2)* ed. by D. Gunopulos, T. Hofmann, D. Malerba, M. Vazirgiannis. Lecture Notes in Computer Science, vol. 6912 (Springer, 2011), pp. 484–500
21. K. Laghmari, C. Marsala, M. Ramdani, Graded multi-label classification: Compromise between handling label relations and limiting error propagation, in *11th International Conference on Intelligent Systems: Theories and Applications (SITA)* (2016), pp. 1–6
22. E. Hüllermeier, J. Fürnkranz, W. Cheng, K. Brinker, Label ranking by learning pairwise preferences. Artif. Intell. **172**(1617), 1897–1916 (2008)
23. J. Fürnkranz, E. Hüllermeier, E. Loza Mencía, K. Brinker, Multilabel classification via calibrated label ranking. Mach. Learn. **73**(2), 133–153 (2008)
24. C. Brinker, E.L. Menca, J. Frnkranz, Graded multilabel classification by pairwise comparisons, in *ICDM* ed. by R. Kumar, H. Toivonen, J. Pei, J.Z. Huang, X. Wu (IEEE Computer Society, 2014), pp. 731–736
25. K. Laghmari, C. Marsala, M. Ramdani, Classification multi-labels graduee apprendre les relations entre les labels ou limiter la propagation d erreur, *Revue des Nouvelles Technologies de l'Information*, vol. Extraction et Gestion des Connaissances, RNTI-E-33 (2017), pp. 381–386
26. G. Tsoumakas, I. Katakis, I. Vlahavas, Mining multi-label data, in *Data Mining and Knowledge Discovery Handbook* (2010), pp. 667–685
27. S. Destercke, *Multilabel Prediction with Probability Sets: The Hamming Loss Case* (Springer International Publishing, Cham, 2014), pp. 496–505
28. S. Godbole, S. Sarawagi, Discriminative methods for multi-labeled classification, in *Advances in Knowledge Discovery and Data Mining: 8th Pacific-Asia Conference, PAKDD 2004, Sydney, Australia, May 26-28, 2004. Proceedings* (Springer, Berlin, Heidelberg, 2004), pp. 22–30
29. I. Pillai, G. Fumera, F. Roli, Designing multi-label classifiers that maximize f measures: State of the art. Pattern Recognit. **61**, 394–404 (2017)
30. M. Kubat, R. Holte, S. Matwin, *Learning When Negative Examples Abound* (Springer, Berlin, Heidelberg, 1997), pp. 146–153
31. K. Trohidis, G. Tsoumakas, G. Kalliris, I.P. Vlahavas, Multi-label classification of music into emotions, in *ISMIR* ed. by J.P. Bello, E. Chew, D. Turnbull (2008), pp. 325–330

32. M.R. Boutell, J. Luo, X. Shen, C.M. Brown, Learning multi-label scene classification. Pattern Recognit. **37**(9), 1757–1771 (2004)
33. A. Elisseeff, J. Weston, A kernel method for multi-labelled classification, in *In Advances in Neural Information Processing Systems 14* (MIT Press, 2001), pp. 681–687
34. A.E. Abele-Brehm, M. Stief, Die prognose des berufserfolgs von hochschulabsolventinnen und -absolventen: Befunde zur ersten und zweiten erhebung der erlanger l'angsschnittstudie Bela-E[predicting career success of university graduates: Findings of the first and second wave of the erlangen longitudinal study Bela-E]. Zeitschrift fr Arbeits- und Organisationspsychologie A&O **48**(1), 4–16 (2004)
35. S. Arctander, *Perfume and Flavor Chemicals: (aroma Chemicals), ser* (Aroma Chemicals. Allured Publishing Corporation, Perfume and Flavor Chemicals, 1969)
36. A. Mauri, V. Consonni, M. Pavan, R. Todeschini, Dragon software: an easy approach to molecular descriptor calculations. MATCH/Commun. Math. Comput. Chem. **56**, 237–248 (2006)

Improving Sparse Representation-Based Classification Using Local Principal Component Analysis

Chelsea Weaver and Naoki Saito

Abstract *Sparse representation-based classification* (SRC), proposed by Wright et al., seeks the sparsest decomposition of a test sample over the dictionary of training samples, with classification to the most-contributing class. Because it assumes test samples can be written as linear combinations of their same-class training samples, the success of SRC depends on the size and representativeness of the training set. Our proposed classification algorithm enlarges the training set by using local principal component analysis to approximate the basis vectors of the tangent hyperplane of the class manifold at each training sample. The dictionary in SRC is replaced by a local dictionary that adapts to the test sample and includes training samples and their corresponding tangent basis vectors. We use a synthetic data set and three face databases to demonstrate that this method can achieve higher classification accuracy than SRC in cases of sparse sampling, nonlinear class manifolds, and stringent dimension reduction.

Keywords Sparse representation · Local principal component analysis
Dictionary learning · Classification · Face recognition · Class manifold

1 Introduction

We are concerned with *classification*, which, in the context of *supervised learning*, is the task of assigning labels to unknown samples given the class information of a training set. It is one of the most important undertakings in pattern recognition and computational intelligence, with applications including the recognition of handwritten digits [16] and face recognition [6, 24, 32]. These tasks are often challenging.

C. Weaver (✉) · N. Saito
Department of Mathematics, University of California, Davis, One Shields Avenue, Davis, CA 95616, USA
e-mail: caweaver@math.ucdavis.edu

N. Saito
e-mail: saito@math.ucdavis.edu

W. Pedrycz and S.-M. Chen (eds.), *Computational Intelligence for Pattern Recognition*, Studies in Computational Intelligence 777,
https://doi.org/10.1007/978-3-319-89629-8_6

For example, in face recognition, the classification algorithm must be robust to within-class variation in properties such as expression, face/head angle, changes in hair or makeup, and differences that may occur in the image environment, most notably, the lighting conditions [24]. Further, in real-world settings, we must be able to handle greatly-deficient training data (i.e., too few or too similar training samples, in the sense that the given training set is insufficient to generalize the data set's class structure) [29], as well as occlusion and noise [32].

In 2009, Wright et al. proposed *sparse representation-based classification* (SRC) [32]. SRC was motivated by the recent boom in the use of sparse representation in signal processing (see, e.g., the work of Candès [4]). The catalyst of these advancements was the discovery that, under certain conditions, the sparsest representation of a signal using an over-complete set of vectors (often called a *dictionary*) could be found by minimizing the ℓ^1-norm of the representation coefficient vector [8]. Since the ℓ^1-minimization problem is convex, this gave rise to a tractable approach to obtaining the sparsest solution.

SRC applies this relationship between the minimum ℓ^1-norm and the sparsest solution to classification. The algorithm seeks the sparsest decomposition of a test sample over the dictionary of training samples via ℓ^1-minimization, with classification to the class whose corresponding portion of the representation approximates the test sample with least error. The method assumes that class manifolds are linear subspaces, so that the test sample can be represented using training samples in its ground truth class. Wright et al. [32] argue that this is precisely the sparsest decomposition of the test sample over the training set. They make the case that sparsity is critical to high-dimensional image classification and that, if properly harnessed, it can lead to superior classification performance, even on highly corrupted or occluded images. Further, good results can be achieved regardless of the choice of image features that are used for classification, provided that the number of retained features is large enough [32]. Though SRC was originally applied to face recognition, similar methods have been employed in clustering [10], dimension reduction [25], and texture and handwritten digit classification [36].

The SRC assumption that class manifolds are linear subspaces is often violated; e.g., facial images that vary in pose and expression are known to lie on nonlinear class manifolds [12, 26]. Additionally, small training set size, one of the primary challenges in face recognition and classification as a whole, can easily make it impossible to represent a given test sample using its same-class training samples, even in the case that the class manifold is linear. However, these reasons alone are not enough to discount SRC even on such data sets, as demonstrated by Wright et al. [32] in experiments on the AR face database [21]. AR contains expression and occlusion variations that suggest the underlying class manifolds are nonlinear, yet SRC often outperformed SVM (support vector machines) on AR for a wide variety of feature extraction methods and feature dimensions [32]. To understand how this is possible, consider that SRC decomposes the test sample over the entire training set, and so components of the test sample not within the span of its ground truth class's training samples may be absorbed by training samples from other classes. A similar fail-safe occurs when the class manifolds (linear or otherwise) are sparsely sampled.

The above discussion, however, illustrates a weakness in SRC. When the algorithm relies on "wrong-class" training samples to partially represent or approximate the test sample, misclassification may ensue, especially when the class manifolds are close together. In the case where class manifolds are nonlinear and/or sparsely sampled, so that it is impossible to accurately approximate the test sample using only the training samples in its ground truth class, this approximation could conceivably be improved if we were able to increase the sampling density around the test sample, "fleshing out" its local neighborhood on the (correct) class manifold. This is the motivation behind this chapter's proposed classification algorithm.

Our contributions in this chapter are the following:

1. We introduce a classification algorithm that improves SRC by increasing the accuracy and locality of the approximation of the test sample in terms of its ground truth class. Our algorithm is designed to increase the training set via nearby (to the test sample) basis vectors of the hyperplanes approximately tangent to the (unknown) class manifolds. This provides the two-fold benefit of counter-balancing the potential sparse sampling of class manifolds (especially in the case that they are nonlinear) and helping to retain more information in few dimensions when used in conjunction with dimension reduction.
2. We state guidelines for the setting of parameters in this algorithm and analyze its computational complexity and storage requirements.
3. We demonstrate that our algorithm leads to classification accuracy exceeding that of traditional SRC and related methods on a synthetic database and three popular face databases. We thoroughly analyze and explain our experimental results (e.g., accuracy, runtime, and dictionary size) of the compared algorithms.
4. We illustrate that the tangent hyperplane basis vectors used in our method can capture sample details lost during principal component analysis in the case of face recognition.

Note that both SRC and the method we use to compute the tangent hyperplane basis vectors have previously been proposed. The novelty of the proposed classification algorithm lies in a solid theoretical foundation for combining these two ideas. This motivating foundation is supported empirically—beyond evidence of increased classification accuracy—in experimental results.[1] Further, by providing thorough guidelines and short-cuts regarding the setting of required parameters, we make it feasible to apply the resulting algorithm in practice.

This paper is organized as follows: In Sect. 2, we discuss work related to our proposed method, and we state SRC in detail in Sect. 3. In Sect. 4, we describe our proposed classification algorithm and discuss its parameters, computational complexity, and storage requirements. We present our experimental results in Sect. 5, and in Sect. 6, we summarize our findings and discuss avenues of future work.

Setup and Notation. We assume that the input data is represented by vectors in $\mathbb{R}^m$ and that dimension reduction, if used, has already been applied. The training set, i.e., the matrix whose columns are the data samples with known class labels,

[1] We are referring to Sect. 5.4.7.

is denoted by $X_{\text{tr}} = [\boldsymbol{x}_1, \ldots, \boldsymbol{x}_{N_{\text{tr}}}] \in \mathbb{R}^{m \times N_{\text{tr}}}$. The number of classes is denoted by $L \in \mathbb{N}$, and we assume that there are N_l training samples in class l, $1 \leq l \leq L$. Lastly, we refer to a given test sample by $\boldsymbol{y} \in \mathbb{R}^m$.

2 Related Work

The approach of using tangent hyperplanes for pattern recognition is not new. When the data is assumed to lie on a low-dimensional manifold, local tangent hyperplanes are a simple and intuitive approach to enhancing the data set and gaining insight into the manifold structure. Our proposed method is very much related to *tangent distance classification* (TDC) [7, 27, 35], which constructs local tangent hyperplanes of the class manifolds, computes the distances between these hyperplanes and the given test sample, and then classifies the test sample to the class with the closest hyperplane. We show in Sect. 5 that our proposed method's integration of tangent hyperplane basis vectors into the sparse representation framework generally outperforms TDC.

On the other hand, approaches to address the limiting linear subspace assumption (i.e., the assumption that class manifolds are linear subspaces) in SRC have been proposed. For example, Ho et al. extended sparse coding and dictionary learning to general Riemannian manifolds [13]. Admittedly only a first step in meeting their ultimate objective, Ho et al.'s work requires explicit knowledge of the class manifolds. This is an unsatisfiable condition in many real-world classification problems and is not a requirement of our proposed algorithm. Alternatively, *kernel methods* have been effective in overcoming SRC's linearity assumption, as nonlinear relationships in the original space may be linear in kernel space given an appropriate choice of kernel [37].

Several "local" modifications of SRC implicitly ameliorate the linearity assumption; in *collaborative neighbor representation-based classification* [30] and *locality-sensitive dictionary learning* (LSDL-SRC) [31], for instance, coefficients of the representation are constrained by their corresponding training samples' distances to the test sample, and so these algorithms need only assume linearity at the local level. Our proposed method is designed to improve not only the locality but also the accuracy of the approximation of the test sample in terms of its ground truth class. Section 5 contains an experimental comparison between our proposed method and LSDL-SRC, as well as a discussion thereof.

Other classification algorithms have been proposed that are similar to ours in that they aim to enlarge or otherwise enhance the training set in SRC. Such methods for face recognition, for example, include the use of virtual images that exploit the symmetry of the human face, as in both the method of Xu et al. [33] and *sample pair based sparse representation classification* [38]. Though visual comparison of these virtual images and our recovered tangent vectors (see Sect. 5.4.7) could be informative, our proposed method can be used for general classification.

Additionally, there have been many local modifications to the sparse representation framework with objectives other than classification. For example, Li et al.'s *robust structured subspace learning* [19] uses the $\ell_{2,1}$-norm for sparse feature extraction, combining high-level semantics with low-level, locality-preserving features. In the feature selection algorithm *clustering-guided sparse structural learning* by Li et al. [18], features are jointly selected using sparse regularization (via the $\ell_{2,1}$-norm) and a non-negative spectral clustering objective. Not only are the selected features sparse; they also are the most discriminative features in terms of predicting the cluster indicators in both the original space and a lower-dimensional subspace on which the data is assumed to lie.

3 Sparse Representation-Based Classification

SRC [32] solves the optimization problem

$$\boldsymbol{\alpha}^* := \arg\min_{\boldsymbol{\alpha}\in\mathbb{R}^{N_{\text{tr}}}} \|\boldsymbol{\alpha}\|_1, \text{ subject to } \boldsymbol{y} = X_{\text{tr}}\boldsymbol{\alpha}. \tag{1}$$

It is assumed that the training samples have been normalized to have ℓ^2-norm equal to 1, so that the representation in Eq. (1) will not be affected by the samples' magnitudes. The use of the ℓ^1-norm in the objective function is designed to approximate the ℓ^0-"norm," i.e., to aim at finding the smallest number of training samples that can accurately represent the test sample $\boldsymbol{y}$. It is argued that the nonzero coefficients in the representation will occur primarily at training samples in the same class, so that

$$\text{class\_label}(\boldsymbol{y}) = \arg\min_{1\leq l\leq L} \left\|\boldsymbol{y} - X_{\text{tr}}\delta_l(\boldsymbol{\alpha}^*)\right\|_2 \tag{2}$$

produces the correct class assignment. Here, δ_l is the indicator function that acts as the identity on all coordinates corresponding to samples in class l and sets the remaining coordinates to zero. In other words, $\boldsymbol{y}$ is assigned to the class whose training samples contribute the most to the sparsest representation of $\boldsymbol{y}$ over the entire training set.

The reasoning behind this is the following: It is assumed that the class manifolds are linear subspaces, so that if each class's training set contains a spanning set of the corresponding subspace, the test sample can be expressed as a linear combination of training samples in its ground truth class. If the number of training samples in each class is small relative to the number of total training samples N_{tr}, this representation is naturally sparse [32].

As real-world data is often corrupted by noise, the constrained ℓ^1-minimization problem in Eq. (1) may be replaced with its regularized version

$$\boldsymbol{\alpha}^* := \arg\min_{\boldsymbol{\alpha}\in\mathbb{R}^{N_{\text{tr}}}} \left\{\frac{1}{2}\|\boldsymbol{y} - X_{\text{tr}}\boldsymbol{\alpha}\|_2^2 + \lambda\|\boldsymbol{\alpha}\|_1\right\}. \tag{3}$$

Here, λ is the trade-off between error in the approximation and the sparsity of the coefficient vector. We summarize SRC in Algorithm 1.

Algorithm 1 Sparse Representation-Based Classification (SRC) [32]

Input: Matrix of training samples $X_{\text{tr}} \in \mathbb{R}^{m \times N_{\text{tr}}}$; test sample $\boldsymbol{y} \in \mathbb{R}^m$; number of classes L; and error/sparsity trade-off λ (optional)
Output: The computed class label of $\boldsymbol{y}$: class_label($\boldsymbol{y}$)
1: Normalize each column of X_{tr} to have ℓ^2-norm equal to 1.
2: Use an ℓ^1-minimization algorithm to solve either the constrained problem (1) or the regularized problem (3).
3: **for** each class $l = 1, \ldots, L$, **do**
4: Compute the norm of the class l residual: $\text{err}_l(\boldsymbol{y}) := \left\| \boldsymbol{y} - X_{\text{tr}} \delta_l(\boldsymbol{\alpha}^*) \right\|_2$.
5: **end for**
6: Classify the test sample $\boldsymbol{y}$ according to class_label($\boldsymbol{y}$) $= \arg\min_{1 \le l \le L}\{\text{err}_l(\boldsymbol{y})\}$.

Remark 1 We briefly note that, in the case that some classes contain very few samples, SRC is not a good candidate for *oversampling*, or using repeated training samples to even out the class count. This is because the linear span of the training samples is invariant to the addition of repeat samples and the classification result will be unaffected. Thus there is no obvious solution to dealing with undersampled classes in SRC.

4 Proposed Algorithm

4.1 Local Principal Component Analysis Sparse Representation-Based Classification

Our proposed algorithm, *local principal component analysis sparse representation-based classification* (LPCA-SRC), is essentially SRC with a modified dictionary. This dictionary is constructed in two steps: (i) an offline phase, and (ii) an online phase.

In the offline phase of the algorithm, we generate new training samples as a means of increasing the sampling density. Instead of the linear subspace assumption in SRC, we assume that class manifolds are well-approximated by local tangent hyperplanes. To generate new training samples, we approximate these tangent hyperplanes at individual training samples using *local principal component analysis* (local PCA), and then add the basis vectors of these tangent hyperplanes (after randomly-scaling and shifting them as described in Step 12 of Algorithm 2 and explained in Sect. 4.3.3) to the original training set. Naturally, the shifted and scaled tangent hyperplane basis vectors (hereon referred to as “tangent vectors”) inherit the labels of their corresponding training samples. The result is an amended dictionary over which a generic test sample can ideally be decomposed using samples that approximate a local patch on the correct class manifold. In the case that the class manifolds are sparsely sampled and/or nonlinear, this allows for a more accurate approximation of $\boldsymbol{y}$ using training samples (and their computed tangent vectors) from the test sample’s ground truth class. Even in the case that class manifolds are linear subspaces, this technique ideally increases the sampling density around $\boldsymbol{y}$ on its (unknown) class manifold so that it may be expressed in terms of *nearby* samples.

In the online phase of LPCA-SRC, this extended training set is "pruned" relative to the given test sample, increasing computational efficiency and the locality of the resulting dictionary. Training samples (along with their tangent vectors) are eliminated from the dictionary if their Euclidean distances to the given test sample are greater than a threshold, and then classification proceeds as in SRC as the test sample is sparsely decomposed (via ℓ^1-minimization) over this local dictionary.

The method in LPCA-SRC has an additional benefit: When SRC is applied to the classification of high-resolution images (e.g., $> O(10^4)$ pixels), some method of dimension reduction is generally necessary to reduce the dimension of the raw samples, due to the high computational complexity of solving the ℓ^1-minimization problem. Basic dimension reduction methods, such as *principal component analysis* (PCA), may result in the loss of class-discriminating details when the PCA feature dimension is small. In Sect. 5.4.7, we show that the tangent vectors computed in LPCA-SRC can contain details of the raw images that have been lost in the dimension reduction process.

Remark 2 This remark serves to draw a distinction between our use of sparse representation and local PCA for classification, and our use of (non-local) PCA to pre-process data samples prior to classification in some of our experiments. Sparse representation and local PCA can themselves be used (separately) for dimension reduction; see, for example, the papers of Qiao et al. [25] and Kambhatla and Leen [14]. We stress that dimension reduction is not the subject of this paper, and in fact we use neither sparse representation nor local PCA to accomplish this task at any point. Instead, we focus on classification and in integrating sparse representation and local PCA towards this purpose. When dimension reduction is used in Sect. 5.4, we use (non-local) PCA simply as a means of pre-processing the data before image classification.

We formally state the offline and online portions of our proposed algorithm in Algorithms 2 and 3, respectively. Obviously, by the definition of "offline phase," the tangent vectors need only be computed once for any number of test samples. More details regarding the user-set parameters d, n and λ are provided in Sects. 4.3.1 and 4.3.2, and an explanation of the pruning parameter r and the tangent vector scaling factor c (in Step 12 of Algorithm 2) are given in Sect. 4.3.3.

Figure 1 illustrates the efficacy of LPCA-SRC's tangent vectors and pruning parameter in the sparse representation framework. The figure shows two classes, represented by the colors red and blue. The training samples in each class are represented by solid colored circles. There is one test sample $\boldsymbol{y}$ displayed in the figure, a member of class 2 (blue) and depicted by a solid blue square. Observe that, before the use of tangent vectors, $\boldsymbol{y}$ is closer to the subspace[2] spanned by $\boldsymbol{x}_1^{(1)}$ and $\boldsymbol{x}_2^{(1)}$ (which are

[2]Technically speaking, we are referring to the affine subspace in this illustration; In SRC, instead the *linear* subspace is used. We have tweaked the algorithm slightly to be able to demonstrate an example in low dimension.

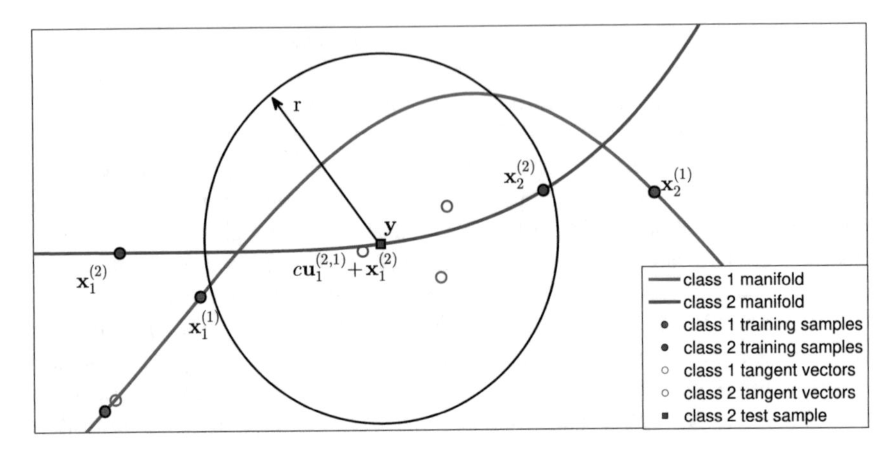

Fig. 1 An example use of LPCA-SRC's tangent vectors and pruning parameter in the SRC framework. Only training samples and tangent vectors relevant to classification of the test sample $\boldsymbol{y}$ have been labeled

class 1 training samples) than the subspace spanned by the class 2 training samples $\boldsymbol{x}_1^{(2)}$ and $\boldsymbol{x}_2^{(2)}$. Thus $\boldsymbol{y}$ would be incorrectly classified by SRC in this scenario.

After the addition of tangent vectors (which are represented by unfilled circles), in particular, the class 2 tangent vector $c\boldsymbol{u}_1^{(2,1)} + \boldsymbol{x}_1^{(2)}$, $\boldsymbol{y}$ is closest to the subspace generated by this tangent vector and $\boldsymbol{x}_2^{(2)}$. Thus the test sample would be correctly classified by LPCA-SRC in this scenario.

The use of the pruning parameter r independently avoids the problem of misclassification. If we consider only samples in the local neighborhood of $\boldsymbol{y}$ (contained in the circle of radius r), the misleading class 1 samples $\boldsymbol{x}_1^{(1)}$ and $\boldsymbol{x}_2^{(1)}$ are eliminated from consideration, leading to the correct classification of $\boldsymbol{y}$.

Thus these two mechanisms in LPCA-SRC—its use of tangent vectors and its localizing pruning parameter—make it especially designed to succeed in these cases of sparse sampling and nonlinear class manifolds in which SRC fails.

4.2 Local Principal Component Analysis

In LPCA-SRC (in particular, Step 5 of Algorithm 2), we use the local PCA technique of Singer and Wu [28] to compute the tangent hyperplane basis $U^{(l,i)}$. We outline our implementation of their method in Algorithm 4. It computes a basis for the tangent hyperplane $T_{\boldsymbol{x}_i}\mathcal{M}$ at a point $\boldsymbol{x}_i$ on the manifold $\mathcal{M}$, where it is assumed that the local neighborhood of $\boldsymbol{x}_i$ on $\mathcal{M}$ can be well-approximated by a tangent hyperplane of some dimension $d < m$. A particular strength of Singer and Wu's method is the weighting

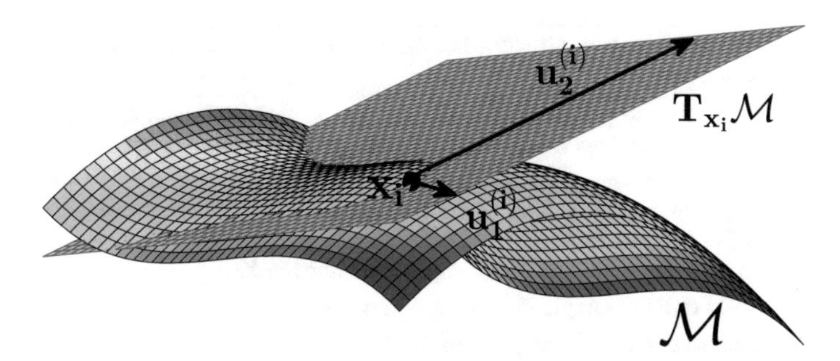

Fig. 2 An illustration of the tangent plane and the tangent basis vectors $\boldsymbol{u}_1^{(i)}$ and $\boldsymbol{u}_2^{(i)}$ at the sample $\boldsymbol{x}_i$ on the manifold $\mathscr{M}$. Here, the intrinsic dimension is $d = 2$

of neighbors by their Euclidean distances to the point $\boldsymbol{x}_i$, so that closer neighbors play a more important role in the construction of the local tangent hyperplane.

A simple illustration of the tangent basis vectors found using local PCA is shown in Fig. 2.

4.3 Remarks on Parameters

In this subsection, we detail the roles of the parameters in LPCA-SRC and suggest strategies for estimating those that must be determined by the user.

4.3.1 Estimate of Class Manifold Dimension and Number of Neighbors

Recall that d is the estimated dimension of each class manifold and n is the number of neighbors used in local PCA. Both parameters must be inputted by the user in our proposed algorithm. The number of samples in the smallest training class, denoted $N_{l_{\min}}$, limits the range of values for d and n that may be used. Specifically,

$$1 \leq d \leq n < N_{l_{\min}} - 1. \tag{4}$$

This follows from the fact that each training sample must have at least $n + 1$ neighbors in its own class, with the dimension d of the tangent hyperplane being bounded above by the number of columns n in the weighted matrix of neighbors B_i. It is important to observe that when the classes are small (as is often the case in face recognition), there are few options for the values of d and n per Eq. (4). Thus these parameters may be efficiently set using cross-validation. This was the method we used to set d and n in the experiments in Sect. 5. We discuss a recommended cross-validation procedure in Sect. 4.3.2.

Algorithm 2 Local PCA Sparse Representation-Based Classification (LPCA-SRC): **OFFLINE PHASE**

Input: $X_{\text{tr}} = [\boldsymbol{x}_1 \ldots, \boldsymbol{x}_{N_{\text{tr}}}] \in \mathbb{R}^{m \times N_{\text{tr}}}$; number of classes L; local PCA parameters d (estimate of class manifold dimension) and n (number of neighbors)

Output: The normalized extended dictionary $D \in \mathbb{R}^{m \times (N_{\text{tr}}(d+1))}$; pruning parameter r

1: Normalize the columns of X_{tr} to have ℓ^2-norm equal to 1.

2: **for** each class $l = 1, \ldots, L$ **do**

3: Let $\mathscr{X}^{(l)}$ be the set of class l training samples contained in X_{tr}.

4: **for** each class l training sample $\boldsymbol{x}_i^{(l)}$, $i = 1, \ldots, N_l$ **do**

5: Approximate the tangent hyperplane of the lth class manifold at $\boldsymbol{x}_i^{(l)}$ as follows:

- Use local PCA in Algorithm 4 with set of samples $\mathscr{X}^{(l)}$ (the samples in the lth class), selected sample $\boldsymbol{x}_i^{(l)}$, and parameters d and n to compute a basis $U^{(l,i)} := [\boldsymbol{u}_1^{(l,i)}, \ldots, \boldsymbol{u}_d^{(l,i)}]$ of an approximate tangent hyperplane at $\boldsymbol{x}_i^{(l)}$ along class l.
- Store the basis $U^{(l,i)}$ and the quantity $r_i^{(l)} := \|\boldsymbol{x}_{i_{n+1}}^{(l)} - \boldsymbol{x}_i^{(l)}\|_2$, the distance between $\boldsymbol{x}_i^{(l)}$ and its $(n+1)$st nearest neighbor in the lth class.

6: **end for**

7: **end for**

8: Define the pruning parameter $r := \text{median}\{r_i^{(l)} \mid 1 \le i \le N_l,\ 1 \le l \le L\}$.

9: Initialize the extended dictionary $D = \emptyset$.

10: **for** each class $l = 1, \ldots, L$ **do**

11: **for** each class l training sample $\boldsymbol{x}_i^{(l)}$, $i = 1, \ldots, N_l$ **do**

12: Set $c := r\gamma$, $\gamma \sim \text{unif}(0, 1)$, and form $\tilde{X}^{(l,i)} := [c\boldsymbol{u}_1^{(l,i)} + \boldsymbol{x}_i^{(l)}, \ldots, c\boldsymbol{u}_d^{(l,i)} + \boldsymbol{x}_i^{(l)}, \boldsymbol{x}_i^{(l)}] \in \mathbb{R}^{m \times (d+1)}$

13: Normalize the columns of $\tilde{X}^{(l,i)}$ to have ℓ^2-norm equal to 1 and add it to the extended dictionary: $D = [D, \tilde{X}^{(l,i)}]$.

14: **end for**

15: **end for**

Remark 3 Interestingly, when cross-validation is used to set d, we find empirically that d is often selected to be smaller than the (expected) true class manifold dimension. Further, in these cases, increasing d from the selected value (i.e., increasing the number of tangent vectors used) does not significantly increase classification accuracy. We expect that the addition of even a small number of tangent vectors (those indicating the directions of maximum variance on their local manifolds, per the local PCA algorithm) is enough to improve the approximation of the test sample in terms of its ground truth class. Additional tangent vectors are often unneeded. Since the value of d largely affects LPCA-SRC's computational complexity and storage requirements, these observations suggest that when the true manifold dimension is large, it is better to underestimate it than overestimate it. Further, setting $d = 1$ can often produce a good result, hence $d = 1$ could be used by default.

There are other methods for determining d besides cross-validation and fixing $d = 1$. One may use the multiscale SVD algorithm of Little et al. [20] or Ceruti et al.'s DANCo (*Dimensionality from Angle and Norm Concentration* [5]). However, in our experiments in Sect. 5, we set d using cross-validation. See Sect. 4.3.2.

Algorithm 3 Local PCA Sparse Representation-Based Classification (LPCA-SRC): **ONLINE PHASE**

Input: Test sample $\boldsymbol{y} \in \mathbb{R}^m$; normalized extended dictionary D; pruning parameter r; estimate of class manifold dimension d; error/sparsity trade-off λ (optional)
Output: The computed class label of $\boldsymbol{y}$: class_label($\boldsymbol{y}$).
1: Normalize $\boldsymbol{y}$ to have $\|\boldsymbol{y}\|_2 = 1$.
2: Initialize the pruned dictionary $D_{\boldsymbol{y}} = \emptyset$ and set $N_{\boldsymbol{y}} = 0$ (# of columns of $D_{\boldsymbol{y}}$).
3: **for** each class $l = 1, \ldots, L$ **do**
4: **for** each class l training sample $\boldsymbol{x}_i^{(l)}$, $i = 1, \ldots, N_l$ **do**
5: **if** $\left\|\boldsymbol{y} - \boldsymbol{x}_i^{(l)}\right\|_2 \leq r$ or $\left\|\boldsymbol{y} - (-\boldsymbol{x}_i^{(l)})\right\|_2 \leq r$ **then**
6: Add the portion $\tilde{X}^{(l,i)}$ of D corresponding to $\boldsymbol{x}_i^{(l)}$ and its tangent vectors to the pruned dictionary: $D_{\boldsymbol{y}} = [D_{\boldsymbol{y}}, \tilde{X}^{(l,i)}]$. Assign the columns of $\tilde{X}^{(l,i)}$ class l labels. Update $N_{\boldsymbol{y}} = N_{\boldsymbol{y}} + (d+1)$.
7: **end if**
8: **end for**
9: **end for**
10: Use an ℓ^1-minimization algorithm to compute the solution to the constrained problem

$$\boldsymbol{\alpha}^* := \arg \min_{\boldsymbol{\alpha} \in \mathbb{R}^{N_{\boldsymbol{y}}}} \left\{\|\boldsymbol{\alpha}\|_1 \text{ s.t. } \boldsymbol{y} = D_{\boldsymbol{y}}\boldsymbol{\alpha}\right\} \tag{4}$$

or the regularized problem

$$\boldsymbol{\alpha}^* := \arg \min_{\boldsymbol{\alpha} \in \mathbb{R}^{N_{\boldsymbol{y}}}} \left\{\frac{1}{2}\|\boldsymbol{y} - D_{\boldsymbol{y}}\boldsymbol{\alpha}\|_2^2 + \lambda\|\boldsymbol{\alpha}\|_1\right\}. \tag{5}$$

11: **for** each class $l = 1, \ldots, L$, **do**
12: Compute the norm of the class l residual: $\text{err}_l(\boldsymbol{y}) := \left\|\boldsymbol{y} - D_{\boldsymbol{y}}\delta_l(\boldsymbol{\alpha}^*)\right\|_2$.
13: **end for**
14: Classify the test sample $\boldsymbol{y}$ according to class_label($\boldsymbol{y}$) $= \arg\min_{1\leq l\leq L}\{\text{err}_l(\boldsymbol{y})\}$.

Remark 4 Certainly, the parameters d and n could vary per class, i.e., d and n could be replaced with d_l and n_l, respectively, for $l = 1, \ldots, L$. In face recognition, however, if each subject is photographed under similar conditions, e.g., the same set of lighting configurations, then we expect that the class manifold dimension is approximately the same for each subject. Further, without some prior knowledge of the class manifold structure, using distinct d and n for each class may unnecessarily complicate the setting of parameters in LPCA-SRC.

4.3.2 Using Cross-Validation to Set Multiple Parameters

On data sets of which we have little prior knowledge, it may be necessary to use cross-validation to set multiple parameters in LPCA-SRC. Since grid search (searching through all parameter combinations in a brute-force manner) is typically expensive, we suggest that cross-validation be applied to the parameters n, λ, and d, consec-

Algorithm 4 Local Principal Component Analysis (Local PCA, adapted from Singer and Wu [28])

Input: Set of samples $\mathscr{X}$; selected sample $\boldsymbol{x}_i \in \mathscr{X}$; dimension of tangent hyperplane d; number of neighbors n

Output: The basis $U^{(l,i)}$ of the approximated tangent hyperplane at the point $\boldsymbol{x}_i$

1: Find the $n+1$ nearest neighbors (with respect to Euclidean distance) of $\boldsymbol{x}_i$ in $\mathscr{X} \backslash \boldsymbol{x}_i$. Store the n nearest neighbors as columns of the matrix $X_i := [\boldsymbol{x}_{i_1}, \ldots, \boldsymbol{x}_{i_n}]$ and use the $(n+1)$st nearest neighbor to define $\varepsilon_{\text{PCA}} := \|\boldsymbol{x}_{i_{n+1}} - \boldsymbol{x}_i\|_2^2$.

2: Form the matrix $\overline{X}_i$ by centering the columns of X_i around $\boldsymbol{x}_i$: $\overline{X}_i := [\boldsymbol{x}_{i_1} - \boldsymbol{x}_i, \ldots, \boldsymbol{x}_{i_n} - \boldsymbol{x}_i]$.

3: Form a diagonal weight matrix D_i based on the distance between each neighbor and $\boldsymbol{x}_i$ as follows: Let
$D_i(j,j) = \sqrt{K\left(\frac{\|\boldsymbol{x}_{i_j} - \boldsymbol{x}_i\|_2}{\sqrt{\varepsilon_{\text{PCA}}}}\right)}, j = 1, \ldots, n$, where K is the Epanechnikov kernel given by
$K(u) := (1 - u^2)\chi_{[0,1]}$.

4: Form the weighted matrix $B_i := \overline{X}_i D_i$.

5: Find the first d left singular vectors of B_i using singular value decomposition. Denote these vectors by $\boldsymbol{u}_1^{(i)}, \ldots, \boldsymbol{u}_d^{(i)}$.

utively in that order as needed.[3] During this process, we recommend holding the error/sparsity trade-off λ (if used) equal to a small, positive value (e.g., $\lambda = 0.001$) and setting $d = 1$ until these parameters' respective values are determined. We justify and detail this approach below.

Our reasons for suggesting this consecutive cross-validation procedure is the following: During experiments, we found that the LPCA-SRC algorithm can be quite sensitive to the setting of n, especially when there are many samples in each training class (since there are many possible values for n). This is expected, as the setting of n affects both the accuracy of the tangent vectors and the pruning parameter r. In contrast, LPCA-SRC is empirically fairly robust to the values of λ and d used, and as mentioned in Remark 3, setting $d = 1$ can result in quite good performance in LPCA-SRC, even when the true dimension of the class manifolds is expected to be larger.

4.3.3 Pruning Parameter

First, we stress that the pruning parameter r is not a user-set parameter. Its value is automatically computed in the offline phase of LPCA-SRC (Algorithm 2). We explain this process here.

Recall that we only include a training sample $\boldsymbol{x}_i^{(l)}$ and its tangent vectors in the pruned dictionary $D_{\boldsymbol{y}}$ if $\boldsymbol{x}_i^{(l)}$ (or its negative) is in the closed Euclidean ball $\overline{B_m(\boldsymbol{y}, r)} \subset \mathbb{R}^m$ with center $\boldsymbol{y}$ and radius r. Thus r is a parameter that prunes the extended dictionary D to obtain $D_{\boldsymbol{y}}$. A smaller dictionary is good in terms of com-

[3]If the constrained optimization problem (Eq. (4)) is used, the error/sparsity trade-off λ is not needed.

putational complexity, as the ℓ^1-minimization algorithm will run faster. Further, we can obtain this computational speedup without (theoretically) degrading classification accuracy: If $\pm\boldsymbol{x}_i^{(l)}$ is far from $\boldsymbol{y}$ in terms of Euclidean distance, then it is assumed that $\pm\boldsymbol{x}_i^{(l)}$ is not close to $\boldsymbol{y}$ in terms of distance along the class manifold. Thus $\boldsymbol{x}_i^{(l)}$ and its tangent vectors should not be needed in the ℓ^1-minimized approximation of $\boldsymbol{y}$.

A deeper notion of the parameter r is to view it as a rough estimate of the local neighborhood radius of the data set. More precisely, r estimates the distance from a sample within which its class manifold can be well-approximated by a tangent hyperplane (at that sample). Given X_{tr} and n, r is automatically computed, as described in Algorithm 2. In words, we set r to be the median distance between each training sample and its $(n+1)$st nearest neighbor (in the same class), where n, the number of neighbors in local PCA, is used to implicitly define the local neighborhood. It follows that r is a robust estimate of the local neighborhood radius, as learned from the training data.

We verified the effectiveness of this automatically-computed parameter by comparing it to the same algorithm but with r set via manual grid search during cross-validation. Though the latter method sometimes resulted in slightly higher accuracy, the saved computational expense of the automated setting of r (as described above) clearly showed it to be an improvement to the overall algorithm.

This also explains our choice for the tangent vector scaling factor $c = r\gamma$ (in Step 12 of Algorithm 2), where $\gamma \sim \text{unif}(0, 1)$. Multiplying each tangent hyperplane basis vector $\boldsymbol{u}_j^{(l,i)}$, $1 \leq j \leq d$, by this scalar and then shifting it by its corresponding training sample $\boldsymbol{x}_i^{(l)}$ helps to ensure that the resulting tangent vector, included in the dictionary $D_{\boldsymbol{y}}$ if $\pm\boldsymbol{x}_i^{(l)}$ is sufficiently close to $\boldsymbol{y}$, lies in the local neighborhood of $\boldsymbol{x}_i^{(l)}$ on the lth class manifold.

Remark 5 If the test sample $\boldsymbol{y}$ is far from the training data, defining r as in Algorithm 2 may produce $D_{\boldsymbol{y}} = \emptyset$, i.e., there may be no training samples within that distance of $\boldsymbol{y}$. Thus to prevent this degenerate case, we use a slightly modified technique for setting r in practice. After assigning the median neighborhood radius $r_1 := \text{median}\{r_i^{(l)} \mid 1 \leq i \leq N_l,\ 1 \leq l \leq L\}$, we define r_2 to be the distance between the test sample $\boldsymbol{y}$ and the closest training sample (up to sign). We then define the pruning parameter $r := \max\{r_1, r_2\}$. In the (degenerate) case that $r = r_2$, the dictionary consists of the closest training sample and its tangent vectors, leading to nearest neighbor classification instead of an algorithm error. However, experimental results indicate that the pruning parameter r is almost always equal to the median neighborhood radius r_1, and so we leave this "technicality" out of the official algorithm statement to make it easier to interpret.

4.4 Computational Complexity and Storage Requirements

In this subsection, we compare the computational complexity and storage requirements of SRC and our proposed algorithm.

4.4.1 Computational Complexity of SRC

When the ℓ^1-minimization algorithm HOMOTOPY [9] is used, it is easy to see that the computational complexity of SRC is dominated by this step. This complexity is $O(N_{\text{tr}}m\kappa + m^2\kappa)$, where κ is the number of HOMOTOPY iterations [34]. HOMOTOPY has been shown to be relatively fast and good for use in robust face recognition [34]. In our experiments, we use it in all classification methods requiring ℓ^1-minimization.

4.4.2 Computational Complexity of LPCA-SRC

The computational complexity of the offline phase in LPCA-SRC (Algorithm 2) is

$$O\Big(m\sum_{l=1}^{L} N_l^2 + N_{\text{tr}}mn\Big), \tag{6}$$

whereas that of the online phase (Algorithm 3) is

$$O\Big(N_{\text{tr}}m + \frac{N_y}{d}\log\left(\frac{N_y}{d}\right) + N_y m\kappa + m^2\kappa\Big). \tag{7}$$

Recall that N_y denotes the number of columns in the pruned dictionary D_y. We note that the offline cost in Eq. (6) is based on the linear nearest neighbor search algorithm for simplicity; in practice there are faster methods. In our experiments, we used ATRIA (*Advanced Triangle Inequality Algorithm* [22]) via the MATLAB TSTOOL functions `nn_prepare` and `nn_search` [23]. The first function prepares the set of class l training samples $\mathscr{X}^{(l)}$ for nearest neighbor search at the onset, with the intention that subsequent runs of `nn_search` on this set are faster than simply doing a search without the preparation function. Other fast nearest neighbor search algorithms are available, for example, *k-d tree* [3]. The cost complexity estimates of these fast nearest neighbor search algorithms are somewhat complicated, and so we do not use them in Eq. (6). Hence, Eq. (6) could be viewed as the worst-case scenario.

Offline and online phases combined, the very worst-case computational complexity of LPCA-SRC is $O(N_{\text{tr}}^4)$, which occurs when the second-to-last term in Eq. (7) dominates: i.e., when (i) $N_y \approx (d+1)N_{\text{tr}}$ (no pruning); (ii) $m \approx N_{\text{tr}}$ (large relative sample dimension); (iii) very large class manifold dimension estimate d, so that d is relatively close to N_{tr} (note that this requires very large N_l for $1 \le l \le L$ by Eq. (4), which implies that L has to be very small); and (iv) $\kappa \approx m$ (many HOMOTOPY iterations). For small κ and N_l, $1 \le l \le L$, and when the pruning parameter r results in small N_y relative to N_{tr}, then the computational complexity reduces to approximately $O(N_{\text{tr}}m)$.

4.4.3 Storage Requirements

The primary difference between the storage requirements for LPCA-SRC and SRC is that the offline phase of LPCA-SRC requires storing the matrix $D \in \mathbb{R}^{m\times(d+1)N_{\text{tr}}}$, which has a factor of $d+1$ as many columns as the matrix of training samples $X_{\text{tr}} \in \mathbb{R}^{m\times N_{\text{tr}}}$ stored in SRC. Hence the storage requirements of LPCA-SRC are at worst $(d+1)$ times the amount of storage required by SRC.

Though this potentially is a large increase, consider that in applications such as face recognition, it is expected that the intrinsic class manifold dimension be small, e.g., 3–5 [17]. Second, as we discussed in Remark 3 in Sect. 4.3.1, it is often sufficient to take d smaller than the actual intrinsic dimension (e.g., $d \in \{1, 2\}$) in LPCA-SRC. This, combined with the assumption that the original training set in SRC is not too large (so that the ℓ^1-minimization problem in SRC can be solved fairly efficiently), suggests that the additional storage requirements of LPCA-SRC over SRC may not deter from the use of LPCA-SRC.

5 Experiments

We tested the proposed classification algorithm on one synthetic database and three popular face databases. For all data sets, we used HOMOTOPY to solve the regularized versions of the ℓ^1-minimization problems, i.e., Eq. (3) for SRC and Eq. (5) for LPCA-SRC, using version 2.0 of the L1 Homotopy toolbox [1].

5.1 Algorithms Compared

We compared LPCA-SRC to the original SRC, SRC_{pruned} (a modification of SRC which we explain shortly), two versions of *tangent distance classification* (our implementations are inspired by Yang et al. [35]), *locality-sensitive dictionary learning SRC* [31], *k-nearest neighbors classification*, and *k-nearest neighbors classification over extended dictionary*.

- SRC_{pruned}: To test the efficacy of the tangent vectors in the LPCA-SRC dictionary, this modification of SRC prunes the dictionary of original training samples using the pruning parameter r, as in LPCA-SRC. $\text{SRC}_{\text{pruned}}$ is exactly LPCA-SRC without the addition of tangent vectors.
- *Tangent distance classification* (TDC1 and TDC2): We compared LPCA-SRC to two versions of tangent distance classification to test the importance of our algorithm's sparse representation framework. Both of our implementations begin by first finding a pruned matrix D_y^{TDC} that is very similar to the dictionary D_y in LPCA-SRC. In particular, D_y^{TDC} can be found using Algorithm 2 and Steps 1–10 in Algorithm 3, *omitting Step 2 in each algorithm.* That is, neither the training

nor test samples are ℓ^2-normalized in the TDC methods; compared to the SRC algorithms, TDC1 and TDC2 are not sensitive to the energy of the samples. We emphasize that the resulting matrix D_y^{TDC} contains training samples that are nearby $\boldsymbol{y}$, as well as their corresponding tangent vectors.

In TDC1, we then divide D_y^{TDC} into the "subdictionaries" $D_y^{(l)}$, where $D_y^{(l)}$ contains the portion of D_y^{TDC} corresponding to class l. The test sample $\boldsymbol{y}$ is next projected onto the space spanned by the columns of $D_y^{(l)}$ to produce the vector $\hat{\boldsymbol{y}}^{(l)}$, and the final classification is performed using

$$\text{class\_label}(\boldsymbol{y}) = \arg \min_{1 \le l \le L} \left\| \boldsymbol{y} - \hat{\boldsymbol{y}}^{(l)} \right\|_2.$$

Our second implementation, TDC2, is similar. Instead of dividing D_y^{TDC} according to class, however, we split it up according to training sample, obtaining the subdictionaries $D_y^{(l,i)}$, where $D_y^{(l,i)}$ contains the original training sample $\boldsymbol{x}_i^{(l)}$ and its tangent vectors. It follows that each subdictionary in TDC2 has $d+1$ columns. The given test sample $\boldsymbol{y}$ is next projected onto the space spanned by the columns of $D_y^{(l,i)}$ to produce $\hat{\boldsymbol{y}}_i^{(l)}$, a vector on the (approximate) tangent hyperplane at $\boldsymbol{x}_i^{(l)}$. The final classification is performed using

$$\text{class\_label}(\boldsymbol{y}) = \arg \min_{1 \le l \le L} \left\{ \min_{1 \le i \le N_l} \left\| \boldsymbol{y} - \hat{\boldsymbol{y}}_i^{(l)} \right\|_2 \right\}.$$

- *Locality-sensitive dictionary learning SRC* (LSDL-SRC): Instead of directly minimizing the ℓ^1-norm of the coefficient vector, LSDL-SRC replaces the regularization term in Eq. (3) of SRC with a term that forces large coefficients to occur only at dictionary elements that are close (in terms of an exponential distance function) to the given test sample. LSDL-SRC also includes a separate dictionary learning phase in which columns of the dictionary are selected from the columns of X_{tr}. We note that though the name "LSDL-SRC" contains the term "SRC," this algorithm is less related to SRC than our proposed algorithm, LPCA-SRC. See Wei et al.'s paper [31] for their reasoning behind this name choice. However, the two algorithms do have very similar objectives, and we thought it important to compare LPCA-SRC and LSDL-SRC in order to validate our alternative approach.
- *k-nearest neighbors classification* (kNN): The test sample is classified to the most-represented class from among the nearest (in terms of Euclidean distance) k training samples (k is odd).
- *k-nearest neighbors classification over extended dictionary* (kNN-Ext): This is kNN over the columns of the (full) extended dictionary that includes the original training samples and their tangent vectors. Samples are not normalized at any stage.

5.2 *Setting of Parameters*

For the synthetic database, we used cross-validation at each instantiation of the training set to choose the best parameters n, λ, and d in LPCA-SRC. (Though the true class manifold dimension is known on this database, we cannot always assume that this is the case.) We optimized the parameters consecutively as described in Sect. 4.3.2. We used the same approach for the parameter λ in SRC, the parameters n and λ in $\text{SRC}_{\text{pruned}}$, and the parameters n and d in the TDC algorithms. Finally, we used a similar procedure for the multiple parameters in LSDL-SRC (including its number of dictionary elements), and we also set k in kNN and kNN-Ext using cross-validation.

Our approach for the face databases was very similar, though in order to save computational costs, we set some parameter values according to previously published works. In particular, we set $\lambda = 0.001$ in LPCA-SRC, SRC, and $\text{SRC}_{\text{pruned}}$, as was used in SRC by Waqas et al. [30]. Additionally, we set most of the parameters in LSDL-SRC to the values used by its authors [31] on the same face databases, though we again used cross-validation to determine its number of dictionary elements.

5.3 *Synthetic Database*

This subsection is organized into two parts: We describe the synthetic database in Sect. 5.3.1, and we present our experimental findings in Sect. 5.3.2. Figures 4 and 5 and Table 3 show the accuracy and runtime results (as well as related information) respectively, for different versions of the synthetic database. A thorough discussion follows. Note that some algorithms from Sect. 5.1 ("Algorithms Compared") have been excluded from these reported findings because of their poor performance, as we explain towards the end of Sect. 5.3.2. We finish this subsection by summarizing our results on the synthetic database.

5.3.1 Database Description

The following synthetic database is easily visualized, and its class manifolds are nonlinear (though well-approximated by local tangent planes) with many intersections. Thus it is ideal for empirically comparing LPCA-SRC and SRC. However, we stress strongly that the classification results on this database (in Sect. 5.3.2) are biased towards the proposed method, as the database structure is specifically designed to illustrate the advantages of LPCA-SRC over SRC. See the results on the face databases in Sect. 5.4 for an unbiased comparison between LPCA-SRC and the methods outlined in Sect. 5.1.

In the synthetic database, class manifolds are sinusoidal waves normalized to lie on S^2, with underlying equations given by

$$
\begin{aligned}
x(t) &= \cos(t+\phi),\\
y(t) &= \sin(t+\phi),\\
z(t) &= A\sin(\omega t).
\end{aligned}
$$

We set $\omega = 3$ and $A = 0.5$, and we varied ϕ to obtain L classes. In particular, we set $\phi = 2\pi/(3l)$ for data in class $1 \le l \le L = 4$. For each training and test set, we generated the same number $N_0 = N_l$, $l = 1, \ldots, L$, of samples in each class by (i) regularly sampling $t \in [0, 2\pi)$ to obtain the points $\boldsymbol{p}(t) = [x(t), y(t), z(t)]^{\mathsf{T}}$; (ii) computing the normalized points $\boldsymbol{p}(t)/\|\boldsymbol{p}(t)\|_2$; (iii) appending 50 "noise dimensions" to obtain vectors in $\mathbb{R}^{53}$; (iv) adding independent random noise to each coordinate of each point as drawn from the Gaussian distribution $\mathcal{N}(0, \eta^2)$; and lastly (v) re-normalizing each point to obtain vectors of length $m = 53$ lying on S^{m-1}. We performed classification on the resulting data samples. Note that the reason why we turned the original $\mathbb{R}^3$ problem into a problem in $\mathbb{R}^{53}$ was because SRC is designed for high-dimensional classification problems [32] and to make the problem more challenging. We emphasize that we did not apply any method of dimension reduction to this database.

Figure 3 shows the first three coordinates of a realization of the training set of the synthetic database. Note that the class manifold dimension is the same for each class and equal to 1. The signal-to-noise ratios (SNRs) are displayed in Table 1 for $N_0 = 25$ and various values of noise level η. These results were obtained by averaging the mean training sample SNR over 100 realizations of the data set.

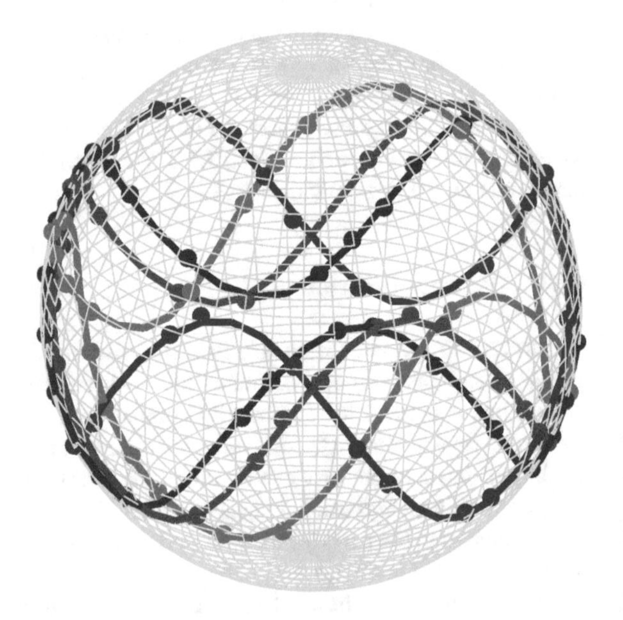

Fig. 3 A realization of the first three coordinates of the synthetic database training set with $N_0 = 25$ and $\eta = 0.01$. Nodes denote training samples; colors denote classes

Table 1 Mean training sample signal-to-noise ratio (in decibels) over 100 realizations of the synthetic database with $N_0 = 25$ and various values of noise level η

$\eta = 0.0001$	$\eta = 0.001$	$\eta = 0.005$	$\eta = 0.01$	$\eta = 0.015$	$\eta = 0.02$	$\eta = 0.03$	$\eta = 0.05$
62.85	42.84	28.86	22.86	19.35	16.89	13.45	9.25

Table 2 Brief descriptions of the parameters relevant to experimental results on the synthetic database

Algorithm parameters	Data set parameters	Output parameters
d, n: Local PCA parameters	N_0: Class size	N: Dictionary size
λ: Error/sparsity trade-off	η: Noise level	t: Time in seconds
r: Pruning parameter (set automatically)		κ: # of Homotopy iterations

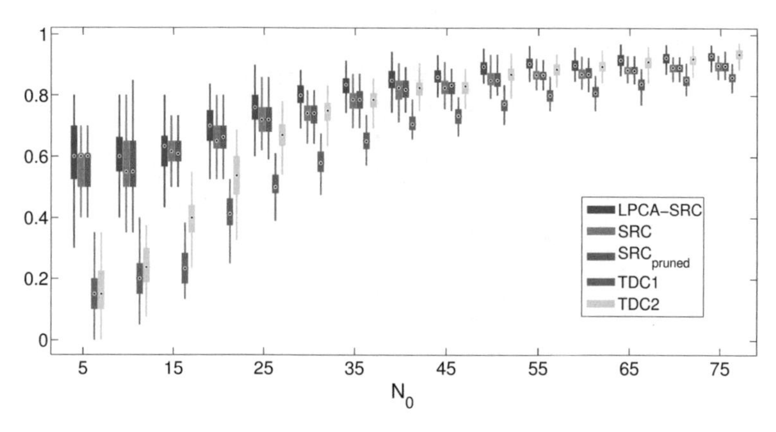

Fig. 4 Box plots of the average classification accuracy (over 100 trials) of competitive algorithms on the synthetic database with varying training class size N_0. We fixed $\eta = 0.001$

5.3.2 Experimental Results

We performed experiments on this database, first varying the number of training samples in each class and then varying the amount of noise. Table 2 contains brief descriptions of the relevant parameters for easy reference; a detailed description of the output parameters is given later on.

The results are presented in Figs. 4, 5 and Table 3; a discussion follows.

Accuracy results for varying class size. Figure 4 shows the average classification accuracy (over 100 trials) of the competitive algorithms as we varied the number of training samples in each class. We fixed the noise level $\eta = 0.001$. LPCA-SRC generally had the highest accuracy. On average, LPCA-SRC outperformed SRC by 3.5%, though this advantage slightly decreased as the sampling density increased and the tangent vectors became less useful, in the sense that there were often already

Table 3 Average runtime in ms (t), dictionary size (N), and number of HOMOTOPY iterations (κ) over 100 trials on the synthetic database with varying training class size N_0. We fixed $\eta = 0.001$

	$N_0 = 5$			$N_0 = 25$			$N_0 = 45$			$N_0 = 65$		
Algorithm	t	N	κ	t	N	κ	t	N	κ	t	N	κ
LPCA-SRC	11.2	56	2	68.8	80	3	115.3	42	3	159.2	30	2
SRC	4.5	20	2	39.9	100	3	104.6	180	3	162.8	260	3
SRC_{pruned}	7.1	20	2	54.1	79	3	130.2	146	3	206.0	201	3
TDC1	10.8	9	N/A	43.6	6	N/A	71.1	5	N/A	92.3	3	N/A
TDC2	19.5	3	N/A	57.0	2	N/A	93.4	2	N/A	125.4	2	N/A

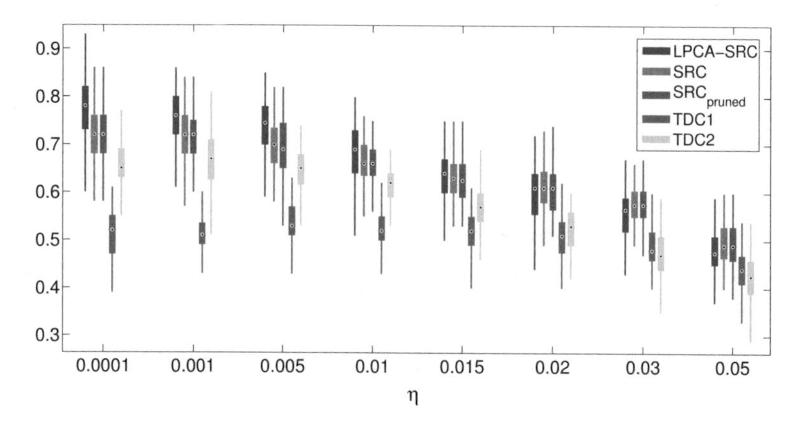

Fig. 5 Box plots of the average classification accuracy (over 100 trials) of competitive algorithms on the synthetic databases with varying noise level η. We fixed $N_0 = 25$

enough nearby training samples in the ground truth class of $\boldsymbol{y}$ to accurately approximate it without the addition of tangent vectors. SRC and $\text{SRC}_{\text{pruned}}$ had comparable accuracy for all tried values of N_0, indicating that the pruning parameter r was effective in removing unnecessary training samples from the SRC dictionary. Further, the increased accuracy of LPCA-SRC over $\text{SRC}_{\text{pruned}}$ suggests that the tangent vectors in LPCA-SRC contributed meaningful class information.

To determine if these results are statistically significant, we performed a Repeated Measures ANOVA test on the results for LPCA-SRC, SRC, and $\text{SRC}_{\text{pruned}}$ as well as a t-test between the results for LPCA-SRC and SRC. The detailed results can be found in Table 9 in Appendix A. In summary, the differences in the accuracies of LPCA-SRC, SRC, and $\text{SRC}_{\text{pruned}}$ are statistically significant for all but $N_0 = 15$, as demonstrated by p-values less than 0.05 for these experiments.

The TDC methods performed relatively poorly for small values of N_0. At low sampling densities, the TDC subdictionaries were poor models of the (local) class manifolds, leading to approximations of $\boldsymbol{y}$ that were often indistinguishable from each other and resulting in poor classification. Both TDC methods improved significantly as N_0 increased, with TDC2 outperforming TDC1 and in fact becoming comparable to LPCA-SRC for $N_0 \geq 60$. We attribute this to the extremely local nature of TDC2: It considers a single local patch on a class manifold at a time, rather than each class as a whole. Hence under dense sampling conditions, TDC2 effectively mimicked the successful use of sparsity in LPCA-SRC.

Accuracy results for varying noise. Figure 5 shows the average classification accuracy (over 100 trials) of the competitive algorithms as we varied the amount of noise. We fixed $N_0 = 25$. LPCA-SRC had the highest classification accuracy for low values of η (equivalently, when the SNR was high), outperforming SRC by as much as nearly 4%. For $\eta \geq 0.015$ (i.e., when the SNR dropped below 20 decibels), LPCA-SRC lost its advantage over SRC and $\text{SRC}_{\text{pruned}}$. This is likely due to noise degrading the accuracy of the tangent vectors. SRC and $\text{SRC}_{\text{pruned}}$ had nearly identical accuracy for all values of η; again, this illustrates that faraway training samples (as defined by the pruning parameter r) did not contribute to the ℓ^1-minimized approximation of the test sample, and the increased accuracy of LPCA-SRC over $\text{SRC}_{\text{pruned}}$ for low noise values demonstrates the efficacy of the tangent vectors in LPCA-SRC in these cases. We briefly note that when we vary the noise level for larger values of N_0, the accuracy of the tangent vectors generally improves. As a result, we see that LPCA-SRC can tolerate higher values of η before being outperformed by SRC and $\text{SRC}_{\text{pruned}}$.

Table 10 in Appendix A contains the p-values for rANOVA and related tests on the accuracy results of LPCA-SRC, SRC, and $\text{SRC}_{\text{pruned}}$, as well as the 5% confidence intervals for the advantage of LPCA-SRC over SRC. These tests concur with the discussion above; LPCA-SRC outperforms SRC for small values of η, there is no clear advantage for $\eta \in \{0.01, 0.015\}$, and SRC outperforms LPCA-SRC for $\eta \geq 0.02$.

TDC2 outperformed TDC1 for all but the largest values of η, though both algorithms were outperformed by the three SRC methods at this relatively low sampling density for the reasons discussed previously. For $\eta \geq 0.03$, TDC2 began performing worse than TDC1. We expect that the local patches represented by the subdictionaries in TDC2 became poor estimates of the (tangent hyperplanes of the) class manifolds as the noise increased, resulting in a decrease in classification accuracy.

Runtime results for varying class size. In Table 3, we display the runtime-related information of the competitive algorithms with varying training class size. (We do not show the runtime results for the case of varying noise; the results for varying class size are much more revealing.) In particular, we report the average runtime (in milliseconds), the number of columns in each algorithm's dictionary (we refer to this as the "size" of the dictionary, as the sample dimension is fixed), and the number of HOMOTOPY iterations. These latter variables are denoted N and κ, respectively. The runtime does not include the time it took to perform cross-validation and is the total time (averaged over 100 trials) of performing classification on the entire database. In the case that the algorithm has separate offline and online phases (e.g., LPCA-SRC), both phases are included in this total. For the TDC methods, we report the average subdictionary sizes,[4] and for conciseness, we display the results for only a handful of the values of N_0. We use "N/A" to indicate that a particular statistic is not applicable to the given algorithm.

The dictionary sizes of LPCA-SRC, SRC, and $\text{SRC}_{\text{pruned}}$ are quite informative. Recall that LPCA-SRC outperformed SRC and $\text{SRC}_{\text{pruned}}$ (by more than 3%) for the shown values of N_0. For $N_0 = 5$, the dictionary in LPCA-SRC was larger than that of the two other methods, adaptively retaining more samples to counter-balance the low sampling density. At large values of N_0, LPCA-SRC took full advantage of the increased sampling density, stringently pruning the set of training samples and keeping only those very close to $\boldsymbol{y}$. Due to the resulting small dictionary, it had comparable runtime to SRC despite its additional cost of computing tangent vectors. In contrast, without the addition of tangent vectors, $\text{SRC}_{\text{pruned}}$ was forced to

[4]Recall that these subdictionaries are the class-specific portions $D_{\boldsymbol{y}}^{(l)}$, $1 \leq l \leq L$, of the main dictionary $D_{\boldsymbol{y}}^{\text{TDC}}$. Thus the values of N for TDC1 and TDC2 are much smaller than those for the other classification methods.

keep a large number of training samples in its dictionary; the cost of the dictionary pruning step resulted in $\text{SRC}_{\text{pruned}}$ running slower than SRC, despite its slightly smaller dictionary. (We note that one might expect that $\text{SRC}_{\text{pruned}}$ would always have a smaller dictionary than LPCA-SRC since it does not include tangent vectors; this is not the case, as the value of the number-of-neighbors parameter n, and hence the pruning parameter r, may be different for the two algorithms.)

The TDC methods ran relatively fast, especially for large values of N_0. This is expected, as these algorithms do not require ℓ^1-minimization.

Summary. The experimental results on the synthetic database show that LPCA-SRC can achieve higher classification accuracy than SRC and similar methods when the class manifolds are sparsely sampled and the SNR is large. In these cases, the tangent vectors in LPCA-SRC help to "fill out" portions of the class manifolds that lack training samples. When the sampling density was sufficiently high, however, we saw that the tangent vectors in LPCA-SRC were less needed to provide an accurate, local approximation of the test sample, and thus LPCA-SRC offered a smaller advantage over SRC and $\text{SRC}_{\text{pruned}}$. Additionally, for higher noise (i.e., low SNR) cases, the computed tangent vectors were less reliable and the classification performance consequently deteriorated. With regard to runtime, LPCA-SRC appeared to adapt to the sampling density of the synthetic database, and though the addition of tangent vectors initially increased the dictionary size in LPCA-SRC, the online dictionary pruning step allowed for runtime comparable to SRC when the class sizes were large.

5.4 *Face Databases*

This subsection is organized as follows:

- We first explain our experimental setup. We describe the different face databases and state the training set sizes in Sect. 5.4.1, and in Sects. 5.4.2 and 5.4.3, we describe the method of dimension reduction used on the raw samples and our approach to handling data samples with occlusion, respectively. Section 5.4.4 simply contains Table 4, which shows brief descriptions of the relevant parameters on the face databases for easy reference.
- We separate our classification results into two parts: Sect. 5.4.5 contains our results on the AR face database, and Sect. 5.4.6 contains our results on the Extended Yale B and ORL face databases. More precisely, Figs. 6, 7 and Table 5 contain the accuracy and runtime results for two versions of the AR face database; Figs. 8, 9 and Tables 6, 7 show the same results for Extended Yale B and ORL. Again, these databases are described in Sect. 5.4.1. The figures and tables in each section are followed by a discussion of their results.
- In Sect. 5.4.7, we offer evidence to support our claim that the tangent vectors in LPCA-SRC can recover discriminative information lost during PCA transforms to low dimensions. We display the PCA-recovered tangent vectors and compare them to the original samples (without PCA transform) as well as the recovered samples (after PCA transform).

- Lastly, Sect. 5.4.8 contains a summary of our experimental findings on the face databases.

5.4.1 Database Description

The *AR Face Database* [21] contains 70 male and 56 female subjects photographed in two separate sessions held on different days. Each session produced 13 images of each subject, the first seven with varying lighting conditions and expressions, and the remaining six images occluded by either sunglasses or scarves under varying lighting conditions. Images were cropped to 165×120 pixels and converted to grayscale. In our experiments, we selected the first 50 male subjects and first 50 female subjects, as was done in several papers (e.g., Wright et al. [32]), for a total of 100 classes. We performed classification on two versions of this database. The first, which we call "AR-1," contains the 1400 un-occluded images from both sessions. The second version, "AR-2," consists of the images in AR-1 as well as the 600 occluded images (sunglasses and scarves) from Session 1.

The *Extended Yale Face Database B* [11] contains 38 classes (subjects) with about 64 images per class. The subjects were photographed from the front under various lighting conditions. We used the version of Extended Yale B that contains manually-aligned, cropped, and resized images of dimension 192×168.

The *Database of Faces* (formerly "The ORL Database of Faces") [2] contains 40 classes (subjects) with 10 images per class. The subjects were photographed from the front against dark, homogeneous backgrounds. The sets of images of some subjects contain varying lighting conditions, expressions, and facial details. Each image in ORL is initially of 92×112 pixels.

Given existing work on the manifold structure of face databases (e.g., that of Saul and Roweis [26], He et al. [12], and Lee et al. [17]), we make the following suppositions: Since images in each class in AR-1 and AR-2 have extreme variations in lighting conditions and differing expressions, the class manifolds of these databases may be nonlinear. Further, the natural occlusions contained in AR-2 make these class manifolds *highly* nonlinear. Alternatively, since the images in each class in Extended Yale B differ primarily in lighting conditions, the class manifolds may be nearly linear. Lastly, since the images in some classes in ORL differ in both lighting conditions and expression, these class manifolds may be nonlinear; however, since the variations are small, these manifolds may be well-approximated by linear subspaces.

With regard to sampling density, we reiterate that Extended Yale B has large class sizes compared to AR and ORL. In our experiments, we randomly selected the same number of samples in each class to use for training, i.e., we set $N_0 \equiv N_l$, $1 \le l \le L$, where N_0 was half the number of samples in each class.[5] We used the remaining samples for testing.

[5]Since the class sizes vary slightly in Extended Yale B, we set $N_0 = 32$ on this database.

5.4.2 Dimension Reduction

To perform dimension reduction on the face databases, we used (global) PCA to transform the raw images to $m_{\text{PCA}} \in \{30, 56, 120\}$ dimensions before performing classification. Similar values for m_{PCA} were used by Wright et al. [32]. For the remainder of this paper, we will refer to the PCA-compressed versions of the raw face images as "feature vectors" and m_{PCA} as the "feature dimension." We note that the data was not centered (around the origin) in the PCA transform space.

5.4.3 Handling Occlusion

Since AR-2 contains images with occlusion, we considered using the "occlusion version" of SRC (with analogous modifications to LPCA-SRC and $\text{SRC}_{\text{pruned}}$) on this database. As discussed by Wright et al. [32], this model assumes that $\boldsymbol{y}$ is the summation of the (unknown) true test sample $\boldsymbol{y}_0$ and an (unknown) sparse error vector. The resulting modified ℓ^1-minimization problem consists of appending the dictionary of training samples with the identity matrix $I \in \mathbb{R}^{m \times m}$ and decomposing $\boldsymbol{y}$ over this augmented dictionary. For more details, see Sect. 3.2 of the SRC paper [32].

However, the context in which Wright et al. use the occlusion version of SRC on the AR database is critically different than our experimental setup here [32]. In the SRC paper, the samples with occlusion make up the test set. In our case, both the training and test set contain samples with and without occlusion. As a consequence, occluded samples in the training set can be used to express test samples with occlusion, and on the other hand, the use of the identity matrix to extend the dictionary in SRC results in too much error allowed in the approximation of un-occluded samples. Correspondingly, we see much worse classification performance in SRC when we use its occlusion version on AR-2. Hence, we stick to Algorithm 1 (the original version of SRC) on all face databases.

5.4.4 Table of Parameters

Table 4 contains brief descriptions of the parameters relevant to the face databases.

5.4.5 AR Face Database Results

Accuracy results on AR. Figures 6 and 7 display the accuracy results over 10 trials for the two versions of AR, respectively. LPCA-SRC had substantially higher classification accuracy than the other methods on both versions of AR with $m_{\text{PCA}} = 30$. This suggests that the tangent vectors in LPCA-SRC were able to recover important class information lost in the stringent PCA dimension reduction. As m_{PCA} increased, however, the methods SRC, $\text{SRC}_{\text{pruned}}$, and LSDL-SRC became more competitive, as

Table 4 Brief descriptions of the parameters relevant to experimental results on the face databases

Algorithm parameters	Data set parameters	Output parameters
d, n: Local PCA parameters	N_0: Class size	N: Dictionary size
λ: Error/sparsity trade-off	m_{PCA}: PCA dimension	t: Time in seconds
r: Pruning parameter (set automatically)		κ: # of Homotopy iterations

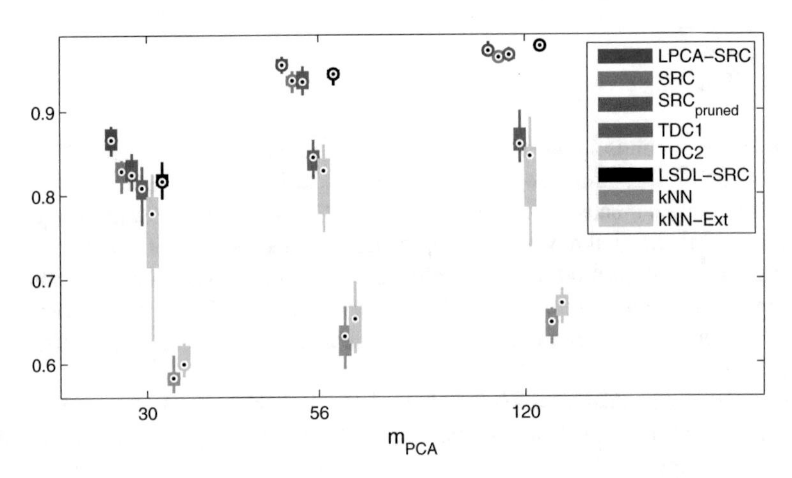

Fig. 6 Box plots of the average classification accuracy (over 10 trials) on the AR-1 face database for different values of m_{PCA}

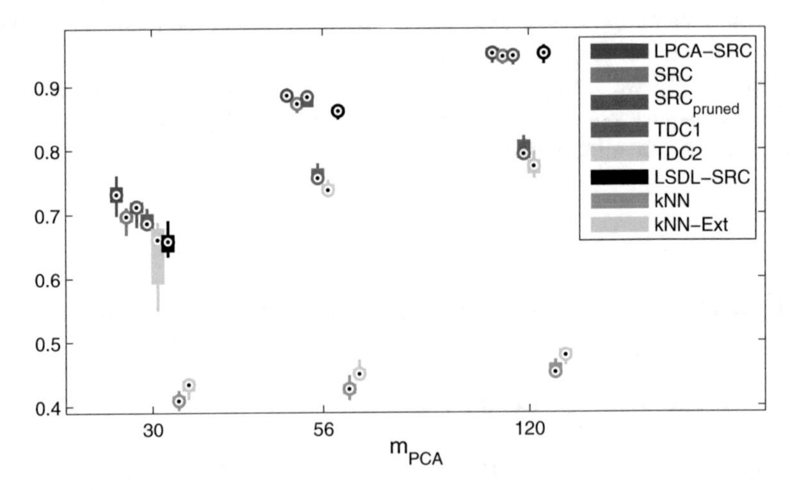

Fig. 7 Box plots of the average classification accuracy (over 10 trials) on the AR-2 face database for different values of m_{PCA}

more discriminative information was retained in the feature vectors and less needed to be provided by the LPCA-SRC tangent vectors. $\text{SRC}_{\text{pruned}}$ had comparable accuracy to SRC, indicating that, once again, training samples could be removed from the SRC dictionary using the pruning parameter r without decreasing classification accuracy. In some cases, the removal of these faraway training samples slightly improved class discrimination.

To test for statistical significance in the differences between accuracy results, we performed a Repeated Measures ANOVA test on LPCA-SRC, SRC, $\text{SRC}_{\text{pruned}}$, and LSDL-SRC as well as two t-tests, one between LPCA-SRC and SRC and the other between LPCA-SRC and LSDL-SRC. The related p-values and confidence intervals are contained in Table 11 in Appendix A. In summary, LPCA-SRC outperforms both methods in a statistically-significant manner, except for LSDL-SRC when $m_{\text{PCA}} = 120$.

For the most part, the other algorithms performed poorly on AR. The exception was LSDL-SRC, which had comparable accuracy to LPCA-SRC for $m_{\text{PCA}} = 120$ (slightly outperforming it for AR-1) and beat SRC on AR-1 for $m_{\text{PCA}} = 56$. However, LSDL-SRC had lower accuracy than the SRC algorithms for $m_{\text{PCA}} = 30$ on both versions of this database. In contrast, the TDC methods performed relatively better for $m_{\text{PCA}} = 30$ than for larger values of m_{PCA} due to their more effective use of tangent vectors at this small feature dimension. Overall, however, their class-specific dictionaries were not as effective on this nonlinear, sparsely sampled database as the multi-class dictionaries of the previously-discussed algorithms. Further, TDC2 often had notably high standard error, presumably because of its sensitivity to the value of the manifold dimension estimate d. This could perhaps be mitigated by using a different cross-validation procedure. Lastly, kNN and kNN-Ext had the lowest classification accuracies, though kNN-Ext offered a slight improvement over kNN. Both methods consistently selected $k = 1$ during cross-validation.

Runtime results on AR. Table 5 displays the average runtime and related results (over 10 trials) of the various classification algorithms for both versions of AR. Again, the runtime does not include the time it took to perform cross-validation and is the total time (averaged over 10 trials) of performing classification on the entire database (offline and online phases both included when applicable). The "dictionary size" N for kNN and kNN-Ext refers to the average size of the set from which the k-nearest neighbors are selected (e.g., for kNN, $N = N_{\text{tr}}$).

The generally large dictionary sizes of LPCA-SRC (and its consequently long runtimes) indicate that minimal dictionary pruning often occurred. Thus LPCA-SRC was generally slower than SRC and $\text{SRC}_{\text{pruned}}$. However, on AR-2 with $m_{\text{PCA}} = 30$, LPCA-SRC was able to eliminate many training samples from its dictionary, due to its effective use of tangent vectors on the (presumably) highly-nonlinear class manifolds of AR-2. At this low feature dimension, the computed tangent vectors contained more class discriminative information than nonlocal training samples, likely allowing for a more accurate—and local—approximation of $\boldsymbol{y}$ on its ground truth class manifold. LPCA-SRC was faster than SRC and $\text{SRC}_{\text{pruned}}$ (which kept a large number of training samples) in this case, and this is impressive, considering

Table 5 Average runtime in ms (t), dictionary size (N), and number of HOMOTOPY iterations (κ) over 10 trials on AR

Algorithm	AR-1								
	$m_{\text{PCA}} = 30$			$m_{\text{PCA}} = 56$			$m_{\text{PCA}} = 120$		
	t	N	κ	t	N	κ	t	N	κ
LPCA-SRC	7253	435	61	12496	676	87	19068	795	112
SRC	6114	700	51	8875	700	72	13574	700	99
$\text{SRC}_{\text{pruned}}$	3763	231	39	5099	226	49	6897	232	60
TDC1	11816	16	N/A	14239	16	N/A	24296	19	N/A
TDC2	8895	5	N/A	16786	5	N/A	36682	5	N/A
LSDL-SRC	7776	440	N/A	8552	470	N/A	9720	490	N/A
kNN	13	700	N/A	18	700	N/A	29	700	N/A
kNN-Ext	102	2170	N/A	132	2240	N/A	253	2660	N/A
Algorithm	**AR-2**								
	$m_{\text{PCA}} = 30$			$m_{\text{PCA}} = 56$			$m_{\text{PCA}} = 120$		
	t	N	κ	t	N	κ	t	N	κ
LPCA-SRC	10533	478	58	35269	1593	10	56169	1690	151
SRC	11394	1000	58	17674	1000	85	27743	1000	121
$\text{SRC}_{\text{pruned}}$	11118	788	54	16631	775	77	24880	767	107
TDC1	20557	25	N/A	27515	26	N/A	43073	26	N/A
TDC2	20930	6	N/A	47571	6	N/A	103796	6	N/A
LSDL-SRC	22698	750	N/A	16337	620	N/A	22191	710	N/A
kNN	15	1000	N/A	21	1000	N/A	37	1000	N/A
kNN-Ext	128	4300	N/A	152	3600	N/A	294	4400	N/A

that LPCA-SRC also outperformed these methods by nearly 4% and more than 2%, respectively.

Despite not requiring ℓ^1-minimization, the TDC methods were often the slowest algorithms on the AR databases. We suspect that this is largely due to the relatively large number of classes in AR—recall that both TDC methods must compute least squares solutions (in TDC2, sometimes many of them) for each class represented in the pruned dictionary D_y^{TDC}. Further, TDC2 selected a relatively large value of d during cross-validation (presumably so that its subdictionaries would contain a wider "snapshot" of the class manifolds), which made it even less efficient. The runtime of LSDL-SRC, unlike those of most of the other algorithms, was fairly insensitive to the feature dimension, and as a result, LSDL-SRC was relatively efficient for $m_{\text{PCA}} \in \{56, 120\}$. However, the expense of its dictionary learning phase for $m_{\text{PCA}} = 30$, at which the ℓ^1-minimization algorithm in the SRC methods could be solved efficiently, resulted in LSDL-SRC's relatively slow runtime. Both kNN methods ran significantly faster than all the other methods.

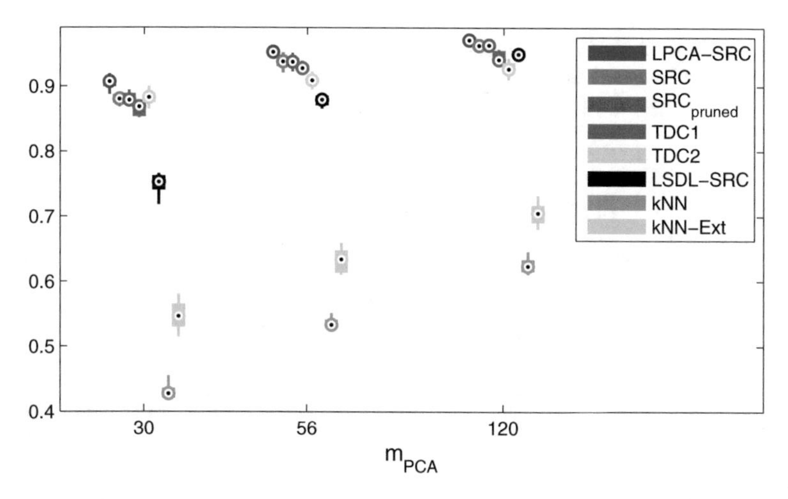

Fig. 8 Box plots of the average classification accuracy (over 10 trials) on the Extended Yale B face database for different values of m_{PCA}

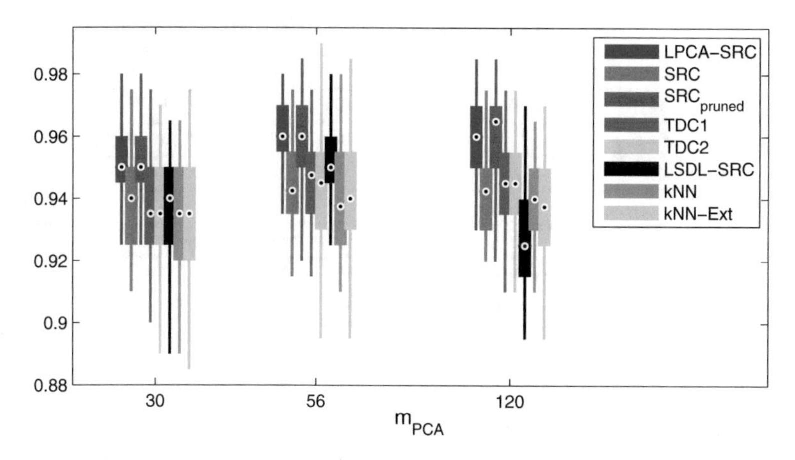

Fig. 9 Box plots of the average classification accuracy (over 50 trials) on the ORL face database for different values of m_{PCA}

5.4.6 Extended Yale Face Database B and Database of Faces ("ORL") Results

Accuracy results on Extended Yale B and ORL. Figure 8 displays the accuracy results for Extended Yale B (over 10 trials), and Fig. 9 displays the accuracy results for ORL (over 50 trials). On Extended Yale B, LPCA-SRC had the highest accuracy for all m_{PCA}, though as we saw on the AR database, this advantage decreased as m_{PCA} increased and SRC became more competitive. SRC and $\mathrm{SRC}_{\mathrm{pruned}}$ had similar accuracy, indicating that training samples excluded from the dictionary via the

Table 6 Average runtime in ms (t), dictionary size (N), and number of HOMOTOPY iterations (κ) over 10 trials on Extended Yale B

	$m_{\text{PCA}} = 30$			$m_{\text{PCA}} = 56$			$m_{\text{PCA}} = 120$		
Algorithm	t	N	κ	t	N	κ	t	N	κ
LPCA-SRC	29204	1922	75	72122	3359	120	141966	3785	182
SRC	15584	1216	62	24697	1216	91	41939	1216	137
$\text{SRC}_{\text{pruned}}$	15915	1111	61	23813	1112	88	40504	1115	131
TDC1	8098	20	N/A	27620	59	N/A	42828	59	N/A
TDC2	11675	6	N/A	23506	6	N/A	56006	6	N/A
LSDL-SRC	67295	1186	N/A	53031	1003	N/A	38731	821	N/A
kNN	17	1216	N/A	26	1216	N/A	49	1216	N/A
kNN-Ext	172	5350	N/A	251	4742	N/A	443	4864	N/A

Table 7 Average runtime in ms (t), dictionary size (N), and number of HOMOTOPY iterations (κ) over 50 trials on ORL

Algorithm	$m_{\text{PCA}} = 30$			$m_{\text{PCA}} = 56$			$m_{\text{PCA}} = 120$		
	t	N	κ	t	N	κ	t	N	κ
LPCA-SRC	539	59	26	730	72	34	1221	111	50
SRC	854	200	40	1337	200	57	2087	200	81
$\text{SRC}_{\text{pruned}}$	254	19	12	343	26	16	530	39	24
TDC1	121	1	N/A	162	1	N/A	344	1	N/A
TDC2	117	3	N/A	233	3	N/A	532	3	N/A
LSDL-SRC	1040	116	N/A	1088	121	N/A	931	102	N/A
kNN	8	200	N/A	8	200	N/A	9	200	N/A
kNN-Ext	25	568	N/A	28	592	N/A	38	568	N/A

pruning parameter r did not provide class information. TDC1 and TDC2 had consistently mediocre performance, neither one outperforming the other over all settings of m_{PCA}, and LSDL-SRC improved as m_{PCA} increased, analogous to its behavior on AR. However, LSDL-SRC was outperformed by LPCA-SRC, even for $m_{\text{PCA}} = 120$, suggesting that the improved approximations in LPCA-SRC via its use of tangent vectors were more effective (even at this high feature dimension) than the procedure in LSDL-SRC. Along these same lines, the tangent vectors in kNN-Ext offered a considerable improvement over kNN, though once again both methods reported lower accuracy than all the other algorithms. As on AR, the kNN methods consistently selected $k = 1$ during cross-validation.

On ORL, LPCA-SRC and $\text{SRC}_{\text{pruned}}$ had comparable accuracy and outperformed SRC. This indicates that: (i) the pruning parameter r in LPCA-SRC and $\text{SRC}_{\text{pruned}}$ was *helpful* to classification (instead of simply being benign); and (ii) the tangent vectors computed in LPCA-SRC were not. With regard to (i), it must be the case that faraway training samples—those in different classes from the test sample—

contributed significantly to the approximation of the test sample in SRC, negatively affecting classification performance. This is an example of *sparsity not necessarily leading to locality* (as it is relevant to class discrimination), as discussed in the LSDL-SRC paper [31]. With regard to (ii), we suspect that the tangent vectors in LPCA-SRC were simply *unneeded* to improve the classification performance on ORL. Though the approximations in SRC contained nonzero coefficients at training samples not in the same class as $\boldsymbol{y}$—presumably because of the sparse sampling and nonlinear structure of the class manifolds—many of these wrong-class training samples could be eliminated simply based on their distance to $\boldsymbol{y}$. This suggests that ORL's class manifolds can be fairly well-separated via Euclidean distance. An additional reason for (ii) was because the PCA transform to the dimensions specified in this experiment did not result in a loss of too much information, at least compared to AR and Extended Yale B. See Table 8 at the end of Sect. 5.4.7 for this comparison.

As we did for the AR face database, we performed statistical analysis on the reported accuracies for Extended Yale B and ORL. The detailed results are contained in Table 12 in Appendix A. In summary, LPCA-SRC outperforms both SRC and LSDL-SRC with 95% confidence in all of these experiments, albeit its lift in accuracy is sometimes small.

All of the remaining methods performed relatively well on ORL. The accuracies of TDC1 and TDC2 were similar and comparable to those of SRC. We ascertained that the success of the TDC methods was not due to their use of tangent vectors but instead the result of their "per-class" approximations of the test sample. This approach was very effective on the (presumably) well-separated class manifolds of ORL. Strikingly, the accuracy of LSDL-SRC was relatively low for $m_{\text{PCA}} = 120$, opposite to the trend we saw on the previous face databases. The performance of LSDL-SRC could be improved for $m_{\text{PCA}} = 120$ on this database if the samples were centered (around the origin) after PCA dimension reduction. However, we confirmed that LDSL-SRC was still outperformed by LPCA-SRC in this case (albeit by a smaller margin), and its performance with centering on the other face databases was much worse than our reported results. In contrast to the results on Extended Yale B, kNN-Ext only provided a slight increase in accuracy over kNN, with the tangent vectors mimicking their unnecessary role in LPCA-SRC on this database. The value $k = 1$ was consistently selected by both kNN and kNN-Ext during cross-validation.

Runtime results on Extended Yale B and ORL. Tables 6 and 7 show the runtime and related results for the Extended Yale B and ORL experiments, respectively. LPCA-SRC had much longer runtimes than SRC on Extended Yale B, especially as m_{PCA} increased. This was due to a combination of large values for d selected during cross-validation and the tangent vectors' decreasing efficacy at larger feature dimensions. However, the dictionary pruning procedure in LPCA-SRC actually eliminated a large number of training samples for all m_{PCA}; once again, the computed tangent vectors contained more class-discriminating information than the eliminated nonlocal training samples, especially at lower feature dimensions for which details provided by these tangent vectors were especially needed. The linearity of the class manifolds of Extended Yale B, combined with this database's relatively dense sampling, lent itself well to the accurate computation of tangent vectors—part of the reason why LPCA-SRC used so many of them. Viewing these points as newly-generated and nearby training samples, LPCA-SRC's boost in accuracy over SRC can be viewed as an argument for locality in classification. We note that we might be able to decrease the value of d in LPCA-SRC while still maintaining an advantage over SRC (see the discussion in Sect. 4.3.1); our cross-validation procedure is designed to obtain the highest accuracy without regard to computational cost.

On Extended Yale B, the TDC methods ran relatively more quickly (compared to the other algorithms) than on AR, presumably due to the much smaller number of classes on this database; both had runtimes typically between those of LPCA-SRC and SRC. Again, we see that LSDL-SRC had a relatively slow runtime for $m_{\text{PCA}} = 30$ and became more competitive as m_{PCA} increased. Though both kNN and kNN-Ext were very fast, the large "dictionary sizes" in kNN-Ext made this algorithm clearly the slower of the two methods.

On ORL, LPCA-SRC and SRC had comparable runtimes, a result of rigorous dictionary pruning in LPCA-SRC. This algorithm and $\text{SRC}_{\text{pruned}}$ retained roughly the same number of training samples in their respective dictionaries, and the latter was notably fast, running in about half the time as SRC. The remaining algorithms were even more efficient. TDC1 and TDC2 had comparable runtimes, both running faster than LSDL-SRC. As before, kNN and kNN-Ext had the fastest runtimes; the former was faster than the latter.

5.4.7 Tangent Vectors and PCA Feature Dimension

In this section, we offer evidence to support our claim that the tangent vectors in LPCA-SRC can recover discriminative information lost during PCA transforms to low dimensions. Thus LPCA-SRC can offer a clear advantage over SRC in these cases, as we saw in experimental results on AR and Extended Yale B.

In Figs. 10, 11 and 12, we display three versions of three example images from AR-1. The first version is the original image (before PCA dimension reduction), the second version is the recovered image from PCA dimension reduction to dimension $m_{\text{PCA}} = 30$, and the third version is the recovered corresponding tangent vector computed in LPCA-SRC. In each case, the tangent vector contains details of the original image not found in the recovered image, supporting our claim that the tangent vectors in LPCA-SRC can recover some (but not all) of the information lost in stringent PCA dimension reduction.

Towards quantifying what we mean by "stringent," Table 8 lists the average energy[6] (over 10 trials) retained in the first m_{PCA} left-singular vectors of the face database training sets, along with the percent improvement in the accuracy of

[6]By "energy," we mean the ratio of the sum of squares of the first m_{PCA} singular values to the sum of squares of all singular values.

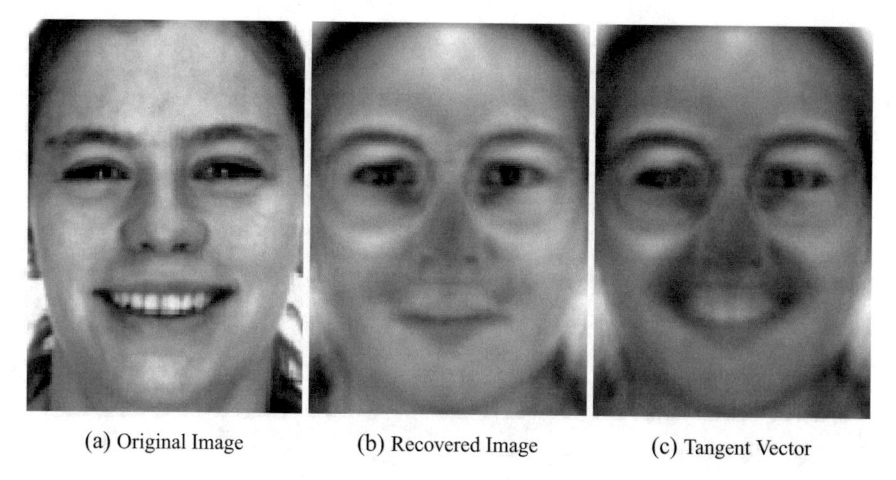

(a) Original Image (b) Recovered Image (c) Tangent Vector

Fig. 10 The tangent vector does a much better job of displaying facial details conveying "happiness" than the recovered image. Images **b** and **c** were recovered from PCA dimension $m_{\text{PCA}} = 30$

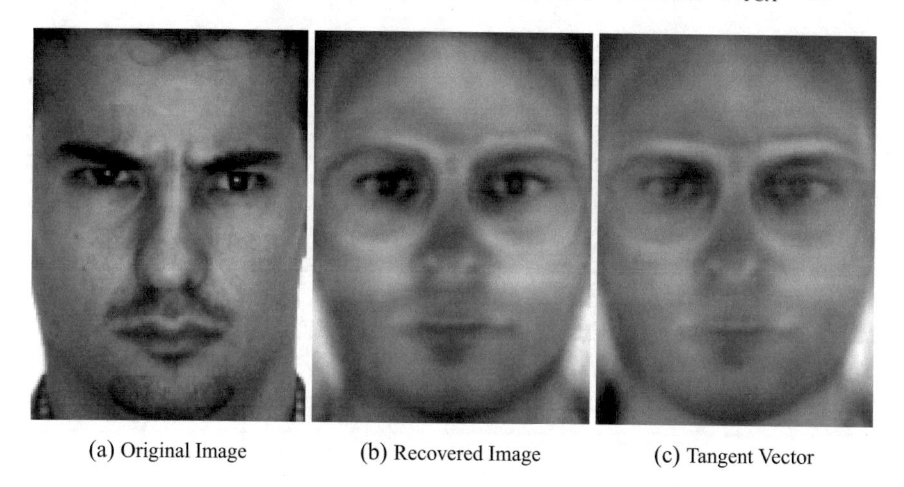

(a) Original Image (b) Recovered Image (c) Tangent Vector

Fig. 11 The tangent vector does a better job of displaying "anger" than the recovered image, most notably in the subject's eyes and eyebrows. Images **b** and **c** were recovered from PCA dimension $m_{\text{PCA}} = 30$

LPCA-SRC over that of SRC and $\text{SRC}_{\text{pruned}}$. Given that the addition of tangent vectors did not increase classification accuracy on ORL, we see a correlation between the efficacy of tangent vectors in LPCA-SRC and the stringency of the PCA dimension reduction.

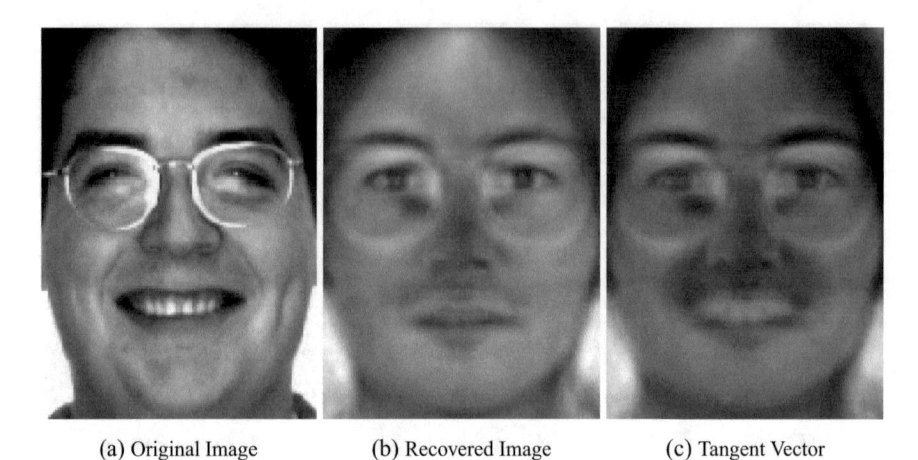

(a) Original Image (b) Recovered Image (c) Tangent Vector

Fig. 12 The tangent vector shows the subject's smile better than the recovered image. Images **b** and **c** were recovered from PCA dimension $m_{\text{PCA}} = 30$

Table 8 Average energy retained in PCA dimension reduction (over 10 trials) to various dimensions m_{PCA} on the face database training sets, as well as the average increase in classification accuracy of LPCA-SRC over SRC and $\text{SRC}_{\text{pruned}}$

Database	$m_{\text{PCA}} = 30$		$m_{\text{PCA}} = 56$		$m_{\text{PCA}} = 120$	
	Energy	% Increased Acc. SRC / $\text{SRC}_{\text{pruned}}$	Energy	% Increased Acc. SRC / $\text{SRC}_{\text{pruned}}$	Energy	% Increased Acc. SRC / $\text{SRC}_{\text{pruned}}$
AR-1	0.4527	3.90/3.86	0.5322	1.87/1.91	0.6522	0.80/0.60
AR-2	0.4137	3.83/2.36	0.4884	1.31/0.63	0.5988	0.62/0.53
Extended Yale B	0.3954	2.46/2.45	0.4803	1.59/1.59	0.6055	0.77/0.74
ORL	0.5385	1.34/0.05	0.6581	1.26/-0.04	0.8487	1.73/0.03

5.4.8 Summary

The experimental results on face databases show that LPCA-SRC can achieve higher accuracy than SRC in cases of low sampling and/or nonlinear class manifolds and small PCA feature dimension. We showed that LPCA-SRC had a significant advantage (in terms of mean accuracy) over SRC and the other algorithms for the small class sizes and nonlinear class manifolds of the AR database when the feature dimension was low. We also showed that LPCA-SRC could improve classification on Extended Yale B and ORL through its use of tangent vectors to provide a local approximation of the test sample and its discriminating pruning parameter, respectively.

The runtime of LPCA-SRC was sometimes much longer than that of SRC, although this was less often seen for small feature dimensions, at which LPCA-SRC tended to excel. The size of the dictionary in LPCA-SRC was observed to be a good predictor of the relationship between the runtimes of LPCA-SRC and SRC, and this could easily be computed (given estimates of the parameters n and d) before deciding between the two methods.

To validate our claim that the tangent vectors in LPCA-SRC can contain information lost in stringent PCA dimension reduction, we provided examples from the AR database. We also compared the energy retained in PCA dimension reduction with the increase in accuracy in LPCA-SRC over SRC and saw that there was a correlation.

6 Further Discussion and Future Work

This paper presented a modification of SRC called *local principal component analysis SRC*, or "LPCA-SRC." Through the use of tangent vectors, LPCA-SRC is designed to increase the sampling density of training sets and thus improve class discrimination on databases with sparsely sampled and/or nonlinear class manifolds. The LPCA-SRC algorithm computes basis vectors of approximate tangent hyperplanes at the training samples in each class and replaces the dictionary of training samples in SRC with a local dictionary (that is constructed based on each test sample) computed from shifted and scaled versions of these vectors and their corresponding training samples. Using a synthetic database and three face databases, we showed that LPCA-SRC can regularly achieve higher accuracy than SRC in cases of sparsely sampled and/or nonlinear class manifolds, low noise, and relatively small PCA feature dimension.

To address the issue of parameter setting, we recommended a consecutive parameter cross-validation procedure and gave detailed guidelines for its use. We also briefly discussed alternative methods for determining the class manifold dimension estimate d. It is important to note that in the case of small training sets, e.g., many face recognition problems, there are few options for the number-of-neighbors parameter n—and consequently for d by Eq. (4)—and so these values can easily be set using cross-validation, as in our experiments. When the training sets are very small (i.e., $N_l = 4$ or 5), one could simply set n to its maximum value, i.e., $n = N_{l_{\min}} - 2$, per Eq. (4). On the other hand, simply setting $d = 1$ may suffice, especially when minimizing algorithm runtime and/or storage requirements is paramount.

One disadvantage of this method is its high computational cost and storage requirements. SRC is already expensive due to its ℓ^1-minimization procedure; in LPCA-SRC, the computation of tangent vectors is added to the algorithm's workload. The size of the dictionary in LPCA-SRC may be larger or smaller than that of SRC, depending on the LPCA-SRC parameters n and d and the effect of the pruning parameter r. Thus LPCA-SRC can be slower or faster than SRC. Further, the storage required by LPCA-SRC is $(d + 1)$ times that of SRC, which may be prohibitive when d is large. As mentioned, simple computations based on the training set could render relative cost and storage estimates of using LPCA-SRC instead of SRC, and a smaller value of d than that found using cross-validation (e.g., $d = 1$) may be used successfully. These estimates can help the user decide between LPCA-SRC and SRC based on their desired balance between accuracy and computational efficiency.

Additionally, as we saw on the synthetic database, the usefulness of the tangent vectors in LPCA-SRC decreases as the noise level in the training data increases. This problem could potentially be alleviated by using the method proposed by Kaslovsky and Meyer [15] to estimate clean points on the manifolds from noisy samples and then computing the tangent vectors at these points. Note that the case of large training sample noise was the only case for which we saw LPCA-SRC not obtain higher accuracy than SRC. Thus LPCA-SRC should be preferred over SRC in low noise scenarios on either small-scale problems (e.g., the size of ORL) or when achieving a modest (e.g., 1–4%) boost in accuracy is worth potentially higher computational cost.

Open questions regarding LPCA-SRC include whether or not the aforementioned general trends hold for different methods of dimension reduction besides PCA. Additionally, one could compare the performance of the "group" or "per-class" methods of the above representation-based algorithms, in which test samples are approximated using class-specific dictionaries (similarly to as in TDC1). Lastly, one could gain insight into the role of ℓ^1-minimization in SRC by comparing LPCA-SRC and $\text{SRC}_{\text{pruned}}$ to versions of these algorithms that replace the ℓ^1-norm with the ℓ^2-norm, analogous to the work of Zhang et al. in their *collaborative representation-based representation* model [39]. This is part of our ongoing work, which we hope to report at a later date.

Acknowledgements C. Weaver's research on this project was conducted with government support under contract FA9550-11-C-0028 and awarded by DoD, Air Force Office of Scientific Research, National Defense Science and Engineering Graduate (NDSEG) Fellowship, 32 CFR 168a. She was also supported by National Science Foundation VIGRE DMS-0636297 and NSF DMS-1418779. N. Saito was partially supported by ONR grants N00014-12-1-0177 and N00014-16-1-2255, as well as NSF DMS-1418779.

A Tests of Statistical Significance

This appendix contains the detailed results for the tests of statistical significance between the most competitive classification algorithms on the experiments presented in Sects. 5.3.2, 5.4.5, and 5.4.6.

Table 9 Tests for statistical significance on the synthetic database for varying N_0: p-values in rAnova, Mauchly, Greenhouse-Geisser, and Huynh-Feldt tests (LPCA-SRC, SRC, and SRC_{pruned}), and 5% confidence interval of the improvement of LPCA-SRC over SRC

N_0	rANOVA	Mauchly	Greenhouse-Geisser	Huynh-Feldt	5% Confidence (LPCA-SRC > SRC)
5	4.1×10^{-5}	0.3598	4.7×10^{-5}	4.1×10^{-5}	[0.0266, 0.0804]
10	2.0×10^{-9}	3.9×10^{-9}	1.2×10^{-7}	1.0×10^{-7}	[0.0276, 0.0604]
15	0.1488	2.3×10^{-10}	0.1610	0.1606	[−0.0025, 0.0265]
20	5.8×10^{-10}	1.1×10^{-8}	3.9×10^{-8}	3.3×10^{-8}	[0.0244, 0.0526]
25	1.7×10^{-8}	1.1×10^{-18}	3.3×10^{-6}	3.1×10^{-6}	[0.0166, 0.0448]
30	2.2×10^{-16}	1.1×10^{-13}	9.6×10^{-14}	7.5×10^{-14}	[0.0349, 0.0611]
35	2.2×10^{-16}	7.0×10^{-10}	4.8×10^{-13}	3.6×10^{-13}	[0.0309, 0.0540]
40	1.1×10^{-6}	5.2×10^{-23}	7.1×10^{-5}	6.8×10^{-5}	[0.0113, 0.0361]
45	2.2×10^{-16}	8.6×10^{-8}	2.2×10^{-16}	6.0×10^{-18}	[0.0248, 0.0395]
50	2.2×10^{-16}	1.9×10^{-17}	2.2×10^{-16}	2.0×10^{-31}	[0.0331, 0.0439]
55	2.2×10^{-16}	1.9×10^{-13}	2.2×10^{-16}	5.0×10^{-37}	[0.0330, 0.0423]
60	2.2×10^{-16}	8.4×10^{-5}	2.2×10^{-16}	3.4×10^{-33}	[0.0258, 0.0356]
65	2.2×10^{-16}	3.2×10^{-10}	2.2×10^{-16}	1.0×10^{-41}	[0.0285, 0.0356]
70	2.2×10^{-16}	3.6×10^{-20}	2.2×10^{-16}	6.1×10^{-38}	[0.0291, 0.0366]
75	2.2×10^{-16}	7.8×10^{-23}	2.2×10^{-16}	1.7×10^{-31}	[0.0264, 0.0343]

A.1 Tests of Statistical Significance for Experiments on the Synthetic Database

Recall that LPCA-SRC, SRC, and SRC_{pruned} were the most competitive algorithms on the synthetic database experiments presented in Sect. 5.3.2. As evidence that LPCA-SRC outperformed SRC in a statistically-significant manner, we performed a Repeated Measures ANOVA test on all three methods as well as a t-test between the results for LPCA-SRC and SRC. The corresponding p-values and confidence intervals are contained in Tables 9 and 10. The columns of these tables are as follows: The value of N_0 (in the case of varying class size) or η (in the case of varying noise level) in the experiment, the p-value for Univariate Type III Repeated-Measures ANOVA Assuming Sphericity, the p-value for Mauchly Tests for Sphericity, the p-values for Greenhouse-Geisser and Huynh-Feldt Corrections for Departure from Sphericity, and the 5% confidence interval for a one-sided t-test of the improvement of LPCA-SRC over SRC. These tests were performed in R with the functions `Anova` (from the car package) and `t.test`.

For all but the Mauchly test, a small p-value indicates that we should reject the null hypothesis, which states that the algorithms have the same average accuracy. For the Mauchly test, a large p-value indicates that the data obeys the sphericity assumption; otherwise, the Greenhouse-Geisser or Huynh-Feldt corrections should

Table 10 Tests for statistical significance on the synthetic database for varying η: p-values in rAnova, Mauchly, Greenhouse-Geisser, and Huynh-Feldt tests (LPCA-SRC, SRC, and SRC_{pruned}), and 5% confidence interval of the improvement of LPCA-SRC over SRC

η	rANOVA	Mauchly	Greenhouse-Geisser	Huynh-Feldt	Confidence (LPCA-SRC > SRC)
0.0001	2.2×10^{-16}	8.1×10^{-21}	6.4×10^{-14}	5.4×10^{-14}	[0.0356, 0.0596]
0.001	2.3×10^{-6}	9.5×10^{-10}	2.9×10^{-5}	2.7×10^{-5}	[0.0135, 0.0428]
0.005	1.4×10^{-9}	5.1×10^{-7}	4.6×10^{-8}	3.8×10^{-8}	[0.0232, 0.0529]
0.01	0.0938	1.3×10^{-21}	0.1180	0.1177	[−0.0073, 0.0271]
0.015	0.9044	9.6×10^{-13}	0.8325	0.8345	[−0.0147, 0.0110]
0.02	0.0027	2.8×10^{-19}	0.0098	0.0096	[−0.0334, −0.0064]
0.03	5.3×10^{-5}	9.0×10^{-17}	0.0006	0.0006	[−0.0388, −0.0109]
0.05	0.0004	9.0×10^{-9}	0.0014	0.0013	[−0.0254, −0.0051]

be used. The confidence intervals can be interpreted as follows: Were we to repeat this experiment, we would expect LPCA-SRC to outperform SRC (with the exception of $N_0 = 15$ and $\eta \geq 0.01$) with the difference in mean accuracies falling within this confidence interval 95 times out of 100.

A.2 Tests of Statistical Significance for Experiments on the Face Databases

To test for statistical significance in the differences between algorithm accuracy on the AR, Extended Yale B, and ORL face databases, we performed Repeated Measures ANOVA tests on LPCA-SRC, SRC, SRC_{pruned}, and LSDL-SRC as well as two t-tests on each database, one between LPCA-SRC and SRC and the other between LPCA-SRC and LSDL-SRC. The related p-values and confidence intervals are contained in Tables 11 and 12. The columns in these tables are as follows: The name of the database, the PCA dimension m_{PCA}, the p-value for Univariate Type III Repeated-Measures ANOVA Assuming Sphericity, the p-value for Mauchly Tests for Sphericity, the p-values for Greenhouse-Geisser and Huynh-Feldt Corrections for Departure from Sphericity, the 5% confidence interval for a one-sided t-test between LPCA-SRC and SRC (LPCA-SRC > SRC), and the 5% confidence interval for a one-sided t-test between LPCA-SRC and LSDL-SRC (LPCA-SRC > LSDL-SRC). For all but the Mauchly test, a small p-value indicates that we should reject the null hypothesis, which states that the algorithms have the same average accuracy. For the Mauchly test, a large p-value indicates that the data obeys the sphericity assumption; otherwise, the Greenhouse-Geisser or Huynh-Feldt corrections should be used. These tests were performed in `R` with the functions `Anova` (from the car package) and `t.test`.

Table 11 Tests for statistical significance on the AR Face Database: *p*-values in rAnova, Mauchly, Greenhouse-Geisser, and Huynh-Feldt tests (LPCA-SRC, SRC, SRC_{pruned}, and LSDL-SRC), and 5% confidence intervals of the improvement of LPCA-SRC over SRC ($t\text{-test}_1$) and over LSDL-SRC ($t\text{-test}_2$)

Database	m_{PCA}	rANOVA	Mauchly	G-G	H-F	$t\text{-test}_1$	$t\text{-test}_2$
AR-1	30	8.5×10^{-12}	0.9182	9.1×10^{-11}	8.5×10^{-12}	[0.0310, 0.0499]	[0.0389, 0.0567]
AR-1	56	8.0×10^{-7}	0.2124	0.0001	1.8×10^{-5}	[0.0145, 0.0226]	[0.0076, 0.0165]
AR-1	120	3.7×10^{-7}	0.0470	7.3×10^{-5}	1.3×10^{-5}	[0.0039, 0.0117]	[−0.0087, −0.0018]
AR-2	30	1.3×10^{-10}	0.0083	5.0×10^{-7}	2.2×10^{-8}	[0.0243, 0.0497]	[0.0544, 0.0952]
AR-2	56	1.8×10^{-5}	0.0651	0.0009	0.0003	[0.0047, 0.0222]	[0.0139, 0.0286]
AR-2	120	0.0115	0.4566	0.0225	0.0115	[0.0009, 0.0095]	[−0.0046, 0.0055]

Table 12 Tests for statistical significance on the Extended Yale B and ORL Face Database: p-values in rAnova, Mauchly, Greenhouse-Geisser, and Huynh-Feldt tests (LPCA-SRC, SRC, $\text{SRC}_{\text{pruned}}$, and LSDL-SRC), and 5% confidence intervals of the improvement of LPCA-SRC over SRC (t-test_1) and over LSDL-SRC (t-test_2)

Database	m_{PCA}	rANOVA	Mauchly	G-G	H-F	t-test_1	t-test_2
Yale B	30	2.2×10^{-16}	0.0014	1.0×10^{-11}	4.8×10^{-13}	[0.0182,0.0300]	[0.1422, 0.1687]
Yale B	56	2.2×10^{-16}	4.2×10^{-5}	4.6×10^{-5}	8.4×10^{-12}	[0.0088,0.0233]	[0.0671, 0.0822]
Yale B	120	9.1×10^{-14}	0.1026	1.5×10^{-9}	6.3×10^{-12}	[0.0052,0.0102]	[0.0190, 0.0192]
ORL	30	2.2×10^{-16}	0.0033	3.9×10^{-12}	1.1×10^{-12}	[0.0085,0.0168]	[0.0092, 0.0192]
ORL	56	4.3×10^{-14}	0.0012	5.7×10^{-12}	1.6×10^{-12}	[0.0121, 0.0201]	[0.0045, 0.0120]
ORL	120	2.2×10^{-16}	3.2×10^{-5}	2.2×10^{-16}	7.8×10^{-35}	[0.0142, 0.0213]	[0.0308, 0.0400]

References

1. M.S. Asif, J. Romberg, L1 Homotopy: A MATLAB toolbox for homotopy algorithms in L1-norm minimization problems. http://users.ece.gatech.edu/~sasif/homotopy/, 2009–2013. Accessed 31 March 2015
2. AT&T Laboratories Cambridge. The Database of Faces. http://www.cl.cam.ac.uk/research/dtg/attarchive/facedatabase.html, 1992–1994. Accessed 26 March 2016
3. J.L. Bentley, Multidimensional binary search trees used for associative searching. Commun. ACM **18**(9), 509–517 (1975). ISSN: 0001-0782, https://doi.org/10.1145/361002.361007
4. E. Candes, Mathematics of sparsity (and a few other things), in *Proceedings of the International Congress of Mathematicians*, Seoul, South Korea, 2014
5. C. Ceruti et al., DANCo: an intrinsic dimensionality estimator exploiting angle and norm concentration. Pattern Recognit. **47**(8), 2569–2581 (2014). ISSN: 0031-3203, https://doi.org/10.1016/j.patcog.2014.02.013
6. H. Cevikalp et al., Two-dimensional subspace classifiers for face recognition. Neurocomputing **72**(4), 1111–1120 (2009). ISSN: 0925-2312, https://doi.org/10.1016/j.neucom.2008.02.015
7. J.-M. Chang, M. Kirby, Face recognition under varying viewing conditions with subspace distance, in *International Conference on Artificial Intelligence and Pattern Recognition (AIPR-09)*, 2009, pp. 16–23. ISBN: 978-1-60651-007-0, https://doi.org/10.1109/ICCV.2005.167
8. D.L. Donoho. For most large underdetermined systems of linear equations the minimal l1-norm solution is also the sparsest solution. Commun. Pure Appl. Math. **59**(6), 797–829 (2006). ISSN: 0010-3640, https://doi.org/10.1002/cpa.20132
9. D.L. Donoho, Y. Tsaig, Fast solution of l1-norm minimization problems when the solution may be sparse. IEEE Trans. Inf. Theory **54**(11), 4789–4812 (2008). ISSN: 0018-9448, https://doi.org/10.1109/TIT.2008.929958
10. E. Elhamifar, R. Vidal, Sparse subspace clustering, in *2009 IEEE Conference on Computer Vision and Pattern Recognition*, June 2009, pp. 2790–2797, https://doi.org/10.1109/CVPR.2009.5206547
11. A.S. Georghiades, P.N. Belhumeur, D.J. Kriegman, From few to many: illumination cone models for face recognition under variable lighting and pose. IEEE Trans. Pattern Anal. Mach. Intell. **23**(6), 643–660 (2001). ISSN: 0162-8828, https://doi.org/10.1109/34.927464
12. X. He et al., Face recognition using Laplacianfaces. IEEE Trans. Pattern Anal. Mach. Intell. **27**(3), 328–340 (2005). ISSN: 0162-8828, https://doi.org/10.1109/TPAMI.2005.55
13. J. Ho, Y. Xie, B.C. Vemuri, On a nonlinear generalization of sparse coding and dictionary learning, in *ICML* (3), vol. 28. *JMLR Proceedings. JMLR.org*, 2013, pp. 1480–1488, http://dblp.uni-trier.de/db/conf/icml/icml2013.html#HoXV13
14. N. Kambhatla, T.K. Leen, Dimension reduction by local principal component analysis. Neural Comput. **9**, 1493–1516 (1997)
15. D.N. Kaslovsky, F.G. Meyer, Non-asymptotic analysis of tangent space perturbation. Inf. Inference **3**(2), 134–187 (2014). ISSN: 2049-8764, https://doi.org/10.1093/imaiai/iau004
16. Y. LeCun et al., Gradient-based learning applied to document recognition. Proc. IEEE **86**(11), 2278–2324 (1998). https://doi.org/10.1109/5.726791
17. K.-C. Lee, J. Ho, D. Kriegman, Acquiring linear subspaces for face recognition under variable lighting. IEEE Trans. Pattern Anal. Mach. Intell. **27**(5), 684–698 (2005). ISSN: 0162-8828, https://doi.org/10.1109/TPAMI.2005.92
18. Z. Li et al., Clustering-guided sparse structural learning for unsupervised feature selection. IEEE Trans. Knowl. Data Eng. **26**(9), 2138–2150 (2014). ISSN: 1041-4347, https://doi.org/10.1109/TKDE.2013.65
19. Z. Li et al., Robust structured subspace learning for data representation. IEEE Trans. Pattern Anal. Mach. Intell. **37**(10), 2085–2098 (2015). ISSN: 0162- 8828, https://doi.org/10.1109/TPAMI.2015.2400461
20. A.V. Little, M. Maggioni, L. Rosasco, Multiscale geometric methods for data sets I: multiscale SVD, noise and curvature. Appl. Comput. Harmon. Anal. (2016). ISSN: 1063-5203, https://doi.org/10.1016/j.acha.2015.09.009

21. A.M Martinez, R. Benavente, The AR Face Database. Technical report 24. Computer Vision Center, June 1998, http://www.cat.uab.cat/Public/Publications/1998/MaB1998
22. C. Merkwirth, U. Parlitz, W. Lauterborn, Fast nearest neighbor searching for nonlinear signal processing. Phys. Rev. E **62**, 2089–2097 (2000). https://doi.org/10.1103/PhysRevE.62.2089
23. C. Merkwirth et al., TSTOOL Homepage, 2009, http://www.physik3.gwdg.de/tstool/index.html. Accessed 6 Feb 15
24. R. Patel, N. Rathod, A. Shah, Comparative analysis of face recognition approaches: a survey. Int. J. Comput. Appl. **57**(17), 50–69 (2012)
25. L. Qiao, S. Chen, X. Tan, Sparsity preserving projections with applications to face recognition. Pattern Recognit. **43**(1), 331–341 (2010). https://doi.org/10.1016/j.patcog.2009.05.005
26. S.T. Roweis, L.K. Saul, Nonlinear dimensionality reduction by locally linear embedding. Science **290**, 2323–2326 (2000). https://doi.org/10.1126/science.290.5500.2323
27. P.Y. Simard et al., Transformation invariance in pattern recognition - tangent distance and tangent propagation, in *Neural Networks: Tricks of the Trade*, 2nd edn, pp. 235–269. ISBN: 978-3-642-35289-8, https://doi.org/10.1007/978-3-642-35289-8_7
28. A. Singer, H.-T. Wu, Vector diffusion maps and the connection Laplacian. Commun. Pure Appl. Math. **65**(8), 1067–1144 (2012). ISSN: 0010-3640, https://doi.org/10.1002/cpa.21395
29. X. Tan et al., Face recognition from a single image per person: a survey. Pattern Recognit. **39**(9), 1725–1745 (2006). ISSN: 0031-3203, https://doi.org/10.1016/j.patcog.2006.03.013
30. J. Waqas, Z. Yi, L. Zhang, Collaborative neighbor representation based classification using l2-minimization approach. Pattern Recognit. Lett. **34**(2), 201–208 (2013). ISSN: 0167-8655, https://doi.org/10.1016/j.patrec.2012.09.024
31. C.-P. Wei et al., Locality-sensitive dictionary learning for sparse representation based classification. Pattern Recognit. **46**(5), 1277–1287 (2013). ISSN: 0031-3203, https://doi.org/10.1016/j.patcog.2012.11.014
32. J. Wright et al., Robust face recognition via sparse representation. IEEE Trans. Pattern Anal. Mach. Intell. **31**(2), 210–227 (2009). https://doi.org/10.1109/TPAMI.2008.79
33. Y. Xu et al., Integrate the original face image and its mirror image for face recognition. Neurocomputing **131**, 191–199 (2014). ISSN: 0925-2312, https://doi.org/10.1016/j.neucom.2013.10.025
34. A.Y. Yang et al., Fast 'ℓ1-minimization algorithms and an application in robust face recognition: A review, in *2010 17th IEEE International Conference on Image Processing*, Sept 2010, pp. 1849–1852, https://doi.org/10.1109/ICIP.2010.5651522
35. J. Yang, K. Zhu, N. Zhong, Local tangent distances for classification problems, in *2012 IEEE/WIC/ACM International Conferences on Web Intelligence and Intelligent Agent Technology (WI-IAT)*, vol. 1, Dec 2012, pp. 396–401, https://doi.org/10.1109/WI-IAT.2012.46
36. J. Yang, J. Wang, T. Huang, Learning the sparse representation for classification, in *2011 IEEE International Conference on Multimedia and Expo (ICME)*, July 2011, pp. 1–6, https://doi.org/10.1109/ICME.2011.6012083
37. J. Yin et al., Kernel sparse representation based classification, in Neurocomputing **77**(1), 120–128 (2012). ISSN: 0925-2312, https://doi.org/10.1016/j.neucom.2011.08.018
38. H. Zhang et al., Sample pair based sparse representation classification for face recognition. Expert Syst. Appl. **45**, 352–358 (2016). ISSN: 0957-4174, https://doi.org/10.1016/j.eswa.2015.09.058
39. L. Zhang, M. Yang, X. Feng, Sparse representation or collaborative representation: which helps face recognition?, in *Proceedings of the 2011 International Conference on Computer Vision* (IEEE Computer Society, 2011), pp. 471–478. ISBN: 978-1-4577-1101-5, https://doi.org/10.1109/ICCV.2011.6126277

Robust Constrained Concept Factorization

Wei Yan and Bob Zhang

Abstract Accurately representing data is a fundamental problem in many pattern recognition and computational intelligence applications. In this chapter, a robust constrained concept factorization (RCCF) method is proposed. RCCF allows the extraction of important information, while simultaneously utilizing prior information when it is available, and is noise invariant. To guarantee data samples share the identical cluster and obtain similar representation in the new laten space, the proposed method uses a constraint matrix that is embodied into the rudimentary concept factorization model. The $L_{2,1}$-norm is used for both the reconstruction function and the regularization, which allows the proposed model to be insensitive to outliers. Furthermore, the $L_{2,1}$-norm regularization assists in the selection of useful information with joint sparsity. An elegant and efficient iterative updating scheme is also introduced with convergence and correctness analysis. Experimental results on commonly used databases in pattern recognition and computational intelligence demonstrate the effectiveness of RCCF.

Keywords Concept factorization · Dimensionality reduction · Clustering

1 Introduction

Obtaining a suitable representation is a fundamental problem for many research areas. For example: machine learning [1], data mining [2, 3], signal processing [4–6], and in particular pattern recognition [7–9], and computational intelligence [10, 11]. Optimal data representation can boost the performance of a learning task by revealing the underlying structure with-in a high-dimensional space. Recently,

W. Yan · B. Zhang (✉)
Department of Computer and Information Science,
University of Macau, Macau, China
e-mail: bobzhang@umac.mo

W. Yan
e-mail: yb67410@umac.mo

W. Pedrycz and S.-M. Chen (eds.), *Computational Intelligence for Pattern Recognition*, Studies in Computational Intelligence 777,
https://doi.org/10.1007/978-3-319-89629-8_7

matrix factorization based methods, including Singular Value Decomposition (SVD) [12], Principal Component Analysis (PCA) [13], Vector Quantization (VQ) [14], Nonnegative Matrix Factorization (NMF) [15–18] and Concept Factorization (CF) [19–22], have been receiving considerable attention as useful techniques for learning meaningful representation.

Generally, the main goal of these methods is to represent the given matrix as a product of two or more matrices. Among them, NMF is superior to PCA, SVD, and VQ for providing meaningful factorization results. Moreover, NMF yields parts-based and sparse representation because the nonnegative constraints allow only additive combinations. Regards to the parts-based representation, there are physiological evidences [23, 24]. However, NMF performs only in original data space. It is an issue about how to successfully apply NMF in reproducing kernel Hilbert space (RKHS), e.g., the transformed data space [20]. Recently, Concept Factorization (CF), an important variation of NMF, which uses linear combination of input data to represent the bases, has been effectively employed in processing real data, such as text and image, due to the fact that CF inherits all the strengths from NMF. Besides this, CF can be employed effectively in the transformed space. When using the CF method in data clustering, each sample is reconstructed as a linear combination of the cluster centers, and each cluster center is expressed as a linear combination of the samples. Here, the task of data clustering can be regarded as finding two sets of coefficients. To further improve the clustering performance of CF, Locally Consistent Concept Factorization (LCCF) [20] preserves the intrinsic structure information of the data set by incorporating the manifold structure into the CF model.

Despite its impressive performances, there are three major drawbacks for basic CF: (1) It is prone to outliers since a few outliers or noisy features with large errors will play a dominative role in the least square error function. Indeed, in many applications, data are additionally corrupted and thus data always contains noisy features or outliers. A potential robust version of CF is needed to deal with these issues. (2) Basic CF does not always result in sparse representation since there is no constraints to manage the sparseness explicitly. That means the representation in the low-dimensional space may still contain redundant and useless information. Generally, adding sparsity regularization is one practical method to control the degree of sparseness in factorization results, but it was designed only for NMF [25, 26]. (3) CF obtains data representation in an unsupervised way. It may not effectively distinguish the constrained data from the unconstrained data. Especially when the prior information is collected and CF does not completely use this information. To bridge this gap, a constrained algorithm, named constrained concept factorization (CCF) [27] is proposed utilizing prior information as a constraint matrix.

However, there is no such a framework that addresses all these drawbacks simultaneously. In this chapter, we propose a robust constrained concept factorization (RCCF) method, which not only makes good use of the available label information, but also addresses noise and learns meaningful information at the same time. Specifically, we utilized the mixed norm $L_{2,1}$-norm instead of the F-norm that is used in basic CF as our loss function, thus improving the robustness of the model such that this new model can effectively deal with outliers and can be employed in pattern recognition and computational intelligence. Then, a constraint matrix, which contains label information, is embedded into the original CF model to guarantee data belong to the same cluster obtain the identical representation in the new representation space. Hence, the learned representation achieves better distinguishing abilities. In addition, the $L_{2,1}$-norm regularization is added in RCCF to obtain sparse results that help select the most relevant information. For optimizing the new model, we derive efficient updating rules that are iterative. At the same time, we analyse the correctness and convergence of the updating rules. Experimental results on three different data sets have shown the effectiveness of RCCF.

The remainder of this chapter is organized as follows. Section 2 proposes the RCCF framework, followed with its updating rules. Section 3 presents the experimental results on three data sets. Finally, we summarize our work.

2 Robust Constrained Concept Factorization (RCCF)

2.1 NMF and CF

Given a matrix $\mathbf{V} = [v_1, v_2, \ldots, v_N] \in \mathbb{R}_+{}^{M\times N}$, N denotes the number of data points and M is the length of the vector. For a dimensionality number K, NMF tries to seek two nonnegative data matrices $\mathbf{W} \in \mathbb{R}_+{}^{M\times K}$ and $\mathbf{H} \in \mathbb{R}_+{}^{K\times N}$ whose product gives an approximation to the input data matrix. The objective function of NMF is:

$$\mathcal{O} = \|\mathbf{V} - \mathbf{W}\mathbf{H}\|_F^2. \tag{1}$$

Since (1) is a nonconvex minimization problem, it is unrealistic to get the optimal solution. However it is convex in W only or H only. Based on this analysis, Lee [15] proposed the following updating rules:

$$\mathbf{W}_{ik} \leftarrow \mathbf{W}_{ik}\frac{(\mathbf{V}\mathbf{H}^{\mathbf{T}})_{ik}}{(\mathbf{W}\mathbf{H}\mathbf{H}^{\mathbf{T}})_{ik}} \tag{2}$$

$$\mathbf{H}_{kj} \leftarrow \mathbf{H}_{kj}\frac{(\mathbf{W}^{\mathbf{T}}\mathbf{V})_{kj}}{(\mathbf{W}^{T}\mathbf{W}\mathbf{H})_{kj}}. \tag{3}$$

In regards to the above solutions, if $\mathbf{W}$ and $\mathbf{H}$ are the solution to (1), $\mathbf{WQ}$ and $\mathbf{HQ}^{-1}$ can also be a solution for any matrix $\mathbf{Q}$, which is positive and diagonal. To ensure it unique, we normalize the solution. In practice, this can be obtained by:

$$\mathbf{W}_{ik} \leftarrow \frac{\mathbf{W}_{ik}}{\sqrt{\sum_i W_{ik}^2}} \tag{4}$$

$$\mathbf{H}_{kj} \leftarrow \mathbf{H}_{kj}\sqrt{\sum_i W_{ik}^2} \tag{5}$$

In [15], Lee gives the proof that the aforementioned updating rules could obtain a local solution of (1).

Xu and Gong [19] modeled the document clustering problem by using two data representations. In the first one, the intrinsic semantics (e.g. clusters) can be represented by related document samples that belong to similar semantics. That is, the entire samples can be used to construct the cluster, and this combination can is linearly. Let V_i denotes the term-frequency vector of sample i, where $i = 1, \ldots, n$, m is the dimensionality and R_c is the centroid of cluster c, where $c = 1, \ldots, k$. The first representation can be defined as:

$$R_c = \sum_i w_{ic} V_i \tag{6}$$

where w_{ic} is nonnegative coefficient that represents the coefficient of data point i relating to cluster c. In the second one, all the clusters can be used to reconstruct the samples. The corresponding weight denotes the coefficient of overlap between the related sample and the cluster. The above two representations can be formulated as:

$$V_i = \sum_c h_{ic} R_c \tag{7}$$

where h_{ic} is the coefficient value that gives the coefficient of overlap between the related sample V_i and the concept cluster R_c. We construct the document matrix $\mathbf{V} = [V_1, V_2, \ldots, V_n] \in \mathbf{R}_+{}^{m\times n}$ with the feature vector of sample i as the ith column. From (6) and (7) we have

$$\mathbf{V} \approx \mathbf{VWH} \tag{8}$$

where $\mathbf{W} = [w_{jk}] \in \mathbf{R}_+{}^{n\times k}$ and $\mathbf{H} = [h_{jk}] \in \mathbf{R}_+{}^{k\times n}$. From Eq. (8), we observe that it can be considered as a factorization process of input sample matrix $\mathbf{X}$ into $\mathbf{X}$, $\mathbf{W}$, and $\mathbf{H}$. With the factorization results, we can find the cluster which are accomplished by constructed by $\mathbf{XW}$. The cluster coefficient of each sample is obtained by finding the $\mathbf{H}$. Thus, we term this process concept factorization (CF). As $k \ll m$ and $k \ll n$, concept factorization leads to low-dimensional representation of the input matrix. This means the object function is defined as:

$$\mathcal{O} = \|\mathbf{V} - \mathbf{VWH}\|_F^2 \tag{9}$$

where $\|\cdot\|_F^2$ is the Frobenius norm of a matrix. Using the formulation (9), the data clustering problem can be solved by finding $\mathbf{W}$ and $\mathbf{H}$ that minimizes the $\mathcal{O}$.

To minimize (9), the multiplicative updating rules are introduced as [19]:

$$w_{nk} \leftarrow w_{nk} \frac{(\mathbf{KH}^{\mathbf{T}})_{nk}}{(\mathbf{KWHH}^{T})_{nk}} \tag{10}$$

$$h_{kn} \leftarrow h_{kn} \frac{(\mathbf{W}^{\mathbf{T}}\mathbf{K})_{kn}}{(\mathbf{W}^{T}\mathbf{KWH})_{kn}}. \tag{11}$$

where $\mathbf{K} = \mathbf{V}^{T}\mathbf{V}$. It is natural to leverage kernel methods on CF. Therefore, CF can make use of kernel methods to improve its performance in real applications. More information can be found in [19].

Concept factorization is an effective tool of data clustering, which is a fundamental topic in data mining. Data mining is about extracting interesting information from raw data. Data clustering aims to efficiently separate a given data set into clusters, which is a kind of key information. Among various clustering methods, concept factorization is widely used since it can provide meaningful clustering results. From the definition of CF, the cluster is constructed by using the input samples. This combination is linearly. The cluster construction as well as the new data representation can be addressed by CF.

2.2 RCCF Model

According to recent semi-supervised algorithms [27–29] a few labeled samples could be used along with the unlabeled samples to improve learning accuracy of unlabeled data. Inspired by previous research CCF [27], we assume that the first l data samples are given label information with c clusters. Then we construct an constraint matrix $\mathbf{C}$, in which $c_{i,j} = 1$ if c_i belongs to the jth class; $C_{i,j} = 0$ otherwise. For example, given n data points, v_1, v_2 and v_3 come from class I, v_4 and v_5 belong to class II, v_6 is labeled with class III. Base on this illustration, the label indicator matrix C can be formulated as follows:

$$\mathbf{C} = \begin{pmatrix} 1 & 0 & 0 \\ 1 & 0 & 0 \\ 1 & 0 & 0 \\ 0 & 1 & 0 \\ 0 & 1 & 0 \\ 0 & 0 & 1 \end{pmatrix}. \tag{12}$$

Based on the indicator matrix C, a label constraint matrix $\mathbf{A}$ is defined as follows:

$$\mathbf{A} = \begin{pmatrix} \mathbf{C}_{l\times c} & 0 \\ 0 & \mathbf{I}_{N-l} \end{pmatrix}, \tag{13}$$

where $\mathbf{I}_{N-l}$ is an identity matrix. The obtained $\mathbf{H}$ of input data points in the new representation space is formulated as $\mathbf{H} = \mathbf{Z}\mathbf{A}^T$. That means if samples v_i and v_j come from the identical category, the ith row and the jth row of $\mathbf{A}$ should be identical, that is $h_i = h_j$, which makes sure that document point with the identical class could obtain the identical low-dimensional representation. Thus, (9) could be reformulated:

$$\min_{\mathbf{W}\geq 0, \mathbf{Z}\geq 0} \|\mathbf{V} - \mathbf{V}\mathbf{W}\mathbf{Z}\mathbf{A}^T\|_F^2 \tag{14}$$

However, F-norm is sensitive to outliers and could be unstable because the error for each sample in the objective function is expressed as squared. The objective function could be dominated by the large errors. To address this drawback, we utilize the mixed norm $L_{2,1}$-norm on the loss function to effectively remove outliers and noise. According to [30], the definition of $L_{2,1}$-norm is:

$$\|\mathbf{U}\|_{2,1} = \sum_{i=1}^{M} \sqrt{\sum_{j=1}^{N} \mathbf{U}_{ji}^2} = \sum_{i=1}^{M} \|u_i\|_2, \tag{15}$$

where u_i is the ith row of $\mathbf{U}$. We rewrite the error function:

$$\begin{aligned} \|\mathbf{V} - \mathbf{V}\mathbf{W}\mathbf{H}\|_{2,1} &= \sum_{i=1}^{N} \sqrt{\sum_{j=1}^{M} (\mathbf{V} - \mathbf{V}\mathbf{W}\mathbf{H})_{ji}^2} \\ &= \sum_{i=1}^{N} \|v_i - \mathbf{V}\mathbf{W}h_i\|. \end{aligned} \tag{16}$$

We can observe that the error for each sample in the new objective function (16) is not of the form x^2, so the large errors because of outliers do not impact the function in (16) dramatically. By employing the $L_{2,1}$-norm as measurement of the reconstruction error, the objective function can be reformulated as

$$\min_{\mathbf{W}\geq 0, \mathbf{Z}\geq 0} \|\mathbf{V} - \mathbf{V}\mathbf{W}\mathbf{Z}\mathbf{A}^T\|_{2,1}. \tag{17}$$

Furthermore, the real data usually contains meaningless features, i.e., not all the features are useful. Although basic CF can lead to sparse results that help extract meaningful features, it does not always result in such representation. Regarding this, we use the $L_{2,1}$-norm regularization term to control row sparsity on the new representation of data to extract informative features. Generally, the group sparsity imposing on the representation matrix $\mathbf{H}^T$ can be represented as follows,

$$\min_{\mathbf{H}\geq 0} \|\mathbf{H}^T\|_{2,1}. \tag{18}$$

As we make $\mathbf{H} = \mathbf{Z}\mathbf{A}^T$, our task is to get the minimum of matrix $\mathbf{A}\mathbf{Z}^T$. Since the constrained matrix $\mathbf{A}$ is given, the task in turn is to find matrix $\mathbf{Z}^T$.

By embedding the constrained matrix into basic CF, and imposing the $L_{2,1}$-norm on both the regularization and reconstruction function, a new model can be obtained as follows,

$$\min_{\mathbf{W}\geq 0,\mathbf{Z}\geq 0} \|\mathbf{V} - \mathbf{V}\mathbf{W}\mathbf{Z}\mathbf{A}^T\|_{2,1} + \alpha\|\mathbf{Z}^T\|_{2,1}, \tag{19}$$

where $\mathbf{V} \in \mathbb{R}_+{}^{M\times N}$, $\mathbf{W} \in \mathbb{R}_+{}^{N\times K}$, $\mathbf{Z} \in \mathbb{R}_+{}^{K\times(N-l+c)}$ and $\mathbf{A} \in \mathbb{R}_+{}^{N\times(N-l+c)}$. In this function, there is only one parameter, e.g., the parameter α. This item plays the role on controlling the sparse regularization.

2.3 Solutions of the RCCF Model

The solutions for the RCCF model via an iterative strategy is given as follows,

$$\mathbf{Z}_{ki} \leftarrow \mathbf{Z}_{ki}\frac{(\mathbf{W}^T\mathbf{V}^T\mathbf{V}\mathbf{D}_1\mathbf{A})_{ki}}{(\mathbf{W}^T\mathbf{V}^T\mathbf{V}\mathbf{W}\mathbf{Z}\mathbf{A}^T\mathbf{D}_1\mathbf{A} + \alpha\mathbf{D}_2\mathbf{Z})_{ki}}, \tag{20}$$

$$\mathbf{W}_{nk} \leftarrow \mathbf{W}_{nk}\frac{(\mathbf{V}^T\mathbf{V}\mathbf{D}_1\mathbf{A}\mathbf{Z}^T)_{nk}}{(\mathbf{V}^T\mathbf{V}\mathbf{W}\mathbf{Z}\mathbf{A}^T\mathbf{D}_1\mathbf{A}\mathbf{Z}^T)_{nk}}, \tag{21}$$

The entries of $\mathbf{D}_1$ and $\mathbf{D}_2$ are defined as:

$$(\mathbf{D}_1)_{ii} = \frac{1}{\|\mathbf{V}_i - \mathbf{V}\mathbf{W}(\mathbf{Z}\mathbf{A}^T)_i\|}, i = 1, 2\ldots, N. \tag{22}$$

$$(\mathbf{D}_2)_{ii} = \frac{1}{\|(\mathbf{Z}^T)_i\|}, i = 1, 2\ldots, K. \tag{23}$$

2.4 RCCF Model Convergence

In this subsection, we give the analysis of the convergence of proposed updating rules with following two Theorems.

Theorem 1 *Obtaining* $\mathbf{Z}$ *utilizing the rule of* (20) *with* $\mathbf{W}$ *being fixed, the objective function of* (19) *is non-increasing,*

$$\begin{aligned} &\|\mathbf{V} - \mathbf{V}\mathbf{W}\mathbf{Z}^{t+1}\mathbf{A}^T\|_{2,1} + \alpha\|(\mathbf{Z}^{t+1})^T\|_{2,1} \\ &-\|\mathbf{V} - \mathbf{V}\mathbf{W}\mathbf{Z}^{t}\mathbf{A}^T\|_{2,1} - \alpha\|(\mathbf{Z}^{t})^T\|_{2,1} \leq 0, \end{aligned} \tag{24}$$

where t *is the number of iteration.*

Theorem 2 *Obtaining* $\mathbf{W}$ *utilizing the solution proposed in* (21) *when* $\mathbf{Z}$ *is fixed, the objective function of* (19) *is nonincreasing,*

$$\|\mathbf{V}-\mathbf{V}\mathbf{W}^{t+1}\mathbf{Z}\mathbf{A}^T\|_{2,1}-\|\mathbf{V}-\mathbf{V}\mathbf{W}^{t}\mathbf{Z}\mathbf{A}^T\|_{2,1}\leq 0, \tag{25}$$

where t *represents the number of iteration.*

We use the following Lemma 1 to prove Theorem 1.

Lemma 1 *With the solution in* (20), *we have the following inequation:*

$$\begin{aligned}
&Tr((\mathbf{V}-\mathbf{V}\mathbf{W}\mathbf{Z}^{t+1}\mathbf{A}^T)\mathbf{D}_1(\mathbf{V}-\mathbf{V}\mathbf{W}\mathbf{Z}^{t+1}\mathbf{A}^T)^T)\\
&+\alpha Tr((\mathbf{Z}^{t+1})^T\mathbf{D}_2\mathbf{Z}^{t+1})\\
&\leq Tr((\mathbf{V}-\mathbf{V}\mathbf{W}\mathbf{Z}^{t}\mathbf{A}^T)\mathbf{D}_1(\mathbf{V}-\mathbf{V}\mathbf{W}\mathbf{Z}^{t}\mathbf{A}^T)^T)\\
&+\alpha Tr((\mathbf{Z}^{t})^T\mathbf{D}_2\mathbf{Z}^{t}).
\end{aligned} \tag{26}$$

Proof Following [31], we introduce an auxiliary function approach to help prove Lemma 1. Firstly, we have

$$\begin{aligned}
J(\mathbf{Z})=\;&Tr((\mathbf{V}-\mathbf{V}\mathbf{W}\mathbf{Z}\mathbf{A}^T)\mathbf{D}_1(\mathbf{V}-\mathbf{V}\mathbf{W}\mathbf{Z}\mathbf{A}^T)^T)\\
&+\alpha Tr(\mathbf{Z}^T\mathbf{D}_2\mathbf{Z}).
\end{aligned} \tag{27}$$

Next we re-express (26) as

$$J(\mathbf{Z}^{t+1})\leq J(\mathbf{Z}^{t}). \tag{28}$$

Base on (27), the following equation can be obtained

$$\begin{aligned}
J(\mathbf{Z})&=Tr(\mathbf{V}\mathbf{D}_1\mathbf{V}^T-2\mathbf{V}\mathbf{D}_1\mathbf{A}\mathbf{Z}^T\mathbf{W}^T\mathbf{V}^T)\\
&+Tr(\mathbf{V}\mathbf{W}\mathbf{Z}\mathbf{A}^T\mathbf{D}_1\mathbf{A}\mathbf{Z}^T\mathbf{W}^T\mathbf{V}^T)+\alpha Tr(\mathbf{Z}^T\mathbf{D}_2\mathbf{Z})\\
&\leq Tr(\mathbf{V}\mathbf{D}_1\mathbf{V}^T-2\mathbf{V}\mathbf{D}_1\mathbf{A}\mathbf{Z}^T\mathbf{W}^T\mathbf{V}^T)\\
&+\sum_{k=1}^{K}\sum_{i=1}^{(N-l+c)}\frac{(\mathbf{S}_1\mathbf{H}'\mathbf{B}_1)_{ki}(\mathbf{H}^2)_{ki}}{\mathbf{H}'_{ki}}\\
&+\sum_{k=1}^{K}\sum_{i=1}^{(N-l+c)}\frac{(\mathbf{S}_2\mathbf{H}'\mathbf{B}_2)_{ki}(\mathbf{H}^2)_{ki}}{\mathbf{H}'_{ki}}\\
&=Tr(\mathbf{V}\mathbf{D}_1\mathbf{V}^T-2\mathbf{V}\mathbf{D}_1\mathbf{A}\mathbf{Z}^T\mathbf{W}^T\mathbf{V}^T)\\
&+\sum_{k=1}^{K}\sum_{i=1}^{(N-l+c)}\frac{(\mathbf{W}^T\mathbf{V}^T\mathbf{V}\mathbf{W}\mathbf{Z}'\mathbf{A}^T\mathbf{D}_1\mathbf{A}+\alpha\mathbf{D}_2\mathbf{Z}')_{ki}(\mathbf{Z}^2)_{ki}}{\mathbf{Z}'_{ki}}\\
&=F(\mathbf{Z},\mathbf{Z}'),
\end{aligned} \tag{29}$$

where $\mathbf{S}_1 = \mathbf{W}^T\mathbf{V}^T\mathbf{V}\mathbf{W}$, $\mathbf{B}_1 = \mathbf{A}^T\mathbf{D}_1\mathbf{A}$, $\mathbf{H} = \mathbf{Z}$, $\mathbf{H}' = \mathbf{Z}'$, $\mathbf{B}_2 = \mathbf{I}$, and $\mathbf{S}_2 = \alpha\mathbf{D}_2$. The equality holds in case of $\mathbf{Z} = \mathbf{Z}'$. The auxiliary function of $J(\mathbf{Z})$ is $F(\mathbf{Z}, \mathbf{Z}')$.

Let

$$\mathbf{Z}^{t+1} = \arg\min_{\mathbf{Z}} F(\mathbf{Z}, \mathbf{Z}^t), \tag{30}$$

we can get

$$J(\mathbf{Z}^{t+1}) = F(\mathbf{Z}^{t+1}, \mathbf{Z}^{t+1}) \leq F(\mathbf{Z}^{t+1}, \mathbf{Z}^t) \leq J(\mathbf{Z}^t), \tag{31}$$

From (31), we can have the provement that $J(\mathbf{Z}^t)$ is non-increasing.

Let $f(\mathbf{Z}) = F(\mathbf{Z}, \mathbf{Z}')$, the gradient of $f(\mathbf{Z})$ is

$$\begin{aligned}\frac{\partial f(\mathbf{Z})}{\partial \mathbf{Z}_{ki}} &= -2(\mathbf{W}^T\mathbf{V}^T\mathbf{V}\mathbf{D}_1\mathbf{A})_{ki} \\ &+ 2\frac{(\mathbf{W}^T\mathbf{V}^T\mathbf{V}\mathbf{W}\mathbf{Z}'\mathbf{A}^T\mathbf{D}_1\mathbf{A} + \alpha\mathbf{D}_2\mathbf{Z}')_{ki}(\mathbf{Z})_{ki}}{\mathbf{Z}'_{ki}}.\end{aligned} \tag{32}$$

The Hessian matrix of $f(\mathbf{Z})$ is

$$\frac{\partial^2 f(\mathbf{Z})}{(\partial \mathbf{Z}_{ki})(\partial \mathbf{Z}_{lj})} = 2\frac{(\mathbf{W}^T\mathbf{V}^T\mathbf{V}\mathbf{W}\mathbf{Z}'\mathbf{A}^T\mathbf{D}_1\mathbf{A} + \alpha\mathbf{D}_2\mathbf{Z}')_{ki}}{\mathbf{Z}'_{ki}}\delta_{ij}\delta_{kl}. \tag{33}$$

Since $f(\mathbf{Z})$ is convex and the second-order derivatives is semi-positive definite, we can obtain the solution for $f(\mathbf{Z})$. By letting (32) be zero, we can obtain the update rule of $\mathbf{Z}$ as:

$$\mathbf{Z}_{ki} \leftarrow \mathbf{Z}'_{ki}\frac{(\mathbf{W}^T\mathbf{V}^T\mathbf{V}\mathbf{D}_1\mathbf{A})_{ki}}{(\mathbf{W}^T\mathbf{V}^T\mathbf{V}\mathbf{W}\mathbf{Z}'\mathbf{A}^T\mathbf{D}_1\mathbf{A} + \alpha\mathbf{D}_2\mathbf{Z}')_{ki}}, \tag{34}$$

Let $\mathbf{Z}^t \leftarrow \mathbf{Z}'$, $\mathbf{Z}^{t+1} \leftarrow \mathbf{Z}$, (34), we can obtain the iterative solution of (20). When we use this strategy to update $\mathbf{Z}$, the objective function of (27) is non-increasing.

Until now, Lemma 1 is proved.

Lemma 2 *In order to finish the proof of this, we refer to the matrix inequality in [32]. If matrices* $\mathbf{S} \geq \mathbf{0}$, $\mathbf{B} \geq \mathbf{0}$, $\mathbf{H} \geq \mathbf{0}$, *the sizes are suitable and* $\mathbf{B} = \mathbf{B}^T$, $\mathbf{S} = \mathbf{S}^T$, *we obtain the matrix inequality:*

$$Tr(\mathbf{H}^T\mathbf{S}\mathbf{H}\mathbf{B}) \leq \sum_{ik}(\mathbf{S}\mathbf{H}'\mathbf{B})\frac{\mathbf{H}^2_{ik}}{\mathbf{H}'_{ik}} \tag{35}$$

Lemma 3 *According to the solution in* (20), *the following in-equation holds*

$$
\begin{aligned}
&\|\mathbf{V}-\mathbf{VWZ}^{t+1}\mathbf{A}^T\|_{2,1}+\alpha\|(\mathbf{Z}^{t+1})^T\|_{2,1}\\
&-\|\mathbf{V}-\mathbf{VWZ}^{t}\mathbf{A}^T\|_{2,1}-\alpha\|(\mathbf{Z}^{t})^T\|_{2,1}\\
&\leq\frac{1}{2}[Tr((\mathbf{V}-\mathbf{VWZ}^{t+1}\mathbf{A}^T)\mathbf{D}_1(\mathbf{V}-\mathbf{VWZ}^{t+1}\mathbf{A}^T)^T)\\
&+\alpha Tr((\mathbf{Z}^{t+1})^T\mathbf{D}_2\mathbf{Z}^{t+1})\\
&-Tr((\mathbf{V}-\mathbf{VWZ}^{t}\mathbf{A}^T)\mathbf{D}_1(\mathbf{V}-\mathbf{VWZ}^{t}\mathbf{A}^T)^T)\\
&-\alpha Tr((\mathbf{Z}^{t})^T\mathbf{D}_2\mathbf{Z}^{t})].
\end{aligned}
\tag{36}
$$

Proof Lemma 3 can be proved with the same method of [33]. Then, we can derive (36).

With the characteristic of $\mathbf{D}_1$ and $\mathbf{D}_2$, we have

$$
\begin{aligned}
&Tr((\mathbf{V}-\mathbf{VWZ}^{t+1}\mathbf{A}^T)\mathbf{D}_1(\mathbf{V}-\mathbf{VWZ}^{t+1}\mathbf{A}^T)^T)\\
&+\alpha Tr((\mathbf{Z}^{t+1})^T\mathbf{D}_2\mathbf{Z}^{t+1})\\
&=\sum_{i=1}^{N}\|\mathbf{V}_i-\mathbf{VW}(\mathbf{Z}^{t+1}\mathbf{A}^T)_i\|^2(\mathbf{D}_1)_{ii}\\
&+\alpha\sum_{i=1}^{K}\|(\mathbf{Z}_i^{t+1})^T\|^2(\mathbf{D}_2)_{ii},
\end{aligned}
\tag{37}
$$

$$
\begin{aligned}
&Tr((\mathbf{V}-\mathbf{VWZ}^{t}\mathbf{A}^T)\mathbf{D}_1(\mathbf{V}-\mathbf{VWZ}^{t}\mathbf{A}^T)^T)\\
&+\alpha Tr((\mathbf{Z}^{t})^T\mathbf{D}_2\mathbf{Z}^{t})\\
&=\sum_{i=1}^{N}\|\mathbf{V}_i-\mathbf{VW}(\mathbf{Z}^{t}\mathbf{A}^T)_i\|^2(\mathbf{D}_1)_{ii}\\
&+\alpha\sum_{i=1}^{K}\|(\mathbf{Z}_i^{t})^T\|^2(\mathbf{D}_2)_{ii}.
\end{aligned}
\tag{38}
$$

The right-hand side (RHS) of (36) becomes

$$
\begin{aligned}
RHS=&\frac{1}{2}\sum_{i=1}^{N}(\|\mathbf{V}_i-\mathbf{VW}(\mathbf{Z}^{t+1}\mathbf{A}^T)_i\|^2(\mathbf{D}_1)_{ii}\\
&-\|\mathbf{V}_i-\mathbf{VW}(\mathbf{Z}^{t}\mathbf{A}^T)_i\|^2(\mathbf{D}_1)_{ii})\\
&+\frac{1}{2}\alpha\sum_{i=1}^{K}(\|(\mathbf{Z}^{t+1})_i^T\|^2(\mathbf{D}_2)_{ii}-\|(\mathbf{Z}^{t})_i^T\|^2(\mathbf{D}_2)_{ii}).
\end{aligned}
\tag{39}
$$

Combining (22) and (23),

$$
\begin{aligned}
RHS = & \frac{1}{2}\sum_{i=1}^{N}\left(\|\mathbf{V}_i - \mathbf{VW}(\mathbf{Z}^{t+1}\mathbf{A}^T)_i\|^2(\mathbf{D}_1)_{ii} - \frac{1}{(\mathbf{D}_1)_{ii}}\right) \\
& + \frac{1}{2}\alpha\sum_{i=1}^{K}(\|(\mathbf{Z}^{t+1})_i^T\|^2(\mathbf{D}_2)_{ii} - \frac{1}{(\mathbf{D}_2)_{ii}}).
\end{aligned} \tag{40}
$$

The left-hand side (LHS) of (36) becomes

$$
\begin{aligned}
LHS = & \|\mathbf{V} - \mathbf{VWZ}^{t+1}\mathbf{A}^T\|_{2,1} + \alpha\|(\mathbf{Z}^{t+1})^T\|_{2,1} \\
& - \|\mathbf{V} - \mathbf{VWZ}^t\mathbf{A}^T\|_{2,1} - \alpha\|(\mathbf{Z}^t)^T\|_{2,1} \\
& = \sum_{i=1}^{N}(\|\mathbf{V}_i - \mathbf{VW}(\mathbf{Z}^{t+1}\mathbf{A}^T)_i\| - \frac{1}{(\mathbf{D}_1)_{ii}}) \\
& + \alpha\sum_{i=1}^{K}(\|(\mathbf{Z}^{t+1})_i^T\| - \frac{1}{(\mathbf{D}_2)_{ii}}).
\end{aligned} \tag{41}
$$

Therefore, we have

$$
\begin{aligned}
& LHS - RHS \\
& = \sum_{i=1}^{N}\frac{-(\mathbf{D}_1)_{ii}}{2}(\|\mathbf{V}_i - \mathbf{VW}(\mathbf{Z}^{t+1}\mathbf{A}^T)_i\| - \frac{1}{(\mathbf{D}_1)_{ii}})^2 \\
& + \sum_{i=1}^{K}\frac{-(\mathbf{D}_2)_{ii}}{2}(\|(\mathbf{Z}^{t+1})_i^T\| - \frac{1}{(\mathbf{D}_2)_{ii}})^2 \leq 0.
\end{aligned} \tag{42}
$$

Until now, the proof of Lemma 3 is accomplished.

With the usage of Lemmas 1–3, the proof of Theorem 1 can be obtained. It means the objective function of (19) is non-increasing under the solution in (20).

We can take the same strategy to prove the Theorem 2, we do not provide details here.

2.5 Correctness of the RCCF Analysis

In the following, we will prove that the proposed algorithms is guaranteed to converge to the Karush-Kuhn-Tucker (KKT) points.

Theorem 3 *Using the updating rule in* (20)*, the obtained solution of* $\mathbf{Z}$ *satisfies the Karush-Kuhn-Tucker condition.*

Proof. The Karush-Kuhn-Tucker condition for $\mathbf{Z}$ with the constrains $(\mathbf{Z})_{ki} \geq 0$, $k = 1, 2, \ldots, K; i = 1, 2, \ldots, (N - l + c)$, is

$$\frac{\partial J(\mathbf{Z})}{\partial(\mathbf{Z})_{ki}}(\mathbf{Z})_{ki} = 0, \forall k, i. \tag{43}$$

The derivative is

$$\frac{\partial J(\mathbf{Z})}{\partial(\mathbf{Z})_{ki}} = -2((\mathbf{W}^T\mathbf{V}^T\mathbf{V}(1 - \mathbf{W}\mathbf{Z}\mathbf{A}^T)\mathbf{D}_1\mathbf{A})_{ki} + \alpha(\mathbf{D}_2\mathbf{Z})_{ki}). \tag{44}$$

Then, the Karush-Kuhn-Tucker condition for $\mathbf{Z}$ is

$$\begin{aligned}
&[-(\mathbf{W}^T\mathbf{V}^T\mathbf{V}\mathbf{D}_1\mathbf{A})_{ki} + (\mathbf{W}^T\mathbf{V}^T\mathbf{V}\mathbf{W}\mathbf{Z}\mathbf{A}^T\mathbf{D}_1\mathbf{A})_{ki} \\
&+ \alpha(\mathbf{D}_2\mathbf{Z})_{ki}](\mathbf{Z})_{ki} \\
&= 0, \forall k, i.
\end{aligned} \tag{45}$$

If the $\mathbf{Z}$ converges under updating rule of (20), the obtained solution $\mathbf{Z}^*$ satisfies

$$\mathbf{Z}^*_{ki} \leftarrow \mathbf{Z}^*_{ki}\frac{(\mathbf{W}^T\mathbf{V}^T\mathbf{V}\mathbf{D}_1\mathbf{A})_{ki}}{(\mathbf{W}^T\mathbf{V}^T\mathbf{V}\mathbf{W}\mathbf{Z}^*\mathbf{A}^T\mathbf{D}_1\mathbf{A} + \alpha\mathbf{D}_2\mathbf{Z}^*)_{ki}}, \tag{46}$$

which can be reformulated as

$$\begin{aligned}
&[-(\mathbf{W}^T\mathbf{V}^T\mathbf{V}\mathbf{D}_1\mathbf{A})_{ki} + (\mathbf{W}^T\mathbf{V}^T\mathbf{V}\mathbf{W}\mathbf{Z}^*\mathbf{A}^T\mathbf{D}_1\mathbf{A})_{ki} \\
&+ \alpha(\mathbf{D}_2\mathbf{Z}^*)_{ki}](\mathbf{Z}^*)_{ki} \\
&= 0, \forall k, i.
\end{aligned} \tag{47}$$

We observe that (47) is the same as (45). This means the learned solution for $\mathbf{Z}^*$ satisfies the Karush-Kuhn-Tucker condition. Until now, we finish the proof.

Theorem 4 *With the solution* $\mathbf{W}$ *under the updating rule of* (21), *the proposed algorithm converges to the Karush-Kuhn-Tucker points.*

In regards to proving Theorem 4, we can take the same strategy to finish it.

3 Experimentation

3.1 Description of the Data

We used three data sets in our experiments. There are two face data sets and one handwritten digits images database. The proposed RCCF method is evaluated on data clustering. Table 1 show details of the selected data sets.

Table 1 Details of the datasets

Datasets	Size	Dimentions	Classes
Yale	165	1024	15
ORL	400	1024	40
MNIST	1000	784	10

Fig. 1 Yale Faces database

Yale Database.[1] The Yale database contains 165 images in gray scale. These images belong to 15 different people. For each person, there are 11 facial images of size 32×32. Each picture is with a different facial expression or configuration. Similar to [34], all samples are normalized in orientation and scale to ensure that two eyes are aligned at the same position. Figure 1 shows some face images from this database.

ORL Database.[2] This database contains 400 gray scale face images of 40 individuals. For each individual, images are in different facial expressions or configurations. All these pictures are collected at different time, varying the lighting. We use the same way as the Yale data set to preprocess this data set. Figure 2 shows some examples from this database.

MNIST Database.[3] The MNIST database contains 10000 images of handwritten digits from 0 to 9 in gray scale. For each subject, there are 1000 images. We resize

[1] http://www.face-rec.org/databases/.

[2] http://www.face-rec.org/databases/.

[3] http://yann.lecun.com/exdb/mnist/.

Fig. 2 ORL Faces database

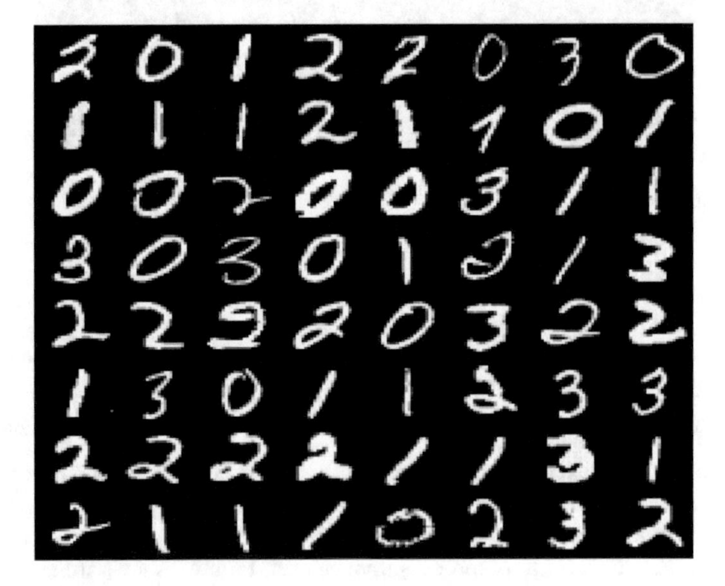

Fig. 3 MNIST database

each image to 16×16, thus the dimensionality of feature vector is 256. Figure 3 provides several example pictures from this data set.

3.2 Evaluation Metrics

Following [34, 35], the normalized mutual information metric (*NMI*) and the accuracy (*AC*) are employed to evaluate the clustering performance. Accuracy reflects the percentage of correctly predicted cluster number. Given a database with n samples, r_i is cluster information provided by the database, for each sample, let l_i be cluster information that we obtain by using different methods. The definition of AC is:

$$AC = \frac{\sum_{i=1}^{n} \delta(r_i, map(l_i))}{n} \tag{48}$$

where $\delta(x, y)$ be set 1 if $x = y$ and $\delta(x, y)$ be set 0 if $x \neq y$, and $map(l_i)$ denotes the permutation mapping function that maps each cluster label l_i to the corresponding label from the data set. We utilize the KM algorithm [36] to obtain the best map.

The normalized mutual information matrix is employed to measure the similarity of two clusters. Let C be the information of clusters achieved from the ground truth and C' obtained from the proposed algorithm. Its mutual information matrix is measured as:

$$MI(C, C') = \sum_{c_i \in C, c_j' \in C'} p(c_i, c_j') \cdot log \frac{p(c_i, c_j')}{p(c_i) \cdot p(c_j')}, \tag{49}$$

where $p(c_i, c_j')$ denotes the joint probability that the chosen sample comes from the cluster c_i and $p(c_j')$ simultaneously. $p(c_i)$ and $p(c_j')$ are the probabilities that a randomly chosen data comes from the clusters c_i and $p(c_j')$, respectively. In these experiments, we used the $NMI(C, C')$, which gets scores ranging from 0 to 1.

$$NMI(C, C') = \frac{MI(C, C')}{max(H(C), H(C'))}, \tag{50}$$

where $H(C)$ and $H(C')$ denote the entropies of C and C'. *NMI* equals to 1 when two selected samples are the same, and it is 0 when these two samples come from two different clusters.

3.3 Experimental Results

Testing was carried out on the proposed RCCF method in terms of clustering performance on three public datasets. At the same time we also make comparisons with related methods as follow:

1. Traditional KMeans clustering method (KMeans for short).
2. Concept-Factorization-based clustering (CF for short) [19].
3. Constrained concept factorization (CCF for short) [27].

The former two methods are unconstrained and the last one is a constrained algorithm.

Experiments are conducted with different cluster numbers K. K takes the value between 2 and 10. For each data set, K categories are selected randomly. Similar to [28], 30% of the data points are extracted to construct the training dataset and the rest for constructing the test dataset. Then, matrix factorization methods are used to obtain low-dimensional representations. The reduced dimensionality is set to be equal to the cluster number K. Once the new representation is obtained, we utilize KMeans by choosing cosine distance to the new representation for data clustering. KMeans is repeated 20 times with various initiations. The result with minimum cost function is recorded to measure accuracy and mutual information. There is an important tunable parameter for the proposed method. To make the experimental results persuasive, we perform grid search in the parameter space for our method and the best results are recorded. In particular, the search of α from 0.1 to 90 was carried out and we set $\alpha = 20, 10, 20$ for the YALE, ORL and MNIST respectively.

The Yale data set clustering results are shown in Table 2. Average *AC* and *NMI* versus *K* can be found in the last row. We can see that RCCF outperforms others most of the time, especially in terms of AC, while comparing to the second best results, i.e., average results in terms of *AC* and *NMI* for CCF, our algorithm RCCF achieves 3.1 percent and 5.3 percent improvements respectively.

Table 3 provides the clustering results for the data set named ORL. In this table it can be observed that RCCF achieves the best results for most cases. RCCF obtains the highest results 8/9 times in *AC* and the highest results 7/9 times in *NMI*. RCCF achieves a 3.7 percent improvement in *AC* and a 3.4 percent improvement in *NMI* on average, compared to the next best method (i.e. CCF)

Using MNIST, the details of the clustering results are given in Table 4. It can be observed from the table that the superiority of our method is obvious when K is

Table 2 Clustering methods' performance on Yale Database

K	Accuracy(%)				Normalized mutual information(%)			
	KMeans	CF	CCF	RCCF	KMeans	CF	CCF	RCCF
2	73.6	85.0	83.2	88.4	25.9	50.1	46.7	59.3
3	70.3	73.0	80.9	82.1	44.1	48.9	61.1	57.2
4	51.6	62.0	69.8	71.0	32.1	38.0	49.6	55.7
5	46.4	57.3	61.5	67.5	38.2	41.3	47.6	56.6
6	49.09	49.5	61.4	62.3	36.3	37.1	50.2	56.3
7	45.8	48.8	55.5	62.9	39.4	39.5	49.4	56.2
8	44.3	48.3	55.1	57.5	41.0	43.1	49.5	54.3
9	43.8	48.4	55.5	56.3	43.4	45.2	53.1	54.9
10	40.6	43.9	52.2	54.7	40.9	41.8	52.0	55.9
Avg.	52.0	57.4	63.9	67.0	37.9	42.8	51.0	56.3

Table 3 Clustering methods' performance on ORL Database

K	Accuracy(%)				Normalized mutual information(%)			
	KMeans	CF	CCF	RCCF	KMeans	CF	CCF	RCCF
2	95.0	87.1	90.7	93.6	76.1	56.7	64.3	75.6
3	60.0	70.4	80.9	83.8	43.2	49.5	62.9	68.7
4	57.5	61.4	65.4	73.9	47.4	50.0	55.8	63.0
5	63.0	57.7	61.7	63.7	52.7	50.3	55.6	55.9
6	56.7	60.5	61.2	63.1	55.2	59.7	59.4	59.1
7	54.5	57.4	62.9	65.3	58.0	58.4	63.8	64.0
8	53.6	58.5	63.8	67.3	59.3	61.2	65.8	67.4
9	59.9	62.2	63.2	68.3	67.0	65.2	65.9	69.6
10	58.3	55.4	62.0	65.9	65.3	63.3	68.1	68.7
Avg.	62.0	63.4	67.9	71.6	58.3	57.2	62.4	65.8

Table 4 Clustering methods' performance on MNIST Database

K	Accuracy(%)				Normalized mutual information(%)			
	KMeans	CF	CCF	RCCF	KMeans	CF	CCF	RCCF
2	90.1	90.1	94.6	98.6	59.9	60.1	73.3	90.6
3	82.2	81.2	82.8	87.0	56.5	54.7	60.0	65.2
4	74.4	70.8	80.9	84.0	53.7	50.9	63.1	69.8
5	67.9	63.2	80.0	86.8	51.5	46.0	62.5	73.7
6	65.9	68.2	77.7	74.6	53.9	52.2	63.1	62.0
7	63.2	61.6	75.1	74.3	53.5	50.2	63.2	61.3
8	57.1	61.2	67.3	68.4	50.9	50.0	57.6	58.0
9	53.9	57.0	67.6	67.6	51.1	48.5	58.5	57.7
10	54.1	53.8	68.9	58.6	50.9	45.8	59.2	50.1
Avg.	67.6	67.4	77.2	77.8	53.5	51.0	62.3	65.4

small. On average, RCCF and CCF have similar performance, however, RCCF still achieves the best results. When matched with the algorithm that performed second best (CCF), RCCF obtains a 3.1 percent improvement in *NMI*,

4 Conclusion

A robust constrained concept factorization (RCCF) method is proposed in this chapter. This new model learns discriminative results since it fully utilizes the labeled information with a constraint matrix. In addition, $L_{2,1}$-norm is applied on both the reconstruction function and the regularization. The $L_{2,1}$-norm based reconstruction function improves the robustness of RCCF, and the $L_{2,1}$-norm regularization is used to select useful information. In order to solve the new model, we have derived an efficient iterative updating algorithm, along with proofs of convergence. Evaluating the proposed method on three data sets showed the superiority of the algorithm as a generalized method in pattern recognition and computational intelligence applications.

Acknowledgements This work is supported by the Science and Technology Development Fund (FDCT) of Macao SAR (124/2014/A3) and the National Natural Science Foundation of China (61602540).

References

1. Y. Bengio, A. Courville, P. Vincent, Representation learning: a review and new perspectives. IEEE Trans. Pattern Anal. Mach. Intell. **35**(8), 1798–1828 (2013)
2. M. Tepper, G. Sapiro, Compressed nonnegative matrix factorization is fast and accurate. IEEE Trans. Signal Process. **64**(9), 2269–2283 (2016)
3. Z. Ma, A.E. Teschendorff, A. Leijon, Y. Qiao, H. Zhang, J. Guo, Variational bayesian matrix factorization for bounded support data. IEEE Trans. Pattern Anal. Mach. Intell. **37**(4), 876–889 (2015)
4. Z. Yang, Y. Xiang, Y. Rong, K. Xie, A convex geometry-based blind source separation method for separating nonnegative sources. IEEE Trans. Neural Netw. Learn. Syst. **26**(8), 1635–1644 (2015)
5. I. Domanov, L.D. Lathauwer, Generic uniqueness of a structured matrix factorization and applications in blind source separation. IEEE J. Sel. Top. Signal Process. **10**(4), 701–711 (2016)
6. X. Fu, W.K. Ma, K. Huang, N.D. Sidiropoulos, Blind separation of quasi-stationary sources: exploiting convex geometry in covariance domain. IEEE Trans. Signal Process. **63**(9), 2306–2320 (2015)
7. Y. Xu, B. Zhang, Z. Zhong, Multiple representations and sparse representation for image classification. Pattern Recognit. Lett. **68**, 9–14 (2015)
8. H. Zhang, J. Yang, J. Xie, J. Qian, B. Zhang, Weighted sparse coding regularized nonconvex matrix regression for robust face recognition. Inf. Sci. **394**, 1–17 (2017)
9. W. Jia, B. Zhang, J. Lu, Y. Zhu, Y. Zhao, W. Zuo, H. Ling, Palmprint recognition based on complete direction representation. IEEE Trans. Image Process. (2017)
10. Y. Xiao, Z. Zhu, Y. Zhao, Y. Wei, S. Wei, X. Li, Topographic nmf for data representation. IEEE Trans. Cybern. **44**(10), 1762–1771 (2014)
11. L. Luo, L. Chen, J. Yang, J. Qian, B. Zhang, Tree-structured nuclear norm approximation with applications to robust face recognition. IEEE Trans. Image Process. **25**(12), 5757–5767 (2016)
12. R.O. Duda, P.E. Hart, D.G. Stork, *Pattern Classification* (Wiley, New York, 2012)
13. I. Jolliffe, *Principal Component Analysis* (Wiley Online Library, Hoboken, 2002)
14. R. Gray, Vector quantization. IEEE Assp Mag. **1**(2), 4–29 (1984)
15. D.D. Lee, H.S. Seung, Learning the parts of objects by non-negative matrix factorization. Nature **401**(6755), 788–791 (1999)
16. Z. Yang, Y. Zhang, W. Yan, Y. Xiang, S. Xie, A fast non-smooth nonnegative matrix factorization for learning sparse representation. IEEE Access **4**, 5161–5168 (2016)
17. X. Zhang, L. Zong, X. Liu, J. Luo, Constrained clustering with nonnegative matrix factorization. IEEE Trans. Neural Netw. Learn. Syst. **27**(7), 1514–1526 (2016)
18. Z. Yang, Y. Xiang, K. Xie, Y. Lai, Adaptive method for nonsmooth nonnegative matrix factorization. IEEE Trans. Neural Netw. Learn. Syst. **28**(4), 948–960 (2017)

19. W. Xu, Y. Gong, Document clustering by concept factorization, in *Proceedings of the 27th Annual International ACM SIGIR Conference on Research and Development in Information Retrieval* (ACM, 2004), pp. 202–209
20. D. Cai, X. He, J. Han, Locally consistent concept factorization for document clustering. IEEE Trans. Knowl. Data Eng. **23**(6), 902–913 (2011)
21. H. Liu, Z. Yang, J. Yang, Z. Wu, X. Li, Local coordinate concept factorization for image representation. IEEE Trans. Neural Netw. Learn. Syst. **25**(6), 1071–1082 (2014)
22. W. Yan, B. Zhang, S. Ma, Z. Yang, A novel regularized concept factorization for document clustering. Knowl.-Based Syst. (2017)
23. S.E. Palmer, Hierarchical structure in perceptual representation. Cogn. Psychol. **9**(4), 441–474 (1977)
24. N.K. Logothetis, D.L. Sheinberg, Visual object recognition. Annu. Rev. Neurosci. **19**(1), 577–621 (1996)
25. P.O. Hoyer, Non-negative matrix factorization with sparseness constraints. J. Mach. Learn. Res. **5**, 1457–1469 (2004)
26. R. Peharz, F. Pernkopf, Sparse nonnegative matrix factorization with 0-constraints. Neurocomputing **80**, 38–46 (2012)
27. H. Liu, G. Yang, Z. Wu, D. Cai, Constrained concept factorization for image representation. IEEE Trans. Cybern. **44**(7), 1214–1224 (2014)
28. H. Liu, Z. Wu, D. Cai, T.S. Huang, Constrained nonnegative matrix factorization for image representation. IEEE Trans. Pattern Anal. Mach. Intell. **34**(7), 1299–1311 (2012)
29. D. Wang, X. Gao, X. Wang, Semi-supervised nonnegative matrix factorization via constraint propagation. IEEE Trans. Cybern. **46**(1), 233–244 (2016)
30. C. Ding, D. Zhou, X. He, H. Zha, R1-PCA: rotational invariant L1-norm principal component analysis for robust subspace factorization, in *Proceedings of the 23rd International Conference on Machine Learning* (ACM, 2006), pp. 281–288
31. D.D. Lee, H.S. Seung, Algorithms for non-negative matrix factorization, in *Advances in Neural Information Processing Systems* (2001), pp. 556–562
32. C.H.Q. Ding, T. Li, M.I. Jordan, Convex and semi-nonnegative matrix factorizations. IEEE Trans. Pattern Anal. Mach. Intell. **32**(1), 45–55 (2010)
33. D. Kong, C. Ding, H. Huang, Robust nonnegative matrix factorization using L21-norm, in *Proceedings of the 20th ACM International Conference on Information and Knowledge Management* (ACM, 2011), pp. 673–682
34. D. Cai, X. He, J. Han, T.S. Huang, Graph regularized nonnegative matrix factorization for data representation. IEEE Trans. Pattern Anal. Mach. Intell. **33**(8), 1548–1560 (2011)
35. D. Cai, X. He, J. Han, Document clustering using locality preserving indexing. IEEE Trans. Knowl. Data Eng. **17**(12), 1624–1637 (2005)
36. L. Lovász, M.D. Plummer, *Matching Theory*, vol. 367 (American Mathematical Society, Providence, 2009)

An Automatic Cycling Performance Measurement System Based on ANFIS

Andre Vieira Pigatto and Alexandre Balbinot

Abstract Cycling is a sport that has grown widely in the last decades, what has attracted the attention of research laboratories seeking to understand the factors that influence the athletes' performance; most of the researches consider only the power transmitted to the wheel to analyze the training level of the cyclist. However, recent studies have shown a weak correlation between the peak power in an indoor test and the times in a time trial competition. Considering this, the following chapter describes the development and evaluation of a new approach to cycling performance measurement, implemented using a three input Adaptive Neuro-Fuzzy System, based on three parameters: the average mechanical power applied by the athlete to the bicycle pedal, the power standard deviation and the effective force bilateral asymmetry index. Data used to develop and train the system were measured using an experimental force platform based on instrumented load cells with built-in conditioning circuit and strain gages that measure and acquire the components of the force that is applied to a road bicycle crank arm during pedaling in real conditions and save them on a SD card. A randomized block experiment design was performed with fifteen cyclists to analyze the significance of the scores obtained using the collected data as input of the ANFIS. ANOVA showed that the subject causes significant variation on the score. The subjects cycling performance score was then determined using the ANFIS; the mean score was $24.7 \pm 18.7\%$ which was considered a consistent result taking into account the volunteers cycling experience. The developed system has shown a promising applicability, since an automatic performance classifier may be a great tool for coaches to objectively compare the performance level among different cyclists or to evaluate the progression of the athlete among different trials.

Keywords Force measuring · Cycling performance · Fuzzy · ANFIS

A. V. Pigatto (✉) · A. Balbinot
Electro-Electronic Instrumentation Laboratory (IEE),
Electrical Engineering Graduate Program (PPGEE) of Federal University of Rio Grande do Sul, UFRGS, Porto Alegre, Brazil
e-mail: pigatto.andre@gmail.com

A. Balbinot
e-mail: alexandre.balbinot@ufrgs.br

W. Pedrycz and S.-M. Chen (eds.), *Computational Intelligence for Pattern Recognition*, Studies in Computational Intelligence 777,
https://doi.org/10.1007/978-3-319-89629-8_8

1 Introduction

Defining the meaning of intelligence is not an easy task; therefore, it still is the focus of many professional researches. Alfred Binet used to define the essential activities of the intelligence as to judge well, to comprehend well and to reason well [1]. In the scope of specialist systems, the intelligence definition is commonly called computational intelligence, which is a field that applies mathematical-computational methods to perform tasks in a faster or more efficient way. Among a wide variety of methods used in the computational intelligence field there are two in particular that stand out, which are the Fuzzy and Neuro-Fuzzy methods.

The Aristotelian culture uses the bivalent logic to solve problems, which means that the membership function that represents it can assume only two values: true or false; zero or one. This approach enables the person to solve most of the daily basis problems, for example, to decide whether it is raining or not, but it has its limitations as it does not tolerate values within the limits of zero to one. However, another approach was introduced by Zadeh [2], the Fuzzy Logic. In a general way, the Fuzzy Logic can be used to solve daily basis problems but using the multivalence concept, where it is possible to comprehend and characterize infinite values within the interval of zero to one. Considering the example of the rain, already presented, there are infinite definitions that could be used to scale the level of rain, rather than just determine if it is raining or not, for example we may represent the zero as not raining, while one represents a storm, and the middle of the scale (0.5) represents a drizzle; somehow it is just a matter of point of view.

Therefore, after this brief introduction, we present a solution proposal for a cycling performance measurement problem where the purely Boolean approach would not be able to provide an adequate solution. The proposed solution is an automatic cycling performance measurement system which uses an Adaptive Neuro Fuzzy System (ANFIS) to recognize the athlete's pedaling pattern, based on the forces that are applied to the pedal, and determine a score that represents the cyclist performance.

In the last two decades, cycling has become one of the most popular sports in the world; some of the adepts use their bicycle as an eco-friendly vehicle to get from place to place, others use it as a training method to enhance their aerobic capacity and a reasonable part of them competes professionally. Considering this, a lot of researches have focused on the study of the characteristics that directly impact on the athlete performance as the physiological condition [3, 4], the bicycle ergonomics and the technique applied during pedaling [4]. One of the ways to analyze the pedaling technique is through the measurement of the components of the forces that are applied to the bicycle pedal during cycling as it reveals important information about the athlete's performance, as the fraction of the resultant force that is effective (is able to rotate the crank arm), the bilateral asymmetry index, the resultant power that is supplied to the transmission of the bicycle and other individual pedaling characteristics [3, 5–7].

The resultant force that is applied to the bicycle pedal can be decomposed into three components perpendicular among each other and referred to the crank arm,

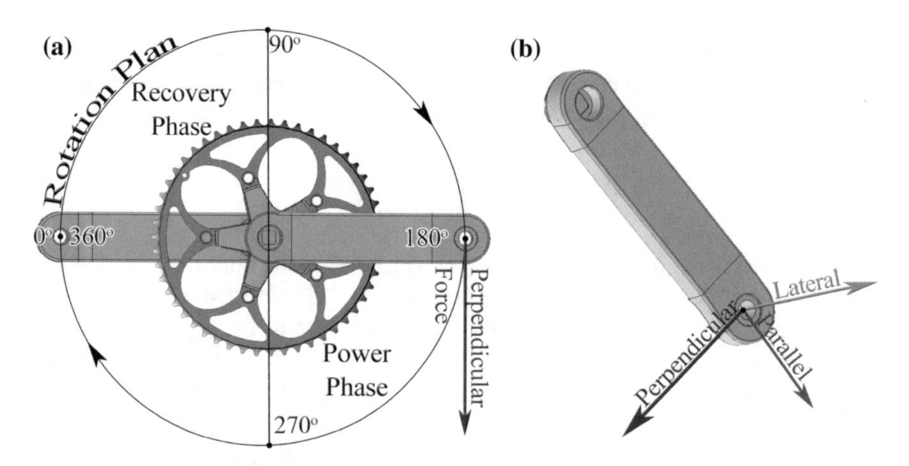

Fig. 1 **a** Pedaling cycle, **b** Components of force

which are: Perpendicular to the crank arm (the only one that can produce mechanical torque), Parallel to the crank arm and Lateral, which is perpendicular to the rotation plan of the crank arm [8]. The pedaling cycle can be divided into two phases [3, 4]: the power phase, which is the part of the cycle where the maximum power is transferred by the cyclist to the bicycle transmission and the recovery phase, where the athlete pulls the pedal upwards to reduce the effect of the leg weight on the movement, in case of a standard pedal is used, or to apply positive effective force, when a clipless pedal is used. Figure 1a shows the pedaling cycle and Fig. 1b shows the force measurement components.

From Fig. 1a it is possible to notice that the only force component that is able to rotate the crank arm is the perpendicular one, of which the maximum value is reached when the pedal is at a 180° angle as almost all the force exerted by the cyclist is applied in the perpendicular direction. As the crank arms are connected by the bottom bracket, each leg is in the opposite phase simultaneously, which implies that the leg that is in the recovery phase should be pulled up, otherwise it will apply a negative effective force, decreasing the amount of torque generated per cycle, thus, affecting the performance of the athlete. As the length of the crank arm and the perpendicular force applied to the pedal is known, the torque can be determined [5, 9]; from the torque and the angular speed, it is possible to determine the power applied to the crank arm, as shown in Eq. (1):

$$P_{out} = (2\pi/60)\, F_{effective} LC \tag{1}$$

where $\boldsymbol{P_{out}}$ is the power output in W, $\boldsymbol{F_{effective}}$ is the effective force, in Newtons, $\boldsymbol{L}$ is the crank arm length, in meters, and $\boldsymbol{C}$ is the cadence, in RPM.

From the individual leg force and power measurement, it is possible to analyze the asymmetry between the inferior members and then help coaches to develop strategies for training and performance improvement [6, 10]. The bilateral asymmetry index

(AI) was determined according to [11], considering that the dominant leg (DO) is the one that applies a higher average resultant force to the bicycle crank arm than the non-dominant leg (ND), as shown in Eq. (2):

$$AI\,(\%) = 100\left(\frac{FDO - FND}{FDO}\right) \tag{2}$$

where $\boldsymbol{AI\,(\%)}$ is the bilateral asymmetry index, $\boldsymbol{FDO}$ is the dominant leg effective force, in Newtons, and $\boldsymbol{FND}$ is the non-dominant leg effective force.

There are several methods to classify the performance of the cyclist, as the VO2 consumption peaks measurement during a trial with power increments every period of time to maximum effort; constant power, where the cyclist maintains a specified mean power to the point of exhaustion [12]; heart rate variability (HRV), combined with kinematic data from the bicycle [13] and via pedaling profile classification, considering the position of the foot and the pedal related to the crank arm during each cycle [14]. However, one of the most usual and effective way of classifying the athlete's performance is analyzing the power generated while cycling [12]. A new approach of cycling performance measurement is introduced in this chapter, which is a score determined through an Adaptive Neuro Fuzzy Inference System (ANFIS), applying the theory proposed by [15], developed based on three input parameters to recognize the pedaling pattern of the athlete: the mean power, the power standard deviation (STD) and the effective force. The main advantage of this new method based on ANFIS is that as the score depends on other variables than the power, it may lead to a better comprehension of the cyclist pedaling technique and the influence of its characteristics on the overall cycling performance. This approach may be useful not only to determine which of the pedaling attributes should be improved with training, based on how much it affects the overall performance, but also to analyze the effectiveness of the coaching strategy through the quantitative comparison of the score obtained before and after the strategy that has been applied.

2 Materials and Methods

The hardware of the system used in this study was already developed and may be seen in more details in [7].

2.1 *Hardware Description*

The measurement system is a road bicycle crank arm based load cell that deforms when a force is applied to the bicycle pedal, unbalancing the cemented strain gage composed Wheatstone bridge. The voltage across the bridge arms is sensed by an instrumentation amplifier and digitalized using the ADC of an ATMEGA 328P

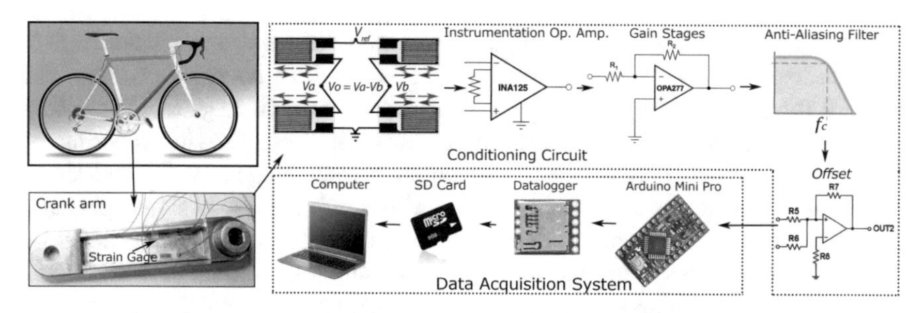

Fig. 2 Measurement system block diagram

Fig. 3 3D model of the crankset: **a** right crank arm up side, **b** right crank arm down side, **c** left crank arm down side and **d** left crank arm up side and Picture of the Measuring System, **e** right crank arm, **f** left crank arm

micro-controller. The digital signal is then recorded on a SD card through a datalogger connected to the serial port of the micro-controller. Figure 2 shows the block diagram of the developed system.

Since the system was developed to be used on track, the experimental load cell was designed to fit a conventional road bike and be ergonomically equivalent to a commercial crankset but with a compartment to fit all the electronics and strain gages. The static simulation and experimental loads were determined considering a cyclist with a body weight of 75 kg, which implied on a maximum loading force or 732.5 N, as the total force applied to the pedal hardly exceeds the athlete's weight [16, 17]. Each crank arm was instrumented with 12 strain gages (HBM 1-LY13-1.5/350), which were positioned according to the tension concentration areas determined by the static simulation results, forming two load cells of full Wheatstone bridges, for Perpendicular force measure (Channel 1 and 4), and four load cells of half bridges, for Lateral (Channel 2 and 5) and Parallel (Channel 3 and 6) force measurement. Figure 3 shows the virtual model of the experimental crank arms and the whole system assembled.

The signal conditioner of each crank arm is composed by three channels, developed following the same topology, where each channel is composed by one instrumentation amplifier (TI—INA125UA), which provides the voltage reference and the first voltage gain stage, four independent low noise operational amplifiers (TI

OPA4227UA), to provide voltage amplification, an offset stage, and a 4th order Butterworth low-pass filter [18, 19]. Two 3.7 V 850 mAh LiPo batteries are connected in series for symmetrical power supply. The preliminary voltage gain of each channel was determined based on the deformation test results and then tuned during the static calibration experiment, ensuring an output voltage range of 0–5 V; the signal conditioning circuit was validated with aid of a Tektronix AFG3052 signal generator and a TDS2001 oscilloscope. To determine the transfer function of each channel, a static calibration experiment was performed applying load in each direction of the force measurement components separately and measuring the voltage output of the respective channel with aid of a Tektronix DMM4050 precision multimeter.

2.2 Data Acquisition and Processing

To acquire and store the output voltage signal of the six channels of the conditioning circuit, an Arduino Mini Pro development platform is used. The routine developed digitalizes the analog signals of all the channels using the ADC (10 bits, input range of 0–5 V) of the ATMEGA 328P micro-controller and then stores them on a SD card, through a datalogger connected to the Arduino's serial port. The signals are stored on a text file formatted as a matrix, on which the first column shows the time of the sample (in ms) and the next 6 columns are the digital value of the voltage of the six channels' outputs of the conditioning circuit; a new line is written every 12 ms, which represents a sampling frequency rate of 83.3 Hz.

All the data acquired during the trials are processed in MATLAB environment; the developed script reads and imports the log files, converts the digital voltage data into the force components data, eliminates the offset voltage and generates all the graphics that show the force, torque, power, cadence and bilateral asymmetry index, in function of time, to the desired segment of the trial. From individual instantaneous power and force measurement data, the average power, its standard deviation and the bilateral asymmetry index, referred to the effective force, are determined for each trial and for the full experiment; the values are then applied to the inputs of an Adaptive Neuro Fuzzy Inference system, implemented using the Fuzzy Logic MATLAB Toolbox, which determines a score that represents the cyclist performance on a scale that varies from 0 to 100%.

2.3 Performance Classification

2.3.1 General Fuzzy Methodology

To classify the performance of the subject, a Fuzzy Inference System (FIS) was developed and implemented in MATLAB Environment, using the concept introduced in [2] which defines a fuzzy set as a set or collection of elements with membership

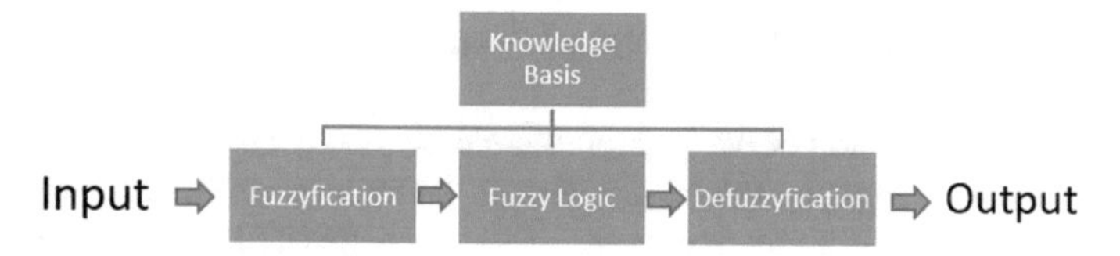

Fig. 4 General Fuzzy Inference System structure

values comprehended in the interval of 0–1, where, differently from the Boolean logic (True or False), it can assume any value within this range. The membership value is characterized by its fuzzy Membership Function (MF), which is modeled in intervals that represent linguistic expressions, e.g., "the value of the power generated by the athlete is good". The FIS used in this study is based on the Takagi-Sugeno Fuzzy Inference System [20] and can be described as shown in Fig. 4, where the knowledge basis represents the application background necessary to develop the system.

The Fuzzyfication interface is the first stage of the FIS and is responsible for the conversion of the scale of input values in the fuzzy scale universe, using the Membership Function. After the conversion, the crisp input values become a set of linguistic expressions, e.g. a 300 W power input value, in this specific system, is defined as an "Excellent" amount of power.

The fuzzy logic is the stage responsible for taking the decision based on a set of rules previously defined by the expert; it is the stage that simulates the logics that are regularly used by people on a daily basis. The rules are determined using the linguistic expressions, e.g. "If Average Power is Poor and Power STD is Poor, then Performance Index is Poor". From the rules set and the heuristic inputs, the fuzzy output is generated using a Strategy (Inference system). This study is based on Sugeno's inference system; the fuzzyfied input values are applied to each rule, using a product implication and applying a sum aggregation, as proposed by [20], resulting in a fuzzy output value.

Differently from Mamdani's Defuzzyfication method, Sugeno's does not have Membership Functions for the Defuzzyfication stage, instead of that, the application of each rule results in a fuzzy singleton, i.e. a well defined value that is unity at a single particular point in the universe of discourse and zero everywhere else. This system was developed using the linear model [20–22]; the rules were formulated as in Eq. (3).

$$IF\, x_1^j\, IS\, A_1^j\, AND\, x_2^j\, IS\, A_2^j \ldots AND\, x_i^j\, IS\, A_i^j\, THEN\; z_j = p_j x_1^j + q_j x_2^j + \cdots + r_j x_i^j + s_j \tag{3}$$

where $\boldsymbol{x}_i^j$ are the inputs of the system, $\boldsymbol{A}_i^j$ are the fuzzy sets in the antecedent and z_j is a crisp function in the consequent, i.e. the result of the application of rule $\boldsymbol{j}$. As seen in Eq. (3), each rule is represented by a linear function of its inputs, where the output of the rule is the result of the first order equation. Therefore, the system output for a set of inputs is the weighted average of all the rules outputs.

2.3.2 Adaptive Neuro-Fuzzy Inference System

ANFIS is the combination of two computational methods, the Artificial Neural Network (ANN) [23, 24] and the Takagi–Sugeno [20] inference system; a hybrid learning algorithm [15, 25, 26] is applied to the ANN to synthesize a FIS system. The combination of the Fuzzy methodology to ANN unifies the FIS advantages of converting the qualitative aspects of human knowledge into quantitative ones with the ANN learning capability, resulting in a powerful tool of supervised learning which enables the user to develop a FIS system in a shorter period of time, as the membership functions are automatically adjusted by the ANN, in addition to reducing the rate of errors in the determination of rules in fuzzy logic [15]. Figure 5 shows the architecture of a two inputs ANFIS system.

As shown in Fig. 5, the architecture of the two-input first-order ANFIS model, with two rules, is composed by five layers, which can be described as follows [27]:

- **Layer 1**: composed by adaptive nodes, where the output of each node is a degree of membership value that is given by the MF; it is the layer responsible for the MF's parameters adjust.
- **Layer 2**: composed only by nonadaptive nodes, where each node is responsible for the multiplication of the incoming signals and for delivering it to the next node, representing the firing strength of each rule. A T-norm operator is usually applied to obtain the output layer.
- **Layer 3**: composed only by nonadaptive nodes; calculates the ratio of the rules' firing strength.
- **Layer 4**: composed by adaptive nodes; applies weights to the inputs' parameters to determine the values of the consequent function of each rule application.
- **Layer 5**: composed by fixed nodes to compute the overall output as a summation of all the incoming signals.

The ANFIS system can be implemented using several programming languages, however, its development can be generally summarized into three steps [27]: the

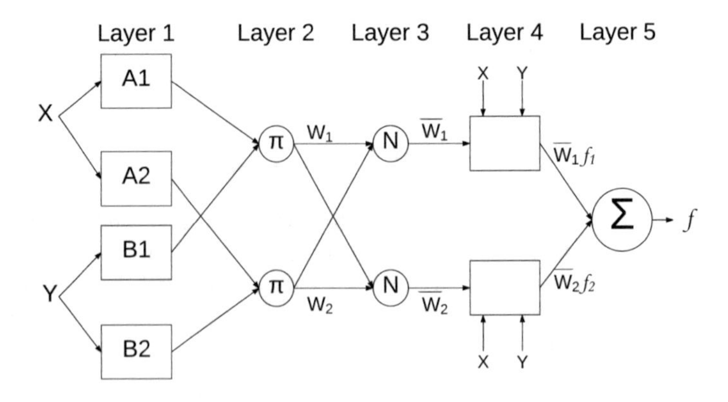

Fig. 5 Two inputs ANFIS architecture

definition of the training and validation data sets, composed by the inputs values and the desired output for each inputs' combination, the definition of the system rules, the training and the validation. During the training phase, each layer of the ANFIS architecture is determined and the weights of the network are tuned, during each training epoch, to decrease the error among the desired outputs and the ANFIS outputs. After that, the validation dataset is used to determine the quality of the training, through the determination of the average error among the validation dataset output and the developed system output; then, if the minimum error requirements are met, the system may be considered ready for use.

2.3.3 Fuzzy Performance Classifier

One of the major advantages of the Fuzzy methodology among a wide variety of classifiers, such as Random Forest [28], Support Vector Machines [29], Multilayer Perceptron [30], Naive Bayes [31] and others, is the possibility of converting qualitative descriptions, used in daily basis, into quantitative definitions, which may facilitate the application of the specialist knowledge to develop the intelligent system [32, 33]. In the Sports area, most of the definitions used by coaches to evaluate the athlete's performance and training are expressed in qualitative terms; as the performance usually is based not only on a single variable, but on a set of characteristics, when developing an automatic classification system, it is desirable to use a methodology that enables the specialist to use his own words and variables on the development of the system. Therefore, this research was developed using the ANFIS classifier, which was developed and implemented in MATLAB Environment. The simplified structure of the developed Fuzzy system is shown in Fig. 6.

As seen in Fig. 6, the system is composed by three inputs, which are represented by the left vertical aligned boxes, where the AVG-POWER is the Average Power input, AVG-ASY is the Average Bilateral Asymmetry index and STD-POWER is the Power standard deviation. A summarized representation of the developed ANFIS

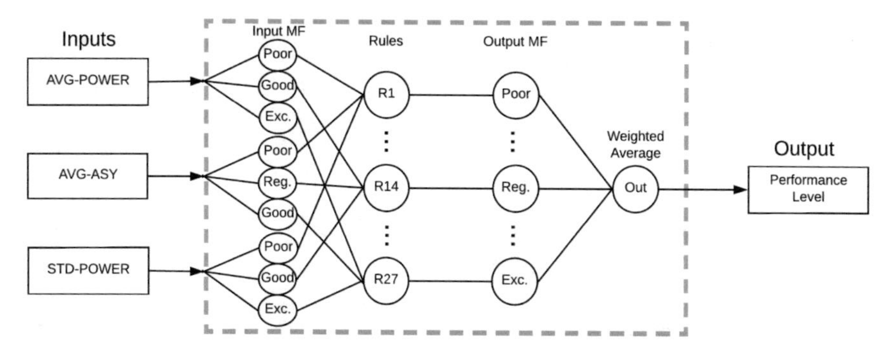

Fig. 6 Developed fuzzy inference system structure

architecture is presented inside the dashed line rectangle and the output is represented by the box denominated Performance Level.

2.3.4 ANFIS Fuzzyfication Interface

The determination of the preliminary values of the input Degrees of Membership (DM) was made based on previous studies for the Average Power MF input, where it was considered a maximum average mechanical power of about 300 W for a regular athlete [5, 9, 10, 17, 34], and for the Bilateral Asymmetry Index [5, 7, 9, 16, 35, 36], as this data have a well-established pattern. The Average Power STD MF values were defined based on the data acquired during the experimental trials conducted with the subjects. As the Fuzzyfication stage is responsible for mapping the crisp input values into the fuzzy values, it is important that the mathematical functions that compose the DM behave as close as possible to the input data that it represents [37]; the data applied to the input of the system follow a normal distribution, thus the DM functions chosen to compose de MF were Gaussian, as in Eq. (4), or Sigmoid, as in Eq. (5). Figure 7 shows the preliminary Membership Functions for all inputs, which represents the fuzzyfication stage of the system developed before training.

$$f\,(x\,|\sigma|\,c) = e^{\frac{-(-x-c)^2}{2\sigma^2}} \tag{4}$$

where $\boldsymbol{c}$ is the position of the center of the peak and $\boldsymbol{\sigma}$ is the standard deviation of the Gaussian function.

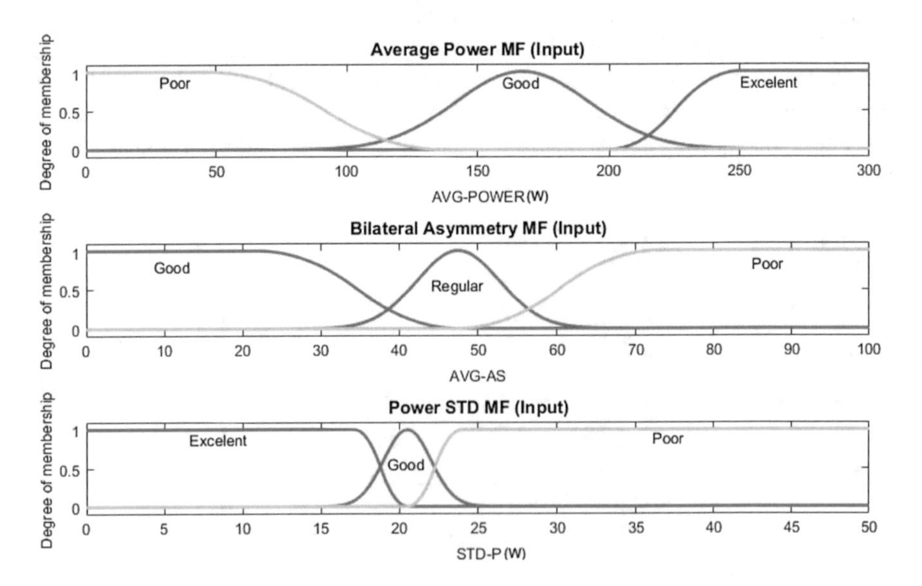

Fig. 7 Fuzzy inputs membership functions

Table 1 Preliminary input degrees of membership parameters

Average power input			
Linguistic expression	Parameters		Function
Excellent	$\alpha = 200$	$b = 250$	Sigmoid
Good	$\sigma = 25.5$	$c = 167$	Gauss
Poor	$\alpha = 47.0$	$b = 137$	Sigmoid
Average bilateral asymmetry input			
Linguistic expression	Parameters		Function
Good	$\alpha = 22.0$	$b = 47.5$	Sigmoid
Regular	$\sigma = 5.31$	$c = 47.5$	Gauss
Poor	$\alpha = 47.5$	$b = 73.0$	Sigmoid
Average power STD			
Linguistic expression	Parameters		Function
Excellent	$\alpha = 17.1$	$b = 20.5$	Sigmoid
Good	$\sigma = 1.49$	$c = 20.5$	Gauss
Poor	$\alpha = 20.5$	$b = 24.0$	Sigmoid

$$f(x|a|b) = \begin{cases} 0, & x \leq a \\ 2\left(\frac{x-a}{b-a}\right)^2, & a < x \leq \frac{a+b}{2} \\ 1 - 2\left(\frac{x-a}{b-a}\right)^2, & \frac{a+b}{2} < x \leq b \\ 1, & x > b \end{cases} \tag{5}$$

where ***a*** is the position of the left extreme of the curve and ***b*** is the position of the right extreme of the Sigmoid function.

Analyzing the DM functions shown in Fig. 7 it is possible to notice that each input Membership Function is composed by three DM functions which map the input crisp value into the fuzzy value defined by the linguistic expression. Table 1 shows the input DM parameters in more details.

The MF parameters shown in Table 1 are the values of the constants of the Eqs. (4) and (5), which determine the position and limits of each function that composes the MF.

2.3.5 ANFIS Logic and Defuzzyfication

Since the ANFIS requires a number of rules equals to the number of possible combinations of the membership functions of all the inputs, the ANFIS Logic was developed considering the specialist's background knowledge and the possible combinations of the system inputs. The rules set is presented in Table 2, where PWR is the average

Table 2 ANFIS rules

Rules	
1.	If PWR is Poor and B.A is Poor and PSTD is Poor then PL is Poor
2.	If PWR is Poor and B.A is Poor and PSTD is Good then PL is Poor
3.	If PWR is Poor and B.A is Poor and PSTD is Exc. then PL is Poor
4.	If PWR is Poor and B.A is Reg. and PSTD is Poor then PL is Poor
5.	If PWR is Poor and B.A is Reg. and PSTD is Good then PL is Poor
6.	If PWR is Poor and B.A is Reg. and PSTD is Exc. then PL is Poor
7.	If PWR is Poor and B.A is Good and PSTD is Poor then PL is Poor
8.	If PWR is Poor and B.A is Good and PSTD is Good then PL is Poor
9.	If PWR is Poor and B.A is Good and PSTD is Exc. then PL is Poor
10.	If PWR is Good and B.A is Poor and PSTD is Poor then PL is Poor
11.	If PWR is Good and B.A is Reg. and PSTD is Poor then PL is Poor
12.	If PWR is Good and B.A is Good and PSTD is Poor then PL is Poor
13.	If PWR is Good and B.A is Poor and PSTD is Good than PL is Reg.
14.	If PWR is Good and B.A is Reg. and PSTD is Good then PL is Reg.
15.	If PWR is Good and B.A is Good and PSTD is Good then PL is Reg.
16.	If PWR is Good and B.A is Poor and PSTD is Exc. then PL is Reg.
17.	If PWR is Good and B.A is Reg. and PSTD is Exc. then PL is Good
18.	If PWR is Good and B.A is Good and PSTD is Exc. then PL is Good
19.	If PWR is Exc. and B.A is Poor and PSTD is Poor then PL is Reg.
20.	If PWR is Exc. and B.A is Reg. and PSTD is Poor then PL is Good
21.	If PWR is Exc. and B.A is Good and PSTD is Poor then PL is Good
22.	If PWR is Exc. and B.A is Poor and PSTD is Good then PL is Good
23.	If PWR is Exc. and B.A is Reg. and PSTD is Good then PL is Good
24.	If PWR is Exc. and B.A is Good and PSTD is Good then PL is V.G.
25.	If PWR is Exc. and B.A is Poor and PSTD is Exc. then PL is V.G.
26.	If PWR is Exc. and B.A is Reg. and PSTD is Exc. then PL is Exc.
27.	If PWR is Exc. and B.A is Good and PSTD is Exc. then PL is Exc.

power during the trial, B.A is the mean bilateral asymmetry index, PSTD is the power standard deviation and PL is the performance Level.

As seen in Table 2 the rules set is composed by 27 rules, which defines the logic of the system based on the fuzzyfied value of each input. It is interesting to point out that although three variables are considered in the determination of the athlete's performance, the two most important ones are the power applied to the pedal and the power std; rules 1–9 logic implies that, regarding the values of the power std and the bilateral asymmetry, if the power is considered poor, the output is poor. This is justified by the fact that, notwithstanding the other fuzzyfied input values are considered good, the performance on track will not be good if the power level is poor. Furthermore, by the analysis of rules 19–21, it is possible to notice that even if the average power level is excellent, if the power std is poor, the output does not reach high values; this is important because it measures the endurance of the athlete, which is essential on track. Therefore, the rules were formulated to combine these three input variables that represent the pedaling characteristics of the athlete in a way that could measure the athlete's training level in a more comprehensive way than the usual techniques based only in the power measurement data.

As the ANFIS is based on Takagi-Sugeno Inference method, the output is the weighted average of the results obtained by the application of each rule to the inputs.

2.4 Experimental Evaluation

To evaluate the behavior of the developed system, a randomized block experiment design was performed indoor using a Btwin IN'Ride 300 magnetic-braked ergometer roll. A total of 300 runs (total distance of 273.2 km, measured using a bicycle speedometer) with a duration of 75 s each were performed by 7 healthy amateur cyclists, 5 cyclists with at least 2 years of experience and 3 professional cyclists; each subject made twenty repetitions of the test, of which ten were performed in the lowest level of magnetic braking (L) and ten in the third level of magnetic braking (2). The sample was one female and 14 males of 29 ± 5 years, a body mass of 73 ± 9 kg and a height of 1.78 ± 0.07 m. The subjects were instructed to adjust the bicycle ergonomics to an equivalent configuration as they use on their own bicycle; the cadence level was freely chosen by each cyclist but the gear was fixed in all trials. An ANOVA test was performed to analyze the significance of the collected data and the subjects' performance scores determined by the developed system and the specialist. From data of the average Power, its STD and the Effective Force bilateral asymmetry, the performance score of each cyclist was determined using the ANFIS. In order to respect personal decision and the Declaration of Helsinki of the World Medical Association, declared consent was obtained from the subjects.

3 Results and Discussion

3.1 Static Calibration Experiment

The static calibration experiment was performed applying known loads to the crank arm in the direction of each force component, for the same range to which the load cell will be used in the application, while the voltage output of each channel was being measured. Channel's 1 (Perpendicular force component measurement) as function of perpendicular force application is shown in Fig. 8. Other channels transfer functions are presented in Table 3.

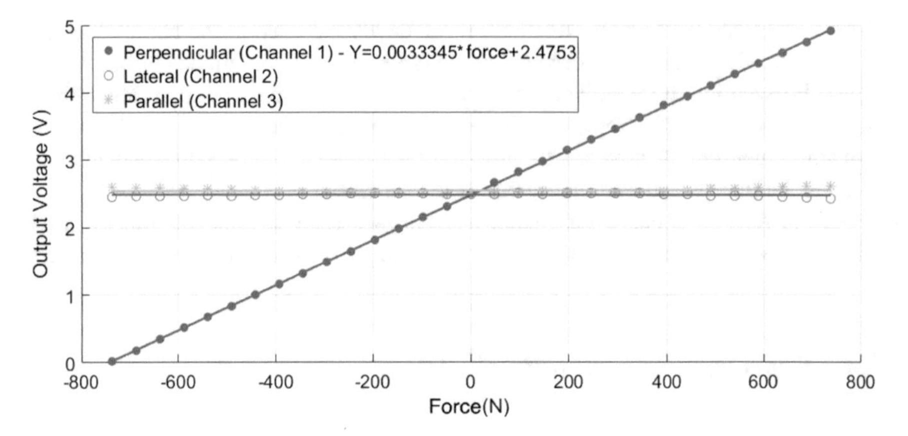

Fig. 8 Voltage output as function of perpendicular force application

Table 3 Load cell voltage output characteristics

Channel	Transfer function (mV)	Linearity error (%)
1	y = 3.334. force + 2475	0.60
2	y = 8.500. force + 2501	0.19
3	y = 1.766. force + 2517	0.41
4	y = 3.303. force + 2508	0.60
5	y = 8.007. force + 2491	0.20
6	y = 1.922. force + 2502	0.44

From Table 3 data it is possible to point out that the mean sensibility is 3.321 mV/N for perpendicular force measure, 8.251 mV/N for lateral force measure and 1.844 mV/N for parallel force measure; the linearity error of all channels is below 0.6%.

Uncertainty analysis was developed using the classical approach, considering the tolerance of the active and passive components of each channel as an uniform distributed function with infinite degrees of freedom [38]. Uncertainty for all channels' sensitivity is below 3.2%.

3.2 Force Components and Power Data

From the force components data acquired during the trials it is possible to analyze the bilateral asymmetry between the forces that are applied by each leg as the cadence kept by the cyclist and other patterns referred to the pedaling technique of the athlete. The instant power produced by the athlete and transmitted to the crank arm is determined based on the effective force component, the cadence and the crank arm length,

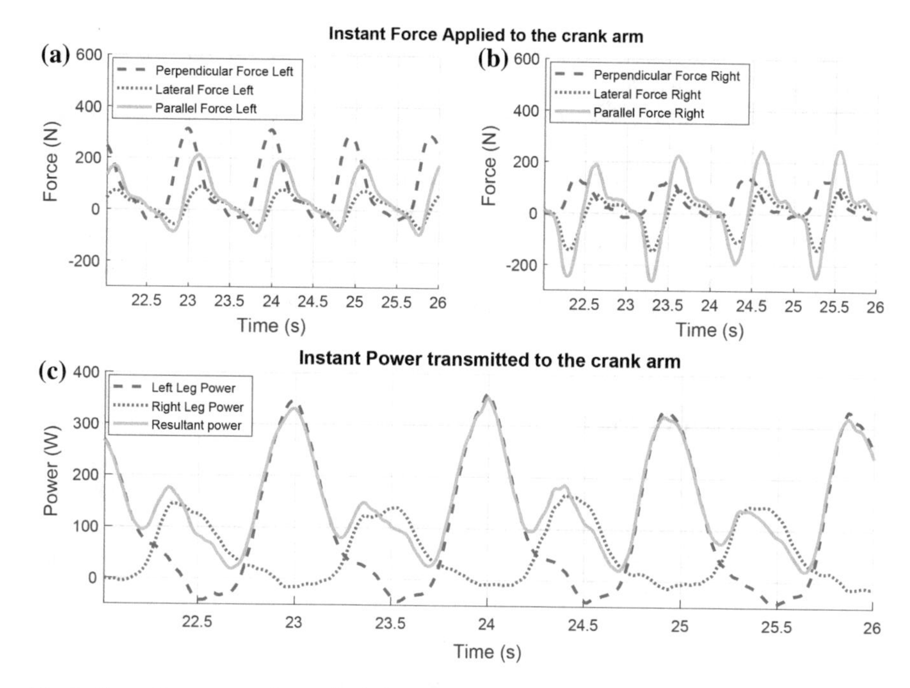

Fig. 9 Force Components data: **a** Left Crank arm, **b** Right Crank arm; Instant Power transmitted to the crank arm (**c**)

as shown in Eq. (1) [5, 9]; the bilateral asymmetry index is determined according to Eq. (2), considering the effective force applied to the crankset by each leg [11]. Figure 9 shows a 4 s duration segment of the data acquired during one of the trials.

Analyzing Fig. 9a, b it is possible to notice the difference between each leg forces application pattern; the maximum effective force applied by the left leg is 314.97 N, while the maximum effective force applied by the right leg is 143.70 N, which leads to a bilateral asymmetry index of 54.4% referred to the peak perpendicular force, considering this segment of the trial. The average perpendicular force is 92.48 N applied by the left leg and 45.04 N applied by the right leg, which leads to an asymmetry index of 51.3% referred to the mean effective force. It is interesting to point out that the different values found for the bilateral asymmetry index, referred to the same force component, show that not only the peak of force varies from one member to another, but also the way the force application is distributed over time, affecting the average force that is able to generate torque to rotate the crank arm. As shown in Fig. 9c the peak of power is 363.48 W, applied by the left leg and 167.03 W applied by the right leg, while the average resultant power is 154.46 W. As the crank arms are connected by the bottom bracket in a 180° angle, each leg is in the opposite phase simultaneously, which implies that a power supplied in the opposite direction reduces the cyclist efficiency as it will be contrary to the rotation of the crankset; in Fig. 9c it is noticeable that a peak of −44 W is applied by the left leg and −16 W by the right leg, which negatively impacts the resultant power, decreasing the overall performance of the athlete.

The differences among the patterns of the forces that are applied by each member may be related to several variables, as the position of the foot in the pedal, the bicycle ergonomics (saddle height, type of pedal used, and other adjusts) and, most importantly, the subject's technique level, which is, in a matter of fact, one of the reasons for the development of this study, as, knowing the pedaling characteristics of the athlete may enable the coach to develop training strategies to improve the cyclist's technique, leading to a performance improvement.

3.3 Experimental Data

From data of the forces, acquired during the trials, and the Eqs. (1) and (2), the average power, its standard deviation and the bilateral asymmetry index were determined. For each trial, a 60 s duration sample of the data was randomly selected and the variables that will be used as input of the classification system were calculated. Figure 10 shows the data of the power and bilateral asymmetry index achieved by each subject during all the trials.

As shown in Fig. 10 it is possible to notice that the average bilateral asymmetry index and the average power vary significantly among trials, considering the same subject, and among subjects. This variation may be explained by the fact that the only controlled factors considered in the experiment design was the subject and the level of magnetic braking; the subjects were instructed to pedal in the highest speed that they could keep during all the trials, i.e., they were able to choose their preferred cadence and speed, leading to high variations of the pedaling characteristics. The maximum average power transmitted to the crank arm was 288.85 W, achieved by

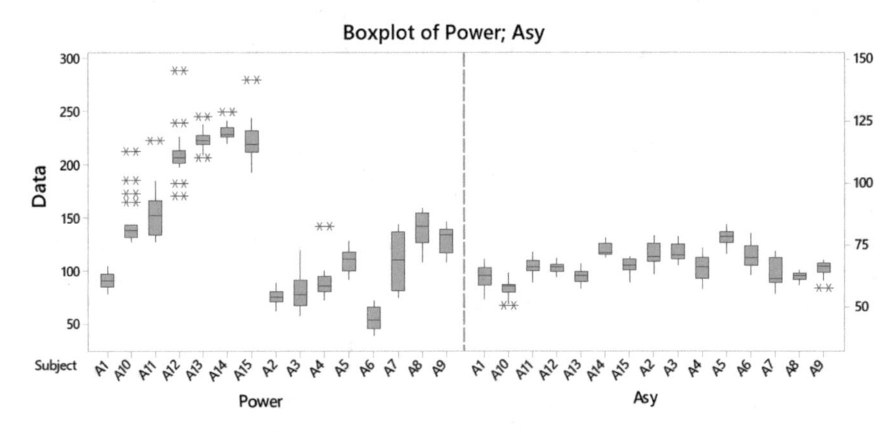

Fig. 10 Average power and bilateral asymmetry index achieved per subject considering all the trials

subject A12 during the first trial, while the minimum average power observed was 38.67 W, reached by subject A6; the mean power workload, considering all the subjects and trials is 137.63 ± 59.60 W, of which 34.82 ± 15.77 W was applied by the right leg and 103.77 ± 44.40 W was applied by the left leg; the average variation among trials considering the same subject is 57.72 ± 28.76 W. The maximum average bilateral asymmetry index value was 83.4%, achieved by subject 5, while the minimum was 50.43% for subject 10; the average bilateral asymmetry index of the group, determined considering all the trials and subjects is 67.0 ± 6.2%. The maximum variation of the average power and bilateral asymmetry observed among trials considering the same subject was 118.1 W and 17.3%, respectively. Previous studies realized with six professional and amateur cyclists have shown a bilateral asymmetry index of about 25% in the beginning of an incremental test, to a work load corresponding to 90% of the cyclist's oxygen consumption [36]. Another research developed with masters' cyclists showed a bilateral asymmetry index in a range of 10– 60%, with a mean value of 30 ± 8%, for a workload of 100–250 W [10]. All the data acquired during the trials were analyzed through an ANOVA test which showed that it is significant.

3.4 ANFIS Evaluation

3.4.1 ANFIS Configuration After Training

The data acquired during the trials were divided into two datasets: the training dataset, which represents 70% of the collected data, and the testing dataset, which represents 30% of the data. The ANFIS training was applied using the Hybrid Learning Algorithm [15, 25, 26]; the training process was developed in 300 epochs. The error measurement methodology adopted was the root mean square error (RMSE) [39–41]; the RMSE reached an average value of 0.646 for the training data (210 data pairs) and 1.09 for the testing data (90 data pairs) while the highest RMSE found was 1.61. Considering the worst case, a maximum error of 1.61 represents a relative output error of 1.61% as it is referred to the Score, which can take values from 0 to 100% in function of the input parameters. Figure 11 shows the ANFIS system output versus the dataset output, i.e. the expected output in function of the input datasets.

As shown in Fig. 11 the ANFIS output values obtained using the testing data as input of the system are very close to the targets, which means that the parameters of the neural network after training was well fitted to the input data, as the ANFIS output, using the dataset as input, follows the same pattern as the scores determined by the specialist.

To ensure a proper validation of the system for the full dataset, a K-Fold cross-validation (K = 10) was developed using the subjects' data as input of the system; the full dataset was split into 10 mutually exclusive subsets of equal size, where the trials that compose each subset were randomly picked. For each iteration of the method, 9 subsets were used for training the model while one was used for testing [42]; the

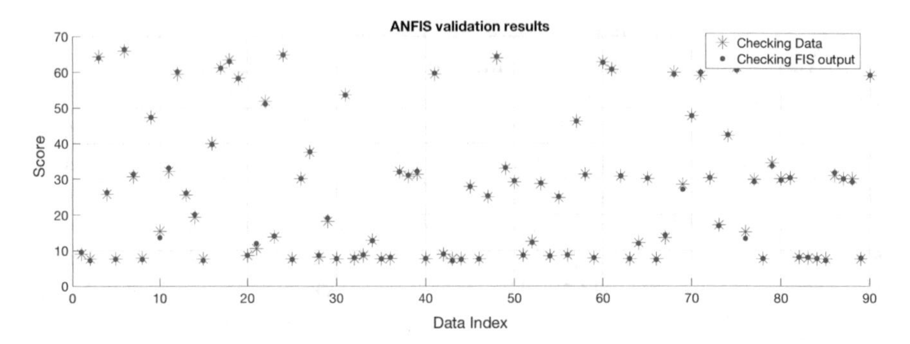

Fig. 11 ANFIS output test realized with the training and checking datasets

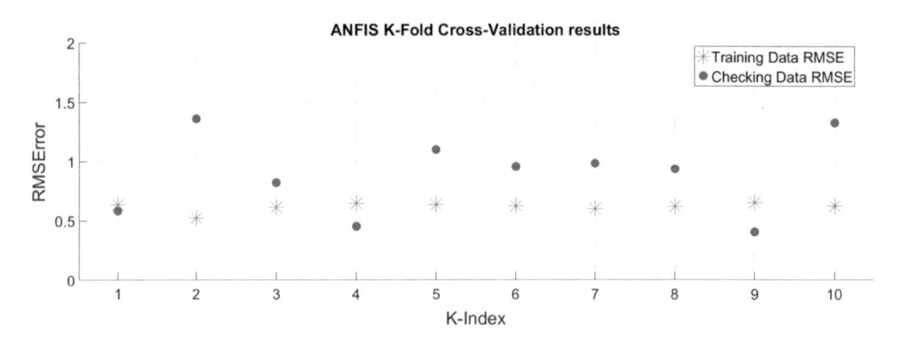

Fig. 12 ANFIS K-fold cross-validation error results

RMSE was computed in each iteration. Figure 12 shows the RMSE obtained by the application of the K-Fold cross-validation method.

Analyzing Fig. 12 it is possible to notice that the maximum RMSE obtained for the checking data was 1.36, while the minimum was 0.405. From the values of the RMSE obtained in each iteration of the K-Fold cross-validation, the mean classification error of the system was determined; the average RMSE is 0.89 ± 0.33. After the training was applied, the developed system was composed by 78 nodes, 108 linear parameters and 18 nonlinear parameters. Since the ANFIS is an adaptive system, its characteristics depend on the training dataset used to determine the weights of the neural network constants, implying that the input membership functions are not fixed. Considering this, the final inputs MF, after the training was applied, are shown in Fig. 13; more details about the functions parameters are presented in Table 4.

From the comparison of Fig. 13 with Fig. 7 it is possible to notice that the main difference among the preliminary inputs MF and the final ones is the range assumed by the variables, which is explained by the adaptation of the Neuro-Fuzzy model to the dataset that were used in the training, i.e. the new MF inputs range is in the same range as the variables used in the training.

Analyzing Tables 1 and 4 it is possible to point out that all the parameters of the Membership Functions remained unaltered except for the average bilateral asym-

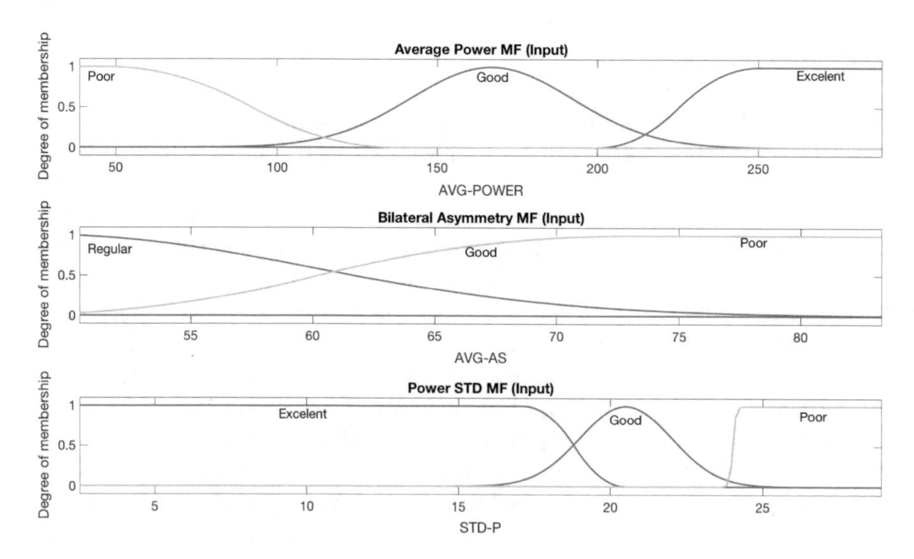

Fig. 13 ANFIS final Input Membership Functions

Table 4 ANFIS Final input degrees of membership parameters

Average power input			
Linguistic expression	Parameters		Function
Excellent	$\alpha = 200$	$b = 250$	Sigmoid
Good	$\sigma = 25.5$	$c = 167$	Gauss
Poor	$\alpha = 47.0$	$b = 137$	Sigmoid
Average bilateral asymmetry input			
Linguistic expression	Parameters		Function
Good	$\alpha = 22.0$	$b = 47.5$	Sigmoid
Regular	$\sigma = 10.6$	$c = 49.2$	Gauss
Poor	$\alpha = 47.5$	$b = 73.0$	Sigmoid
Average power STD			
Linguistic expression	Parameters		Function
Excellent	$\alpha = 17.1$	$b = 20.5$	Sigmoid
Good	$\sigma = 1.49$	$c = 20.5$	Gauss
Poor	$\alpha = 20.5$	$b = 24.0$	Sigmoid

metry input, in which the MF parameters that represent the "Regular" linguistic expression have changed from $\sigma = 5.31$ and $c = 47.5$ to $\sigma = 10.6$ and $c = 49.2$.

3.4.2 Fuzzy System Evaluation

The purpose of the developed cycling performance classifier is to implement a new metric that enables the coach to predict the potential results that would be achieved by the athlete in a real competition based on the results of an indoor trial. To accomplish that, not only the power transmitted to the bicycle pedal is considered but also its application uniformity, through the consideration of the power STD and the pedaling characteristics, through the bilateral asymmetry index. Therefore, the values that were used as input of the performance classification system were determined based on the data acquired during the trials and are shown in columns one to four of Table 5; the scores obtained by each cyclist, using the data as input of the ANFIS, for the considered trials, are shown in column five. To properly evaluate the quality of the model prediction, not only the error obtained with the training data should be considered, but also the ability of the model to predict the output correctly when different data than the one used in building the model are applied to its inputs [43]. Therefore, the system was tested only with the data that were not used to train the ANFIS; the values presented in Table 5 represent only 30% of the trials, of which the average power workload of the group was 123.46 ± 57.749 W, the mean bilateral asymmetry index was $65.98 \pm 5.4\%$ and the average power standard deviation was 8.930 ± 5.495 W.

Analyzing the results presented in Table 5, it is possible to notice a clear difference among the scores obtained by the amateur, the more experienced and the professional cyclists; subjects 1 to 7, which declared themselves as amateur cyclists, scored values from 7.5 to 12%, which is considered a poor performance level, using the linguistic

Table 5 Data acquired per subject during all trials and performance scores

Subject	Average power (W)	Power STD (W)	Average bilateral asymmetry (%)	Score (%)
1	97.81	59.11	6.4	11.2
2	68.85	74.01	4.9	7.7
3	72.76	73.24	8.9	7.9
4	82.95	64.12	13.6	8.2
5	108.86	77.49	12.9	13.4
6	63.47	67.21	6.9	7.5
7	85.22	62.10	16.0	8.6
8	123.46	62.10	9.1	28.3
9	133.84	67.35	6.9	29.4
10	135.74	58.34	4.5	34.4
11	154.15	66.93	17.8	29.9
12	207.22	64.24	17.9	34.2
13	223.01	63.01	5.9	59.6
14	230.19	72.03	7.2	62.6
15	217.29	67.18	14.7	53.2

term defined in the universe of discourse, while the volunteers from 8 to 12, which declared that they use to cycle on a daily basis, reached values around 28.3–34.4%, which is considered a regular performance level, thus, the subjects from 13 to 15, the professional athletes, reached values from 53.2 to 62.6%, which is considered a very good performance level. Therefore, the scores obtained during the trials were considered compatible with the experience level declared by the subjects. Taking into consideration that there were groups of volunteers with significantly different levels of cycling experience participating on the study, it is considered to be an interesting result.

The main goal of the developed system is to introduce a new point of view over the cycling performance measurement that not only considers the average or peak power reached by the cyclist in an indoor trial to determine the training level of the athlete, but to develop a method that analyzes the athlete's pedaling pattern leading to a more representative result, as it considers the consistency of the cyclist's technique and the possibility of improving it. For instance, comparing the score obtained by subject 12 (34.2) with the one reached by subject 10 (34.4) it is considered that subject 10 achieved a better performance in the test, however, if the only variable that was analyzed to determine the performance was the power, subject 12 would be considered more trained than the other, as the average power applied to the pedal by subject 12 (207.22 W) is higher than the one applied by subject 10 (135.74 W). This result is explained by the fact that even though subject 12 applied more power to the pedal, his power STD (17.9 W) was higher than subject's 10 (4.5 W). Taking into consideration that all the trials were developed in the same environmental conditions, using the same equipment, it is considered that, although subject 12 developed more power, subject 10 kept it more constant, which may be interpreted as a better performance, as it can be related to a higher physiological condition or a better pedaling technique than the other.

Previous studies conducted with elite sprint cyclists have found that there is no statistically significant difference between the values of the maximum power and torque among indoor and outdoor trials [44, 45]. Thus, the correlation between the peak power (PPO) on a maximal aerobic power indoor test, carried out with aid of an ergometer, and the PPO on a 16.1 km Time Trial (TT) endurance test outdoor trial is strong and positive ($r = 0.99$, $P < 0.001$), however, the correlation among the peak power and the TT times is weak ($r = 0.46$, $P < 0.05$) [46]. Another study conducted with adolescent cyclists showed that although the maximal and mean power measured in an indoor trial and on track are strongly correlated, the power output results were consistently higher in the laboratory trials [47]. Therefore, it is noticeable that the peak of power data obtained in an indoor trial is not enough to classify the athlete's performance or to predict the results that will be achieved on track, which corroborates with the conclusion that another pedaling characteristics, rather than just the power, should be considered to determine the overall performance score of the athlete. Taking into account that the approach proposed in this study combines the computational intelligence and the specialist knowledge to determine the performance score of the cyclist based on the pedaling pattern, it is considered that it may be a great tool for coaches to objectively compare the performance progression of the athlete among trials, before and after the training strategy has been applied, and for performance analysis among different athletes.

3.4.3 Statistical Analysis

The score and power data were examined on Minitab 17 Stats tool, which showed that the data significance can be analyzed through an Analysis of Variance as its residuals follow a normal distribution and the premise of equal variances is valid [48]. The behavior of the residuals distribution was visually checked, through its histogram, and examined using a Ryan-Joiner test, which corroborated the conclusion that it follows a normal distribution, since $RJ = 0.977$, for a P-Value > 0.100. Considering this, the significance of the collected data was analyzed through an ANOVA test, developed using a 99% confidence interval (F-Table $\alpha = 0.01$), considering three controlled factors: the subject, factor A, of 15 levels, the classification methodology, factor B, of two levels (the score determined by the specialist and the score obtained through the developed system) and the level of magnetic breaking applied on the training roll (denominated as gear), factor C, of two levels. ANOVA showed that the only controlled factors that significantly impact the average power and the score response variables is the subject (Power F-Value = 601.88, Score F-Value = 209.18, P-Value = 0.000) and the gear (Power F-Value = 93.71, Score F-Value = 60.09, P-Value = 0.000), i.e. the data are significant. Since the applied test is a hypothesis test, whose results show that all the means are equal or at least one mean is different than the others, it is interesting to apply a multiple means comparison test to analyze which of the means are significantly different; Fisher's multiple means comparison test showed that all means of the score and power are significantly different (F-Table $\alpha = 0.01$). Figure 14 shows the main effects plot for the score.

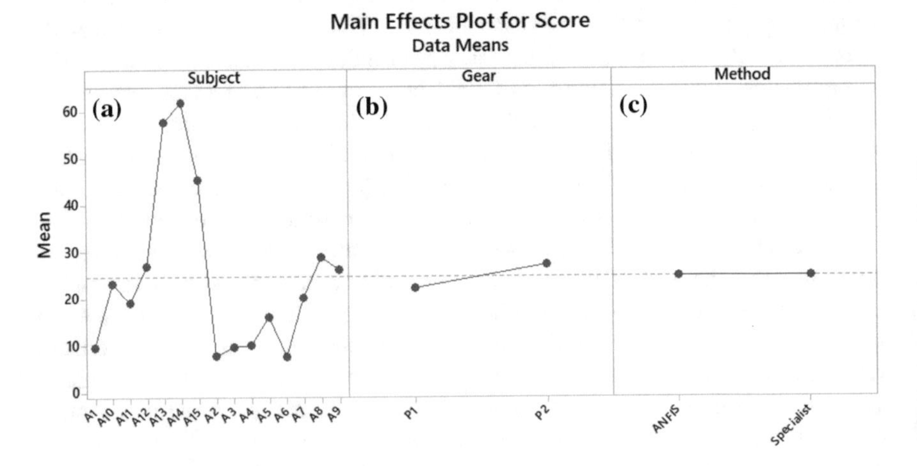

Fig. 14 Main effects plot for Score in function of: **a** Subject, **b** Gear and **c** Method

Analyzing Fig. 14 (a) it is possible to notice the subject's influence over the score values, e.g. while level A1 of the subjects' controlled factor results in a mean output value of 9.74%, level A14 results in a mean output of 61.8%, (b) from the gear's main effects plot it is possible to view a significant difference between the mean score obtained in levels P1 and P2, (c) from the method's main effects plot it is noticeable, as expected, that the method has no significant influence over the score. As the ANOVA showed that there is no significant difference among the score values determined by the specialist and the ones determined by the developed system, using the same data as input, it is considered that the intelligent system was able to recognize the pedaling patterns shown by the cyclists.

4 Conclusion

The aim of this study was to develop an intelligent automatic system to measure, analyze and evaluate the pedaling technique of a cyclist based on the forces that are applied to the bicycle pedal during cycling. To accomplish that, an instrumented load cell was developed and built from scratch, with mechanical and dimensional characteristics equivalent to a standard crankset used in road cycling competitions, but with the ability to measure the components of the forces that are applied to the bicycle pedal with a linearity error under 0.6% and an uncertainty under 3.2% referred to the sensitivity of the channel. To evaluate the system a randomized block Experiment was developed; ANOVA showed that all the data acquired during the trials were significant. Data comparison with results obtained by other studies developed using similar methods showed consistent similarity among this study results and the ones found on the literature. To evaluate the cyclist performance an Adaptive Neuro Fuzzy Inference system was developed and trained with the data acquired during the trials, previously analyzed by the specialist; after training, the system showed a maximum average error of 0.65% for the checking data and 1.1% for the training data.

Based on the data acquired during the trials, the average power, its STD and the bilateral asymmetry index were determined and used as input of the classifying system; the group reached an average score of 24.7% $\pm$ 18.7%. Taking into account that there were subjects with different levels of training and that the scores reflected the declared skills of each subject of the group, it was considered that the developed system showed consistent results. The combination of the forces data, acquired using the instrumented crankset system, with the score generated by the classification system, seems to be a powerful tool for helping coaches to evaluate the performance level of different athletes or the progression of the cyclist among different trials.

From the statistical analysis, it was shown that the ANFIS was able to recognize the pedaling patterns of the cyclist, as the ANOVA showed that there is no statistically significant difference among the scores determined by the specialist and the ones obtained with the developed system, using the same inputs. Analyzing the score means through the Fisher's method, it was possible to notice that all the score means are significantly different among the subjects. Therefore, it was considered that the use of a computational method based on Fuzzy to develop the intelligent performance analysis system was essential to apply the knowledge of the specialist, which is based on linguistic expressions.

References

1. A. Binet, T. Simon, New methods for the diagnosis of the intellectual level of subnormals (L'Ann´ee Psych., 1905), pp. 191–244, in *The Development of Intelligence in Children* (The Binet-Simon Scale), 1916, pp. 37–90
2. L. Zadeh, Fuzzy sets. Inf. Control **8**(3), 338–353 (1965)
3. A. Schmidt, *Competitive Cycling* (Meyer & Meyer Sport, 2014)
4. S. Sovndal, *Cycling Anatomy* (Human Kinetics, 2009)
5. C.D. Lazzari, A. Balbinot, Wireless Crankarm dynamometer for cycling. Sensors & Transducers **128**(5), 39–54 (2011)
6. R.R. Bini, F.P. Carpes, *Biomechanics of Cycling* (Springer International Publishing, Cham, 2014)
7. A.V. Pigatto, A. Balbinot, G.W. Favieiro, K.O. Moura, A new crank arm based load cell, with built-in conditioning circuit and strain gages, to measure the components of the force applied by a cyclist, in *2016 38th Annual International Conference of the IEEE Engineering in Medicine and Biology Society (EMBC)* (IEEE, Orlando, 2016), pp. 1983–1986
8. M. Ericson, R. Nisell, Efficiency of pedal forces during ergometer cycling. Int. J. Sports Med. **09**(02), 118–122 (1988)
9. A. Balbinot, C. Milani, J.D.S.B. Ahia Nascimento, A new crank arm-based load cell for the 3D analysis of the force applied by a cyclist. Sensors (Basel, Switzerland) **14**(12), 22921–22939 (2014)
10. W.M. Bertucci, A. Arfaoui, G. Polidori, Analysis of the pedaling biomechanics of Master's cyclists: a preliminary study. J. Sci. Cycling **1**(2), 42–46 (2012)
11. P. Chavet, M. Lafortune, J. Gray, Asymmetry of lower extremity responses to external impact loading. Hum. Mov. Sci. **16**(4), 391–406 (1997)
12. C.D. Paton, W.G. Hopkins, Tests of cycling performance. Sports Med. (Auckland, N.Z.) **31**(7), 489–496 (2001)
13. S. Sudin, A.Y. Md Shakaff, F. Aziz, F.S. Ahmad Saad, A. Zakaria, A.F. Salleh, Track cyclist performance monitoring system using wireless sensor network, in *Regional Conference on Science, Technology and Social Sciences (RCSTSS 2014)* (Springer Singapore, Singapore, 2016), pp. 123–131
14. J.Y. Xu, X. Nan, V. Ebken, Y. Wang, G.J. Pottie, W.J. Kaiser, Integrated Inertial sensors and mobile computing for real-time cycling performance guidance via pedaling Profile Classification. IEEE J. Biomed. Health Inf. **19**(2), 440–445 (2015)
15. J.-S. Jang, ANFIS: adaptive-network-based fuzzy inference system. IEEE Trans. Syst. Man Cybern. **23**(3), 665–685 (1993)
16. W. Smak, R.R. Neptune, M.L. Hull, The influence of pedaling rate on bilateral asymmetry in cycling. J. Biomech. **32**(9), 899–906 (1999)
17. D.J. Sanderson, E.M. Hennig, A.H. Black, The influence of cadence and power output on force application and in-shoe pressure distribution during cycling by competitive and recreational cyclists. J. Sports Sci. **18**(3), 173–181 (2000)
18. S. Franco, *Design with Operational Amplifiers and Analog Integrated Circuits*, 4th edn (McGraw-Hill, 2014)

19. B. Robert, *Northrop, Introduction to Instrumentation and Measurements*, 3rd edn. (CRC Press, Boca Raton, 2014)
20. T. Takagi, M. Sugeno, Fuzzy identification of systems and its applications to modeling and control. IEEE Trans. Syst. Man Cybern. **SMC-15**(1), 116–132 (1985)
21. M. Sugeno, T. Takagi, Derivation of fuzzy control rules from human operator control actions.pdf, in *IFAC*, 1983, pp. 55–60
22. M. Sugeno, G.T. Kang, Structure identification of fuzzy model. Fuzzy Sets Syst. **28**(1), 15–33 (1988)
23. A. Jain, J. Mao, K. Mohiuddin, Artificial neural networks: a tutorial. Computer **29**(3) 31–44 (1996). arXiv:1411.3159v1
24. A.P. Engelbrecht, *Computational Intelligence* (Wiley, 2007)
25. J. Jang, Fuzzy modeling using generalized neural networks and Kalman filter algorithm, in *Proceedings of the 9th National Conference on Artificial Intelligence*, vol. 91, 1991, pp. 762–767
26. J.S.R. Jang, C.T. Sun, Neuro-fuzzy modeling and control. Proc. IEEE **83**(3), 378–406 (1995)
27. J.S.R. Jang, C.T. Sun, E. Mizutani, *Neuro-fuzzy and Soft Computing; A Computational Approach to Learning and Machine Intelligence* (Prentice Hall, Upper Saddle River, 1997)
28. L. Breiman, *Random Forest* (Statistics Department University of California Berkeley, 2001), pp. 1–35
29. A. Ben-Hur, J. Weston, *A User's Guide to Support Vector Machines*, vol. 609, in Methods Molecular Biology (Springer, Clifton, NJ, 2010), pp. 223–239
30. M. Minsky, S. Papert, *Perceptrons*, vol. 522, expanded edn. (MIT Press, Cambridge, MA, 1988), p. 20
31. K. Larsen, Generalized Naive Bayes classifiers. ACM SIGKDD Explor. Newslett. **7**(1), 76–81 (2005)
32. A. Sala, P. Albertos, *Fuzzy Logic Controllers: Advantages and Drawbacks* (1998)
33. A.I. Al-Odienat, A.A. Al-Lawama, The advantages of PID fuzzy controllers over the conventional types. Am. J. Appl. Sci. **5**(6), 653–658 (2008)
34. V.-C. Omar, D. Rafael, C. Vinicius, B. Alexandre, Complete factorial design experiment for 3D load cell instrumented crank validation, in *2015 37th Annual International Conference of the IEEE Engineering in Medicine and Biology Society (EMBC)* (IEEE, 2015), pp. 3655–3658
35. F.P. Carpes, M. Rossato, I.E. Faria, C.B. Mota, Bilateral pedaling asymmetry during a simulated 40-km cycling time-trial. J. Sports Med. Phys. Fitness **47**(1), 51–57 (2007)
36. F.P. Carpes, M. Rossato, I. Faria, C.B. Mota, During an incremental exercise cyclists improve bilateral pedaling symmetry. Brazilian J. Biomotricity **2**(3), 155–159 (2008)
37. E. Dadios (Ed.), *Fuzzy Logic—Controls, Concepts, Theories and Applications* (InTech, 2012)
38. ISO, Evaluation of measurement data Guide to the expression of uncertainty in measurement. International Organization for Standardization Geneva ISBN 50, 134 (2008)
39. D. Azeez, M.A.M. Ali, K.B. Gan, I. Saiboon, Comparison of adaptive neuro-fuzzy inference system and artificial neutral networks model to categorize patients in the emergency department. SpringerPlus **2**(1), 416 (2013)
40. H.M. Azamathulla, A. Ab. Ghani, S.Y. Fei, ANFIS-based approach for predicting sediment transport in clean sewer. Appl. Soft Comput. J. **12**(3), 1227–1230 (2012)
41. Shahriar Jahan Hossain, Adaptive neuro-fuzzy inference system (ANFIS) based surface roughness prediction model for ball end milling operation. J. Mech. Eng. Res. **4**(3)
42. R. Kohavi, A Study of Cross-validation and bootstrap for accuracy estimation and model selection, in *Appears in the International Joint Conference on Artificial Intelligence (IJCAI)*, 1995
43. S. Haykin, *Neural Networks and Learning Machines*, vol. 3 (Pearson Prentice Hall, New Jersey, USA, 2008)
44. A.S. Gardner, J.C. Martin, D.T. Martin, M. Barras, D.G. Jenkins, Maximal torque and power-pedaling rate relationships for elite sprint cyclists in laboratory and field tests. Eur. J. Appl. Physiol. **101**(3), 287–292 (2007)

45. B. Karsten, S. Jobson, J. Hopker, A. Jimenez, C. Beedie, High Agreement between laboratory and field estimates of critical power in cycling. Int. J. Sports Med. **35**(04), 298–303 (2013)
46. J. Balmer, R.C.R. Davison, S.R. Bird, Peak power predicts performance power during an outdoor 16.1-km cycling time trial. Med. Sci. Sports Exerc. **32**(8), 1485–1490 (2000)
47. A. Nimmerichter, C.A. Williams, Comparison of power output during ergometer and track cycling in adolescent cyclists. J. Strength Condition. Res. **29**(4), 1049–1056 (2015)
48. D.C. Montgomery, *Design and Analysis of Experiments*, vol. 2, 8th edn. (Wiley, 2012)

Fuzzy Classifiers Learned Through SVMs with Application to Specific Object Detection and Shape Extraction Using an RGB-D Camera

Chia-Feng Juang and Guo-Cyuan Chen

Abstract In several studies, fuzzy classifiers (FCs) have been shown to achieve higher generalization ability when learned through support vector machines (SVMs) compared to learning through neural networks that aim to minimize only the training error. This chapter introduces the learning of FCs using linear SVMs. Two types of FCs with zero-order and high-order Takagi-Sugeno (TS)-type consequents are considered. Given a number of rules, the antecedent parameters in the two FCs are determined using a self-splitting clustering (SSC) algorithm. Regarding the consequent parameter optimization, this chapter first describes the basic concept of linear SVMs followed by its application to the learning of the consequent parameters to endow the FCs with high generalization ability. These two types of FCs are subsequently applied to object detection and shape extraction using a red-green-blue-depth (RGB-D) camera. In this application, after the detection of an object using a color-feature-based FC, the depth information from the camera is used to extract the shape of an object. A histogram-based shape feature is proposed for improving the object detection performance. The performance of the proposed classification approach is evaluated through the detection of different objects and comparisons with various object detection approaches.

Keywords Fuzzy classifiers · Support vector machines · Neural fuzzy systems Object detection · Object shape extraction

1 Introduction

In contrast to other popular machine learning (ML)-based classifiers, such as support vector machines (SVMs) and neural networks (NNs), fuzzy classifiers (FCs) have the

C.-F. Juang (✉) · G.-C. Chen
Department of Electrical Engineering, National Chung Hsing University, Taichung, Taiwan
e-mail: cfjuang@dragon.nchu.edu.tw

G.-C. Chen
e-mail: mikechen@gmail.com

W. Pedrycz and S.-M. Chen (eds.), *Computational Intelligence for Pattern Recognition*, Studies in Computational Intelligence 777,
https://doi.org/10.1007/978-3-319-89629-8_9

potential for model interpretability [1], which helps to bridge the gap between ML and human understanding. In addition, because of the local mapping property of each rule, clustering algorithms can be directly employed to determine the structure and initial antecedent parameters of an FC [2–6]. These unsupervised clustering learning algorithms help build a good initial FC model for subsequent parameter optimization, which results in small model size and facilitates the quick establishment of a well-performing model without the collection of a large training data set.

Because of the onerous amount of work that is required to handcraft the optimal fuzzy rules in an FC, many automatic learning approaches for building FCs have been proposed. Among the many FC learning approaches, two popular ones are neural FCs (NFCs) [2–4, 6] and evolutionary FCs (EFCs) [7, 8]. The learning approaches in both NFCs and EFCs are based on empirical risk error minimization [9]. With the consideration of structural risk minimization, the learning of FCs using SVMs has been proposed [5, 10–14]. In this learning approach, SVMs are employed to find the parameters in an FC based on soft margin minimization [15]. SVMs with different types of kernels have been proposed for learning FCs. In learning FCs using Gaussian-kernel-based SVMs [10, 11], a support vector (SV) corresponds to a zero-order Takagi-Sugeno (TS)-type fuzzy rule, and Gaussian kernels are regarded as Gaussian membership functions. A new SVM kernel is defined as the product of a linear kernel and a Gaussian kernel and applied to learn FCs [12]. In this learning method, an SV corresponds to a first-order TS-type fuzzy rule. In these nonlinear kernel-based learning methods, the constructed FCs suffer the problem of large model size due to the large number of SVs. To address this problem, in linear-kernel-based SVM learning of FCs [13, 14], clustering algorithms have been proposed to determine the structure of an FC to build a small FC model size. Linear SVMs are only applied to optimize the consequent parameters of the constructed FC. This chapter introduces the use of this learning technique in designing zero-order and high-order TS-type FCs. The zero-order FC that is learned through the self-splitting clustering (SSC) algorithm and SVM (FC-SSCSVM) [13] and the high-order FC with SV learning in expanded high-order consequent space (SFC-SVHC) [14] are described as examples.

This chapter applies the two FCs FC-SSCSVM and SFC-SVHC to detect a known, specific object that contains multiple colors with non-homogeneous distributions in complex backgrounds using a red-green-blue-depth (RGB-D) image that was obtained using a Kinect camera. Detection primarily consists of two stages: feature extraction and classification. Table 1 shows different types of features and classifiers that have been proposed for the object detection task and the method proposed in this chapter. One category of features consists of those that are extracted from the edges and/or corners of an object [16–20], such as Haar-like wavelet features [16], scale-invariant feature transform (SIFT) [17], histogram of oriented gradients (HOG) [18], and shape information [19, 20]. Another category of features consists of those that are extracted from the color information of an object, such as color histograms [21, 22] and a two-phase color feature that is composed of the entropy of the composing color (ECC) and entropies of geometric color distributions (EGCD) [13]. In contrast to the above two handcrafted feature categories, one popular and powerful feature extraction method is to use a deep convolutional neural network (DCNN) to

Table 1 Different types of features and classifiers for the object detection task and the proposed method

Studies	Features			Classification
	Color	Edges/corners	Hybrid	
Previous work	Color histograms [21, 22], ECC+EGCD [13]	Haar-like wavelet [16], SIFT [17], HOG [18], shape information [19, 20]	Convolution feature [23, 24], hand-crafted+convolution feature [26]	Cascaded Adaboost [16], template matching [17, 19, 21], SVMs [18, 20], FCs [13, 22], NNs [23, 24, 26]
Proposed method	Hybrid feature: Stage 1: ECC+EGCD, Stage 2: Shape contour from depth image			Stage 1: SFC-SVHC Stage 2: template matching

automatically extract features [23, 24]. Since different types of features (including colors and shapes) are extracted, the classification accuracy of this type of feature is high. For complex classification problems, a huge set of training data is generally required for the DCNN to achieve good performance. If only a small set of training data is available, the technique of transfer learning could be employed to improve classification performance [24, 25]. Most DCNNs suffer from heavy computational loads. Though specialized hardware, such as graphics processing units, have been developed to reduce runtime, the resulting implementation costs increase. For these problems, one promising approach is the combination of the handcrafted features and the DCNN-based features [26], where the former are responsible for quickly filtering out uninteresting candidates and the latter for making the final decision.

The contributions of this chapter are threefold. First, this chapter introduces a unified and systematic way of applying the SSC and linear SVM to respectively determine the antecedent and consequent parameters of TSK-type FCs with different orders. The zero-order FC (FC-SSCSVM) [13] and the high-order FC (SFC-SVHC) [14] are taken as examples of this design approach. Second, this paper applies the SFC-SVHC to detect objects using the two-phase color feature of the ECC and EGCD for object detection in color images. This new computer vision application was not studied in the previous study of the SFC-SVHC [14]. Based on the two-phase color feature, this chapter studies the performance comparisons of the FC-SSCSVM and SFC-SVHC in the detection problem. Third, different from the use of only color features for object detection in [13], this chapter proposes a new method to extract object shape features from the depth image following the color-based object detection. A new contour-based shape feature is proposed in this method. Experimental results verify the effectiveness of this shape extraction method.

This chapter is organized as follows. Section 2 describes the functions in the two types of FCs and their structure learning. Section 3 describes the basic concept of linear SVMs followed by the details of their application to learn the consequent parameters of the two FCs. Section 4 describes the object detection process based on the two-phase color feature and an FC. Section 5 introduces the shape-based object detection method. Section 6 presents experimental results and performance comparisons of various detection methods. Finally, conclusions are summarized in Sect. 7.

2 Fuzzy Classifiers and Structure Learning

2.1 Fuzzy Classifiers

This section describes the functions of zero-order and high-order TS-type FCs [13, 14]. The ith rule in the zero-order FC is described as follows [13]:

$$\text{Rule}\, i : \text{If}\, x_1 \text{ is } A_{i1} \text{ and}, \ldots, \text{ and}\, x_n \text{ is } A_{in} \text{ then } y' \text{ is } a_i, \tag{1}$$

where n is the input feature dimension, A_{ij} is a fuzzy set, and a_i is a crisp value. Using a Gaussian membership function with center m_{ij} and width σ_i, the firing strength $\mu_i(\vec{x})$ is given as follows:

$$\mu_i(\vec{x}) = \prod_{j=1}^{n} \exp\left\{\frac{(x_j - m_{ij})^2}{\sigma_i^2}\right\} = \exp\left\{-\frac{||\vec{x} - \vec{m}_i||^2}{\sigma_i^2}\right\}, \tag{2}$$

where $\vec{x} = [x_1, x_2, \cdots, x_n]^T$ and $\vec{m}_i = [m_{i1}, m_{i2}, \cdots, m_{in}]^T$. The FC output after a simple defuzzification operation is

$$y' = \sum_{i=1}^{r} a_i \cdot \mu_i(\vec{x}) = \left\langle \vec{a},\, \vec{\mu}(\vec{x}) \right\rangle, \tag{3}$$

where r is the number of rules, $\vec{a} = [a_1, \ldots, a_r]$, and $\vec{\mu}(\vec{x}) = [\mu_1(\vec{x}), \ldots, \mu_r(\vec{x})]$. The two-class decision function of the FC is

$$f(\vec{x}) = \text{sign}\left(\sum_{i=1}^{r} a_i \cdot \mu_i(\vec{x}) + b\right), \tag{4}$$

where b is a threshold.

The high-order FC expands the rule-mapped space in a first-order TS-type FC [14]. In a first-order FC, each rule maps an input vector $\vec{x}$, $x_j \in [-1,\ 1]$ to a new vector $\vec{g}_i = [\mu_i x_0, \ \ldots, \ \mu_i x_n] \in \mathbb{R}^{n+1}$ with $x_0 := 1$ in the consequent space. Using a trigonometric function transformation, the high-order FC further expands vector $\vec{g}_i$ to a new vector $\vec{\Phi}_i$, which is given as follows:

$$\vec{\Phi}_i = [\mu_i x_0,\ \sin(\pi\ \mu_i x_0),\ \cos(\pi\ \mu_i x_0),\ \ldots,\ \mu_i x_n,\ \sin(\pi\ \mu_i x_n),\ \cos(\pi\ \mu_i x_n)] \in \mathbb{R}^{3(n+1)}. \tag{5}$$

Based on the simple weighted-sum defuzzification function with weighting vector $\hat{\vec{a}}_i \in \mathbb{R}^{3(n+1)}$ for rule i, the FC output is given as follows:

$$y' = \sum_{i=1}^{r} \left\langle \hat{\vec{a}}_i,\ \vec{\Phi}_i(\vec{x}) \right\rangle = \sum_{i=1}^{3rn+1} a_i \cdot \phi_i(\vec{x}) = \left\langle \vec{a},\ \vec{\phi}(\vec{x}) \right\rangle, \tag{6}$$

where $\vec{a} = [a_1, \ldots, a_{3rn}]$ and $\vec{\phi}(\vec{x}) = [\phi_1(\vec{x}),\ \ldots,\ \phi_{3rn+1}(\vec{x})]$. In two-class classification problems, the decision function of the FC is

$$f(\vec{x}) = \operatorname{sign}\left(\sum_{i=1}^{3rn+1} a_i \cdot \phi_i(\vec{x}) + b \right). \tag{7}$$

2.2 *Structure Learning*

The FCs use the self-splitting clustering (SSC) algorithm [13] to find antecedent parameters m_{ij} and σ_i in (2). The SSC algorithm starts with one cluster and generates new clusters using iterative splitting operations. To generate a new cluster from $\hat{r}$ existing clusters, the SSC algorithm splits the cluster with the largest variance (denoted as cluster I) into two new clusters. The top two input samples having the shortest Euclidean distances to the center of cluster I are selected to be the initial centers of the two new clusters. After the splitting process, the k-means algorithm is applied to identify the new centers of all clusters. The SSC algorithm stops when $\hat{r}$ is equal to an assigned cluster number r. Finally, m_{ij} and σ_i are assigned to be the center and standard deviation of the ith cluster, respectively.

3 Parameter Learning Through Linear Support Vector Machines

The consequent parameters a_i of the two types of FCs are optimized by a linear SVM to minimize a soft margin. Given N labeled training data $(\vec{x}_k,\ y_k)$, $y_k \in \{+1, -1\}$, the linear SVM aims to find a hyperplane decision function [15]

$$f(\vec{x}) = \text{sign}\left(\vec{w}^T \vec{x} + b\right) \tag{8}$$

to minimize a soft margin:

$$\begin{array}{ll} \text{Min}_{\vec{w},\xi_k} & \frac{1}{2}\vec{w}^T \vec{w} + C\sum_{k=1}^{N} \xi_k \\ \text{Subject to } y_k\left(\vec{w}^T \vec{x}_k + b\right) & \geq 1 - \xi_k \end{array}, \tag{9}$$

where $\xi_k \geq 0$ is a slack variable and C is a positive coefficient. The optimal decision function is found to be

$$f(\vec{x}) = \text{sign}\left(\sum_{k=1}^{N} y_k\alpha_k < \vec{x},\ \vec{x}_k > +b\right), \tag{10}$$

where the training samples for which $\alpha_k \neq 0$ are called SVs. According to [27], the minimization of the soft margin tends to minimize the bound on the structural risk. Thus, the obtained decision function should improve the generalization ability.

To apply the linear SVM to find the parameters a_i in a zero-order FC, the original training data $\left(\vec{x}_k,\ y_k\right)$ are transformed to $\left(\vec{\mu}(\vec{x}_k),\ y_k\right)$ in the rule-mapped space. Equation (3) shows the linear relationship between the FC output y' and the transformed input data $\vec{\mu}(\vec{x})$. According to (9), a soft margin is formulated in terms of the free parameters a_i as follows:

$$\begin{array}{ll} \text{Min}_{\vec{a},\xi_k} & \frac{1}{2}\vec{a}^T \vec{a} + C\sum_{k=1}^{N} \xi_k \\ \text{Subject to } y_k\left(\vec{a}^T \vec{\mu}(\vec{x}_k) + b\right) & \geq 1 - \xi_k \end{array}. \tag{11}$$

According to (10), the optimal decision function of the zero-order FC is [13]:

$$\begin{aligned} f(\vec{x}) &= \text{sign}\left(\sum_{k=1}^{N} y_k\alpha_k < \vec{\mu}(\vec{x}),\ \vec{\mu}\left(\vec{x}_k\right) > +b\right) \\ &= \text{sign}\left(\sum_{i=1}^{r}\left(\sum_{k=1}^{N} y_k\alpha_k\mu_i\left(\vec{x}_k\right)\right)\mu_i(\vec{x}) + b\right), \end{aligned} \tag{12}$$

where α_k is optimized through the linear SVM. It follows from the equivalence of (4) and (12) that

$$a_i = \sum_{k=1}^{N} y_k \alpha_k \mu_i \left(\vec{x}_k\right) = \sum_{k \in SVs}^{N} y_k \alpha_k \mu_i (\vec{x}_k), \quad i = 1, \ldots, r. \tag{13}$$

Similarly, for the high-order FC, the original training data $\left(\vec{x}_k, y_k\right)$ are transformed to $\left(\vec{\phi}(\vec{x}_k), y_k\right)$ in the expanded space. According to (10), the optimal decision function of the high-order FC is written as follows [14]:

$$\begin{aligned} f(\vec{x}) &= \text{sign}\left(\sum_{k=1}^{N} y_k \alpha_k < \vec{\phi}(\vec{x}), \vec{\phi}(\vec{x}_k) > +b\right) \\ &= \text{sign}\left(\sum_{i=1}^{r}\left(\sum_{k=1}^{N} y_k \alpha_k \phi_i \left(\vec{x}_k\right)\right) \phi_i(\vec{x}) + b\right), \end{aligned} \tag{14}$$

where α_k is optimized through the linear SVM. It follows from the equivalence of (6) and (14) that

$$a_i = \sum_{k=1}^{N} y_k \alpha_k \phi_i \left(\vec{x}_k\right) = \sum_{k \in SVs}^{N} y_k \alpha_k \phi_i \left(\vec{x}_k\right) \cdot \quad i = 1, \ldots, 3nr + 1. \tag{15}$$

4 Color Features for Object Detection

Figure 1 shows the flowchart of the detection process. The FCs are applied to detect a specific object using the two-phase color feature [13]. An object in a given image is detected using a scanning window with a size of $w_1 \times w_2$ pixels and a stride of s pixels. The multi-scale scan search method with a scale factor of $\alpha = 0.8$ is applied to find objects with different sizes. For each scanning window, the first phase uses the ECC feature to quickly filter out non-objects. The second phase filters the candidates generated from phase one by using the EGCD and an FC. Extraction of the two features is briefly described as follows.

4.1 Entropy of the Composing Color (ECC) Feature

The objective of the ECC feature is to extract the information of the composing color of an object. To this end, the ECC feature is computed from the color histograms of all the training patterns of an object of interest in the HS space. To obtain the color

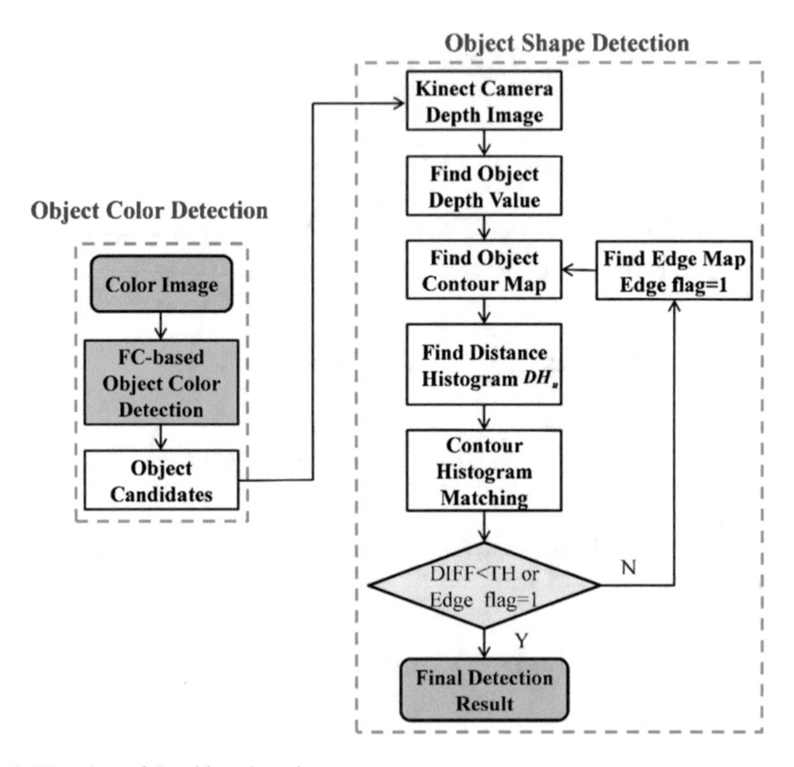

Fig. 1 Flowchart of the object detection process

histograms, the SSC algorithm is applied to partition the HS into $C_{\max}$ clusters. In this application, the criterion for determining which cluster should be split is based on the number of object pixels instead of the variance in each cluster. This criterion is selected so that the pixels of each object will be evenly distributed among the clusters to ensure that the color entropy of an object will be greater than that of a non-object. Figure 2 shows the partition results in a scaled HS space that are obtained by using the SSC algorithm for the five different objects considered in this chapter. The blue points in each HS space represent the HS values of all the pixels belonging to the training objects (i.e., the distribution of the training pixels in the HS space). The black lines and red points in the HS space represent the boundaries and centers of different clusters, respectively. The clusters are formed so that the pixels in the training objects are evenly distributed over the clusters. The determination of the cluster number $C_{\max}$ for each object is described in Sect. 6.

For a scanning window, the ECC feature determines the entropy of the color histograms in the $C_{\max}$ clusters. This entropy is computed as follows:

$$H_{ECC} = -\sum_{i=1}^{C_{\max}} p_{hs}^{i} \log_2 p_{hs}^{i}, \tag{16}$$

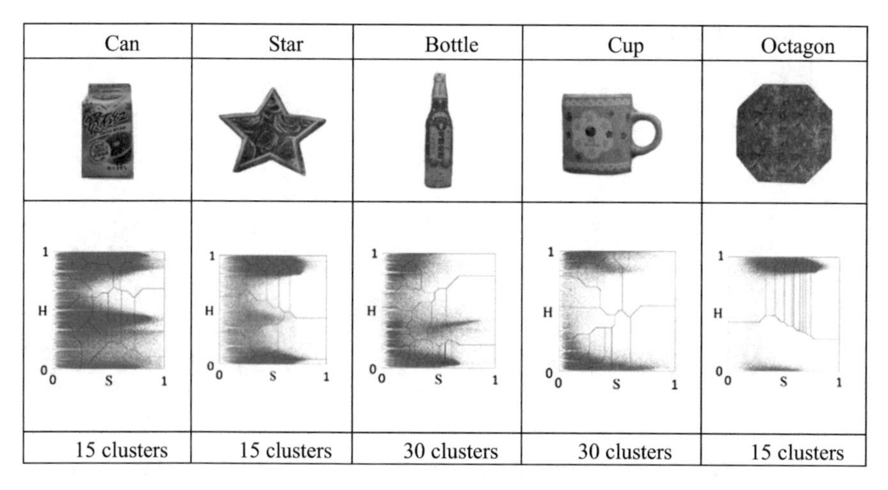

Fig. 2 HS partition results that were obtained by using the SSC algorithm for different objects. The blue and red points are the pixels of the object and the cluster centers, respectively

where

$$p_{hs}^{i} = \frac{T_i}{w_1 \times w_2}, \ i = 1, \ldots, C\max. \tag{17}$$

In this equation, T_i is the count of pixels in cluster C_i. Since the pixels in an object are expected to be evenly distributed over the clusters, the H_{ECC} value of an object is generally higher than a non-object. For example, Fig. 3 shows the histogram of the H_{ECC} values of some cans and non-cans, where it is observed that the H_{ECC} values of most cans are larger than non-cans. Therefore, a test pattern is regarded as an object candidate if $H_{ECC} > \theta_1$, where θ_1 is a threshold that is determined according to a validation set of images.

4.2 Entropies of Geometric Color Distributions (EGCD) Feature

The objective of the EGCD feature is to extract the information of the geometrical distributions of the composing colors of an object. To this end, the EGCD feature is computed based on the entropy of the geometric color histogram of the *i*th composing color of an object. A scanning window is divided into $N = (w_1/s) \times (w_2/s)$ non-overlapping sub-blocks, where the size of each sub-block is $s \times s$ pixels. The total number of pixels in the sub-block k of cluster (color) i is denoted as t_i^k, where $\sum_{k=1}^{N} t_i^k = T_i$. The entropy value H_{EGCD}^{i} of the distribution of composing color i is given as follows:

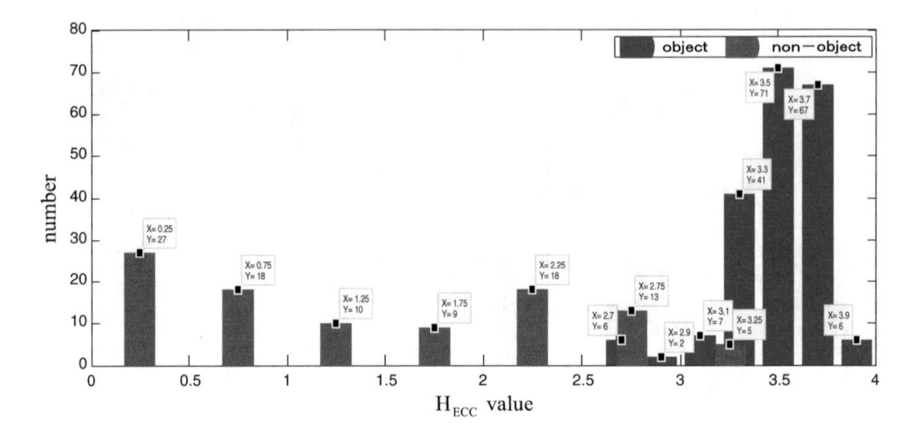

Fig. 3 Histogram of the H_{ECC} values of some objects (cans) and non-objects, where the range of each bin is 0.5

$$H^i_{EGCD} = -\sum_{k=1}^{N} \frac{t_i^k}{T_i} \log_2 \frac{t_i^k}{T_i}, \quad i = 1, \ldots, C_{\max}. \tag{18}$$

These $C_{\max}$ EGCD feature values are fed to an FC to filter object candidates.

After the scanning and detection process, the center of the scanning window that detects an object is recorded. Three 3×3 dilation operations are applied to the recorded centers. A minimum enclosing rectangle (MER) of each connected region is found. The MERs whose sizes are greater than the training pattern size of the detected object are regarded as object candidates.

5 Shape Extraction and Shape-Based Detection

5.1 Shape and Feature Extraction

After the color-based detection process, the object candidates are further filtered using the shape features extracted from the depth image. The object shape extraction process consists of two phases. The first phase finds the depth value of the object center. All pixels of the object candidate are mapped to the depth image. Next, a histogram of these mapped depths is devised. The depth x_d with the maximum histogram value is found to be the depth of the object. Figure 4a shows the detection of a cup using the color features. Figure 4b shows the pixels of the object candidate. Figure 4c shows the depth image, where the background regions with a long distance and the regions not covered by the depth sensor (the lower part) are shown in white. Figure 4d shows the mapping results of Fig. 4b, c. The depth $x_d = 204$ with the maximum histogram value is taken as the depth of the object. To identify the region of

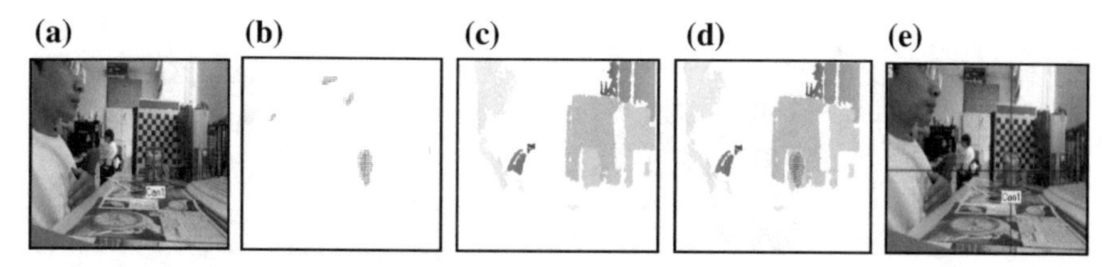

Fig. 4 **a** Object color detection result. **b** Object candidate pixels. **c** The depth image. **d** The mapping result. **e** The final detection result and the extracted can shape

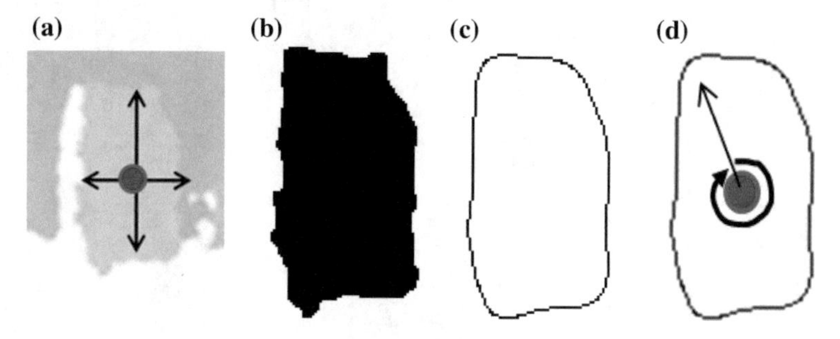

Fig. 5 **a** Center of an object candidate and the four-directional extension from the center. **b** The segmented object area. **c** The smoothed contour. **d** The distance from the object center to each of the contour points

the object from the depth image, the depth range is set to $x_d \pm D_{range}$, where $D_{range} = 3$ is the object depth filtering threshold. The new object center is recomputed using the average of the object candidate pixels whose depths are within the range $x_d \pm D_{range}$. A four-directional extension that starts from the new center is then used to find the object region. For a pixel of a detected object, the next extension direction stops when the depth of the pixel in that direction is not within the depth range. The extension is followed by a morphological opening operation to smooth the shape contour. Figure 5a shows the center of the object candidate in Fig. 4 and the four-directional extension. Figure 5b shows the segmented object region that was obtained using the depth image. Figure 5c shows the smoothed shape contour.

After finding the smoothed contour, the next step is to find the Euclidean distance D_i from the object center $(\bar{X}, \bar{Y})$ to each contour pixel (x_i, y_i), as shown in Fig. 5d. A low-pass filter is applied to the distance curve and the smoothed distance value is calculated as follows:

$$\bar{D}_i = \frac{1}{25} \sum_{j=-12}^{12} D_{i+j}, \ i = 1, \ldots, m. \tag{19}$$

Figure 6a shows the contours of different detected objects. Figure 6b shows the distance curves of the objects. The distance range is evenly divided into 20 bins. The normalized distance histogram DH_u, $u = 1, \ldots, 20$, which counts the total number

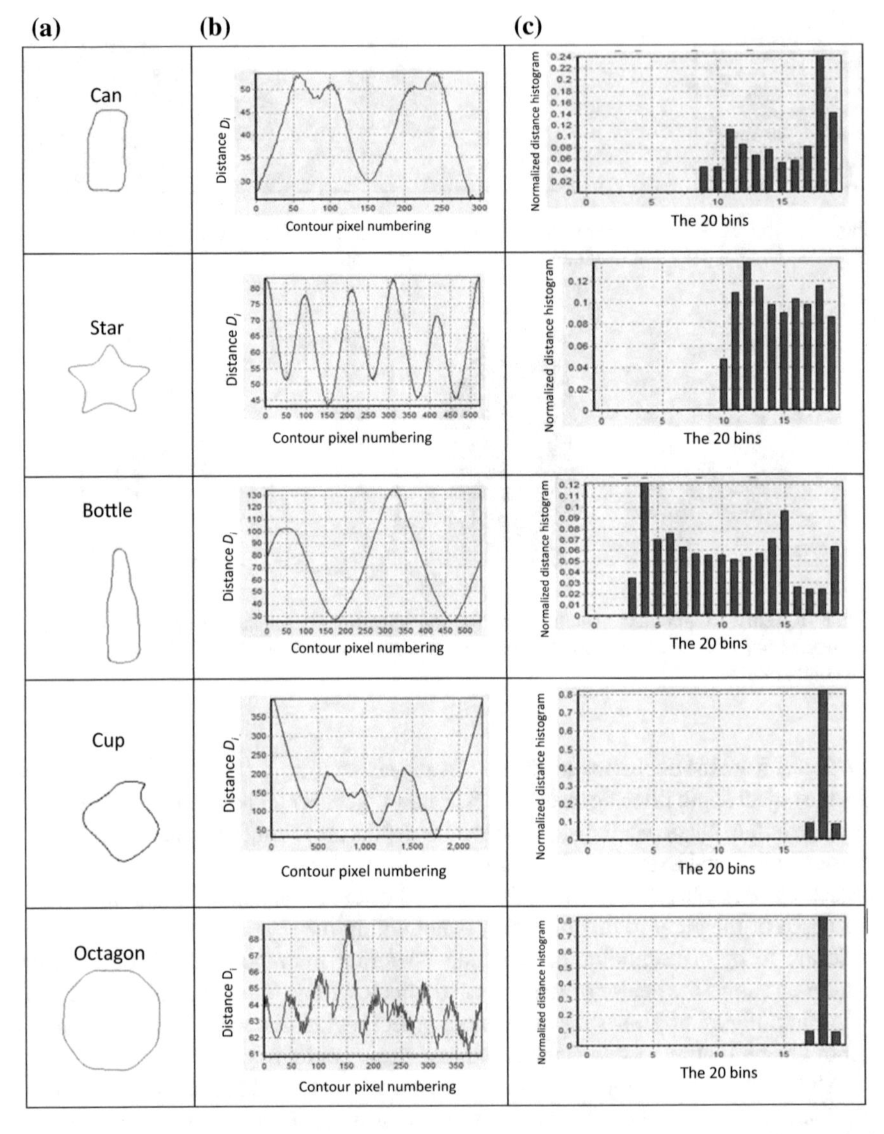

Fig. 6 **a** Object shape contours. **b** The distance curves of the objects. **c** The distance histograms

of distance values that fall into each bin divided by the total number of contour pixels, is obtained. Figure 6c shows the distance histograms of the objects. The distance histogram is scale-invariant and rotation-invariant and is therefore suitable for detecting objects with different scales and orientations.

5.2 Shape Extraction and Final Object Detection

Based on the normalized distance histogram DH_u, the template matching method is used to filter the object candidates. Twenty templates are collected for different views of the object. For a test pattern, the minimum difference between its normalized distance histogram DH_u^{test} and the distance-histogram templates $DH_u^{template,j}$, $j = 1, \ldots, 20$ is determined using the following equation:

$$Diff = \min_j \left(\sum_{u=1}^{20} \left| DH_u^{test} - DH_u^{template,j} \right| \right). \tag{20}$$

If $Diff$ is smaller than a pre-defined threshold TH, then the candidate is regarded as an object. If the difference between the object and the background depths is small or the object is tilted, the object shape region extracted from the depth image may be incorrect. For this reason, if $Diff$ is greater than TH, then a more complex shape extraction process is proposed. If $Diff > TH$, the depth image is transformed into an edge image by using Canny edge detection. The four-directional extension, starting from the depth-based center, is applied to the edge map to segment the object, and the extension continues until the edge is reached. The above shape contour extraction and matching process is reapplied to the newly segmented region to filter out non-objects.

Figure 7 shows a detection example when $Diff > TH$. Figure 7a shows the detection of a bottle using the color features. Figure 7b shows the depth image, where the depths of the bottle and the background are close. Figure 7c shows the segmented region that was obtained using the depth range, where part of the background is improperly segmented as the object. Figure 7d shows the edge map and the four-directional extension from the object center on the edge map. Figure 7e shows the smothered shape contour and region. Figure 7f shows the final detection result. When multiple candidates are generated from the color-based detection method, the shape-based detection method helps find the correct one. Figure 8 shows two examples of this case, where the shape-based detection approach helps successfully detect the object and find its shape. Instead of a single stage of detection based on the conjunction of the color and shape features, a two-stage detection method is proposed in this chapter. In this two-stage method, most non-object regions scanned by the window are filtered out using the color feature in the first stage, which saves the time required to extract the shape feature of these non-objects.

6 Experiments

The performances of the two FCs and the object detection system are evaluated through the detection of five objects: a can, a star, a bottle, a cup, and an octagon. These objects contain multi-color distributions with different shapes. All of the images were collected from the Kinect device and are of size 640×480 pixels. The

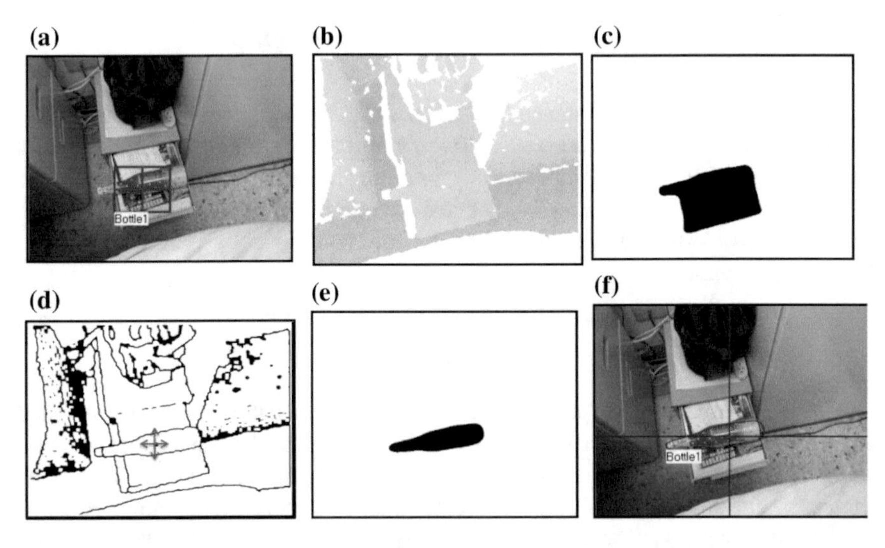

Fig. 7 **a** Object color detection result. **b** Depth image. **c** The edge map. **d** The segmented object shape region that was obtained by using the depth range, where the background is also incorrectly segmented. **e** The segmented object region that was obtained by using the edge map. **f** The final detection result

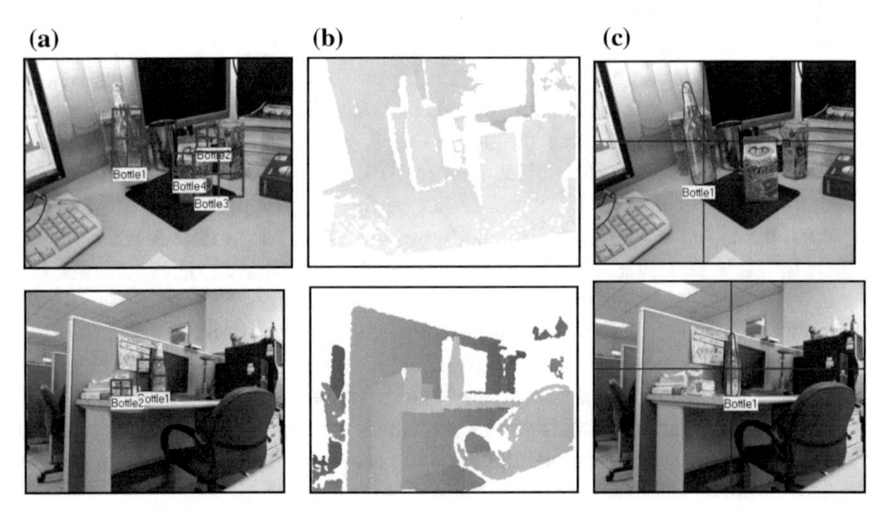

Fig. 8 **a** Original color images and the detected bottle candidates using the color features. **b** The depth images. **c** The correctly detected bottles and the extracted shapes

positive training patterns were collected from different views of an object. The negative training patterns were randomly selected from backgrounds. Table 2 shows the details of each experiment. Denote the total number of true objects, correctly detected objects, and detections as N_T, N_C, and N_D, respectively. The detection performance is evaluated in terms of detection rate (DR $= N_C/N_T$), precision (PR $= N_C/N_D$), and DR + PR − 1 [28]. A higher value of DR+PR-1 indicates a better detection performance.

The sizes of the object pattern, non-object pattern, and scanning window were identically set to the minimum size of the objects of interest to be detected in an image. A larger scanning window than this assignment may miss the objects smaller than the scanning window. A smaller scanning window than this assignment increases the detection time. In addition, when the scanning window size is too small, the features extracted from the low image resolution may perform poorly. Training patterns consist of 200 objects and 100 non-objects. There were 100 test images in each experiment. Another 100 cross-validation (CV) images were collected for selecting system parameters, including r, $C_{\max}$ in the SSC algorithm, the ECC threshold value, the coefficient C in the linear SVM for FC training, and the threshold TH of *Diff* in (20). For the FCs, r was selected from the set $\{5, 10, 15\}$. Table 3 shows the selected number of rules in the FCs in each experiment. The SFC-SVHC used five rules in all experiments. A larger number of rules in the SFC-SVHC tended to decrease the detection performance in the CV images mainly due to the overtraining problem. The results also showed that the FC-SSCSVM used a larger number of rules than the SFC-SVHC in each experiment, since the former used simpler, lower order consequent parts than the latter. Four out of the five FC-SSCSVMs in the experiments used 10 rules. Because of the overtraining problem, the use of the larger number of 15 rules did not show better detection performance in the CV images.

The coefficient C was selected from the set $\{2^0, 2^1, \ldots, 2^{10}\}$. For the HS partitioning problem, $C_{\max}$ was selected from the set $\{15,\ 30,\ 45\}$. This chapter set the threshold TH to 0.38 in all experiments. A higher value of TH means a candidate has a higher chance of being detected as an object. Thus, this setting will concurrently increase the DR and decrease the PR (more false alarms). In contrast, a smaller value of TH concurrently decreases the DR and increases the PR. Because of the trade-off between DR and PR, the threshold $TH = 0.38$ that achieved the maximum DR+PR-1 in the 100 CV images was selected.

6.1 Detection Performance

For the detection using the color feature from an RGB image, the detected objects and the ground truths are marked by rectangles. For the detection using the combination of color and shape features from the RGBD image, the detected objects and the ground truths are marked by the regions bounded by the object shapes. Table 3 shows the detection performances for the five objects in Experiments 1 to 5 of the two FCs using only the color features and the combination of the color and shape features. The results showed that the incorporation of the shape feature improved the detection performance, with the sole exception of Experiment 1. In this experiment, the detection rate using both the color and the shape features was only slightly lower than that using the color feature. In addition, the proposed shape-based method had the advantage of finding the orientation of the object in this experiment.

Table 2 Color images of the objects, their shapes, and some of the training patterns in different experiments

Experiments (Pattern size)	Objects	Object shape	Training patterns (object)	Training patterns (non-object)	#Color cluster
Exp. 1. Can (45 × 90)		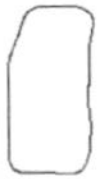			15
Exp. 2. Star (45 * 45)			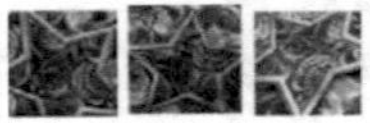		15
Exp. 3. Bottle (45 * 45)					30
Exp. 4. Cup (45 * 45)			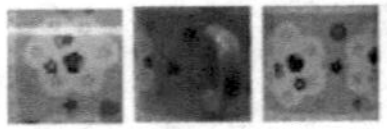		30
Exp. 5. Octagon (45 * 45)					15

Table 3 Detection performances of the two types of FCs using color and color + shape features in Experiments 1 to 5. The highest DR + PR-1 value in each experiment is marked in boldface

Experiments	Classifiers	#rules	Color			Color + Shape		
			DR(%)	PR(%)	DR + PR -1	DR(%)	PR(%)	DR + PR -1
1 (Can)	SFC-SVHC	5	99	93.4	0.924	96	98.0	0.940
	FC-SSCSVM	10	99	97.1	**0.961**	96	98.0	0.940
2 (Star)	SFC-SVHC	5	100	65.4	0.654	95	85.6	**0.806**
	FC-SSCSVM	10	100	60.6	0.606	93	85.3	0.783
3 (Bottle)	SFC-SVHC	5	84	28.7	0.127	89	85.6	0.746
	FC-SSCSVM	10	86	41.4	0.274	91	93.8	**0.848**
4 (Cup)	SFC-SVHC	5	96	57.8	0.538	89	87.3	**0.763**
	FC-SSCSVM	15	80	45.5	0.255	85	84.2	0.692
5 (Octagon)	SFC-SVHC	5	98	88.3	0.863	93	100	**0.930**
	FC-SSCSVM	10	98	87.5	0.855	92	100	0.920

For the two FCs, the high-order SFC-SVHC outperformed the zero-order FC-SSCSVM in three of the five experiments on the RGB images when using the same color features. The FC-SSCSVM outperformed the SFC-SVHC on the RGBD images only in Experiment 1. Regarding the FC structure, the SFC-SVHC used five rules in all experiments, while the FC-SSCSVM did not. The FC-SSCSVM has been shown to outperform Gaussian kernel-based SVMs, multilayer perceptron (MLP), and neural fuzzy classifiers in terms of overall classification performance in object detection problems using color features [13]. The SFC-SVHC has been shown to outperform different NFCs, EFCs, and statistical classifiers in terms of overall classification performance on 20 benchmark classification problems [14]. Therefore, this chapter focuses only on studying the detection performances of the FC-SSCSVM and SFC-SVHC in the application examples.

Figure 9 shows the successful detection results for some of the test images in each experiment using the SFC-SVHC. Figure 10 shows some false alarms and missed detections, where the latter were mainly caused by excessive or insufficient illumination. The test images contained objects with different distances to the camera, with all distances within the depth sensor range of the Kinect for Windows. In the experiments, the influence of the distance between the object and the camera was minor. However, if the object was sufficiently far from the camera such that the size of the object was smaller than the scanning window, then missed detections may occur.

Fig. 9 Detection results of some of the test images in different experiments

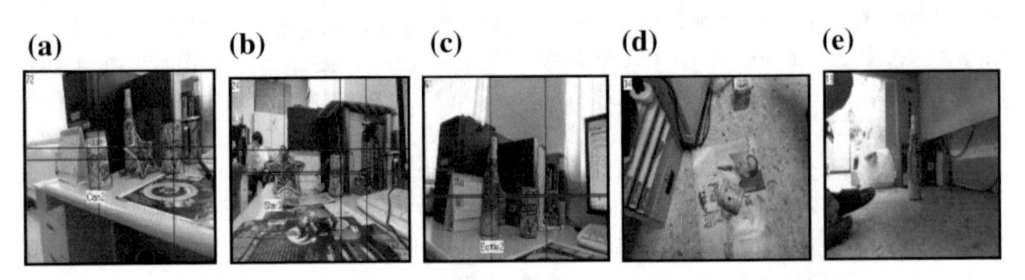

Fig. 10 **a–c** False alarms and **d**, **e** missed detections for some test images

6.2 Comparisons with Other Detection Methods

For comparison, Table 4 shows the detection performances of the proposed SFC-SVHC-based detection method and four other methods. The first method is a histogram-based template matching (H-TM) method [21]. The method uses the color histogram feature and template-matching classification.

Table 4 Detection performances of different methods, where the highest DR+PR-1 value in each experiment is marked in boldface

Experiments	Methods	DR(%)	PR(%)	DR+PR-1
1 (Can)	SFC-SVHC	96	98.0	**0.940**
	H-TM	83	51.6	0.346
	TFS-SVMPC	98	96.0	**0.940**
	CM	69	4.10	−0.269
	SFC-SVHC+CM	94	88.7	0.827
2 (Star)	SFC-SVHC	95	85.6	**0.806**
	H-TM	85	32.0	0.17
	TFS-SVMPC	97	50.5	0.475
	CM	84	9.6	−0.064
	SFC-SVHC+CM	89	58.2	0.472
3 (Bottle)	SFC-SVHC	89	85.6	**0.746**
	H-TM	57	16.2	−0.268
	TFS-SVMPC	89	45.9	0.349
	CM	49	9.0	−0.42
	SFC-SVHC+CM	74	27.2	0.012
4 (Cup)	SFC-SVHC	89	87.3	**0.763**
	H-TM	60	25.1	−0.149
	TFS-SVMPC	86	67.7	0.537
	CM	50	5.40	−0.446
	SFC-SVHC+CM	87	52.4	0394
5 (Octagon)	SFC-SVHC	93	100.0	**0.93**
	H-TM	96	80.0	0.76
	TFS-SVMPC	97	55.4	0.524
	CM	65	2.4	−0.326
	SFC-SVHC+CM	79	68.1	0.471

The second method consists of two stages [22]. The first stage generates object candidates by feeding color histograms of a candidate region to a FC. The FC used is the TS-type fuzzy system learned through an SVM in the principal component space (TFS-SVMPC). The second stage filters the candidates by feeding the ten-dimensional HS values from five locally partitioned regions to another TFS-SVMPC.

The third method uses the chamfer matching (CM) method [29] to detect an object using the shape information from a depth image. The Kinect depth image is transformed into an edge map using the Canny edge detection method. The CM method is applied to the edge map by minimizing a generalized distance between the edges of the map and the edge templates of the object [30]. The matching is performed on a pyramid of images with the same scene but in different resolutions.

The fourth method combines the color-based SFC-SVHC detection approach and the CM method. The first phase uses the color-based SFC-SVHC detection method to generate object candidates. The second phase uses the CM method to filter the candidates.

Table 4 shows that the SFC-SVHC-based detection method achieved the highest DR+PR-1 value. The detection performance of directly applying the CM method to the depth image was poor. When using the same color-based SFC-SVHC method in phase one, the proposed method outperformed the fourth method, which used the CM method in phase two. This comparison shows the superiority of the proposed shape extraction and detection method over the CM method.

Experimental results show that the proposed method generalizes well in the test images when using only a small set of 300 training images. The DCNN has been shown to achieve remarkable performance in different image processing problems when a large set of training data is collected. However, the computational load of a well-trained DCNN is generally high. As the suggested combination approach in [26], the proposed method can be combined with a well-trained DCNN in which the former is responsible for quickly filtering out non-objects and the latter for making the final decision.

In contrast to the 3D templates that generally suffer from the enormous complexities of 3D shapes in geometric spaces [31], the contour-distance based shape feature is simple in its implementation. In addition, the above experimental results show the proposed shape feature is more discriminative than the well-known edge-based feature. The experiments contain objects in different positions and lighting conditions to show the robustness of the proposed method to the two factors. The experiments focus on the detection performance of the images captured from real environments. Therefore, detection performance for the images with artificial noise is not studied. For objects with close shapes, the color feature helps discriminate the right one. When affine transformations or in and/or out plane transformations kick in, the contours of the objects may be different from those found in ideal viewing angles in the templates. These shape transformations may cause missed detections when using the proposed method.

Finally, it is worth noting that the incorporation of principal component space (PCA) for feature extraction could improve the accuracy of the SVM hyperplane [22]. The technique of PCA or linear discriminant analysis can be introduced into the FC-SSCSVM and SFC-SVHC for possible improvement of the detection performance.

7 Conclusions

This chapter describes two types of SVM-trained FCs, known as the zero-order FC-SSCSVM and the high-order SFC-SVHC, and their applications to object detection. In structure learning, the two FCs use the VSSC algorithm to automatically assign proper fuzzy set positions and shapes in the input space. In parameter learning, the use of the linear SVM helps to determine a set of consequent parameters with high generalization ability. The primary difference between the two FCs is that the zero-order TS-type rules used in the FC-SSCSVM makes it more interpretable than the SFC-SVHC. However, the SFC-SVHC shows better classification performance than the FC-SSCSVM because of the functional expansion of the rule-mapped space to a higher-order space. The choice between the FC-SSCSVM and the SFC-SVHC depends on whether higher interpretability or accuracy of the classification problem is preferred. In the FC-based object detection applications, in addition to the color features, object depth and shape extraction using the RGBD camera are introduced. The contour-distance histogram feature helps improve the detection performance. Experimental results on the detection of five different objects show the effectiveness of the FC-based detection method.

References

1. J. Casillas, O. Cordón, F. Herrera, L. Magdalena, *Interpretability Issues in Fuzzy Modeling (Studies in Fuzziness and Soft Computing)* (Springer, Berlin, Germany, 2003)
2. C.F. Juang, C.T. Lin, An on-line self-constructing neural fuzzy inference network and its applications. IEEE Trans. Fuzzy Syst. **6**(1), 12–32 (1998)
3. G.D. Wu, P.H. Huang, A maximizing-discriminability-based self-organizing fuzzy network for classification problems. IEEE Trans. Fuzzy Syst. **18**(2), 362–373 (2010)
4. C.F. Juang, T.C. Chen, W.Y. Cheng, Speedup of implementing fuzzy neural networks with high-dimensional inputs through parallel processing on graphic processing units. IEEE Trans. Fuzzy Syst. **19**(4), 717–728 (2011)
5. W.Y. Cheng, C.F. Juang, A fuzzy model with online incremental SVM and margin-selective gradient descent learning for classification problems. IEEE Trans. Fuzzy Syst. **22**(2), 324–337 (2014)
6. M. Pratama, J. Lu, G. Zhang, Evolving type-2 fuzzy classifier. IEEE Trans. Fuzzy Syst. **24**(3), 574–589 (2016)
7. A. Orriols-Puig, J. Casillas, E. Bernadó-Mansilla, Genetic-based machine learning systems are competitive for pattern recognition. Evol. Intel. **1**(3), 209–232 (2008)
8. P. Ducange, G. Mannarà, F. Marcelloni, R. Pecori, M. Vecchio, M, A novel approach for internet traffic classification based on multi-objective evolutionary fuzzy classifiers, in *Proceedings of the IEEE International Conference on Fuzzy Systems*, Italy, 2007, pp. 1–6
9. N. Cristianini, J.S. Taylor, *An Introduction to Support Vector Machines and Other Kernel-based Learning Methods* (Cambridge University Press, Cambridge UK, 2000)
10. Y. Chen, J.Z. Wang, Support vector learning for fuzzy rule-based classification systems. IEEE Trans. Fuzzy Syst. **11**(6), 716–728 (2003)
11. C.T. Lin, C.M. Yeh, S.F. Liang, J.F. Chung, N. Kumar, Support-vector-based fuzzy neural network for pattern classification. IEEE Trans. Fuzzy Syst. **14**(1), 31–41 (2006)
12. C.F. Juang, S.H. Chiu, S.W. Chang, A self-organizing TS-type fuzzy network with support vector learning and its application to classification problems. IEEE Trans. Fuzzy Syst. **15**(5), 998–1008 (2007)
13. G.C. Chen, C.F. Juang, Object detection using color entropies and a fuzzy classifier. IEEE Comput. Intell. Mag. **8**(1), 33–45 (2013)
14. G.C. Chen, C.F. Juang, An accuracy-oriented self-splitting fuzzy classifier with support vector learning in high-order expanded consequent space. Appl. Soft Comput. **15**(1), 231–242 (2014)

15. B. Schölkopf, A.J. Smola, *Learning With Kernels: Support Vector Machines, Regularization, Optimization, and Beyond* (MIT Press, Cambridge, MA, 2002)
16. R. Lienhart, J. Maydt, An extended set of Haar-like features for rapid object detection, in *Proceedings of the* 2005 *IEEE International Conference Image Processing*, 2002, pp. 900–903
17. D.G. Lowe, Distinctive image features from scale-invariant keypoints. Int. J. Comput. Vision **60**(2), 91–110 (2004)
18. N. Dalal, B. Triggs, Histograms of oriented gradients for human detection, in *Proceedings of the* IEEE International Conference on *Computer Vision Pattern Recognition*, 2005, pp. 886–893
19. K. Schindler, D. Suter, Object detection by global contour shape. Pattern Recognit. **41**(12), 3763–3748 (2008)
20. M. Marszalek, C. Schmid, Accurate object recognition with shape masks. Int. J. Comput. Vis. **97**(2), 191–209 (2012)
21. T. Kawanishi, H. Murase, S. Takagi, M. Werner, Dynamic active search for quick object detection with pan-tilt-zoom camera, in *Proceedings of the International Conference* on *Image Processing*, vol. 3, 2001, pp. 716–719
22. C.F. Juang, G.C. Chen, A TS fuzzy system learned through a support vector machine in principal component space for real-time object detection. IEEE Trans. Ind. Electron. **59**(8), 3309–3320 (2012)
23. A. Krizhevsky, I. Sutskever, G.E. Hinton, ImageNet classification with deep convolutional neural networks, in *Advances in Neural Information Processing Systems (NIPS)*, 2012, pp. 1–9
24. S. Ren, K. He, R. Girshick, J. Sun, Faster R-CNN: towards real-time object detection with region proposal networks. IEEE Trans. Pattern Anal. Mach. Intell. **39**(6), 1137–1149 (2017)
25. S.J. Pan, Q. Yang, A survey on transfer learning. IEEE Trans. Knowl. Data Eng. **22**(10), 1345–1359 (2010)
26. J. Cao, Y. Pang, X. Li, Learning multilayer channel features for pedestrian detection. IEEE Trans. Image Process. **26**(7), 3210–3220 (2017)
27. N. Cristianini, J.S. Taylor, *An Introduction to Support Vector Machines and Other Kernel-based Learning Methods* (Cambridge University Press, Cambridge, U.K, 2000)
28. M.F. Wu, W.C. Huang, C.F. Juang, K.M. Chang, C.Y. Wen, Y.H. Chen, C.Y. Lin, Y.C. Chen, C.C. Lin, A new method for self-estimation of the severity of obstructive sleep apnea using easily available measurements and neural fuzzy evaluation system. IEEE J. Biomed. Health Inf. **21**(6), 1524–1532 (2017)
29. G. Borgefors, Hierarchical chamfer matching: a parametric edge matching algorithm. IEEE Trans. Pattern Anal. Mach. Intell. **10**(6), 849–865 (1988)
30. P.H. Lee, Y.L. Lin, S.C. Chen, C.H. Wu, C.C. Tsai, Y.P. Hung, Viewpoint-independent object detection based on two-dimensional contours and three-dimensional sizes. IEEE Trans. Intell. Transp. Syst. **12**(4), 1599–1608 (2011)
31. W. Hu, S.C. Zhu, Learning 3D object templates by quantizing geometry and appearance spaces. IEEE Trans. Pattern Anal. Mach. Intell. **37**(6), 1190–1205 (2015)

Particle Swarm Optimization Based HMM Parameter Estimation for Spectrum Sensing in Cognitive Radio System

Yogesh Vineetha, E. S. Gopi and Shaik Mahammad

Abstract Spectrum Estimation has emerged as the major bottleneck for the development of advanced technologies (IoT and 5G) that demand for a unperturbed continuous availability of the spectrum resources. Opportunistic dynamic access of spectrum by unlicensed users when the licensed user is not using the resources is seen as a solution to the pressing issue of spectrum scarcity. The idea proposed for spectrum estimation is to model the Cognitive Radio (CR) network as Hidden Markov Model (HMM). The spectral estimation is done once in a frame. 100 such frames with 3000 slots each is considered for performing the experiment, assuming that the PU activity is known for a fraction of 3.33% of the slots i.e., for 100 slots. The parameters of the HMM are estimated by maximizing the generating probability of the sequence using the Particle Swarm Optimization (PSO). For the typical values of the network parameters, the experiments are performed and the results are presented. A novel sum squared error minimization based "Empirical Match" algorithm is proposed for an improved latent sequence estimation.

Keywords Computational intelligence · Particle swarm optimization · Empirical match algorithm · Cognitive radio · Spectrum estimation · Hidden Markov model

---

Y. Vineetha · E. S. Gopi (✉) · S. Mahammad
Department of Electronics and Communication Engineering,
National Institute of Technology Trichy, Tiruchirapalli
620015, Tamil Nadu, India
e-mail: esgopi@nitt.edu

Y. Vineetha
e-mail: yogesh.vineetha25@gmail.com

S. Mahammad
e-mail: mahammad.shaik2@gmail.com

W. Pedrycz and S.-M. Chen (eds.), *Computational Intelligence for Pattern Recognition*, Studies in Computational Intelligence 777,
https://doi.org/10.1007/978-3-319-89629-8_10

1 Introduction

A fast pace development of wireless technologies such as Internet Of Things and the emerging 5G technology demand for a continuous availability of spectrum resources for all its users. However, the major bottleneck is the spectrum scarcity problem that might be a consequence of either poor throughput of the network due to congestion or lack of available spectrum resources [5, 8, 9]. The Cognitive Radio technology aims at improving the spectrum utilization and network throughput by enabling the unlicensed users (Secondary Users (SU)) to access the resources of the licensed user (Primary User (PU)) whenever the PU is not utilizing the allocated resources. For an extensive utilization of the resources, the SU must learn the behavior or trend of how the PU is utilizing the resources. Physical spectrum sensing based on energy detection is performed in-order to know the presence of licensed user. However owing to the erroneous channel conditions due to external noise and interference from other users, it is highly possible that the results of physical spectrum sensing are not reliable. The following cases may arise:

1. If the PU is idle, and the physical spectrum sensing decides that the PU is active - **False alarm**, then it will lead to under-utilization of the spectrum resources, since neither the PU is using the channel nor the SU can make use of the free channel owing to false detection. This will have an adverse impact on the throughput and efficiency of the channel.
2. If the PU is busy and the SU senses the channel to be free - **Miss detection**, then the PU and the SU will simultaneously attempt to transmit data i.e., channel contention thereby resulting in congestion.

Hence, the SU cannot rely on the outcome of physical spectrum sensing. A more accurate estimate of PU activity is needed for opportunistic spectrum access by the SU to utilize the spectrum resources without leading to congestion in network and also increasing the network throughput.

2 Problem Formulation

The SU performs the physical spectrum sensing at sensing slots that are uniformly distributed over time. Let the PU spectrum access be represented by the random vector $\overline{\mathbf{S}}$. The outcome of the random vector in each sensing slot can be either 0 (PU is inactive) or 1 (PU is active). The SU performs spectrum sensing based on energy detection. Let $\overline{V}$ be a random vector that denotes the outcome of physical spectrum sensing, the outcome of the random vector will be a binary sequence. Under ideal conditions the outcome $\overline{V}$ is same as $\overline{S}$. However, in real time the result of the physical spectrum sensing and the actual PU activity are seldom in coherence. Hence, the problem under consideration is to estimate the PU activity as accurately as possible, given the SU observation sequence. The solution to this problem is to model the CR network as HMM [1] and extract the hidden network parameters which will further aid in estimating the PU activity more precisely.

2.1 Hidden Markov Model

The Hidden Markov Models (HMM) belongs to the class of mixture models, where the latent variables are discrete and belong to a finite set. The HMM consists of two stochastic processes of which one is hidden - latent. The other stochastic process is a result of the hidden process and is referred to as observation sequence. The hidden process in case of HMM is a markov process. A markov process of first order can be defined as the one in which the next state of the process depends only on the current state of the process and is independent of all the past history of the process. As mentioned earlier, the latent states of the HMM form the markov chain of first order [2, 3].

The latent state of the HMM at nth instant can be represented by a variable S_n that takes a value from a discrete set of k values and yields an observation V_n, where, k is the number of distinct states involved in the markov process. The latent process of the HMM is characterized with the transition probability defined as $p_{ij} = P(s_{n+1} = j/s_n = i)$, where $i, j \in k$. The observation process is a result of the hidden latent process that is generated as a result of emission of observations from the latent states. The emission of the observations depends on the emission probabilities defined as $h_{ij} = P(v_n = j/s_n = i)$. Each possible value of latent state is associated with a prior probability $\pi_k = P(s_1 = k)$, such that, $\sum_k \pi_k = 1$. The transition probabilities and emission probabilities of a two state HMM can be represented in the form of tables as in Fig. 1. In the Fig. 1, S_i and S_a are the two distinct states in the markov process, where subscript $'i'$ and $'a'$ indicate inactive and active in accordance with the application in CR network. Similarly, V_i and V_a implies the observation result, where, i and a hold the same meaning.

A HMM can be completely described by a model defined by $\lambda = [\pi \quad A \quad O]$, where, A is a matrix representing all possible transition probabilities between different states and O represents all the possible emission probabilities.

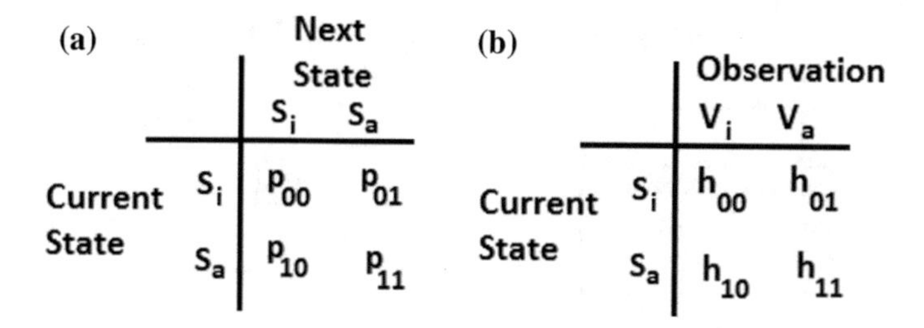

Fig. 1 **a** Transition probability table. **b** Observation probability table

Regarding the problem of spectrum estimation, the only information available with the SU is the erroneous observation sequence $\overline{V}$. The HMM model corresponding to the CR Spectrum estimation problem is as shown in Fig. 2. Using $\overline{V}$, the HMM model (i.e., $\lambda = [\pi \quad A \quad O]$) is to be estimated. Using the estimated HMM model parameters, further, the PU activity sequence is to be estimated. Hence, the Spectrum Estimation problem in hand can be broadly split into two tasks, where the solution of first problem paves way to solve the second problem.

1. **Task 1** To estimate the model λ
2. **Task 2** To estimate the hidden PU activity $\overline{S}$.

A generalized flow chart illustrating the steps employed for solving spectrum estimation problem is as in Fig. 3.

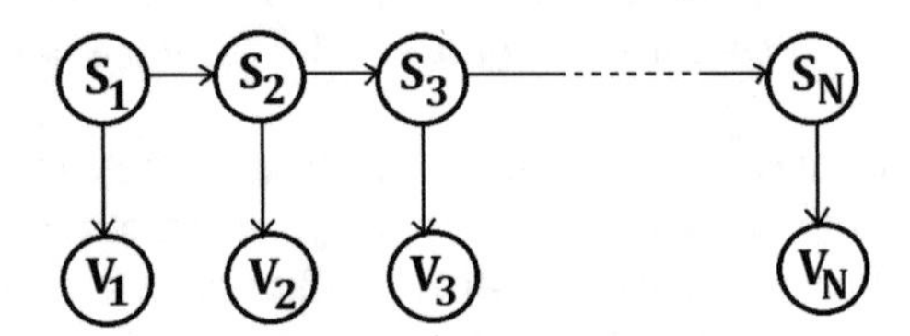

Fig. 2 Hidden Markov model of SU in cognitive radio network

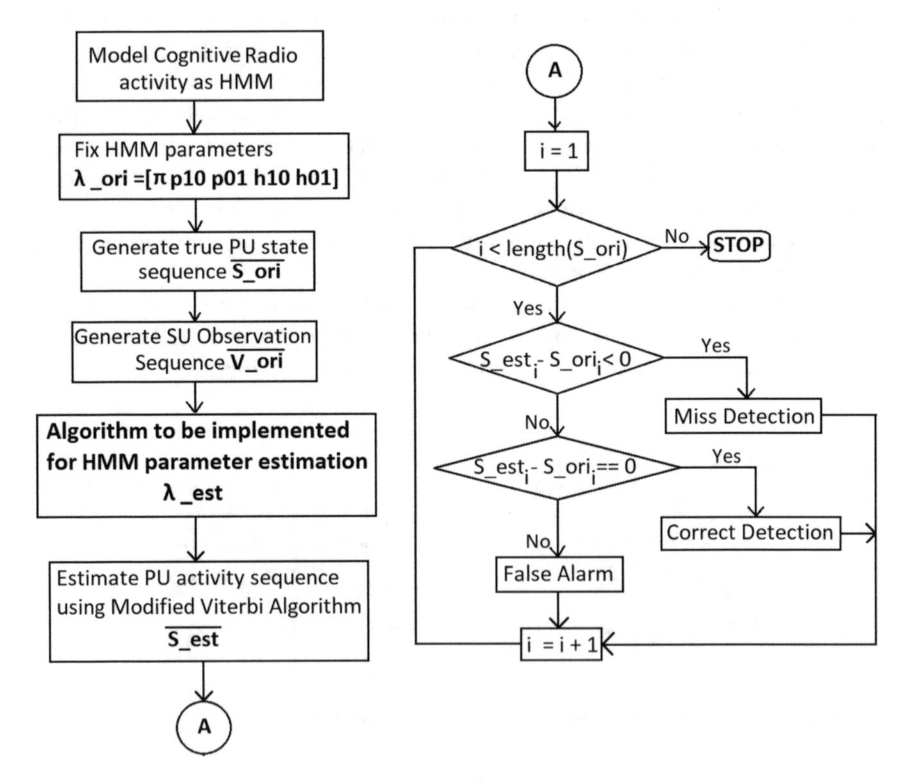

Fig. 3 Generalized steps involved in spectrum estimation problem

3 PSO Based Estimation of Hidden Parameters of Network - Task 1

The first problem associated with the spectrum estimation is to estimate the hidden parameters of the network (CR modeled as HMM). This can be solved conventionally by using Expectation Maximization (EM) algorithm. The algorithm aims at finding the solution that maximizes the probability of generation of model.

3.1 Expectation Maximization

The EM algorithm [4] is an iterative algorithm that tries to adjust the model parameters such that the probability of generation of the random vector $\overline{V}$, given the model parameters increases. It provides the Maximum likelihood solution to the problem. The generation probability of the model can be described by (1)

$$G = (\pi_0)^{1-s_1}(1-\pi_0)^{s_1} \prod_{i=1}^{i=N-1} p_{s_i s_{i+1}} h_{s_{i+1} v_{i+1}} \qquad (1)$$

The EM algorithm can be summarized as follows

1. Initialize the model with random parameters
2. Run EM algorithm (Refer Appendix) till convergence (200 iterations)
3. Repeat steps 1 and 2 for 10 times
4. From output in step 3, choose the best solution (model parameters that give maximum probability of generation) as λ_{est}.

However, there are a few drawbacks associated with the EM algorithm

1. The EM algorithm takes large time for convergence, i.e., slow convergence
2. The solution obtained using EM depends on the initialization, a bad initialization will result in convergence of EM algorithm to a local maxima.

3.2 PSO Based Proposed Technique

The use of CI technique - Particle Swarm Optimization (PSO) can be used to overcome the disadvantages of conventional EM algorithm. Particle Swarm Optimization (PSO) is a biologically inspired computational intelligence algorithm. Swarm here refers to a group of living objects such as a flock of birds, school of fishes etc. Each bird in the flock is technically referred to as particle in the swarm. The aim of the algorithm is to emulate the biological behavior of birds in the way they inter-communicate to reach their home(destination). It has been studied that the final decision of the bird about the direction in which it has to fly is based on the individual decision of that bird (Local Decision) and the decision of the flock (Global Decision). The objective is to minimize the distance of the bird from their current location to their final destination. Hence in general, PSO is a minimization algorithm. Each particle in the swarm is a potential solution to the problem [6, 7].

PSO Objective Function

The PSO is modeled to find the solution to HMM that maximizes the probability of generation of the model. The probability of generation of the model is formulated in terms of two iteratively updated variables. A forward probability variable $F(r, t)$ is considered, which takes care of generation of the random sequence $\overline{V}$, upto the rth frame, with the condition that the rth frame gets generated from tth state. The forward probability variable can be written in a recursive manner as,

- Initialization: $F(1, 0) = \pi_0 * P(V_1/S_1 = 0)$ and $F(1, 1) = \pi_1 * P(V_1/S_1 = 1)$
- Recursive equation

$$\begin{aligned} F(r, t = i) = F(r-1, t = 0) \times p_{0,i} \times P(V_r/t = i) + \\ F(r-1, t = 1) \times p_{1,i} \times P(V_r/t = i); \quad \forall i = 1 \; or \; 0 \end{aligned} \tag{2}$$

- Terminate when r = N (final state/ final slot)

A backward probability vector $B(r, t)$ is used, which governs the generation of random vector $\overline{V}$ from $(r + 1)$th frame to Nth frame, under the condition that the rth frame gets generated from tth state. Backward probability can be written in a recursive manner as

- Initialization: $B(N, 0) = 1$ and $B(N, 1) = 1$
- Recursive equation

$$\begin{aligned} B(r, t = i) = B(r+1, t = 0) \times p_{i,0} \times P(V_{r+1}/t = 0) + \\ B(r+1, t = 1) \times p_{i,1} \times P(V_{r+1}/t = 1); \quad \forall i = 1 \; or \; 0 \end{aligned} \tag{3}$$

- Terminate when $r = 1$.

The generation probability of HMM can be written in terms of $F(r, t)$ and $B(r, t)$ as

$$G = \sum_{t=0}^{1} F(r, t) \times B(r, t); r \in 1, 2, 3, \ldots, N \tag{4}$$

The objective function of PSO is hence formulated using (4) as $J = \frac{1}{G}$.

PSO Algorithm

The outcome of the vector $[\pi_0 \; p_{10} \; p_{01} \; h_{10} \; h_{01}]$ (hidden parameters) is treated as the position of the particle. The distance from the destination position is treated as the objective function $J = \frac{1}{G}$. It is noted that the elements of the vector are probabilities and hence the range is restricted between 0 to 1. The PSO algorithm is adopted that minimizes the objective function J is as given below.

1. Initialize the positions of the particles $a_1, a_2, \ldots a_N$. (with the elements of the vector ranging from 0 to 1).
2. initialize the tentative next positions of the birds $b_1, b_2, \ldots b_N$ (with the elements of the vector ranging from 0 to 1). Compute the corresponding cost function associated with the corresponding particles as $J_1, J_2, \ldots J_N$.
3. Compute $t = arg_i min J_i$.
4. Identify the next set of locations as follows.
   $c_i = |a_i + \alpha_1 \times (b_i - a_i) + \alpha_2 \times (b_t - a_i)|$
   if $(c_i >= 1)$, c_i is randomly chosen with the elements ranging from 0 to 1.
5. Assign $a_i = c_i$.
6. Compute the cost function associated with the corresponding particles $c_1, c_2, \ldots c_N$ as $K_1, K_2, \ldots K_N$.
7. If $J_i > K_i$, then $b_i = c_i$ else $b_i = b_i$.
8. Repeat the steps for finite number of iterations and the best particle's position corresponding to the lowest functional value J is declared as the estimated hidden parameter.

An illustration of how particles move in PSO occurs is as in Fig. 4. In the figure, the boundary of box is having range from -1 to $+1$, because, the particle elements are probabilities that can take a valid value in range 0 to 1. 3 particles are considered.

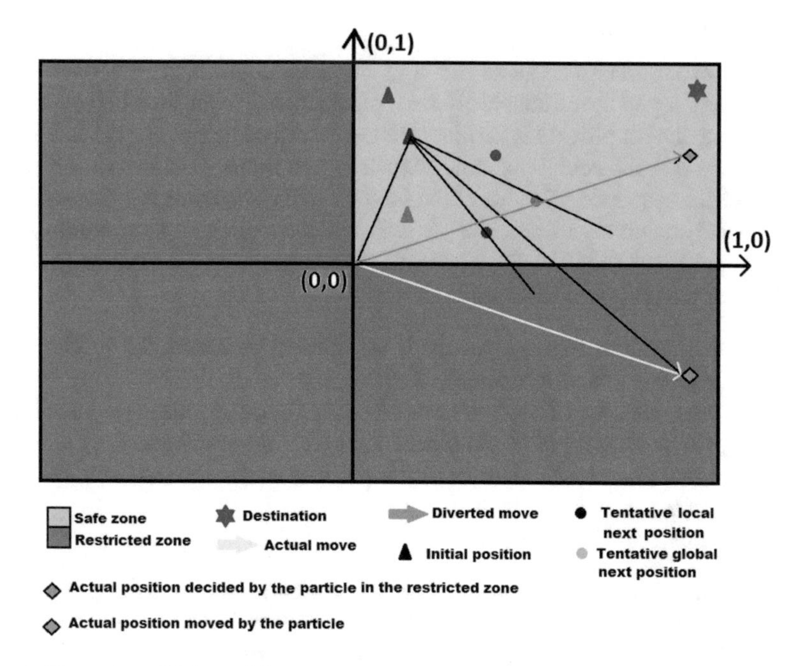

Fig. 4 Illustration of PSO particle movement

In the illustration, the triangles represent the initial position of the particles and the circles represent the tentative next position (local decision) of the particles. The global decision of the flock (green circle) can be considered as the one that is nearest to the destination (purple star). For understanding the particle motion in space, let us consider the maroon color particle (triangle) movement. The particle movement is governed by a linear combination of the particle's local decision (maroon circle) and the flock global decision (green circle). Also, in the box, the portion colored green is the safe zone which corresponds to a valid probability value and red color corresponds to the restricted zone - invalid probability values. A case of the linear combination resulting in the location of the particle's next position (orange diamond) in the restricted zone is considered for illustration. In such case, the particle is flipped back to a position in the safe zone (blue diamond). A case that the particle position outside the box boundary might also arise. In such a case, the particle is positioned in a random location within the safe zone and iterations continue.

Since it is assumed that the PU activity is known for first 100 slots out of 3000 slots in a frame, the information is utilized to estimate the parameters of the HMM parameters. The initialization of PSO is made using the known information.

4 Proposed Technique to Estimate the Outcome of Random Vectors - Task 2

Empirical Match Algorithm

Estimation of the the outcome of the random vector $\overline{\mathbf{S}}$ given the outcome for the random vector $\overline{\mathbf{V}}$ and the estimated hidden parameters λ_{est} from Task 1 is proposed to be done using the Empirical Match algorithm as described below. For the estimation, it is assumed that the actual outcome of the random vector $\overline{\mathbf{S}}$ is known for $1/30$th (3.33%) of the sequence. The rest of the 96.67% of the sequence is estimated using the empirical Match algorithm. The objective of the algorithm is to minimize the Sum Squared Error (SSE) of the stochastic parameters used for estimating $\overline{S}$. The algorithm of the proposed technique is as follows

1. Trend of transitions in the sequence is assumed to be known for $1/30$th of the sequence (say n slots information is known)
2. For the next outcome of random vector $\overline{V}$, i.e., V_{n+1}, the SSE is calculated considering the possibility of generation of V_{n+1} from $S_{n+1} = 0$ and $S_{n+1} = 1$.
3. The S_{n+1} that gives lower value of SSE is considered and the estimated sequence $\overline{S_{est}}$ is updated.
4. Repeat steps 2 and 3 till $n = N$.

The flow chart of the algorithm is as in Fig. 5. The algorithm tries to track closely the changes in the parameters while estimating the activity of the PU in every successive slot, there by providing a reliable estimate of the PU activity.

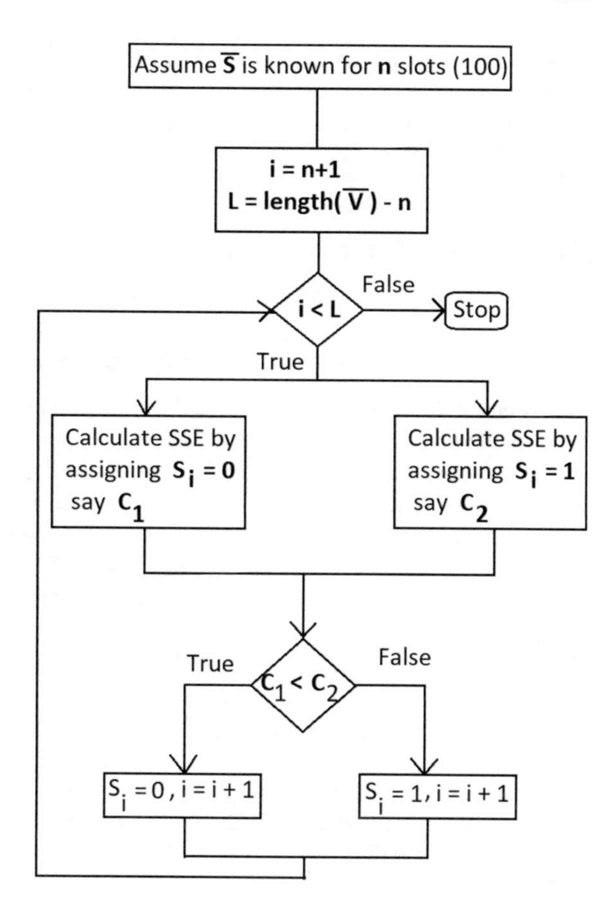

Fig. 5 Flowchart illustrating the empirical match algorithm

5 Experimental Setup and Results

The spectrum sensing is performed at uniform intervals of times referred to as sensing slots or just slots. The sensing results (1 or 0 based whether the PU is active or inactive in the slot respectively) in the slots are considered as the outcome of random vector $\overline{S}$. It is assumed that the CR network parameters donot cange over 300000 slots (Stationary process). For the purpose of experiment, the spectrum sensing data is arranged into a 100×3000 matrix. Each row of the matrix is referred to as a frame. It is assumed that activity of the PU is known for 3.33% of slots per frame, i.e., 100 slots per frame. HMM model of the CR network is considered and the model parameters are the set of transition and emission probabilities of the HMM model represented by the vector $\lambda = [\pi, p_{10}, p_{01}, h_{10}, h_{01}]$. The definition of the elements of the HMM model is provided in Sect. 2. The elements of vector λ reflects to the PU activity $-[\pi, p_{10}, p_{01}]$ and the erroneous channel conditions $-[h_{10}, h_{01}]$. The PU is assumed not to change its state within a slot and also it is assumed that the probability that the PU continues to be active/inactive for some slots continuously, once it becomes active/inactive is high i.e., p_{11} *and* p_{00} is high compared to p_{10} *and* p_{01}. The choice of the emission probabilities h_{01} *and* h_{10} is made randomly, a high value of which indicates bad channel conditions and hence a high probability of error in sensed sequence and vice-versa.

Estimation of PU Activity Using Conventional and Proposed Techniques

The spectrum estimation is performed using

1. Conventional method Expectation Maximization followed by Empirical match algorithm
2. Proposed method using PSO followed by Empirical match algorithm.

Experiments were performed for 6 different sets of typical combinations of CR network parameters (HMM parameters). The performance of the proposed algorithms is compared with the conventional solution. The result is oriented on estimating the PU activity. The solution obtained using the Proposed CI technique and the conventional method is compared with the result of the physical spectrum sensing which clearly illustrates the need for the proposed algorithm. The comparison is done in terms of percentage of match of the estimated sequence with the actual PU activity sequence. Also, the percentage of miss detection and false alarm is compared which is an indication of reduced number of errors in the estimated sequence. The spectrum estimation was performed using the conventional EM algorithm and the PSO algorithm followed by empirical match algorithm and results are tabulated in Table 1.

The PSO algorithm with 500 particles was run for 10 iterations and the convergence graph is as shown in Fig. 6. Also an illustration of the HMM parameter estimation using PSO is shown in Fig. 7 for one set of typical network parameters (refer to set 1 in Table 1). The solution obtained using PSO is compared with the original parameters as well as with the initial estimate of parameters using the known

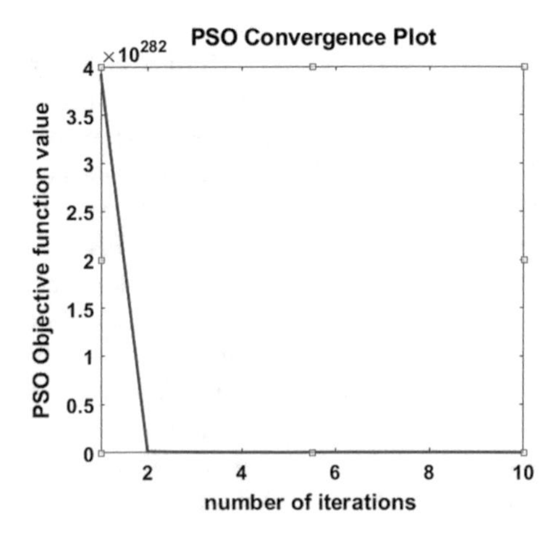

Fig. 6 PSO Convergence Plot

Table 1 Comparison of performance of conventional algorithm and proposed algorithm with the physical spectrum sensing result in estimating the PU activity for different sets of typical CR network parameters

Parameter sets	True network parameters					Observations							
	β_0	p_{10}	p_{01}	h_{10}	h_{01}	Match			Miss detection		False alarm		
						Physical sensing	Without CI (using EM)	With CI (using PSO)	Without CI (using EM)	With CI (using PSO)	Without CI (using EM)	With CI (using PSO)
Set 1	0.4900	0.2000	0.2200	0.8500	0.7700	**18.8817**	24.0630	**81.4850**	39.4520	**7.8963**	36.4850	**10.6187**
Set 2	0.2200	0.1600	0.2500	0.7800	0.8500	**19.1513**	61.6260	**80.9130**	21.1910	**10.4267**	17.1830	**8.6603**
Set 3	0.6200	0.2000	0.1200	0.5100	0.9000	**24.8067**	40.5517	**75.9867**	0.6023	**17.9533**	58.8460	**6.0600**
Set 4	0.1800	0.2000	0.2300	0.6600	0.8900	**23.3490**	56.8863	**75.4627**	17.2067	**13.6070**	25.9070	**10.9303**
Set 5	0.6200	0.2000	0.1200	0.5100	0.9000	**24.7680**	62.1423	**75.6580**	35.6203	**18.1743**	2.2373	**6.1677**
Set 6	0.6000	0.1000	0.1500	0.8100	0.7400	**21.8277**	48.2133	**75.4773**	27.9520	**15.0710**	23.8347	**9.4517**

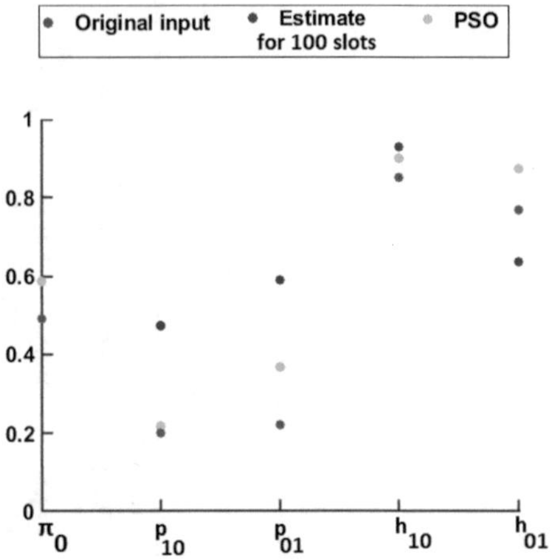

Fig. 7 Parameter estimation using PSO (set 1 in Table 1)

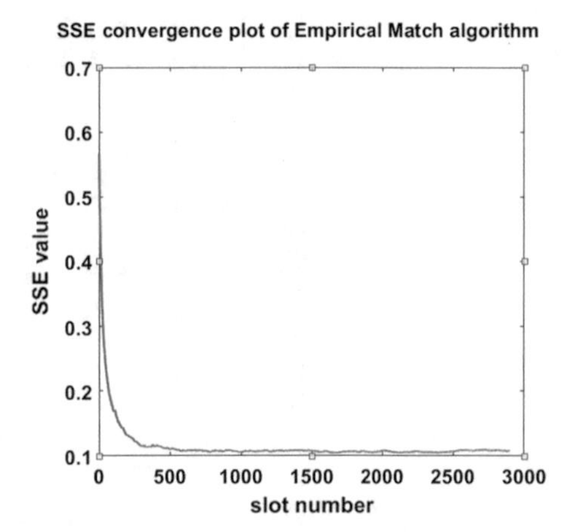

Fig. 8 Convergence plot of empirical match algorithm (SSE minimization)

information (100 slots). It can be seen that, for all parameters, the PSO solution tries to converge closer to the original parameter value (Fig. 7).

The Empirical match algorithm is executed once the parameters are estimated using the PSO algorithm and Expectation maximization. The assumption that the PU activity is known for 3.33% of slots holds true for empirical match algorithm. Hence, using Empirical match algorithm, the rest 2900 slot activity of PU is estimated. Estimation of PU activity in every slot beginning from the 101th slot to the 3000th slot is done with the objective of reducing the Sum Square Error (SSE) between the estimated and obtained parameters. The Convergence plot of SSE in Empirical Match algorithm is as shown in Fig. 8.

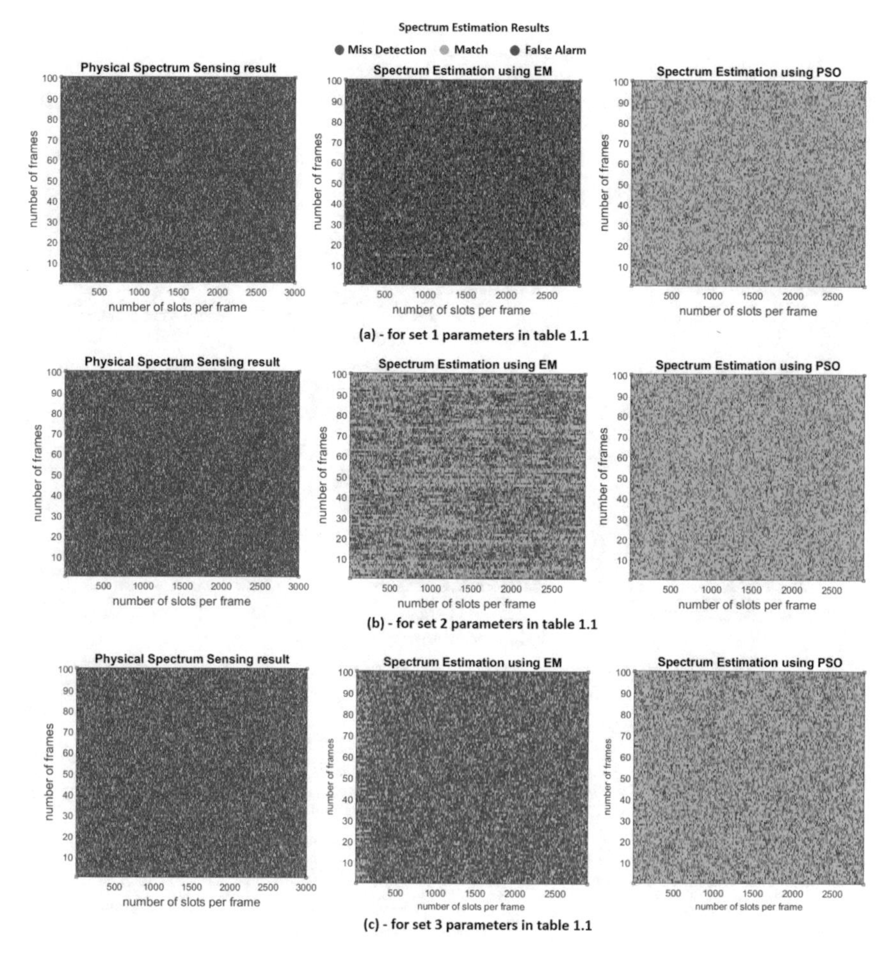

Fig. 9 Illustration of comparison of physical spectrum sensing with the spectrum estimation using EM and PSO

An illustration of the results of physical spectrum sensing, estimation using EM followed by empirical match and estimation using PSO followed by empirical match is shown in Figs. 9 and 10. The figures correspond to the various sets of network parameters considered for performing the experiments. The figures follow a color code RGB, where, Red and Blue are used to represent the mismatch between the estimated sequence and the actual PU activity sequence. Amongst Red and Blue. Red indicates miss detection and Blue indicates false alarm. The green color is used to indicate the match which is the main focus of the experiments.

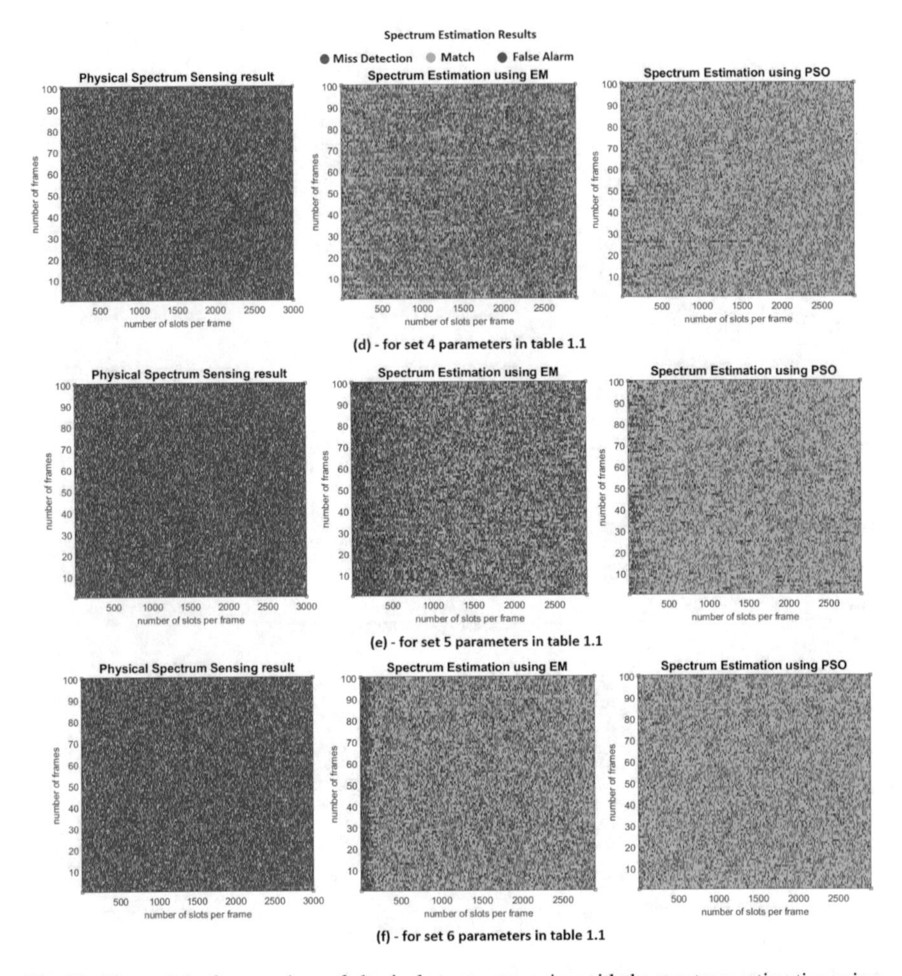

Fig. 10 Illustration of comparison of physical spectrum sensing with the spectrum estimation using EM and PSO

6 Conclusion

The chapter proposes a computational intelligence based solution for spectrum estimation in cognitive radio networks. The solution proposed uses the Particle Swarm Optimization (PSO) followed by the novel Empirical Match algorithm. The futility of the physical spectrum sensing for opportunistic spectrum access can be overcome by using the proposed CI based technique. An average improvement of 55.36% over the physical spectrum sensing is obtained by the use of PSO followed by empirical match which accounts for additional match of one lakh sixty six thousand slots. This implies that the SU can better utilize the spectrum, thereby improving the spectrum utilization and network throughput. The proposed CI based algorithm was compared with the conventional solution to the problem using the expectation maximization followed by empirical match algorithm. It can be seen from Table 1 that the CI based solution to spectrum estimation problem outperforms the convention solution to problem as well as the physical spectrum sensing method (Energy Detection).

7 Future Scope

The experiments are performed based on the assumption that the PU activity is known for 3.33% of total time i.e., 100 slots. Scope for further reduction in the amount of known information can be seen. It is believed that the spectral estimation match can be improved by increasing the number of states in the HMM.

8 Appendix

Expectation Maximization Algorithm The EM algorithm used in our algorithm is as follows.

Consider the HMM with N observations. Let the observation sequence be defined as $\overline{V} = \left[V_1 V_2 \ldots V_r V_{r+1} \ldots V_N\right]$, where r is the index variable, $r = 1, 2, \ldots N$.

Also, since the Latent states of the HMM are governed by the PU activity, the Latent state can be in either of the two states, let the state of the PU be denoted by a binary variable, $'t'$.

Let $F(r, t)$ denote the forward probability variable and $B(r, t)$ denote the backward probability variable.
F(r,t) means the probability of generating the observation sequence till the rth bit, with the condition that the rth bit is generated from PU being in tth state. A recursive formula for forward probability can be written as in Eq. 5. The initializations being: $F(1, 0) = \pi_0 * P(V_1/S_1 = 0)$ and $F(1, 1) = \pi_1 * P(V_1/S_1 = 1)$

$$\begin{aligned} F(r, t = i) = F(r-1, t = 0) \times p_{0,i} \times P(V_r/t = i) + \\ F(r-1, t = 1) \times p_{1,i} \times P(V_r/t = i); \quad \forall i = 1 \ or \ 0 \end{aligned} \tag{5}$$

Similarly, $B(r, t)$ denote the probability of generating the observation sequence from $(r + 1)$th bit till end, with the condition that the rth bit is generated from the PU being in state t. A recursive relation can be developed for finding the backward probability as in (6). The initializations are $B(N, 0) = 1$ and $B(N, 1) = 1$

$$\begin{aligned} B(r, t = i) = B(r+1, t = 0) \times p_{i,0} \times P(V_{r+1}/t = 0) + \\ B(r+1, t = 1) \times p_{i,1} \times P(V_{r+1}/t = 1); \quad \forall i = 1 \ or \ 0 \end{aligned} \tag{6}$$

Let $M(V_r, t = i)$ denote the fraction of V_rth bit being generated by the PU in state $t = i$, i∈(1,0). Then $M(V_r, t)$ can be obtained as

$$M(V_r, t = i) = \frac{F(V_r, t = i) \times B(V_r, t = i)}{F(V_r, t = 0) \times B(V_r, t = 0) + F(V_r, t = 1) \times B(V_r, t = 1)} \tag{7}$$

Let $Q(V_r, i, j)$ denote the fraction of bit V_r being generated as a result of transition of PU from state i to state j

$$Q(V_r, t = i) = \frac{F(V_r, i) \times p_{ij} \times P(V_{r+1}/t = j) \times B(V_{r+1}, j)}{\sum_{i=0}^{1} \sum_{j=0,1} F(V_r, i) \times p_{ij} \times P(V_{r+1}/t = j) \times B(V_{r+1}, j)} \tag{8}$$

The HMM model parameters can be estimated to maximize the probability of generation of the observation given the stochastic HMM parameters. The different stochastic parameters namely, the transition probability, observation probability and the prior probability can be derived using (7) and (8).

References

1. K.W. Choi, E. Hossain, Estimation of primary user parameters in cognitive radio systems via hidden Markov model. IEEE Trans. Signal Process. **61**(3) (2013)
2. C.M. Bishop, *Pattern Recognition and Machine Learning* (Springer, Berlin, 2006)
3. S. Shimpi, V. Patil, Hidden Markov model as classifier: a survey. Elixir (2013)
4. L.R. Rabiner, A Tutorial on hidden Markov models and selected applications in speech recognition. Proc. IEEE **77**(2) (1989)
5. E. Hossain, D. Niyato, Z. Han, *Dynamic Spectrum Access and Management in Cognitive Radio Networks* (Cambridge University Press, Cambridge, 2009)
6. E.S. Gopi, *Mathematical Summary for Digital Signal Processing Applications with MATLAB* (Springer, Berlin, 2010)
7. S. Rao, *Engineering Optimization: Theory and Practice*, 4th edn. (Wiley, New Jersey, 2009)
8. L. Lu, X. Zhou, U. Onunkwo, G.Y. Li, Ten years of research in spectrum sensing and sharing in cognitive radio. EURASIP J. Wirel. Commun. Netw. (2012)
9. L. Csurgai-Horwath, J. Bito, in *Primary and secondary user activities for cognitive wireless network*, 11th International Conference on Telecommunications - ConTEL (2011)

Computational Intelligence for Pattern Recognition in EEG Signals

Aunnoy K Mutasim, Rayhan Sardar Tipu, M. Raihanul Bashar, Md. Kafiul Islam and M. Ashraful Amin

Abstract Electroencephalography (EEG) captures brain signals from Scalp. If analyzed and patterns are recognized properly this has a high potential application in medicine, psychology, rehabilitation, and many other areas. However, EEG signals are inherently noise-prone, and it is not possible for human to see patterns in raw signals most of the time. Application of appropriate computational intelligence is must to make sense of the raw EEG signals. Moreover, if the signals are collected by a consumer grade wireless EEG acquisition device, the amount of interference is ever more complex to avoid, and it becomes impossible to see any sorts of pattern without proper use of computational intelligence to discover patterns. The objective of EEG based Brain-Computer Interface (BCI) systems is to extract specific signature of the brain activity and to translate them into command signals to control external devices or understand human brains action mechanism to stimuli. A typical BCI system is comprised of a Signal Processing module which can be further broken down into four submodules namely, Pre-processing, Feature Extraction, Feature Selection and Classification. Computational intelligence is the key to identify and extract features also to classify or discover discriminating characteristics in signals. In this chapter we present an overview how computational intelligence is used to discover patterns in brain signals. From our research we conclude that, since EEG signals are

A. K. Mutasim · R. S. Tipu · M. R. Bashar · M. A. Amin (✉)
Computer Vision & Cybernetics Group,
Department of CSE, Independent University, Bangladesh, Dhaka, Bangladesh
e-mail: aminmdashraful@iub.edu.bd

A. K. Mutasim
e-mail: aunnoy@iub.edu.bd

R. S. Tipu
e-mail: 1330418@iub.edu.bd

M. R. Bashar
e-mail: 1320454@iub.edu.bd

M. K. Islam
Computer Vision & Cybernetics Group,
Department of EEE, Independent University, Bangladesh, Dhaka, Bangladesh
e-mail: kafiul_islam@iub.edu.bd

W. Pedrycz and S.-M. Chen (eds.), *Computational Intelligence for Pattern Recognition*, Studies in Computational Intelligence 777,
https://doi.org/10.1007/978-3-319-89629-8_11

the outcome of a highly complex non-linear and non-stationary stochastic biological process which contain a wide variety of noises both from internal and external sources; thus, the use of computational intelligence is required at every step of an EEG-based BCI system starting from removing noises (using advanced signal processing techniques such as SWTSD, ICA, EMD, other than traditional filtering by identifying/exploiting different artifact/noise characteristics/patterns) through feature extraction and selection (by using unsupervised learning like PCA, SVD, etc.) and finally to classification (either supervised learning based classifier like SVM, probabilistic classifier like NB or unsupervised learning based classifiers like neural networks namely RBF, MLP, DBN, k-NN, etc.). And the usage of appropriate computational intelligence significantly improves the end results.

Keywords Computational intelligence · Pattern recognition
Electroencephalography (EEG) · Brain-computer interface (BCI) · Stationary wavelet transform (SWT) · SWTSD · PCA · LDA · SVD · Supervised learning
Neural networks · Deep belief network (DBN) · Convolution neural network (CNN) · Event related potential (ERP) · Fast Fourier transform (FFT) · Motor imagery (MI) · Naïve Bayes (NB) · Support vector machine (SVM) · Video category classification (VCC)

1 Introduction

Being one of the most natural parts in human-computer interaction (HCI), Brain-Computer Interfaces (BCIs) have shown great promise for the physically disabled people or people with severe neuromuscular disorders [1, 2]. According to several studies, signals recorded from the brain can become a substitute for any job that requires muscle control or movement [3]. There are a number of methods, such as electroencephalography (EEG), functional MRI (fMRI), electrocorticography (ECoG), calcium imaging, magnetoencephalography (MEG), functional near-infrared spectroscopy (fNIRS), etc., using which such brain signals can be captured.

Electroencephalography (Scalp EEG) signals, which are small amounts of electromagnetic waves emitted by the neurons in the brain [4], are one of the most popularly used signal acquisition techniques in the existing BCI systems due to their non-invasiveness, easy to use, reasonable temporal resolution and cost effectiveness compared to other brain signal recording methods [2]. As far as EEG recordings are concerned, the signals are the outcome of a highly complex non-linear and non-stationary stochastic biological process which contain a wide variety of noises both from internal and external sources. Thus, the use of computational intelligence is required at every step of an EEG-based BCI system starting from removing noises (using advanced signal processing techniques such as SWTSD, ICA, EMD, other than traditional filtering by identifying/exploiting different artifact/noise characteristics/patterns) through feature extraction and selection (by using unsupervised learning like PCA, SVD, etc.) and finally for classification (either supervised learning

Table 1 Rhythms and their traits of EEG signal

Rhythm	Bandwidth	Traits
Delta (δ)	[0.5–3] Hz	This activity occurs in unconscious, anesthetized or in deep sleep stage but almost no activity when in wake state
Theta (θ)	[4–7] Hz	Associated with emotional pressure, deep physical relaxation, and/or deep meditation
Alpha (α)	[8–13] Hz	This occurs in rest state. Thinking, blinking, etc. makes alpha waves disappear
Beta (β)	[14–30] Hz	Beta waves are generated when a person is receiving sensory stimulation, attentive or thinking actively
Gamma (γ)	[31–50] Hz	It is related to perceptual and cognition activity; selective attention can also trigger this

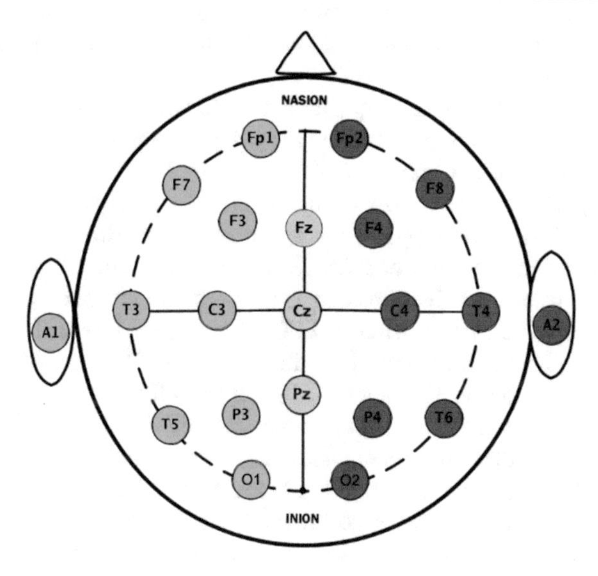

Fig. 1 The 10–20 international system

based classifier like SVM, probabilistic classifier like NB or unsupervised learning based classifiers like neural networks namely RBF, MLP, DBN, k-NN, etc.).

EEG signals can be broken down into five main rhythms based on their frequency range: delta (δ), theta (θ), alpha (α), beta (β) and gamma (γ) [4, 5]. A brief description of the EEG rhythms and traits are shown in Table 1.

EEG relies on the averaging of the responses of many neurons [6]. It is non-invasive where signal acquiring electrodes are positioned on the scalp according to the standard 10–20 international system [7] (see Fig. 1) to ensure reproducibility among studies.

Every electrode in the 10–20 system has a unique identity that identifies which lobe and hemisphere of the brain does one particular electrode correspond to. The letters F, T, C P, and O stand for frontal, temporal, central, parietal, and occipital lobes respectively. Right hemisphere electrode positions are referred with even numbers (between 2 and 8) whereas odd numbers (between 1 and 7) correspond to the left hemisphere. Electrodes positioned on the midline are referred as a "z" (zero) [8]. This means that each of the electrodes provides information to a particular area of the brain. However, this highly depends on the accuracy of the electrodes' placement.

One of the biggest disadvantages of EEG signals is that they are highly susceptible to noise mainly because of its non-invasive nature. These noises, often termed as artifacts, are influenced by extraneous signals, for example electromyography (EMG)—electrical signals originating from muscles in the face and scalp instead of signals originating in the brain [9] and electrooculography (EOG)—noise generated from eye movements/blinking [10]. Also motion artifact is a major source of noise in EEG due to physical movement of the subject [11]. Fortunately, there have been a significant number of researches done to utilize advanced signal processing techniques to overcome these noises [11–15].

There are several EEG signal acquisition devices typically used by researchers in the literature [16]. They are g.USBamp [17, 18], g.BSamp [19], and g.BCIsys [20] made by g.tec in Austria, Cerebus [21–23] made by Black-Rock Microsystems in USA, SynAmps 2 [24–26] made by Compumedics Neuroscan in Australia, wireless Emotiv EPOC [27–30] made by Emotiv Systems in USA, BrainNet-36 [31], ANT-Neuro [32], FlexComp Infiniti encoder [33], etc. In the recent past, a whole new domain for BCI researchers have opened up thanks to the advent of low-cost, easy to use portable dry/wet electrode wireless EEG recording devices such as NeuroSky's MindWave [34], InteraXon's Muse [35], Emotiv EPOC [27], etc. which have been used by researchers in several studies [4, 36–38] as well.

The objective of BCI systems is to extract specific signature of the brain activity and to translate them into command signals to control external devices (see Fig. 2) [39]. These features can be P300 evoked potentials, event-related potentials (ERPs) recorded on the cortex, slow cortical potentials (SCPs), sensorimotor rhythms acquired from the scalp, neuronal action potentials recorded within the cortex, etc.

Computational Intelligence is mainly involved in the Signal Processing module in Fig. 2 which can be broken down into four submodules [2]:

- Pre-processing—removal of noises/artifacts from the EEG signals,
- Feature Extraction—extracting features from the EEG signals,
- Feature Selection—selecting only the features that contains most of the information and
- Classification—deciding to which group does this set of EEG signals correspond to.

Researchers often skip the Feature Selection submodule [40–46] because, this step is only useful when the size or the dimensions of the features extracted by step (ii) is quite large. Large feature sets correspond to slower execution time making several BCI systems completely useless, especially online BCI systems. Thus, the Feature Selection step is used as a dimensionality reduction step to speed up computational time.

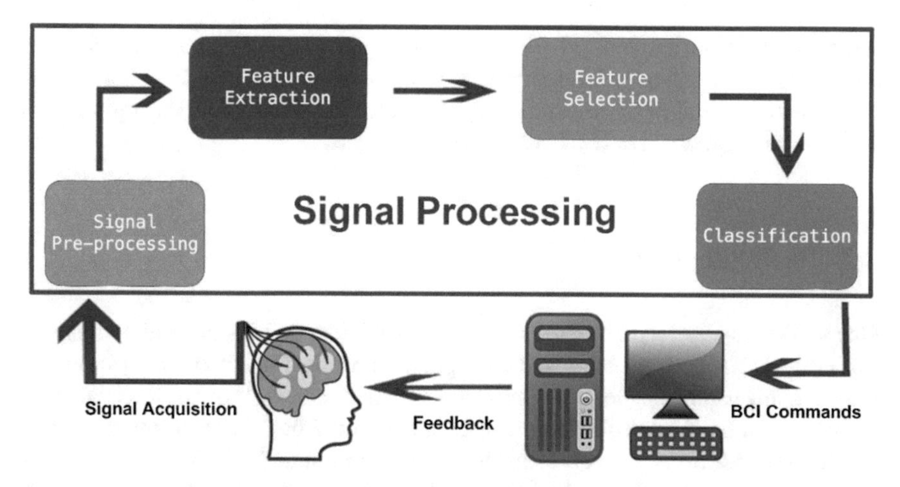

Fig. 2 A general description of a BCI system. The signal processing module can be divided into four submodules: pre-processing, feature extraction, feature selection and classification

In this chapter, we first present a thorough review of several articles for different BCI paradigms. Our focus is on algorithms used by researchers for each of the submodules of the Signal Processing module of a BCI system to solve a particular problem. Then, we analyze different contemporary algorithms for each submodule of a Signal Processing module on two datasets we acquired from:

- 19 college-aged young adults using Emotiv EPOC [27] at a sampling frequency of 128 Hz and
- 19 college-aged young adults using the Muse headband [35] at a sampling frequency of 220 Hz

where each of the participants was shown three different types of videos [47].

2 Use of Computational Intelligence in Different BCI Applications

Based on brain activity patterns, there are mainly four types of EEG-based BCI systems [16]—event related desynchronization/synchronization (ERD/ERS) [48], steady-state visual evoked potential (SSVEP) [2], event-related potential (ERP) [49], and slow cortical potential (SCP) [50]. Except for SCP, the other three are most popular among researchers [51–53].

These EEG-based BCI paradigms have led to many BCI applications. Emotion classification [36, 54–56], cognitive task classification [38, 57], P300 spellers [58–62] and others [63, 64] as an alternative and augmentative communication (AAC) platform [65], brain-controlled wheelchair [66–69], controlling a robot [70–73], rehabilitation of locked-in patients [74–78], neuro-prosthesis [79–82], gaming [83, 84], etc. are just a few examples. In this section, we will discuss about different pattern recognition techniques used by researchers for the detection of the three most prominent brain activity patterns i.e. ERD/ERS, SSVEP and ERP.

2.1 Motor Imagery

One of the most researched domain in ERD/ERS based BCIs are Motor Imagery (MI) [85–87]. MI corresponds to the imagination of moving a body part (for example right/left hand, tongue, both feet, etc.) without actually moving it. Oscillatory activities can be observed in different locations in the brain's sensorimotor cortex for different MI tasks. The objective is to classify such activities to be able to recognize the underlying MI task performed [88]. To achieve this, researchers in the past have experimented with various algorithms to improve the efficiency of the system as much as possible. A summary of different techniques used by researchers is presented in Table 2.

Band-pass filtering the EEG data from 0.5 to 30 Hz, Hamedi et al. [40] implemented Integrated EEG (IEEG) and Root Mean Squares (RMS) as feature extraction algorithms and Radial Basis Function (RBF) Neural Networks and Multilayer Perceptron (MLP) as classifiers for three class (right/left hand and tongue movement) MI classification. Comparing these algorithms with Support Vector Machine (SVM) classifier and Willison Amplitude (WAMP) feature extraction algorithm, it was illustrated that SVM performs better with regards to accuracy and time taken for training and WAMP was more suitable than RMS and IEEG.

Chatterjee et al. [89] classified the BCI competition II [94] MI dataset of left and right-hand movements with the accuracy of 85% and 85.71% for SVM and MLP respectively. They achieved this result by applying wavelet-based energy-entropy method as the feature extraction technique and average power-based feature provided better ROC area than the statistical feature. Their data were filtered using an elliptic band-pass filter on the range 0.5 to 30 Hz.

An et al. [90] in their paper also used an elliptic band-pass filter to attenuate signals in the range of 8 to 30 Hz and used Neuroscan software to remove EOG artifacts. They found that deep belief network (DBN) gives a 4–6% better performance compared to SVM when DBN was constructed with the combination of Restrict Boltzmann Machine (RBM), Adaboost algorithm and Contrastive Divergence (CD) for 8 hidden layers. Number of nodes had no effect, but subject's concentration and status played an important part in the performance of the classifier.

In a study [88] on BCI competition IV dataset 2b and competition II dataset III [94], the authors applied a combination of convolutional neural network (CNN) and stacked autoencoders (SAE) model and achieved an accuracy of 90.0% whereas the winner algorithm achieved 89.3% accuracy. According to kappa value, 9% improvement was achieved using this deep learning approach than the BCI competition winner algorithm.

Table 2 Summary of algorithms used by researchers for MI-based BCI

Authors	Pre-processing	Feature extraction	Feature selection	Classifier
Hamedi et al. [40]	Band-pass filter, 0.5–30 Hz	IEEG, RMS and WAMP		MLP, RBF Neural networks and SVM
Chatterjee et al. [89]	Elliptic band-pass filter, 0.5–30 Hz	Wavelet based energy-entropy method	Average power band	SVM and MLP
An et al. [90]	Elliptic band-pass filter, 8–30 Hz			DBN and SVM
Tabar et al. [88]				CNN and SAE
Kevric et al. [41]	Band-pass filter, 0.05–200 Hz MSPCA as noise removal technique	EMD, DWT and WPD		k-NN
Hsu et al. [91]	Gaussian filter	CWT along with student's two-sample t-statistics	GA	SVM
Li et al. [92]	Band-pass filter, 0.05–200 Hz		Mean, standard deviation, skewness, maximum, minimum and kurtosis	CC-LR
Zhang et al. [93]	Band-pass filter (0.5–100 Hz) and a 50 Hz notch filter	CSP		SBLFB

Kevric et al. [41] presented a comparison among three feature extraction methods— Discrete Wavelet Transform (DWT), Wavelet Packet Decomposition (WPD), and Empirical Mode Decomposition (EMD). The maximum average accuracy of 92.8% was achieved with the combination of Multiscale Principal Component Analysis (MSPCA) as noise removal technique, higher-order statistical features extracted from WPD sub-bands and k-nearest neighbour (k-NN) as the classifier. EEG data were band-pass filtered from 0.05 to 200 Hz.

Hsu et al. [91] classified 10 subjects' motor imagery data with SVM, genetic algorithm (GA) as feature selection method and student's two-sample t-statistics and continuous wavelet transform (CWT) as feature extraction method. They achieved an average classification accuracy of 86.7%. Gaussian filter was used in order to smooth the power spectrum data.

In [92], a modified cross-correlation based logistic regression (CC-LR) algorithm was used on three statistical feature sets consisting of mean, standard deviation, skewness, maximum, minimum and kurtosis as six features for BCI competition III dataset IVa and IVb [94]. Their algorithms provide better accuracy in three out of five subjects when compared with eight other known algorithms and the difference between proposed method accuracy and BCI competition III winner algorithm is 0.3. Digitized data of 1000 Hz was band-pass filtered between 0.05 to 200 Hz with a 16-bit accuracy.

Zhang et al. [93] achieved 81.7% accuracy (with ±15.1 standard deviation) and computational time of less than 10 seconds by implementing sparse Bayesian learning of frequency bands (SBLFB). They extracted features via common spatial pattern (CSP) and achieved better results when this combination was compared with other proposed methods implemented on the BCI Competition IV IIb dataset [94]. A band-pass filter was applied (0.5–100 Hz) with a 50 Hz notch filter.

2.2 *Steady State Visual Evoked Potential*

When the flickering frequency of the visual stimuli matches the frequency of the firing frequency of the visual cortex's neurons, the resulting brain signals are called Steady State Visual Evoked Potential (SSVEP) [95, 96]. SSVEP is identifiable in the range 5–60 Hz and is a very useful BCI tool due to its quite low signal to noise ratio (SNR). SSVEP can easily be identified in EEG signals and therefore it is possible to classify various kinds of visual stimuli. Researchers in the past have experimented with various techniques to classify these stimuli with competitive results.

Chen et al. [45] proposed a SSVEP-based single-channel BCI system using control algorithm and fuzzy tracking for amyotrophic lateral sclerosis (ALS) patient. Fuzzy control algorithm achieved average recognition rate of 96.97% compared to 94.9% achieved by canonical correlation analysis (CCA). In their proposed BCI system, they used fast Fourier transform (FFT) as feature extraction algorithm and in the pre-processing module, to extract data in the range 4–60 Hz, a 2nd-order Butterworth band-pass filter.

Maronidis et al. [97] proposed the use of Subclass Marginal Fisher Analysis (SMFA) to detect SSVEP and compared its result with CCA and Multiple Linear Regression (MLR) for different number of trials and channels. In both the settings, SMFA achieved better results than the other two algorithms. Authors used a 3rd degree band-pass Butterworth Infinite Impulse Response (IIR) filter (6–80 Hz) in the pre-processing module.

Kalaganis et al. [46] experimented with error-related potentials in SSVEP-based BCI system. Authors implemented Minimum Covariance Determinant (MCD) as an outlier detection algorithm or to remove noisy data, Common Spatial Patterns (CSP) as feature extraction technique, SVM, Random Forrest (RF) and Adaboost as classifiers. In comparison between SVM, RF and Adaboost, RF provides better average accuracy (0.8187) and recall rate (0.5633).

In the study conducted by Friman et al. [98], the authors achieved an average classification accuracy of 84% with the minimum energy method as classifier which takes

Table 3 Summary of algorithms used by researchers for SSVEP-based BCI

Authors	Pre-processing	Feature extraction	Feature selection	Classifier
Chen et al. [45]	2nd-order Butterworth band-pass filter, 4–60 Hz	FFT		Fuzzy control algorithm and CCA
Maronidis et al. [97]	3rd degree band-pass Butterworth IIR filter, 6–80 Hz			SMFA, CCA and MLR
Kalaganis et al. [46]	MCD	CSP		SVM, RF and Adaboost
Friman et al. [98]	Autoregressive model			Minimum energy method
Carvalho et al. [99]	Butterworth band-pass (5–60 Hz) and notch filtered (58–62 Hz)	Bank of filters, Welch's method and short-term Fourier transform	Incremental wrapper, Pearson's method and Davies-Bouldin index	LDA, SVM and ELM

about 4 msec computational time. Autoregressive model was implemented to calculate the noise level in SSVEP signal. In [99], authors compared between three feature extraction, feature selection and classification techniques for SSVEP-based BCI system. They implemented bank of filters, Welch's method and short-term Fourier transform as feature extraction methods, incremental wrapper, Pearson's method and Davies-Bouldin index as feature selection methods and support vector machine (SVM), linear discriminant analysis (LDA), and extreme learning machine (ELM) as classifiers on band-pass Butterworth (5–60 Hz) and notch filtered (58–62 Hz) EEG signal. LDA provides a better classification accuracy with Welch's method and incremental wrapper as feature extraction and feature selection methods respectively. Table 3 summarizes the algorithms used by different studies to classify SSVEP from EEG signals.

2.3 Event Related Potentials

The very small voltages in the brain structure generated due to the occurrence of certain events or stimuli are known as event-related potentials (ERPs) [100]. These fluctuations in the brain signal are evoked by and is also time-locked to a motor, sensory or cognitive event. Among several types of ERPs, namely N100 or N1, N200 or N2, P100 or P1, P200 or P2, etc., the P300 or P3 is the largest ERP component which gets triggered during an oddball paradigm. This oddball paradigm is one in which a participant is presented with a series of events which can be classified into two groups—frequently presented class and a rarely occurring class. The infrequent event generates a positive deflection (or a P300 peak) in the scalp voltage about 300 msec after stimulus presentation [101].

This P300 ERP has contributed substantially in the development of several EEG-based BCI applications. P300 Spellers [58–62], Brain Painting [102, 103], controlling

Table 4 Summary of algorithms used by researchers for ERP-based BCI

Authors	Pre-processing	Feature extraction	Feature selection	Classifier
Speier et al. [42]	Band-pass filter, 0.1–60 Hz	Ordinary least-squares regression [108]		SWLDA
Chaurasiya et al. [106]	8-order Chebyshev Type I band-pass filter, 1–10 Hz	Concatenation of six samples of all 64 channels	Binary Differential Evolution	Weighted ensemble of SVMs (WESVM)
Pinegger et al. [109]	Band-pass filter between 0.5 and 100 Hz and a 50-Hz notch filter	FFT		SWLDA
Li et al. [43, 44]	Band-pass filter, 0.01–30 Hz and a regression analysis algorithm to remove EOG artifacts	Wavelet decomposition and reconstruction		SVM and BLDA [43] SVM ensemble [44]
Kulasingham et al. [107]	Butterworth 4th order band-pass filter, 1–30 Hz	Filtered 1 second epochs		SVM ensemble

a virtual environment [104], gaming [105], etc. are just a few examples. For such applications the proper detection of the P300 peak, like any other pattern recognition problem involves pre-processing, feature extraction, feature selection, and classification.

Typically, band-pass filters are used on raw EEG signals to extract data in the range 0.1–30 Hz [101]. Although, Speier et al. [42] and Chaurasiya et al. [106] used substantially different high cut-off frequencies of 60 Hz and 10 Hz respectively and were able to achieve very good results. Filtered raw EEG data as features are not uncommon for P300 Spellers [106, 107]. However, sophisticated methods like ordinary least-squares regression [108] or conventional methods involving wavelet transforms [43, 44] can also be found in the literature. Feature selection, as discussed before, are used only when the size of the dataset is quite big and therefore, out of the articles summarized in Table 4, only one paper used feature selection methods [106].

Currently, stepwise linear discriminant analysis (SWLDA) and SVM ensembles are the two classifiers dominating the detection of the P300 wave in the literature [42, 44, 106, 107, 109]. In [43], the authors experimented with a different classifier, Bayesian Linear Discriminate Analysis (BLDA), and noted that although increasing the training set size decreases the difference in results between BLDA and SVM, the results of SVM in P300-speller with familiar face model by utilizing a small training set is better than that of BLDA.

3 An Experiment with State-of-the-Art Algorithms—Video Category Classification

In this section, we experimented with several state-of-the-art algorithms (discussed in the previous section) for each of the submodules of the Signal Processing module of a BCI system (see Sect. 1) on two datasets we acquired using two EEG signal acquisition devices (Muse headband [35] and Emotiv EPOC [27]) where each of the participants were shown videos of three different genres. Our objective is to passively classify which type of video a person is watching from their Scalp EEG signals as this is the fundamental step of our long-term goal of building a BCI based passive video recommender system [47]. This data with preliminary code is downloadable from [110].

3.1 Experimental Setup and Data Acquisition Techniques

EEG Recordings

As previously mentioned, Muse headband by InteraXon [35] and Emotiv EPOC [27] by Emotiv Systems were used to record electroencephalogram (EEG) data to create two datasets. These off-the-shelf wireless devices have been used previously in several studies as well [4, 36–38]. Muse is a dry electrode EEG recording device with 5 channels (TP9, AF7, AF8 and TP10 with reference channel at FPz) and the Emotiv EPOC is a wet electrode device with 16 channels (AF3, F7, F3, FC5, T7, P7, O1, O2, P8, T8, FC6, F4, F8 and AF4 with two reference channels at P3 and P4) arranged according to the international 10–20 system. Recording sampling frequency of the Muse and Emotiv EPOC were 220 Hz and 128 Hz respectively and the data were wirelessly transmitted to a computer via Bluetooth.

Demographics of Subjects

23 (15 males and 8 females) and 44 (32 males and 12 females) college-aged young adults contributed to dataset 1 (dataset created using the Muse headband) and dataset 2 (dataset created using Emotiv EPOC) respectively. The subjects had no personal history of mental or neurological disorders and had either normal or corrected-normal vision. The whole experiment for each of the subjects were also recorded using a webcam. We discarded data of 3 male and 1 female subjects from dataset 1 as after analysing these videos, we identified that one or more artifacts (excessive blinking, hand or body movements, etc. even after being instructed to move as less as possible) were excessively present in the signal. For this reason, we also selected 19 subjects with the same male to female ratio (12 males and 7 females) from dataset 2 as well to keep the comparisons between the two datasets legitimate.

All the participants signed informed consent forms prior to the study. The 19 selected participants for each of the datasets 1 and 2 had maximum, minimum,

Table 5 Details of the video clips

No.	Video title	Genre	Year
1	Birds-of-Paradise Project Introduction	Calm, Informative	2012
2	Doctor Strange Official Trailer 2	Fictional	2016
3	The Present—Official	Emotional	2016

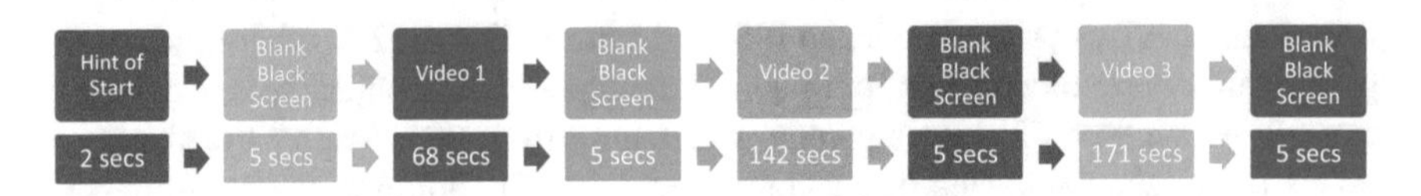

Fig. 3 EEG data collection protocol for video category classification from EEG data

average, and standard deviation age of 26, 20, 22.5 and 1.35 and 23, 19, 21.2 and 1.32 respectively and all the 38 participants were right-handed.

Experimental Setup

Three different types of videos were shown to the participants (see Table 5): 1. Calming and informative, 2. Fictional and 3. Emotional. The criteria of choosing these three videos can be found in [47]. A five second blank black screen were shown between each of the three videos and also, at the beginning and at the end of the whole experiment. To give a hint of start, a message stating "The video will start in 5 seconds" was shown for two seconds at the very beginning (see Fig. 3). The compiled experimental video (accessible online in [111]) was of 6 minutes 43 seconds and the total experimental procedure including device setup took about 10 minutes per subject. The stimuli were presented on a 21.5-inch LED monitor with 60 Hz refresh rate.

3.2 Experimental Study and Findings

Algorithms and Methods

In this section, we list out all the algorithms we experimented with for the Pre-processing, Feature Extraction, Feature Selection, and Classification submodules of the Signal Processing module of a BCI system to observe the best algorithm combination that achieves the highest accuracy in predicting the category of video a person is watching.

Pre-processing: As the three videos presented as stimuli were of different lengths, to classify without biasness, we selected one minute of raw EEG data from each of these videos—the part involved with the main story line of the video. The last minute of the first video states most of the information, one minute in the exact middle of the

Table 6 Illustration of SWT coefficients in relation to EEG rhythms in different frequency bands for dataset-1 using MUSE

SWT Coef (Level = 5, Fs = 220)	D1	D2	D3	D4	D5	A5
Freq band (Hz)	55–110	27.5–55	13.75–27.5	6.825–13.75	3.9125–6.825	0–3.9125
EEG rhythm		Gamma	Beta	Alpha	Theta	Delta

Table 7 Illustration of SWT coefficients in relation to EEG rhythms in different frequency bands for dataset-2 using Emotiv EPOC

SWT Coef (Level = 4, Fs = 128)	D1	D2	D3	D4	A4
Freq band (Hz)	32–64	16–32	8–16	4–8	0–4
EEG rhythm	Gamma	Beta	Alpha	Theta	Delta

second video comprises of the main climax and/or story and the last minute of the third video reveals the emotional climax and thus, we selected raw EEG data from these parts.

After the extraction of these one-minute data, we carried out experiments following three different approaches. We did not use any artifact removal techniques in our first approach, i.e. used raw data. In our second approach, to remove artifacts, we used Stationary wavelet transform (SWT) based denoising and as our third artifact removal technique we used an extended SWT technique were we first applied SWT following which we eliminated all data whose absolute difference was above 2 standard deviation from the mean (SWTSD).

SWT-based denoising was chosen in order to correct stereotyped artifacts such as muscle artifacts (EMG), motion artifacts, blinking and lateral eye movement artifacts (EOG). We chose SWT as it is better than DWT (Discrete Wavelet Transform) because of its transitional invariance (e.g. slight change in signal does not change the wavelet coefficients much and thus doesn't introduce much variations in energy distribution in different wavelet levels) [112]. A 5-level and 4-level SWT with Haar as mother (aka basis) wavelet has been applied on the EEG signals recorded from Muse (Fs = 220 Hz) and Emotiv EPOC (Fs = 128 Hz) headbands respectively. After the application of SWT, the output consists of final approximate coefficients (a_5/a_4) which represent distinct low frequency bands and a series of detail coefficients ($d_1 - d_5/d_1 - d_4$) which are the values of high frequency bands (see Tables 6 and 7). To remove artifacts from EEG signal, the updated universal threshold [113, 114] was applied on different scales of wavelet coefficients. Finally, by applying inverse SWT with Garrote threshold function as used in [113, 114], the artifact-reduced EEG data are reassembled using the latest set of wavelet coefficients.

After applying an artifact removal technique, we experimented with two basic family of filters namely, Finite and Infinite Impulse Response (FIR and IIR) filters to band-pass filter out EEG signals in the range 5–48 Hz which also removed EOG artifacts as they are low frequency signals (less than 4 Hz) [115]. In addition, the

Table 8 Different filters with their design specifications

Filters	Type	Specifications
FIR	i. Least Squares (FLS) ii. Equiripple (FE)	Order: 256 for dataset 1
		Order: 128 for dataset 2
		Sample rate: 220 Hz for dataset 1
		Sample rate: 128 Hz for dataset 2
		Stopband frequency 1: 4 Hz
		Passband frequency 1: 5 Hz
		Stopband frequency 2: 48 Hz
		Passband frequency 2: 50 Hz
IIR	iii. Chebyshev II (Stopband ripple) (ICS2) iv. Chebyshev I (Passband ripple) (ICS1) v. Elliptic (IE)	Order: Automatic
		Sample rate: 220 Hz for dataset 1
		Sample rate: 128 Hz for dataset 2
		Stopband frequency 1: 4 Hz
		Passband frequency 1: 5 Hz
		Stopband frequency 2: 48 Hz
		Passband frequency 2: 50 Hz

Table 9 List of feature extraction methods with their default parameters

Algorithm	DWT	FFT	PWelch	PYAR	STFT
Parameters	Wav.: db1 Dec.: 5 for dataset 1 Dec.: 4 for dataset 2	Nfft: 512 Freq. range: 0–110 Hz for dataset 1 Freq. range: 0–64 Hz for dataset 2	Nfft: 512 Freq. range: 0–110 Hz for dataset 1 Freq. range: 0–64 Hz for dataset 2	Nfft: 512 Freq. range: 0–110 Hz for dataset 1 Freq. range: 0–64 Hz for dataset 2 Order of AR model: 20	Nfft: 512 Freq. range: 0–110 Hz for dataset 1 Freq. range: 0–64 Hz for dataset 2

selected bandwidth of the mentioned filter also inherently removes the power line interference (i.e. 50 Hz in our recording location) and its harmonics, thus Notch filter was not used in the preprocessing stage. We designed two FIR filters and three IIR filters. Table 8 presents their detailed configurations.

Feature Extraction: The objective of this submodule is extracting useful features from the filtered EEG data which are to be used by the Classification step. There exist several feature extraction algorithms among which we selected: Discrete Wavelet Transform (DWT), Fast Fourier Transform (FFT), Welch Spectrum (PWelch), Yule—AR Spectrum (PYAR) and Short Time Fourier Transform (STFT). Table 9 presents the parameters chosen for each of these algorithms.

Feature Selection: Reduction of the dimensions of the features extracted in the last step can substantially reduce the execution time with pretty less or ignorable change in classification accuracy. For our problem, we chose two of the most popular feature selection algorithms—Principal Component Analysis (PCA) and Singular Value Decomposition (SVD).

Classification: Six very different type of classifiers in design and architecture were chosen for the classification submodule—Adaboost (AB), Support Vector Machines (SVMs), Multi-Class Linear Discriminant Analysis (MLDA), Multiple Linear Regression (MLR), Naïve Bayes (NB) and Decision Trees (MLTREE). Parameters for SVM were chosen kernel = linear and C = 1. An ensemble of 100 weak classifiers were used in Adaboost. The default parameters implemented by the MATLAB's Statistics and Machine Learning toolbox were used as parameters for all the other classifiers.

The 10-fold Cross-Validation approach which in our case is also Leave-One-Out Cross-Validation (LOOCV) was used as an evaluation criterion for classification accuracy. We implemented the subject-specific approach in which the classifier is trained and tested using the data of the same subject, i.e. we divided the data of one subject into 10 epochs (6 second epochs), trained the classifier with 9 of them and tested with the remaining one and the whole procedure was repeated 10 times.

A computer with 3.4 GHz processor (Intel Core i7) and 16 GB memory were used to run all the experiments and they were implemented using the EEG processing toolbox developed by Oikonomou et al. [116].

Experimental Results and Discussion

Since, it is impossible to report the results of all the combinations of algorithms (300 combinations for each artifact removal technique, i.e. 900 combinations) we chose in the previous section, based on our preliminary results, except for the artifact removal techniques, we selected two top performing algorithms from each of the submodules. Thus, as filters we selected FLS and ICS1, PYAR and PWelch as feature extraction techniques, both PCA and SVD as feature selection methods and NB and SVM as classifiers. Tables 10 and 11, for dataset 1 (data acquired using Muse headband) and 2 (data acquired using Emotiv EPOC) respectively, presents the results achieved for each of the combination of algorithms when different artifact removal techniques were applied for all the channels of Muse (TP9, AF7, AF8 and TP10) and corresponding closely located channels of Emotiv EPOC (T7, AF3, AF4 and T8).

Artifact Removal Techniques: For dataset 1 (Table 10), an increase of 3.1% in average accuracy can be observed when SWT (57.8%) was applied compared to the average accuracy when raw data (54.7%) were used. The mixture of SWT followed by SD (SWTSD) was able to achieve even better average accuracy of 61.9% with a difference of 4.1 and 7.2% with SWT and raw data respectively.

Table 10 Average accuracies for each of the combination of algorithms when different artifact removal techniques were applied for all the channels of Muse

Artifact Removal Techniques				Raw					SWT					SWTSD				
Algorithms				Channel					Channel					Channel				
Pre-processing	Feature Extraction	Feature Selection	Classifier	TP9	AF7	AF8	TP10	Average	TP9	AF7	AF8	TP10	Average	TP9	AF7	AF8	TP10	Average
FLS	PWelch	PCA	SVM	44.6	58.8	67.5	47.9	54.7	52.1	60.9	69.6	53.7	59.1	64.9	62.5	71.2	59.3	64.5
FLS	PYAR	PCA	SVM	47.7	59.6	68.4	47.9	55.9	58.2	64.2	73.7	57.5	63.4	63.5	64.4	71.6	64.6	66.0
FLS	PWelch	PCA	NB	43.5	55.6	60.4	47.9	51.8	49.8	54.6	60.5	49.6	53.6	59.1	58.8	66.5	54.2	59.6
FLS	PYAR	PCA	NB	42.6	47.2	58.8	46.7	48.8	51.1	50.4	55.8	44.9	50.5	54.4	46.3	63.3	48.2	53.1
ICSI	PWelch	PCA	SVM	43.0	61.1	67.7	48.2	55.0	50.4	62.8	69.3	48.1	57.6	63.5	62.1	70.5	59.1	63.8
ICSI	PYAR	PCA	SVM	46.8	61.9	70.0	50.2	57.2	52.3	62.5	71.9	57.2	61.0	59.8	64.4	72.5	64.4	65.3
ICSI	PWelch	PCA	NB	46.8	54.0	61.8	48.9	52.9	48.2	56.7	58.4	51.4	53.7	54.6	56.1	61.9	55.1	56.9
ICSI	PYAR	PCA	NB	40.5	52.8	54.9	46.8	48.8	41.2	44.4	55.3	48.6	47.4	50.7	51.6	65.3	52.3	55.0
FLS	PWelch	SVD	SVM	42.6	57.4	66.7	48.9	53.9	48.8	61.8	69.5	51.6	57.9	57.5	58.6	70.2	55.1	60.4
FLS	PYAR	SVD	SVM	47.0	56.8	64.9	47.4	54.0	54.0	63.3	71.4	57.0	61.4	57.7	63.7	68.2	61.6	62.8
FLS	PWelch	SVD	NB	53.2	60.0	65.6	52.3	57.8	52.8	60.0	71.9	53.3	59.5	60.2	64.0	74.4	58.9	64.4
FLS	PYAR	SVD	NB	51.1	60.7	69.5	54.6	58.9	56.0	64.2	74.9	55.8	62.7	62.5	62.6	77.7	60.5	65.8
ICSI	PWelch	SVD	SVM	44.7	58.1	63.9	49.5	54.0	49.6	63.7	70.0	47.4	57.7	53.5	61.8	70.7	56.5	60.6
ICSI	PYAR	SVD	SVM	44.6	60.4	69.8	47.5	55.6	50.2	60.0	69.3	56.0	58.9	56.1	64.0	71.9	63.7	63.9
ICSI	PWelch	SVD	NB	53.7	61.6	66.0	50.7	58.0	51.8	61.2	69.3	54.2	59.1	57.9	62.3	73.9	56.7	62.7
ICSI	PYAR	SVD	NB	50.4	61.6	66.5	54.4	58.2	53.2	61.2	73.5	55.6	60.9	60.2	66.1	74.2	59.8	65.1
Average				46.4	58.0	65.1	49.4	54.7	51.2	59.5	67.8	52.6	57.8	58.5	60.6	70.3	58.1	61.9

Table 11 Average accuracies for each of the combination of algorithms when different artifact removal techniques were applied for 4 channels of Emotiv EPOC which are close correspondence with the channels of Muse

Artifact Removal Techniques				Raw					SWT					SWTSD				
Algorithms				Channel					Channel					Channel				
Pre-processing	Feature Extraction	Feature Selection	Classifier	T7	AF3	AF4	T8	Average	T7	AF3	AF4	T8	Average	T7	AF3	AF4	T8	Average
FLS	PWelch	PCA	SVM	46.5	51.1	42.5	53.5	48.4	51.2	48.2	50.7	52.6	50.7	58.9	51.6	46.8	58.1	53.9
FLS	PYAR	PCA	SVM	51.4	47.7	42.5	55.6	49.3	50.0	45.6	42.5	52.1	47.5	63.0	52.6	48.4	60.7	56.2
FLS	PWelch	PCA	NB	51.8	44.7	36.1	50.2	45.7	52.8	47.0	42.8	46.1	47.2	52.5	48.1	45.3	53.5	49.8
FLS	PYAR	PCA	NB	49.8	41.6	40.5	49.5	45.4	52.5	46.7	38.9	44.4	45.6	54.0	51.6	44.9	53.0	50.9
ICSI	PWelch	PCA	SVM	47.4	58.2	49.3	54.7	52.4	49.3	55.4	50.7	52.6	52.0	63.2	59.5	55.1	66.7	61.1
ICSI	PYAR	PCA	SVM	50.2	58.4	51.6	59.3	54.9	45.8	58.1	47.5	55.8	51.8	62.3	58.1	53.7	61.9	59.0
ICSI	PWelch	PCA	NB	51.9	50.5	45.1	53.2	50.2	52.3	45.3	45.4	51.1	48.5	57.0	52.3	49.5	55.6	53.6
ICSI	PYAR	PCA	NB	54.9	49.3	47.0	53.5	51.2	49.6	47.7	44.7	55.3	49.3	54.6	53.0	50.7	59.3	54.4
FLS	PWelch	SVD	SVM	38.6	42.3	40.5	46.3	41.9	45.8	41.2	45.1	45.4	44.4	51.8	41.9	38.6	50.0	45.6
FLS	PYAR	SVD	SVM	37.0	42.6	36.1	50.4	41.5	46.3	42.6	37.2	49.3	43.9	46.3	41.6	43.9	51.6	45.8
FLS	PWelch	SVD	NB	43.7	43.7	37.9	47.4	43.2	45.3	45.6	41.1	48.9	45.2	51.1	47.0	46.1	54.9	49.8
FLS	PYAR	SVD	NB	42.5	46.3	43.7	48.4	45.2	50.9	47.5	46.0	48.1	48.1	50.0	46.8	48.8	54.6	50.0
ICSI	PWelch	SVD	SVM	37.4	45.4	38.6	42.8	41.1	40.2	40.2	41.2	46.7	42.1	50.5	45.4	47.9	51.9	48.9
ICSI	PYAR	SVD	SVM	37.5	37.7	39.8	47.5	40.7	36.7	42.6	37.9	47.0	41.1	50.0	43.3	42.5	53.2	47.2
ICSI	PWelch	SVD	NB	42.1	44.9	42.5	47.9	44.3	43.2	46.3	45.3	47.4	45.5	54.6	51.2	50.2	57.7	53.4
ICSI	PYAR	SVD	NB	41.8	45.3	43.3	46.8	44.3	41.9	45.4	43.3	50.0	45.2	53.5	51.1	53.9	61.8	55.0
Average				46.9	45.3	50.4	42.3	46.2	46.6	47.1	49.6	43.8	46.8	54.6	49.7	47.9	56.5	52.2

Similar results can also be observed for dataset 2 (Table 11). Although, the introduction of SWT (46.8%) slightly improved the average classification accuracy compared to raw data (46.2%), SWTSD (52.2%) substantially improved the results by 6.0%.

The results achieved from EEG data of both the devices infer the fact that, EEG signals are highly prone to artifacts and therefore, appropriate usage of artifact removal technique(s) can significantly improve classification accuracy. For the video category classification (VCC) problem, based on our results we can conclude that, our new method SWTSD performs better than the conventional artifact removal technique SWT. It is important to note that this does not however infer that SWTSD will

Table 12 (**a**) Average accuracies for each of the combination of algorithms for all the channels of Muse when feature selection techniques were not applied. (**b**) Average accuracies for each of the combination of algorithms for 4 channels of Emotiv EPOC when feature selection techniques were not applied

(a)

Artifact Removal Techniques			SWTSD				
Algorithms			Channel				
Pre-Processing	Feature Extraction	Classifier	TP9	AF7	AF8	TP10	Average
FLS	PWelch	SVM	58.8	60.5	66.3	62.8	62.1
FLS	PYAR	SVM	61.6	62.6	69.1	61.8	63.8
FLS	PWelch	NB	55.3	57.0	63.2	50.2	56.4
FLS	PYAR	NB	58.9	58.2	64.7	55.4	59.3
ICSI	PWelch	SVM	61.1	58.2	69.5	60.5	62.3
ICSI	PYAR	SVM	60.0	65.4	70.0	61.9	64.3
ICSI	PWelch	NB	52.3	58.6	64.0	52.1	56.8
ICSI	PYAR	NB	53.3	56.1	66.0	52.8	57.1
Average			57.7	59.6	66.6	57.2	60.3

(b)

Artifact Removal Techniques			SWTSD				
Algorithms			Channel				
Pre-Processing	Feature Extraction	Classifier	T7	AF3	AF4	T8	Average
FLS	PWelch	SVM	55.1	51.9	52.5	54.6	53.5
FLS	PYAR	SVM	62.3	48.6	52.1	58.6	55.4
FLS	PWelch	NB	59.6	55.4	53.3	60.7	57.3
FLS	PYAR	NB	60.4	56.0	53.0	58.1	56.8
ICSI	PWelch	SVM	62.3	58.8	57.4	61.2	59.9
ICSI	PYAR	SVM	63.9	60.5	57.0	64.0	61.4
ICSI	PWelch	NB	64.4	62.6	60.7	64.2	63.0
ICSI	PYAR	NB	63.5	63.3	60.5	68.4	63.9
Average			61.4	57.1	55.8	61.2	58.9

perform better than SWT for other types of studies (e.g. MI, SSVEP, ERP, etc.) as well.

Impacts of usage of Feature Selection Algorithms: Our analysis will now concentrate on the results of SWTSD only as it is the better performing artifact removal technique. Table 12 present the results for dataset 1 and 2 respectively when feature selection techniques were not used.

As expected, when feature selection techniques were not used, an increase in average execution time per subject was observed for both the datasets. For dataset 1, the average execution time increased from 4.03 msec to 6.73 msec (increased by 67.0 percent) and for dataset 2, the average execution time almost doubled from 3.03 msec to 5.96 msec (increased by 96.7 percent).

When all the data are used for classification, the classification accuracy is expected to be higher compared to when feature selection methods are applied before classification. Although, as per Tables 11 and 12b, this is the case when Emotiv EPOC's data were used (average accuracy of 52.2% and 58.9% with and without feature selection methods respectively), however, slightly different results can be observed when Muse's data were used, i.e. average accuracy decreased from 61.9% to 60.3% when feature selection methods were not used (see Tables 10 and 12a).

This decrease in accuracy for Muse can be explained by the differences in sampling rate of the two devices (128 Hz for Emotiv EPOC and 220 Hz for Muse). The number of components chosen by the dimensionality reduction algorithms for both the datasets remained the same and so, feature selection algorithms had more options to choose from for dataset 1 than for dataset 2 and therefore, the number of redundant features selected for dataset 1 are less as well. Also, the selected features probably had less noise compared to the original data and hence, the classification accuracy improved.

Channel Selection: To identify which channel is most suitable for the VCC problem, for the sake of simplicity, our analysis will now concentrate on the results when feature selection algorithms were applied to the datasets (Tables 10 and 11). Average accuracies of channels TP9 (58.5%) and TP10 (58.1%) of the Muse headband are very similar with just a 0.4% difference. The results improve even further to 60.6% when data of channel AF7 were used. A significant increase in average accuracy can be observed for channel AF8 (70.3%) located at the right dorsolateral prefrontal cortex. As videos have the potential to evoke working memory in participants, one possible reason for such an increase of about 10% for this particular channel can be explained by the findings of [117] where the authors conclude that right dorsolateral prefrontal cortex is heavily involved with spatial working memory related tasks. There can be several other explanations for this abrupt increase in average accuracy which include emotions triggered by different videos in subjects, attentiveness, etc.

Comparing the results with the electrodes T7 (54.6%) and T8 (56.5%) of Emotiv EPOC, average accuracies were somewhat similar to that of the Muse headband for the electrodes TP9 (58.5%) and TP10 (58.1%) compared to the electrodes located at the frontal lobe. First of all, unlike Muse headband, the average accuracies deteriorated substantially for the electrodes AF3 (49.7%) and AF4 (47.9%) in comparison with the electrodes located at the temporal lobe (T7 and T8). Secondly, the difference between the average accuracies of AF7 (60.6%) and AF3 (49.7%) was 10.9% and between AF8 (70.3%) and AF4 (47.9%) a huge difference of 22.4% can be observed.

As reported in several studies [118–121], the performance of Emotiv EPOC compared to other EEG signal acquisition devices, is not up to the mark. This might be because, as per Fakhruzzaman et al. [122], Emotiv EPOC is a not a medical grade device, i.e. it is a consumer grade device and the all-size all-fit concept of this device is not so good as it sounds. Other than these reasons, the deterioration in average accuracy, specifically for the frontal lobe electrodes, can also be explained by the difference in spatial position of the two electrodes of the devices (Figs. 4 and 5). AF7 and AF8 of the Muse headband is located on the forehead whereas AF3 and AF4 of Emotiv EPOC is positioned on or above the hairline on the forehead depending on the size of the forehead of different individuals. This obstruction of hair for channels AF3 and AF4 makes them much more susceptible to artifacts compared to the channels AF7 and AF8 which are placed right on top of the skin and therefore, results of frontal lobe channels of Emotiv EPOC are worse than that of Muse.

Table 13 provides the average accuracies achieved by each of the 16 combinations of algorithms for all the channels of Emotiv EPOC when SWTSD artifact removal technique was used. From all the results of the Muse headband (Table 10) and Emotiv EPOC (Table 13) the only channel that exceeded the minimal BCI performance criteria of 70% [123] was when data of channel AF8 of the Muse headband were used leading us to conclude that this is the most suitable channel and the Muse headband is the better device for the VCC problem. The highest average classification accuracy achieved by this channel was 77.7% (4.83 msec average total execution time per

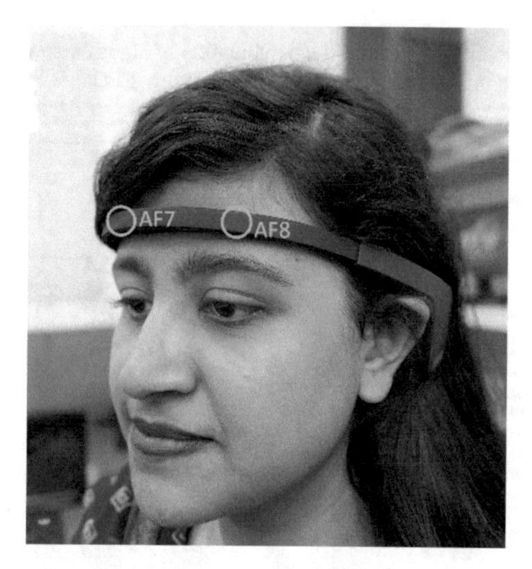

Fig. 4 AF7 and AF8 channel locations of the Muse headband

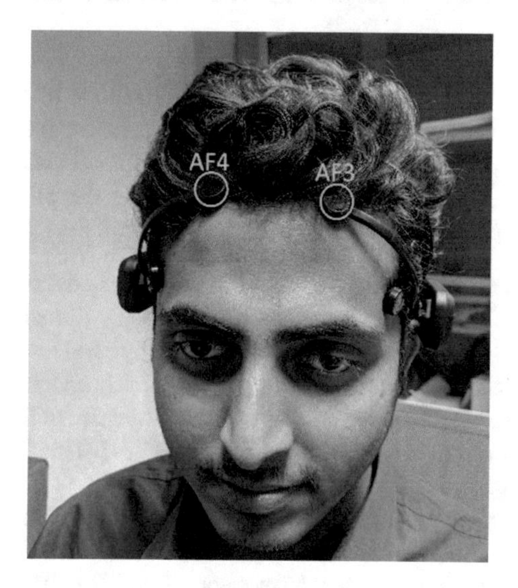

Fig. 5 AF3 and AF4 channel locations of Emotiv EPOC

subject) and the combination of algorithms responsible were SWTSD and FLS for the pre-processing submodule, PYAR for the feature extraction submodule and SVD and NB for the submodules feature selection and classification respectively. Even though none of the channels of Emotiv EPOC achieved the minimal BCI performance criteria of 70% [123], the channel whose results were closest to it was T8 with highest average accuracy of 66.7% (2.96 msec average total execution time per subject) when SWTSD and ICS1, PWelch, PCA and SVM were used.

Table 13 Average accuracies for each of the combination of algorithms when SWTSD were used as an artifact removal technique for all the channels of Emotiv EPOC

Artifact Removal Techniques				SWTSD														
Algorithms				Channels														
Pre-processing	Feature Extraction	Feature Selection	Classifier	AF3	F7	F3	FC5	T7	P7	O1	O2	P8	T8	FC6	F4	F8	AF4	Average
FLS	PWelch	PCA	SVM	51.6	52.1	54.4	55.3	58.9	54.0	50.5	47.2	52.6	58.1	53.2	49.3	58.2	46.8	53.0
FLS	PYAR	PCA	SVM	52.6	50.4	57.4	51.9	63.0	55.6	54.6	50.9	53.0	60.7	53.7	56.1	53.2	48.4	54.4
FLS	PWelch	PCA	NB	48.1	47.4	44.9	44.0	52.5	49.5	43.0	45.6	50.7	53.5	49.8	43.2	46.0	45.3	47.4
FLS	PYAR	PCA	NB	51.6	41.8	44.7	44.9	54.0	45.1	49.1	45.4	47.4	53.0	47.0	44.2	42.8	44.9	46.9
ICSI	PWelch	PCA	SVM	59.5	56.1	63.0	60.9	63.2	56.7	55.4	52.8	56.5	66.7	58.1	57.4	54.6	55.1	58.3
ICSI	PYAR	PCA	SVM	58.1	51.1	63.9	57.0	62.3	55.8	56.1	55.3	56.0	61.9	62.8	57.5	49.1	53.7	57.2
ICSI	PWelch	PCA	NB	52.3	49.5	59.8	57.2	57.0	48.8	52.5	49.3	50.7	55.6	50.5	50.2	42.3	49.5	51.8
ICSI	PYAR	PCA	NB	53.0	48.9	54.6	54.9	54.6	45.1	52.8	51.2	51.1	59.3	47.5	50.7	46.1	50.7	51.5
FLS	PWelch	SVD	SVM	41.9	43.0	46.1	39.1	51.8	39.8	37.7	41.8	46.3	50.0	41.9	42.5	43.7	38.6	43.2
FLS	PYAR	SVD	SVM	41.6	39.8	46.8	38.1	46.3	43.3	44.9	43.7	46.8	51.6	41.6	44.7	43.0	43.9	44.0
FLS	PWelch	SVD	NB	47.0	43.5	50.4	44.4	51.1	45.3	47.0	50.7	51.1	54.9	45.1	45.1	43.2	46.1	47.5
FLS	PYAR	SVD	NB	46.8	45.4	48.4	43.5	50.0	45.1	48.6	47.9	47.7	54.6	44.2	46.1	43.5	48.8	47.2
ICSI	PWelch	SVD	SVM	45.4	43.3	52.5	45.1	50.5	42.6	42.1	47.9	46.7	51.9	47.0	47.2	46.5	47.9	46.9
ICSI	PYAR	SVD	SVM	43.3	40.5	52.1	44.7	50.0	42.1	48.4	44.9	46.7	53.0	46.8	41.9	37.9	42.5	45.4
ICSI	PWelch	SVD	NB	51.2	48.9	53.3	48.2	54.6	46.8	50.5	55.6	53.0	57.7	48.2	49.8	46.1	50.2	51.0
ICSI	PYAR	SVD	NB	51.1	47.0	52.1	51.4	53.5	49.6	55.1	54.7	52.5	61.8	49.8	53.7	47.7	53.9	52.4
Average				49.7	46.8	52.8	48.8	54.6	47.8	49.3	49.1	50.5	56.5	49.2	48.7	46.5	47.9	49.9

The results of the channels located at the occipital lobe of Emotiv EPOC were surprisingly low. Other than the limitations of Emotiv EPOC mentioned before, this may be because that although exposure of videos triggers visual evoked potential (VEP) in the brain, other parts of the brain including the prefrontal dorsolateral cortex are more involved or is activated when such stimuli are presented.

Future Works: There are several areas we can work on in the future to improve our results. For example, the order of the IIR filters and the dimensions of the feature selection algorithms are being selected automatically by MATLAB. Optimizing these parameters will have an impact on the results. We used data epochs of 6 seconds which is a big epoch size for EEG related studies as the stationarity of the EEG signals with increasing epoch duration is expected to disappear [124]. This is a very crucial area which we hope to address in the future.

The relevant frequency bands for MI (7–30 Hz, mu and beta bands) [125], ERP (< 4 Hz, delta band) [126] and SSVEP (12–18 Hz) [127] based BCIs are well known by researchers. Analyzing a relevant frequency band for the VCC problem was beyond the scope of this study. As discussed in our previous work [47], we hope to target high-frequency gamma oscillations as they are heavily involved in working memory load related activity [128–130] and in activities requiring cross-modal sensory processing—perception combined from two separate senses, for example from sound and sight [131, 132].

One category of Machine Learning algorithms, neural networks, especially Deep Learning algorithms which is the recent hype among Machine Learning researchers, was not used in this study. As Deep Learning algorithms compared to conventional Machine Learning algorithms are performing much better in almost all type of studies including EEG-based BCIs [88, 90], we believe that using such algorithms will improve our results substantially. Also, apart from SWT, there exists several other artifact removal tools in the literature, e.g. Empirical Mode Decomposition (EMD), Adaptive filtering, Independent Component Analysis (ICA), etc. which we hope to apply on the VCC study as well [133].

In addition, in this study, we have used a single feature (either PWelch or PYAR) during feature extraction step. However, features extracted using combination of different statistical and non-statistical features found in time, frequency and wavelet domain [133] (e.g. standard deviation, variance, entropy, kurtosis, skewness, periodicity, maximum or minimum power in all three domains, AR features, line length, NEO, FFT-based features, etc.) with different weights concatenated into a single feature vector might enhance the classification accuracy significantly, which is also one of our future works.

4 Conclusion

This chapter attempted to address the existing computational intelligence techniques for pattern recognition in one of the EEG-based BCI applications, i.e. Video Category Classification (VCC) and their accuracies, challenges and suitability for such application.

Based on results found from experiments on VCC and reports from other studies, computational intelligence in BCI systems is problem or application specific and depends on several factors. For example, as reported in the previous section, data acquired from two different EEG signal acquisition devices (Muse headband and Emotiv EPOC) for the same experiment resulted in considerably different results. The correct choice of the relevant frequency band (e.g. SSVEP, 12–18 Hz [127]) also plays a crucial role in the end results.

In accordance with other studies [36, 134], it is also found that proper usage and optimization of artifact removal techniques significantly improves classification accuracies. Similarly, depending on the BCI paradigm and type, appropriate selection of feature extraction, feature selection and classification algorithms will also have a positive impact on the results.

We hope and believe that the rapid progress in technology, both hardware (e.g. better signal acquisition devices, better computer and smart phone hardware, etc.) and software (better Machine Learning techniques such as Deep Learning, continuous improvement of artifact removal techniques, etc.), will in the near future improve the accuracy and feasibility of BCI systems to a level at which these systems can be deployed in real world scenarios (e.g. online BCIs) resulting in a better lifestyle of the physically disabled people and also will increase the quality of life for all people across the world.

Appendix

Abbreviation	Definition
AAC	Alternative and Augmentative Communication
AB	Adaboost
ALS	Amyotrophic lateral sclerosis
BCI	Brain-Computer-Interface
BLDA	Bayesian Linear Discriminative Analysis
CCA	Canonical Correlation Analysis
CC-LR	Cross-Correlation based Logistic Regression
CD	Contrastive Divergence
CNN	Convolution Neural Network
CSP	Common Spatial Pattern
CWT	Continuous Wavelet Transform
DBN	Deep Belief Network
DWT	Discrete Wavelet Transform
ECoG	Electrocorticography
EEG	Electroencephalography
ELM	Extreme Learning Machine
EMD	Empirical Mode Decomposition
EMG	Electromyography
EOG	Electrooculography
ERD	Event Related Desynchronization
ERP	Event Related Potential
ERS	Event Related Synchronization
FE	FIR Equiripple
FFT	Fast Fourier Transform
FIR	Finite Impulse Response
FLS	FIR Least Squares
fMRI	Functional Magnetic Resonance Imaging
fNIRS	Functional Near-Infrared Spectroscopy
GA	Genetic Algorithm
HCI	Human-Computer Interaction
ICA	Independent Component Analysis
ICS1	IIR Chebyshev I
ICS2	IIR Chebyshev II
IE	IIR Elliptic
IEEG	Integrated Electroencephalography

Abbreviation	Definition
IIR	Infinite Impulse Response
k-NN	K- Nearest Neighbor
LDA	Linear Discriminant Analysis
LOOCV	Leave-One-Out Cross-Validation
MCD	Minimum Covariance Determinant
MEG	Magnetoencephalography
MI	Motor Imagery
MLDA	Multi-Class Linear Discriminant Analysis
MLP	Multilayer Perceptron
MLR	Multiple Linear Regression
MLTREE	Decision Tree
MSPCA	Multi Scale Principal Component Analysis
NB	Naïve Bayes
PCA	Principal Component Analysis
PWelch	Welch Spectrum
PYAR	Yale - AR Spectrum
RBF	Radial Basis Function
RBM	Restrict Boltzmann Machine
RF	Random Forest
RMS	Root Mean Squares
ROC	Receiver Operating Characteristic
SAE	Stacked Autoencoders
SBLFB	Sparse Bayesian Learning of Frequency Bands
SCP	Slow Cortical Potentials
SMFA	Subclass Marginal Fisher Analysis
SNR	Signal to Noise Ratio
SSVEP	Steady State Visual Evoked Potential
STFT	Short Time Fourier Transform
SVD	Singular Value Decomposition
SVM	Support Vector Machine
SWLDA	Stepwise Linear Discriminant Analysis
SWT	Stationary Wavelet Transform
SWTSD	Stationary Wavelet Transform with Standard Deviation
VCC	Video Category Classification
VEP	Visual Evoked Potential
WAMP	Willison Amplitude
WESVM	Weighted Ensemble of SVMs
WPD	Wavelet Packet Decomposition

References

1. J.R. Wolpaw, N. Birbaumer, W.J. Heetderks, D.J. McFarland, P.H. Peckham, G. Schalk, E. Donchin, L.A. Quatrano, C.J. Robinson, T.M. Vaughan, Brain-computer interface technology: a review of the first international meeting. IEEE Trans. Rehab. Eng. **8**(2), 164–173 (2000)
2. V.P. Oikonomou, G. Liaros, K. Georgiadis, E. Chatzilari, K. Adam, S. Nikolopoulos, I. Kompatsiaris, Comparative evaluation of state-of-the-art algorithms for SSVEP-based BCIs, 2016. arXiv:1602.00904
3. J.R. Wolpaw, N. Birbaumer, D.J. McFarland, G. Pfurtscheller, T.M. Vaughan, Brain–computer interfaces for communication and control. Clin. Neurophysiol. **113**(6), 767–791 (2002)
4. N.H. Liu, C.Y. Chiang, H.C. Chu, Recognizing the degree of human attention using EEG signals from mobile sensors. Sensors **13**(8), 10273–10286 (2013)
5. P. Campisi, D. La Rocca, G. Scarano, EEG for automatic person recognition. Computer **45**(7), 87–89 (2012)
6. G.A. Light, L.E. Williams, F. Minow, J. Sprock, A. Rissling, R. Sharp, N.R. Swerdlow, D.L. Braff, Electroencephalography (EEG) and event-related potentials (ERPs) with human participants. Curr. Protocols Neurosci., 6–25 (2010)
7. V. Jurcak, D. Tsuzuki, I. Dan, 10/20, 10/10, and 10/5 systems revisited: their validity as relative head-surface-based positioning systems. Neuroimage **34**(4), 1600–1611 (2007)
8. Wikipedia contributors, 10-20 system (EEG), Wikipedia, The Free Encyclopedia, https://en.wikipedia.org/w/index.php?title=10-20_system_(EEG)&oldid=748809006. Accessed 10 Nov 2016
9. D. Szafir, B. Mutlu, Pay attention!: designing adaptive agents that monitor and improve user engagement, in *Proceedings of the SIGCHI Conference on Human Factors in Computing Systems*. ACM, pp. 11–20
10. H.A.T. Nguyen, J. Musson, F. Li, W. Wang, G. Zhang, R. Xu, C. Richey, T. Schnell, F.D. McKenzie, J. Li, EOG artifact removal using a wavelet neural network. Neurocomputing **97**, 374–389 (2012)
11. G. Bonmassar, P.L. Purdon, I.P. Jääskeläinen, K. Chiappa, V. Solo, E.N. Brown, J.W. Belliveau, Motion and ballistocardiogram artifact removal for interleaved recording of EEG and EPs during MRI. Neuroimage **16**(4), 1127–1141 (2002)
12. A. Delorme, T. Sejnowski, S. Makeig, Enhanced detection of artifacts in EEG data using higher-order statistics and independent component analysis. Neuroimage **34**(4), 1443–1449 (2007)
13. J.A. Urigüen, B. Garcia-Zapirain, EEG artifact removal—state-of-the-art and guidelines. J. Neural Eng. **12**(3), 031001 (2015)
14. T.P. Jung, S. Makeig, C. Humphries, T.W. Lee, M.J. Mckeown, V. Iragui, T.J. Sejnowski, Removing electroencephalographic artifacts by blind source separation. Psychophysiology **37**(2), 163–178 (2000)
15. J. Hu, C.S. Wang, M. Wu, Y.X. Du, Y. He, J. She, Removal of EOG and EMG artifacts from EEG using combination of functional link neural network and adaptive neural fuzzy inference system. Neurocomputing **151**, 278–287 (2015)
16. X. Mao, M. Li, W. Li, L. Niu, B. Xian, M. Zeng, G. Chen, Progress in EEG-based brain robot interaction systems. Comput. Intell. Neurosci. (2017)
17. M. Alimardani, N. Shuichi, H. Ishiguro, The effect of feedback presentation on motor imagery performance during BCI-teleoperation of a humanlike robot, in *2014 5th IEEE RAS & EMBS International Conference on Biomedical Robotics and Biomechatronics* (IEEE, 2014), pp. 403–408
18. F. Duan, D. Lin, W. Li, Z. Zhang, Design of a multimodal EEG-based hybrid BCI system with visual servo module. IEEE Trans. Auton. Mental Dev. **7**(4), 332–341 (2015)
19. D. Coyle, J. Garcia, A.R. Satti, T.M. McGinnity, EEG-based continuous control of a game using a 3 channel motor imagery BCI: BCI game, in 2011 *IEEE Symposium on Computational Intelligence, Cognitive Algorithms, Mind, and Brain (CCMB)* (IEEE, 2011), pp. 1–7

20. P. Gergondet, S. Druon, A. Kheddar, C. Hintermüller, C. Guger, M. Slater, Using brain-computer interface to steer a humanoid robot, in *2011 IEEE International Conference on Robotics and Biomimetics (ROBIO)* (IEEE, 2011), pp. 192–197
21. J. Zhao, W. Li, X. Mao, M. Li, SSVEP-based experimental procedure for brain-robot interaction with humanoid robots. JoVE (J. Vis. Exp.) **105**, e53558–e53558 (2015)
22. M. Li, W. Li, H. Zhou, Increasing N200 potentials via visual stimulus depicting humanoid robot behavior. Int. J. Neural Syst. **26**(01), 1550039 (2016)
23. W. Li, M. Li, W. Li, Independent component analysis-based channel selection to achieve high performance of N200 and P300 classification, in *2015 IEEE 14th International Conference on Cognitive Informatics & Cognitive Computing (ICCI* CC)* (IEEE, 2015), pp. 384–389
24. L. Yao, J. Meng, X. Sheng, D. Zhang, X. Zhu, A novel calibration and task guidance framework for motor imagery BCI via a tendon vibration induced sensation with kinesthesia illusion. J. Neural Eng. **12**(1), 016005 (2014)
25. H. Wang, T. Li, Z. Huang, Remote control of an electrical car with SSVEP-based BCI, in *2010 IEEE International Conference on Information Theory and Information Security (ICITIS)* (IEEE, 2010), pp. 837–840
26. C.Y. Chen, C.W. Wu, C.T. Lin, S.A. Chen, A novel classification method for motor imagery based on Brain-Computer Interface, in *2014 International Joint Conference on Neural Networks (IJCNN)* (IEEE, 2014), pp. 4099–4102
27. Emotiv EPOC - 14 Channel Wireless EEG Headset, in *Emotiv*, https://www.emotiv.com/epoc/ . Accessed 13 Aug 2017
28. A. Güneysu, H.L. Akin, An SSVEP based BCI to control a humanoid robot by using portable EEG device, in *2013 35th Annual International Conference of the IEEE Engineering in Medicine and Biology Society (EMBC)* (IEEE, 2013), pp. 6905–6908
29. H.L. Jian, K.T. Tang, Improving classification accuracy of SSVEP based BCI using RBF SVM with signal quality evaluation, in *2014 International Symposium on Intelligent Signal Processing and Communication Systems (ISPACS)* (IEEE, 2014), pp. 302–306
30. O. Çağlayan, R.B. Arslan, Humanoid robot control with SSVEP on embedded system, in *Proceedings of the 5th International Brain-Computer Interface Meeting: Defining the Future*, June 2013, pp. 260–261
31. S.M.T. Müller, T.F. Bastos-Filho, M. Sarcinelli-Filho, Using a SSVEP-BCI to command a robotic wheelchair, in *2011 IEEE International Symposium on Industrial Electronics (ISIE)* (IEEE, 2011), pp. 957–962
32. R.C. Panicker, S. Puthusserypady, Y. Sun, An asynchronous P300 BCI with SSVEP-based control state detection. IEEE Trans. Biomed. Eng. **58**(6), 1781–1788 (2011)
33. W. Song, X. Wang, S. Zheng, Y. Lin, Mobile robot control by BCI based on motor imagery, in *2014 Sixth International Conference on Intelligent Human-Machine Systems and Cybernetics (IHMSC)*, vol. 2 (IEEE, 2014), pp. 383–387
34. MindWave, in *MindWave*. http://store.neurosky.com/pages/mindwave. Accessed 13 Aug 2017
35. MUSE ™ | Meditation Made Easy, in *Muse: The Brain Sensing Headband*. http://www.choosemuse.com/. Accessed 13 Aug 2017
36. A. Jalilifard, E.B. Pizzolato, M.K. Islam, Emotion classification using single-channel scalp-EEG recording, in *38th Annual International Conference of the Engineering in Medicine and Biology Society (EMBC 2016)* (IEEE Press, Orlando, FL), pp 845–849. https://doi.org/10.1109/embc.2016.7590833
37. M.S.Z. Nine, M. Khan, B. Poon, M.A. Amin, H. Yan, Human computer interaction through wireless brain computer interfacing device, in *9th International Conference on Information Technology and Applications (ICITA 2014)*, 2014
38. S.K. Paul, M.Z. Nine, M. Hasan, M.A. Amin, Cognitive task classification from wireless EEG, in *International Conference on Brain Informatics and Health (BIH 2015)*. LNCS, vol. 9250 (Springer, London, 2015), pp 13–22. https://doi.org/10.1007/978-3-319-23344-4_2
39. G. Schalk, D.J. McFarland, T. Hinterberger, N. Birbaumer, J.R. Wolpaw, BCI2000: a general-purpose brain-computer interface (BCI) system. IEEE Trans. Biomed. Eng. **51**(6), 1034–1043 (2004)

40. M. Hamedi, S.H. Salleh, A.M. Noor, I.M. Rezazadeh, Neural network-based three-class motor imagery classification using time-domain features for BCI applications, in *Region 10 Symposium* (IEEE, 2014), pp. 204–207
41. J. Kevric, A. Subasi, Comparison of signal decomposition methods in classification of EEG signals for motor-imagery BCI system. Biomed. Signal Process. Control **31**, 398–406 (2017)
42. W. Speier, A. Deshpande, L. Cui, N. Chandravadia, D. Roberts, N. Pouratian, A comparison of stimulus types in online classification of the P300 speller using language models. PloS one **12**(4), e0175382 (2017)
43. Q. Li, S. Ma, K. Shi, N. Gao, Comparing the classification performance of Bayesian linear discriminate analysis (BLDA) and support vector machine (SVM) in BCI P300-speller with familiar face paradigm, in *International Congress on Image and Signal Processing, BioMedical Engineering and Informatics (CISP-BMEI)*, (IEEE, 2016), pp. 1476–1481
44. Q. Li, K. Shi, S. Ma, N. Gao, Improving classification accuracy of SVM ensemble using random training set for BCI P300-speller, in *2016 IEEE International Conference on Mechatronics and Automation (ICMA)* (IEEE, 2016), pp. 2611–2616
45. Y.J. Chen, S.C. Chen, I.A. Zaeni, C.M. Wu, Fuzzy tracking and control algorithm for an SSVEP-based BCI system. Appl. Sci. **6**(10), 270 (2016)
46. F. Kalaganis, E. Chatzilari, K. Georgiadis, S. Nikolopoulos, N. Laskaris, Y. Kompatsiaris, An error aware SSVEP-based BCI
47. A.K. Mutasim, R.S. Tipu, M.R. Bashar, M.A. Amin, Video Category Classification Using Wireless EEG, in *International Conference on Brain Informatics (BI 2017)* (Springer, 2017)
48. C. Neuper, M. Wörtz, G. Pfurtscheller, ERD/ERS patterns reflecting sensorimotor activation and deactivation. Progr. Brain Res. **159**, 211–222 (2006)
49. E. Sellers, Y. Arbel, E. Donchin, BCIs that uses event related potentials, in *Brain-Computer Interfaces: Principles and Practice*, ed. by J. Wolpaw, E.W. Wolpaw (2012)
50. G. Garipelli, R. Chavarriaga, J. del R Millán, Single trial analysis of slow cortical potentials: a study on anticipation related potentials. J. Neural Eng. **10**(3), 036014 (2013)
51. B.Z. Allison, J. Jin, Y. Zhang, X. Wang, A four-choice hybrid P300/SSVEP BCI for improved accuracy. Brain-Comput. Interfaces **1**(1), 17–26 (2014)
52. M. Wang, I. Daly, B.Z. Allison, J. Jin, Y. Zhang, L. Chen, X. Wang, A new hybrid BCI paradigm based on P300 and SSVEP. J. Neurosci. Methods **244**, 16–25 (2015)
53. L.W. Ko, S.C. Lin, M.S. Song, O. Komarov, Developing a few-channel hybrid BCI system by using motor imagery with SSVEP assist, in *2014 International Joint Conference on Neural Networks (IJCNN)*, (IEEE, 2014), pp. 4114–4120
54. M. Li, B.L. Lu, Emotion classification based on gamma-band EEG, in *Annual International Conference of the IEEE Engineering in Medicine and Biology Society, 2009. EMBC 2009* (IEEE, 2009), pp. 1223–1226
55. M. Murugappan, N. Ramachandran, Y. Sazali, Classification of human emotion from EEG using discrete wavelet transform. J. Biomed. Sci. Eng. **3**(04), 390 (2010)
56. W.L. Zheng, J.Y. Zhu, Y. Peng, B.L. Lu, EEG-based emotion classification using deep belief networks, in *2014 IEEE International Conference on Multimedia and Expo (ICME)* (IEEE, 2014), pp. 1–6
57. G.F. Wilson, F. Fisher, Cognitive task classification based upon topographic EEG data. Biol. Psychol. **40**(1), 239–250 (1995)
58. W. Speier, C. Arnold, N. Pouratian, Integrating language models into classifiers for BCI communication: a review. J. Neural Eng. **13**(3), 031002 (2016)
59. L.A. Farwell, E. Donchin, Talking off the top of your head: toward a mental prosthesis utilizing event-related brain potentials. Electroencephalogr. Clin. Neurophysiol. **70**(6), 510–523 (1988)
60. C. Guger, S. Daban, E. Sellers, C. Holzner, G. Krausz, R. Carabalona, F. Gramatica, G. Edlinger, How many people are able to control a P300-based brain–computer interface (BCI)? Neurosci. Lett. **462**(1), 94–98 (2009)
61. G. Townsend, B.K. LaPallo, C.B. Boulay, D.J. Krusienski, G.E. Frye, C. Hauser, N.E. Schwartz, T.M. Vaughan, J.R. Wolpaw, E.W. Sellers, A novel P300-based brain–computer interface stimulus presentation paradigm: moving beyond rows and columns. Clin. Neurophysiol. **121**(7), 1109–1120 (2010)

62. R. Fazel-Rezai, W. Ahmad, P300-based brain-computer interface paradigm design, in *Recent Advances in Brain-Computer Interface Systems* (InTech, 2011)
63. P. Nuyujukian, J.C. Kao, S.I. Ryu, K.V. Shenoy, A nonhuman primate brain-computer typing interface. Proc. IEEE **105**(1), 66–72 (2017)
64. C. Pandarinath, P. Nuyujukian, C.H. Blabe, B.L. Sorice, J. Saab, F.R. Willett, L.R. Hochberg, K.V. Shenoy, J.M. Henderson, High performance communication by people with paralysis using an intracortical brain-computer interface. eLife **6**, e18554 (2017)
65. D.E. Thompson, S. Blain-Moraes, J.E. Huggins, Performance assessment in brain-computer interface-based augmentative and alternative communication. Biomed. Eng. **12**(1), 43 (2013)
66. T. Carlson, G. Monnard, J.D.R. Millán, Vision-based shared control for a BCI wheelchair. Int. J. Bioelectromagn. **13**(EPFL-ARTICLE-168977), 20–21 (2011)
67. R. Singla, A. Khosla, R. Jha, Influence of stimuli color on steady-state visual evoked potentials based BCI wheelchair control. J. Biomed. Sci. Eng. **6**(11), 1050 (2013)
68. T. Carlson, J.D.R. Millan, Brain-controlled wheelchairs: a robotic architecture. IEEE Robot. Autom. Mag. **20**(1), 65–73 (2013)
69. N.R. Waytowich, D.J. Krusienski, Development of an extensible SSVEP-BCI software platform and application to wheelchair control, in *2017 8th International IEEE/EMBS Conference on Neural Engineering (NER)* (IEEE, 2017), pp. 259–532
70. E. Tidoni, P. Gergondet, G. Fusco, A. Kheddar, S.M. Aglioti, The role of audio-visual feedback in a thought-based control of a humanoid robot: a BCI study in healthy and spinal cord injured people. IEEE Trans. Neural Syst. Rehab. Eng. **25**(6), 772–781 (2017)
71. S. Inoue, Y. Akiyama, Y. Izumi, S. Nishijima, The development of BCI using alpha waves for controlling the robot arm. IEICE Trans. Commun. **91**(7), 2125–2132 (2008)
72. D.J. McFarland, J.R. Wolpaw, Brain-computer interface operation of robotic and prosthetic devices. Computer **41**(10) (2008)
73. E. Hortal, D. Planelles, A. Costa, E. Iáñez, A. Úbeda, J.M. Azorín, E. Fernández, SVM-based Brain-Machine Interface for controlling a robot arm through four mental tasks. Neurocomputing **151**, 116–121 (2015)
74. S.L. Norman, M. Dennison, E. Wolbrecht, S.C. Cramer, R. Srinivasan, D.J. Reinkensmeyer, Movement anticipation and EEG: implications for BCI-contingent robot therapy. IEEE Trans. Neural Syst. Rehab. Eng. **24**(8), 911–919 (2016)
75. C. Guger, W. Coon, J. Swift, B. Allison, G. Edlinger, A motor rehabilitation BCI with multimodal feedback in chronic stroke patients (P5. 300). Neurology **88**(16 Suppl), P5–300 (2017)
76. M. Petti, D. Mattia, F. Pichiorri, J. Toppi, S. Salinari, F. Babiloni, L. Astolfi, F. Cincotti, A new descriptor of neuroelectrical activity during BCI-assisted motor imagery-based training in stroke patients, in *2014 36th Annual International Conference of the IEEE Engineering in Medicine and Biology Society (EMBC)* (IEEE, 2014), pp. 1267–1269
77. A. Frisoli, C. Loconsole, D. Leonardis, F. Banno, M. Barsotti, C. Chisari, M. Bergamasco, A new gaze-BCI-driven control of an upper limb exoskeleton for rehabilitation in real-world tasks. IEEE Trans. Syst. Man Cybern. Part C (Appl. Rev.) **42**(6), 1169–1179 (2012)
78. W. Wang, J.L. Collinger, M.A. Perez, E.C. Tyler-Kabara, L.G. Cohen, N. Birbaumer, S.W. Brose, A.B. Schwartz, M.L. Boninger, D.J. Weber, Neural interface technology for rehabilitation: exploiting and promoting neuroplasticity. Phys. Med. Rehab. Clin. North America **21**(1), 157–178 (2010)
79. G.R. Muller-Putz, G. Pfurtscheller, Control of an electrical prosthesis with an SSVEP-based BCI. IEEE Trans. Biomed. Eng. **55**(1), 361–364 (2008)
80. A. Jackson, C.T. Moritz, J. Mavoori, T.H. Lucas, E.E. Fetz, The Neurochip BCI: towards a neural prosthesis for upper limb function. IEEE Trans. Neural Syst. Rehab. Eng. **14**(2), 187–190 (2006)
81. T. Pailla, W. Jiang, B. Dichter, E.F. Chang, V. Gilja, ECoG data analyses to inform closed-loop BCI experiments for speech-based prosthetic applications, in *2016 IEEE 38th Annual International Conference of the Engineering in Medicine and Biology Society (EMBC)*. (IEEE, 2016), pp. 5713–5716

82. K.D. Katyal, M.S. Johannes, S. Kellis, T. Aflalo, C. Klaes, T.G. McGee, M.P. Para, Y. Shi, B. Lee, K. Pejsa, C. Liu, A collaborative BCI approach to autonomous control of a prosthetic limb system, in *2014 IEEE International Conference on Systems, Man and Cybernetics (S*MC) (IEEE, 2014), pp. 1479–1482
83. I. Martišius, R. Damaševičius, A prototype SSVEP based real time BCI gaming system. Comput. Intell. Neurosci. **2016**, 18 (2016)
84. E.M. Holz, J. Höhne, P. Staiger-Sälzer, M. Tangermann, A. Kübler, Brain–computer interface controlled gaming: evaluation of usability by severely motor restricted end-users. Artif. Intell. Med. **59**(2), 111–120 (2013)
85. V. Kaiser, I. Daly, F. Pichiorri, D. Mattia, G.R. Müller-Putz, C. Neuper, Relationship between electrical brain responses to motor imagery and motor impairment in stroke. Stroke **43**(10), 2735–2740 (2012)
86. C.Y. Lin, W.F. Chiang, S.C. Yang, S.Y. Huang, Combining event-related synchronization and event-related desynchronization with fuzzy C-Means to classify motor imagery-induced EEG signals, in *Proceedings of the 2nd International Conference on Intelligent Technologies and Engineering Systems (ICITES2013)* (Springer, Cham, 2014), pp. 1039–1045
87. T. Kasahara, K. Terasaki, Y. Ogawa, J. Ushiba, H. Aramaki, Y. Masakado, The correlation between motor impairments and event-related desynchronization during motor imagery in ALS patients. BMC Neurosci. **13**(1), 66 (2012)
88. Y.R. Tabar, U. Halici, A novel deep learning approach for classification of EEG motor imagery signals. J. Neural Eng. **14**(1), 016003 (2016)
89. R. Chatterjee, T. Bandyopadhyay, EEG based motor imagery classification using SVM and MLP, in *2016 2nd International Conference on Computational Intelligence and Networks (CINE)*, (IEEE, 2016), pp. 84–89
90. X. An, D. Kuang, X. Guo, Y. Zhao, L. He, A deep learning method for classification of EEG data based on motor imagery, in *International Conference on Intelligent Computing* (Springer, Cham, 2014), pp. 203–210
91. W.Y. Hsu, Single-trial motor imagery classification using asymmetry ratio, phase relation, wavelet-based fractal, and their selected combination. Int. J. Neural Syst. **23**(02), 1350007 (2013)
92. Y. Li, P.P. Wen, Modified CC-LR algorithm with three diverse feature sets for motor imagery tasks classification in EEG based brain–computer interface. Comput. Methods Programs Biomed. **113**(3), 767–780 (2014)
93. Y. Zhang, Y. Wang, J. Jin, X. Wang, Sparse Bayesian learning for obtaining sparsity of EEG frequency bands based feature vectors in motor imagery classification. Int. J. Neural Syst. **27**(02), 1650032 (2017)
94. BCI Competitions. 2008. http://www.bbci.de/competition/. Accessed 6 Oct 2017
95. G. Bin, X. Gao, Y. Wang, B. Hong, S. Gao, VEP-based brain-computer interfaces: time, frequency, and code modulations [Research Frontier]. IEEE Comput. Intell. Mag. **4**(4) (2009)
96. K.B. Ng, A.P. Bradley, R. Cunnington, Stimulus specificity of a steady-state visual-evoked potential-based brain–computer interface. J. Neural Eng. **9**(3), 036008 (2012)
97. A. Maronidis, V.P. Oikonomou, S. Nikolopoulos, I. Kompatsiaris, Steady state visual evoked potential detection using Subclass Marginal Fisher analysis, in *2017 8th International IEEE/EMBS Conference on Neural Engineering (NER)* (IEEE, 2017), pp. 37-41
98. O. Friman, I. Volosyak, A. Graser, Multiple channel detection of steady-state visual evoked potentials for brain-computer interfaces. IEEE Trans. Biomed. Eng. **54**(4), 742–750 (2007)
99. S.N. Carvalho, T.B. Costa, L.F. Uribe, D.C. Soriano, G.F. Yared, L.C. Coradine, R. Attux, Comparative analysis of strategies for feature extraction and classification in SSVEP BCIs. Biomed. Signal Process. Control **21**, 34–42 (2015)
100. S. Sur, V.K. Sinha, Event-related potential: an overview. Ind. Psychiatry J. **18**(1), 70 (2009)
101. R. Fazel-Rezai, B.Z. Allison, C. Guger, E.W. Sellers, S.C. Kleih, A. Kübler, P300 brain computer interface: current challenges and emerging trends. Front. Neuroeng. **5** (2012)
102. A. Kübler, S. Halder, A. Furdea, A. Hösle, Brain painting-BCI meets art, in *Proceedings of the 4th International Brain-Computer Interface Workshop and Training Course* (University of Graz, Graz, 2008), pp. 361–366

103. J.I. Münßinger, S. Halder, S.C. Kleih, A. Furdea, V. Raco, A. Hösle, A. Kübler, Brain painting: first evaluation of a new brain–computer interface application with ALS-patients and healthy volunteers. Front. Neurosci. **4** (2010)
104. G. Edlinger, C. Holzner, C. Groenegress, C. Guger, M. Slater, Goal-oriented control with brain-computer interface, in *International Conference on Foundations of Augmented Cognition* (Springer, Berlin, Heidelberg, 2009), pp. 732–740
105. G. Edlinger, C. Guger, Social environments, mixed communication and goal-oriented control application using a brain-computer interface, in *Universal Access in Human-Computer Interaction. Users Diversity*, 2011, pp. 545–554
106. R.K. Chaurasiya, N.D. Londhe, S. Ghosh, Binary DE-based channel selection and weighted ensemble of SVM classification for novel brain–computer interface using Devanagari script-based P300 speller paradigm. Int. J. Hum.-Comput. Interact. **32**(11), 861–877 (2016)
107. J.P. Kulasingham, V. Vibujithan, W.A.R. Kithmini, Y.V.A.C. Kumara, A.C. De Silva, P300 speller for local languages using Support Vector Machines, in *2016 IEEE International Conference on Information and Automation for Sustainability (ICIAfS)* (IEEE, 2016), pp. 1–5
108. W. Speier, J. Knall, N. Pouratian, Unsupervîsed training of brain-computer interface systems using exnectation maximization, in *2013 6th International IEEE/EMBS Conference on Neural Engineering (NER)* (IEEE, 2013), pp. 707–710
109. A. Pinegger, J. Faller, S. Halder, S.C. Wriessnegger, G.R. Müller-Putz, Control or non-control state: that is the question! An asynchronous visual P300-based BCI approach. J. Neural Eng. **12**(1), 014001 (2015)
110. http://www.cvcrbd.org/data. Accessed 20 Jan 2018
111. Experiment 1 version 2 (2016), https://www.youtube.com/watch?v=elTcEnCOMc0&feature=youtu.be. Accessed 30 July 2017
112. H. Guo, C.S. Burrus, Convolution using the undecimated discrete wavelet transform, in *1996 IEEE International Conference on Acoustics, Speech, and Signal Processing, 1996. ICASSP-96. Conference Proceedings*, vol. 3 (IEEE, 1996), pp. 1291–1294
113. M.K. Islam, A. Rastegarnia, A.T. Nguyen, Z. Yang, Artifact characterization and removal for in vivo neural recording. J. Neurosci. Methods **226**, 110–123 (2014)
114. M.K. Islam, A. Rastegarnia, Z. Yang, A wavelet-based artifact reduction from scalp eeg for epileptic seizure detection. IEEE J. Biomed. Health Inf. **20**(5), 1321–1332 (2016)
115. S. Boudet, L. Peyrodie, P. Gallois, C. Vasseur, A global approach for automatic artifact removal for standard EEG record, in *Proceedings of the 28th IEEE EMBS Annual International Conference, New York City, USA*, 2006, pp. 5719–5722
116. V.P. Oikonomou, G. Liaros, K. Georgiadis, E. Chatzilari, K. Adam, S. Nikolopoulos, I. Kompatsiaris, *EEG Processing Toolbox* (2016)https://github.com/MAMEM/eeg-processing-toolbox. Accessed 14 Nov 2016
117. G. Giglia, F. Brighina, S. Rizzo, A. Puma, S. Indovino, S. Maccora, R. Baschi, G. Cosentino, B. Fierro, Anodal transcranial direct current stimulation of the right dorsolateral prefrontal cortex enhances memory-guided responses in a visuospatial working memory task. Funct. Neurol. **29**(3), 189 (2014)
118. M. van Vliet, A. Robben, N. Chumerin, N.V. Manyakov, A. Combaz, M.M. Van Hulle, Designing a brain-computer interface controlled video-game using consumer grade EEG hardware, in *2012 ISSNIP Biosignals and Biorobotics Conference (BRC)* (IEEE, 2012), pp. 1–6
119. Y. Liu, X. Jiang, T. Cao, F. Wan, P.U. Mak, P.I. Mak, M.I. Vai, Implementation of SSVEP based BCI with Emotiv EPOC, in *2012 IEEE International Conference on Virtual Environments Human-Computer Interfaces and Measurement Systems (VECIMS)* (IEEE, 2012), pp. 34–37
120. D. Matthieu, C. Thierry, P. Mathieu, A P300-based quantitative comparison between the Emotiv EPOC headset and a medical EEG device. Int. J. Biomed. Eng. **12**(56), 201 (2013)
121. M. Duvinage, T. Castermans, M. Petieau, T. Hoellinger, G. Cheron, T. Dutoit, Performance of the Emotiv Epoc headset for P300-based applications. Biomed. Eng. **12**(1), 56 (2013)
122. M.N. Fakhruzzaman, E. Riksakomara, H. Suryotrisongko, EEG wave identification in human brain with Emotiv EPOC for motor imagery. Proc. Comput. Sci. **72**, 269–276 (2015)

123. A. Myrden, T. Chau, Effects of user mental state on EEG-BCI performance. Front. Hum. Neurosci. **9** (2015)
124. L. Kipiński, R. König, C. Sielużycki, W. Kordecki, Application of modern tests for stationarity to single-trial MEG data. Biol. Cybern. **105**(3–4), 183–195 (2011)
125. I.I. Goncharova, D.J. McFarland, T.M. Vaughan, J.R. Wolpaw, EMG contamination of EEG: spectral and topographical characteristics. Clin. Neurophysiol. **114**(9), 1580–1593 (2003)
126. G. Schalk, J. Mellinger, General-purpose software for brain-computer interface research, data acquisition, stimulus presentation, and brain monitoring, in *A Practical Guide to Brain–Computer Interfacing with BCI2000* (Springer Science & Business Media, 2010)
127. R. Kuś, A. Duszyk, P. Milanowski, M. Łabęcki, M. Bierzyńska, Z. Radzikowska, M. Michalska, J. Żygierewicz, P. Suffczyński, P.J. Durka, On the quantification of SSVEP frequency responses in human EEG in realistic BCI conditions. PloS One **8**(10), e77536 (2013)
128. M.W. Howard, D.S. Rizzuto, J.B. Caplan, J.R. Madsen, J. Lisman, R. Aschenbrenner-Scheibe, A. Schulze-Bonhage, M.J. Kahana, Gamma oscillations correlate with working memory load in humans. Cerebral Cortex **13**(12), 1369–1374 (2003)
129. D.E. Linden, N.N. Oosterhof, C. Klein, P.E. Downing, Mapping brain activation and information during category-specific visual working memory. J. Neurophysiol. **107**(2), 628–639 (2012)
130. F. Roux, M. Wibral, H.M. Mohr, W. Singer, P.J. Uhlhaas, Gamma-band activity in human prefrontal cortex codes for the number of relevant items maintained in working memory. J. Neurosci. **32**(36), 12411–12420 (2012)
131. N. Kanayama, A. Sato, H. Ohira, Crossmodal effect with rubber hand illusion and gamma-band activity. Psychophysiology **44**(3), 392–402 (2007)
132. M.A. Kisley, Z.M. Cornwell, Gamma and beta neural activity evoked during a sensory gating paradigm: effects of auditory, somatosensory and cross-modal stimulation. Clin. Neurophysiol. **117**(11), 2549–2563 (2006)
133. M.K. Islam, A. Rastegarnia, Z. Yang, Methods for artifact detection and removal from scalp EEG: a review. Neurophysiologie Clinique/Clinical Neurophysiology **46**(4), 287–305 (2016)
134. M.K. Islam, Artifact characterization, detection and removal from neural signals. Doctoral dissertation (2015)

Neural Network Based Physical Disorder Recognition for Elderly Health Care

Sriparna Saha and Raktim Das

Abstract This chapter presents a novel approach to recognize fourteen gestures which would help us recognize muscle and joint pains in human body. The subject is acknowledged in a twenty joint skeletal form with the help of Microsoft's Kinect. In order to extract features from the subject, those twenty body joints are worked on and ten features in the form of distances and angles are calculated. For recognizing unknown sample gestures, a neural network with Levenberg-Marquardt learning rule has been used. The explained methodology provides an accuracy of 88.19%, which is relatively higher than other algorithms previously used in elderly healthcare domain.

Keywords Gesture recognition · Healthcare · Kinect · Neural network

1 Introduction

Gesture is always more effective way to convey ideas across human-to-human and human-to-computer when verbal language is absent [1]. People use gestures even while speaking. Conventionally humans and machines communicate using standard keys/buttons provided for users to input like in case of keyboards, mouse, joystick, etc. [1]. But these modes of inputs are inefficient and do not serve the purpose in case of people with disability. In such cases, identifying these gestures can be employed as inputs to the machines for better human-computer interaction.

S. Saha (✉)
Computer Science and Engineering Department, Maulana Abul Kalam
Azad University of Technology, West Bengal, India
e-mail: sahasriparna@gmail.com

R. Das
Standard Chartered - Global Business Services, Chennai, India
e-mail: misterraktim@gmail.com

W. Pedrycz and S.-M. Chen (eds.), *Computational Intelligence for Pattern Recognition*, Studies in Computational Intelligence 777,
https://doi.org/10.1007/978-3-319-89629-8_12

Currently we have the technology to capture a human body in 3-D space. This can be done with the help of different human motion sensing devices. One example of this kind of device is the Microsoft's Kinect sensor [2–4]. Complex body gestures related to different physical disorders, can be successfully identified using the Kinect. This device uses the inbuilt RGB camera and 3D depth sensor to map the human body to its skeletal form. Due to the low cost, this device is used widely in many application areas.

The body gestures taken into account for this chapter are the symptoms due to the physical disorders for muscle and joint pains shown by elderly persons. General causes of these disabilities may be from injury, fatigue, and aging. These disorders go to advanced stage due to negligence, bad habits and aging of the disabled persons. So the explained home monitoring system can be utilized as an alternative for the troublesome process of visiting hospitals on a frequent basis [1].

There are various research proposed previously for the purpose of gesture recognition in elderly healthcare domain, where Microsoft Kinect Sensor is used for gathering the gesture related information. Desk jobs demands long working hours in the same sitting posture, this results in deterioration in the functioning of tendons and joints of the persons working. In [5], the authors have touched upon a technique which will help in recognizing the symptoms of physical disorders at an early stage. Recognising these symptoms involves principal component analysing for linear dimensionality reduction and fuzzy c-means algorithm. Another work in [6] describes similar type of work for young person by calculating Euclidean distances from each frame and ReliefF algorithm is used to remove space complexity. The classification is done using fuzzy k-nearest neighbour classifier. Parajuli et al. has put forth a method for monitoring senior health using Kinect sensor [7]. The authors have approximated the gestures when elders are likely to fall by measuring gait. The recognition stage takes the help of support vector machine. In the current social model where a couple are both out working elderly healthcare is a major concern. To detect the fall of an elder, ensemble decision tree is being used along with the Kinect sensor in [8]. Yu et al. has presented an interesting approach to analyze children tantrum behaviour [9]. The paper exploits medical knowledge and questionnaire based attitude investigation. For dimensionality reduction, principal component analysis is applied and Euclidean distance is employed to estimate the proximity between behaviours like push, shout and attack. Finally k-means clustering is implemented.

In this chapter, we have explained how a real-time home-monitoring system can be build which is useful in alarming the subjects in case the muscle and joint pains are noticed using Kinect. Generally, the majority of daily life activities are performed by elder persons while sitting on a chair. Thus, fourteen distinct body gestures associated with muscle and joint pains are taken into account while the subject is sitting on a chair. After gathering the gestures of the subjects in the form of joint coordinates using Kinect sensor, those are being worked upon to extract ten features from each gesture. Next, the classification is carried out using Levenberg-Marquardt optimization based neural network. In this process, assistance can be provided to elderly people from their homes itself by monitoring their day-to-day activities. Whenever the system detects any physical disorders by examining the gestural features, an alarm is generated and

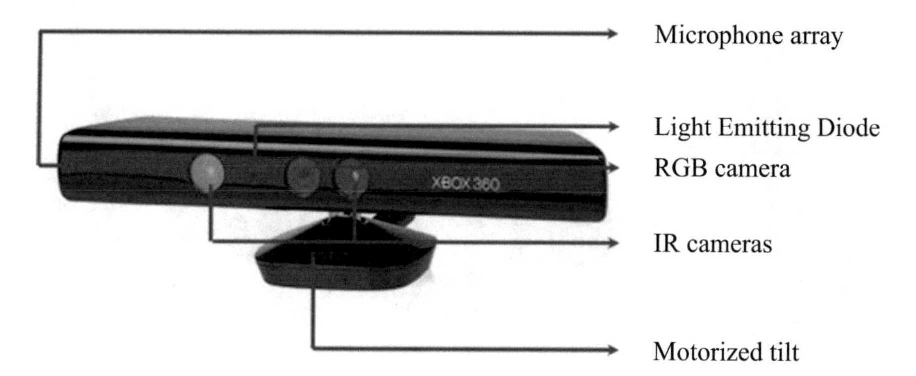

Fig. 1 Kinect sensor

the subject is recommended to do specific exercises based on the recognized disorder. But if the same disorder occurs persistently in a specific subject for a long time, the subject is advised to consult with the doctors.

In Sect. 2, a brief introduction to some preliminary ideas is provided. Sections 3 and 4 elaborates about the proposed work and experimental results respectively. Section 5 concludes the chapter following with Matlab codes in Sect. 6.

2 Preliminary Knowledge

In this section, a brief introduction to Kinect sensor, physical disorders and neural network are given.

2.1 Microsoft's Kinect Sensor

Kinect is manufactured by Microsoft mainly for real-time gaming purposes [2–4]. But this sensor has tremendous scope for home monitoring. It looks like a webcam with one RGB camera, one IR (infra-red) emitter and one IR receiver (Fig. 1). Based on the data captured, x, y and depth co-ordinates are measured using RGB camera IR camera respectively (Fig. 2). Based on these two information, human body is skeletonised into twenty 3D body joints (Fig. 3). This device can record data within a distance of 1.2–3.5 m [10, 11]. The Kinect sensor can work twenty four hours in a day and in almost all lighting conditions (except for fluorescent light). Also the subject's dress does not affect the recognition process (until he/she is wearing fully black dress).

Fig. 2 RGB image with its corresponding body joints as recognized using Kinect sensor

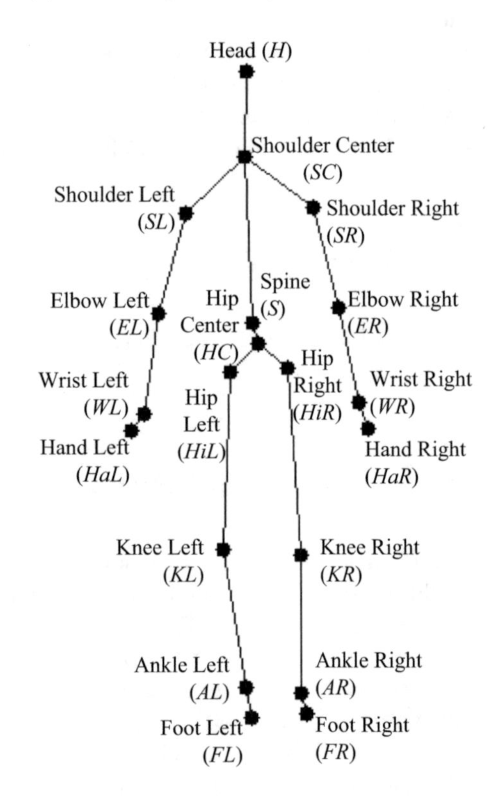

Fig. 3 Twenty body joints as captured using Kinect sensor

2.2 Considered Muscle and Joint Pains

Though the definition of old age cannot be governed by any hard boundary, but after consulting with eminent doctors, the subjects with more than 40 years age are considered as elderly people. From this age, they tend to show some symptoms related to muscle and joint pains. If the disorders can be taken care of at an early stage, the disorders cannot go to an advanced stage. For this chapter, fourteen gestures are taken into account while the subject is sitting. The descriptions about the disorders are given in Table 1.

2.3 Neural Network with Levenberg-Marquardt Optimization

The current chapter briefly discusses the feed forward neural network. A feed forward neural network is made up of three layers namely an input layer, few hidden layers and an output layer, and a weighted sum of input flows in forward one direction only (Fig. 4) [1]. Although a number of hidden layers can be used, one hidden layer with several neurons can fit any input-output mapping [1]. If satisfactory results are not obtained, the number of neurons present in the hidden layer may change. Best output from this method is obtained when ten neurons in one hidden layer is utilised. A mapping between input domain with output domain is attained using a random combination of weights which is used so that error between target and output is reduced. Back-propagation algorithm is another model where the weight adaption happens from the last to first layer (Fig. 5) [1]. Among different weight adaptation techniques for implementing back propagation algorithm, Levenberg-Ma rquardt optimization (LM-NN) [12, 13] has been considered as it excels in gradient descent search and conjugate gradient (or quadratic approximation) methods for medium-sized problems.

The drawback of gradient descent learning [13] is that it converges very slowly. This is because it does not take fixed size step towards negative gradient of the error function, but it adopts minute step size which is roughly fixed times the negative gradient.

$$w_{i+1} = w_i - \eta \nabla E(w) \tag{1}$$

The result of this is fast convergence in sharp neighbourhood (large gradient) and dim motion in valley neighbourhood (small gradient) on the error surface (Fig. 6). The optimization technique can be speeded up by using the curvature information.

Table 1 The addressed fourteen gestures related to elderly healthcare

Disorder name	Skeleton				Description
	RGB image		Skeletal image		
	Pain at left side	Pain at right side	Pain at left side	Pain at right side	
Lumbar spondylosis					The inconvenience to joint is caused due to degenerative alteration to the lumbar spine. The subject suffers from pain in the lower back position. The subject massages the lower back by her left hand in response to the pain
Tennis elbow					Tennis elbow is a condition which affects the tendon and causes pain over the lateral aspect of the elbow due to strain of the tedious origin
Plantar fasciitis					The person experiences pain in ankle due to strain of the plantar fascia as seen in the image beside
Osteo-arthritis knee					Joint problem at knee, happens due to negative implications resulting in pain and stiffness of the joint

(continued)

Table 1 (continued)

Cerviacal spondylosis					Negative changes in the cervical spine leads to the disorder. The patient gets relief from the pain by holding their neck and tilting it
Osteo-arthritis hand			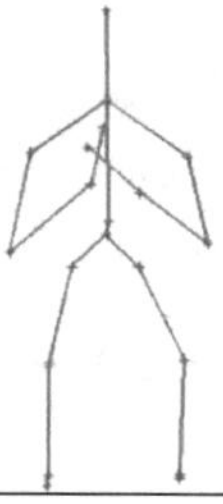	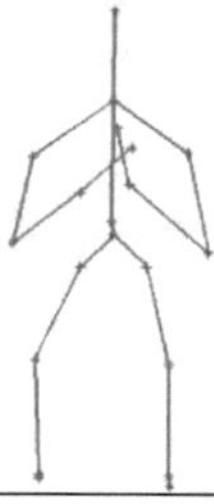	The degenerative change of the joints at palm and fingers, leading to pain and stiffness in the hand. To relieve the pain, patient normally massages the painful joints in hand with the other arm
Frozen shoulder					Degenerative changes in the shoulder resulting in pain and stiffness leads to the disorder

Both the gradient and curvature can be obtained from the second order information but it is costly for reckoning. Hence, most of the techniques rely on approximating the gradient by first order derivative and the curvature by function evaluation.

The error surface is described by mean squared error function in (2) where the mean is done over the input and output pairs.

$$E\,(w) = \left\langle (f\,(x; w) - y)^2 \right\rangle \tag{2}$$

where x is the input vector to the neural net,

w is the weight matrix of the interconnections,
$f(.)$ gives the output vector from the neural net,
y is the target vector for the neural net, and
$E(.)$ is the error which is a function of weights throughout the training phase.

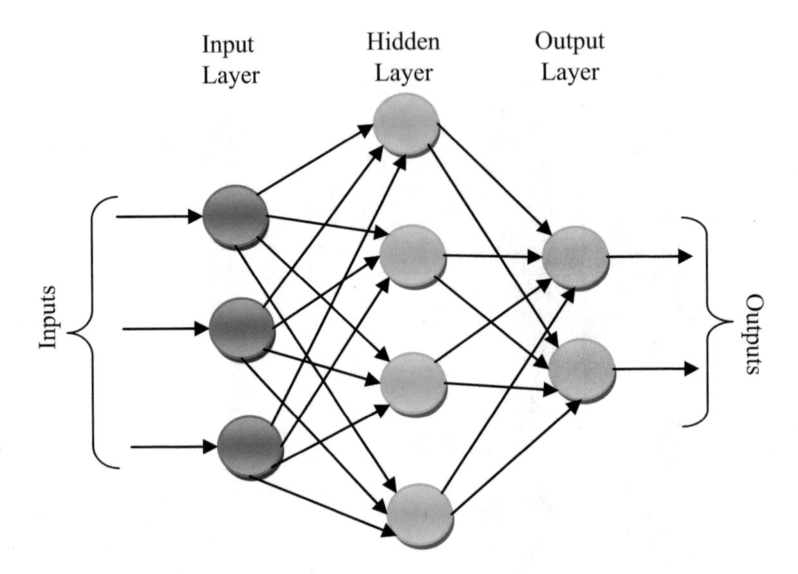

Fig. 4 Feed forward neural network

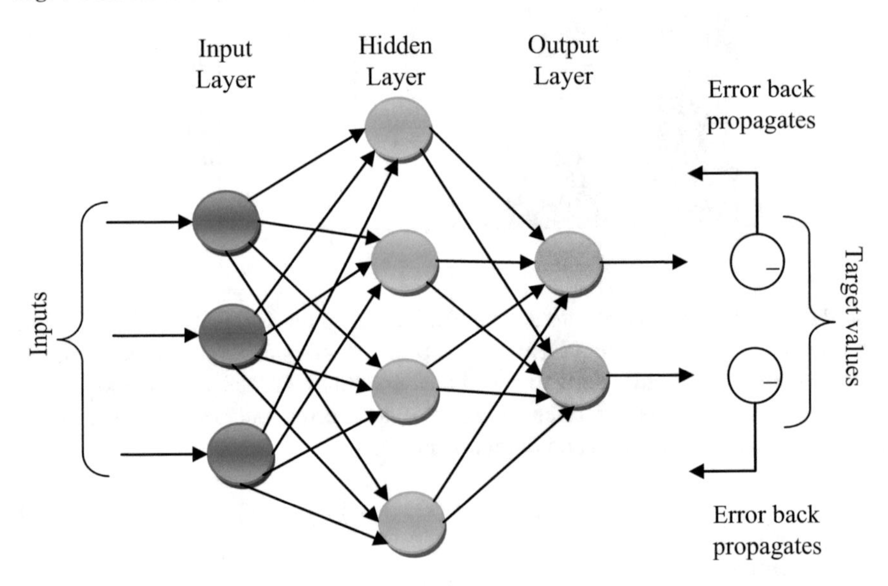

Fig. 5 Back propagation neural network

When E is a quadratic expression, $f(.)$ is a linear model from which the minima can be directly evaluated without exploring the most exorbitant descent search. Thus, estimating $f(.)$ as linear, the weight adjustment rule can be a little modified where d is a derivative and H is an estimate of Hessian matrix [13] obtained by taking mean of the first order derivative.

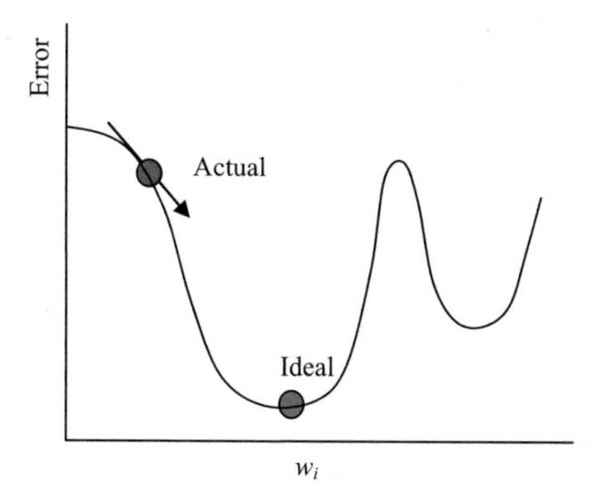

Fig. 6 Convergence using gradient descent learning

$$w_{i+1} = w_i - H^{-1}d \tag{3}$$

However, it may not be correct to treat E as a q all along the error surface except near the minima. So, Levenberg merged the two algorithms where at first the minimum is loosely acquired by using the gradient descent learning and then the quadratic approximation is applied to fine-tune the previous result. Here, weights and all parameters are arbitrarily initialized, and the output and respective error are evaluated. The Levenberg proposed optimization rule (4) is then applied. An enhancement in the error implies the quadratic approximation is running out and so λ is increased to implement gradient descent. Similarly, when a reduction in error means minimum is nearby and so, λ is decreased. The weight adaptation continues to iterate until error is within a dictated limit.

$$w_{i+1} = w_i - (H + \lambda\, I)^{-1}\, d \tag{4}$$

Marquadt noted that gradient descent search [13] dominates when λ is large. So, the original Levenberg equation was changed to yield the concluding Levenberg-Marquardt rule (5) where instead of the identity matrix the diagonal of the approximated Hessian matrix is used.

$$w_{i+1} = w_i - (H + \lambda\, diag\,[H])^{-1}\, d \tag{5}$$

3 Proposed Work

The block diagram of the elderly healthcare related gesture recognition scheme is shown in Fig. 7. Let, G (= 14) be the total number of physical disorders considered here. For recognition of an unknown gesture u, a training dataset is constructed by

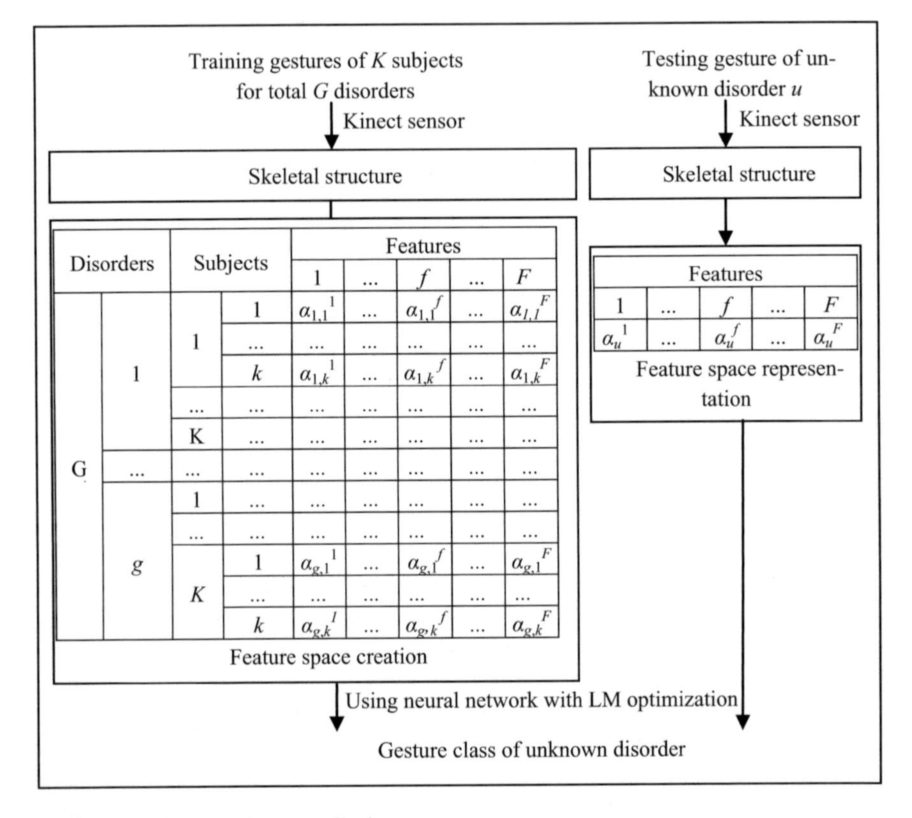

Fig. 7 Block diagram of the described system

D (HaL, HaR)	D (EL, ER)	D (KL, KR)	A (WL, EL, SL)	A (WR, ER, SR)	A (AL, KL, HiL)	A (AR, KR, HiR)	A (SC, S, HC)	A (SL, HiL, KL)	A (SR, HiR, KR)

Fig. 8 Feature vector

taking gestural data from K different subjects of a specific disorder g, where $g \in [1, G]$. For each kth gesture for $k \in [1, K]$, F number of total features (α) are extracted. The meaning of fth feature for $f \in [1, F]$ is given in Fig. 8, 9. As the Kinect sensor is able to capture the human body using twenty 3D joint co-ordinates, using this as the raw information, features are mined which can be used to distinguish between a normal gesture and a gesture revealing physical disorder. To construct the feature vector, ten features are extracted as given in Fig. 8, where α can be D and A implying denotes Euclidean distance and angle features respectively (Fig. 9). The meanings of the joint names are already provided in Fig. 3. Hence, a $1 \times F$ (=10) feature vector is prepared for every frame captured using Kinect sensor.

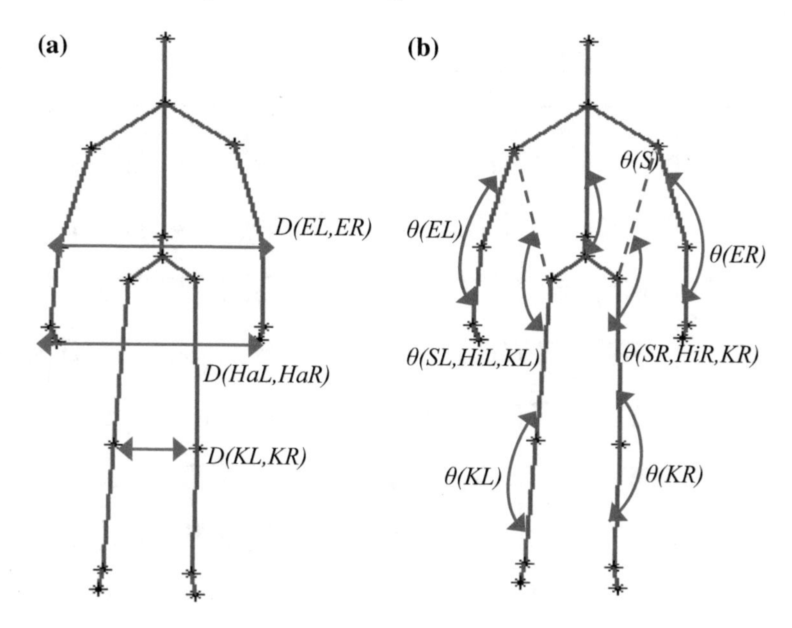

Fig. 9 Features extracted: (a) distance and (b) angle

Table 2 Preparation of training dataset

	Dataset number	No. of subjects (male)	No. of subjects (female)	Age group
Training dataset	1	14	16	21 ± 3 years
	2	18	12	27 ± 2 years
	3	22	8	33 ± 4 years
Testing dataset		37	43	27 ± 6 years

Figure 7 also pictorially depicts the philosophy of identifying an unknown gesture, which is passed through the feature extraction step. The testing gesture is then recognized with the help of neural network, already explained in Sect. 2.3.

4 Experimental Results

Three datasets have been prepared for this work, with thirty subjects ($K = 30$) in each dataset as provided in Table 2. The feature vector corresponding to a gesture of every particular disease is enlisted in each of the columns of Table 5.1. The numbering of the gestures in Table 5.1 is same as that in Fig. 5.1. Table 3.

Table 3 Sample feature vectors from training dataset 1

	Features									
	$\alpha^{1}_{g,21}$	$\alpha^{2}_{g,21}$	$\alpha^{3}_{g,21}$	$\alpha^{4}_{g,21}$	$\alpha^{5}_{g,21}$	$\alpha^{6}_{g,21}$	$\alpha^{7}_{g,21}$	$\alpha^{8}_{g,21}$	$\alpha^{9}_{g,21}$	$\alpha^{10}_{g,21}$
For $g=1$	0.39	0.52	0.24	133.92	65.51	127.99	127.80	141.27	133.46	143.68
For $g=3$	0.29	0.32	0.23	143.80	57.61	127.70	126.16	136.52	136.46	134.10
For $g=5$	0.43	0.48	0.30	87.01	139.51	124.71	104.22	164.35	104.17	109.46
For $g=7$	0.2353	0.32	0.32	144.34	118.86	108.16	121.66	162.17	107.22	104.85
For $g=9$	0.53	0.38	0.27	135.06	27.61	129.03	126.68	129.19	123.09	126.48
For $g=11$	0.05	0.40	0.27	80.41	81.59	117.96	118.59	129.46	142.48	132.70
For $g=13$	0.34	0.30	0.27	91.19	119.55	126.99	123.65	134.46	127.42	129.33

The explained work is compared with four other well-known techniques for the performance analysis. The other existing algorithms are ensemble decision tree (EDT) [8], type-1 fuzzy classifier (T1FS) [14], support vector machine (SVM) [7] and *k*-nearest neighbour (*k*NN) [15]. EDT classifier is based on adaptive boosting principle by taking maximum iterations as 100. T1FS algorithm measures the support of the feature vector based on Gaussian membership curves. SVM algorithm uses a radial basis function kernel whose kernel parameter has a value 1 and the cost value of 100 is tuned in the classifier. For *k*NN, the value of *k* is taken as 5 and Euclidean distance based similarity measure with majority voting determines the class of the unknown gesture.

All the stated algorithms are multiclass in nature except for SVM, which is innately binary. The performance analysis is carried out based on positive predicted value (PPV), negative predicted value (NPV), sensitivity, specificity, accuracy, average error rate (AER) and F1 score (F1 S) as given in (6–12). Here, *TP*, *TN*, *FP* and *FN* stand for true positive, true negative, false positive and false negative respectively. The comparison for each training dataset for all the performance metrics are given in Figs. 10, 11, 12 From the three figures, it is evident that LMA-NN is the best choice for physical disorder recognition for elderly healthcare.

$$\text{PPV} = \frac{TP}{TP + FP} \tag{6}$$

$$\text{NPV} = \frac{TN}{TN + FN} \tag{7}$$

$$\text{Sensitivity} = \frac{TP}{TP + FN} \tag{8}$$

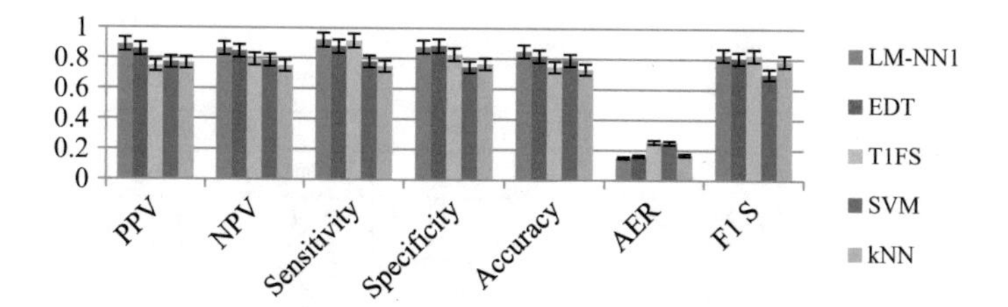

Fig. 10 Comparison of five algorithms for elderly healthcare for training dataset 1

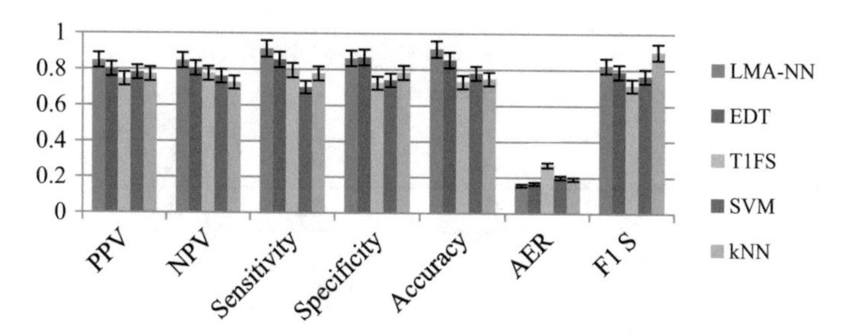

Fig. 11 Comparison of five algorithms for elderly healthcare for training dataset 2

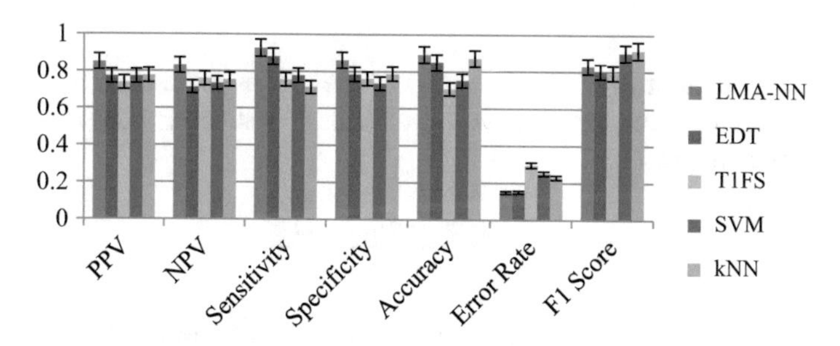

Fig. 12 Comparison of five algorithms for elderly healthcare for training dataset 3

$$\text{Specificity} = \frac{TN}{TN + FP} \tag{9}$$

$$\text{Accuracy} = \frac{TP + TN}{TP + TN + FP + FN} \tag{10}$$

$$\text{AER} = \frac{FP + FN}{TP + TN + FP + FN} \tag{11}$$

$$\text{F1 S} = 2 \times \frac{\text{Precision} \times \text{Recall}}{\text{Precision} + \text{Recall}} \tag{12}$$

To statistically validate the work using neural network for elderly healthcare, three tests are considered. The first one is McNemar's Test. Here, P and Q are the two competitor algorithms with same training dataset. Again, let n_{01} is the number of cases wrongly classified by P but not by Q, and n_{10} is the number of cases wrongly

Table 4 Performance analysis using McNemar's test

Algorithm = Q	Algorithm P = LMA-NN											
	Dataset 1				Dataset 2				Dataset 3			
	n_{01}	n_{10}	Z	Comment	n_{01}	n_{10}	Z	Comment	n_{01}	n_{10}	Z	Comment
EDT	314	417	14.23	Reject	421	416	0.01	Accept	471	401	5.45	Reject
T1FS	583	374	45.20	Reject	293	399	15.93	Reject	225	411	53.81	Reject
SVM	513	398	14.26	Reject	351	425	6.86	Reject	297	378	9.48	Reject
kNN	294	408	18.18	Reject	352	427	7.02	Reject	274	380	16.85	Reject

Table 5 Performance analysis using Friedman and Iman-Davenport tests

Algorithms	Dataset 1	Dataset 2	Dataset 3	R_a	Friedman Test	
					χ^2	Comment
LMA-NN	1	2	1	1.33	9.71	Reject
EDT	2	1	2	1.33		
T1FS	4	4	3	3.67		
SVM	3	3	4	3.33		
kNN	5	5	5	5.00		

classified by Q but not by P. So according to the null hypothesis considering both classifiers have the same error rate, the McNemar's statistic Z obeys a χ^2 with 1 degree of freedom [16].

$$Z = \frac{(|n_{01} - n_{10}| - 1)^2}{n_{01} + n_{10}} \tag{13}$$

From Table 4, it can be observed that the null hypothesis is rejected where $Z > 3.84$, as 3.84 is the threshold value of the chi square distribution at probability of 0.05.

The next statistical analysis is carried out using Friedman Test. Here, let r_a^b is the ranking of the *Accuracy* obtained by the ath algorithm ($1 \le a \le A$) for the bth dataset ($1 \le b \le B$). The best and worst of all classifiers is given ranks of 1 and B respectively. Table 5 gives the idea about Friedman rankings [17].

$$R_a = \frac{1}{B} \sum_{b=1}^{B} r_a^b \tag{14}$$

$$\chi^2 = \frac{12B}{A(A+1)} \left[\sum_{a=1}^{A} R_a^2 - \frac{A(A+1)^2}{4} \right] \tag{15}$$

For the current work, $B = 3$ and $A = 5$. In Table 5, the null hypothesis is rejected, as $\chi_F^2 = 9.71$ is greater than the threshold value (i.e., 9.49) of the χ^2 distribution for $A - 1 = 4$ degrees of freedom at probability of 0.05 [18].

The last analysis is using Iman-Davenport Statistical Test. It is based on F distribution with $(A-1)$ and $(A-1)\times(B-1)$ degrees of freedom [17].

$$F = \frac{(B-1)\times\chi^2}{B\times(A-1)-\chi^2} \tag{16}$$

It is evident that the null hypothesis is rejected, as $F=8.50$ is greater than the critical value (i.e., 5.05) of the F distribution for $A-1=4$ and $(A-1)\times(B-1)=8$ degrees of freedom at probability of 0.05 [18].

5 Conclusion and Future Work

The elderly healthcare system for home monitoring is quite a novel and user-friendly method of recognizing physical disorders related to joint and muscle pains. The work is carried out after consulting with several doctors for preparation of the datasets. There is a closed loop between the subject and the Kinect sensor interfacing computer. Whenever any ambiguity is detected in the normal day-to-day life of the subject, alarm is generated. If that physical disorder is continued for several hours in a day, specific exercise videos are shown to the subjects.

Though the work is mainly demonstrated for elderly people, but it is equally important for young individuals working in multi-national companies. Due to the sedentary working environments in the offices, certain muscle and joint fatigues are developed in the employees. If early detection of those physical disorders can be done, then it will be beneficial for the employees and in turn total company health will be improved. As Kinect sensor only detects the skeleton of the subject, thus privacies of the subjects are persevered. The work can be utilised for other areas, like e-learning of several dances and sign languages and also training in several sports.

Kinect sensor does not require refresh time and can run throughout a day in most lighting conditions. But only disadvantage of using Kinect sensor is that its limited range as it uses the IR. In the future, we should delve into several other data acquisition techniques that can subdue the limitations stated above with introducing new gestures covering more physical disorders.

6 Matlab Codes

The input file 'video_1.txt' contains the twenty body joints 3D information. Here the feature extraction procedure is demonstrated using following Matlab code.

```
dims=3; % as 3D view of the subject is captured using Kinect sensor
joints=20; % as human body is approximated using 20 joints
fid = fopen('video_1.txt'); % input file is named as 'video_1.txt'
A = fscanf(fid, '%g', [1 inf]); % scanning of input file using a pointer
P=1;
total=length(A);
frames=total/(dims*joints); % number of frames in the input file is calcu-
lated
X=zeros(frames,joints,dims); % initializing
for i=1:frames
   for j=1:joints
      for k=1:dims
         X(i,j,k)=A(P); % a 3D matrix is formed where x, y and z dimensions
have assigned in such a fashion that for any specific frame, its joints val-
ues can be easily noted for all three dimensions
         P=P+1;
      end
   end
end
fclose(fid);

for iter=(frames):(frames) % for a particular frame 'iter', skeletal struc-
ture is obtained
   figure(iter)
   hold on
   axis equal
   for i=1:joints
   scatter3(X(iter,i,1),X(iter,i,2),X(iter,i,3),'*b'); % 20 joints are depicted
using black stars
   end
x=[ 4;  3;  3;  5;  6;  7;  9; 10; 11;  3; 2;  1;  1; 13; 14; 15; 17; 18; 19];
y=[ 3;  5;  9;  6;  7;  8; 10; 11; 12;  2; 1; 13; 17; 14; 15; 16; 18; 19; 20];
for i=1:length(x)
    plot3([X(iter,x(i),1) X(iter,y(i),1)],[X(iter,x(i),2)
 X(iter,y(i),2)],[X(iter,x(i),3) X(iter,y(i),3)],'Color','r','LineWidth',1.5); % to
 connect the joints accordingly
```

```
end
end

dist1=zeros(1,frames); % initializing
dist2=zeros(1,frames);
dist3=zeros(1,frames);
angle1=zeros(1,frames);
angle2=zeros(1,frames);
angle3=zeros(1,frames);
angle4=zeros(1,frames);
angle5=zeros(1,frames);
angle6=zeros(1,frames);
angle7=zeros(1,frames);

for i=1:frames % calculation of feature values for all the frames
  dist1(i)=sqrt(sum((X(i,8,:)-X(i,12,:)).^2));
  dist2(i)=sqrt(sum((X(i,10,:)-X(i,6,:)).^2));
  dist3(i)=sqrt(sum((X(i,14,:)-X(i,18,:)).^2));
  v1=[(X(i,5,1)-X(i,6,1)) (X(i,5,2)-X(i,6,2)) (X(i,5,3)-X(i,6,3))];

  v2=[(X(i,7,1)-X(i,6,1)) (X(i,7,2)-X(i,6,2)) (X(i,7,3)-X(i,6,3))];
  angle1(i)=(atan2(norm(cross(v1,v2)), dot(v1,v2)))*180/pi;

  v1=[(X(i,9,1)-X(i,10,1)) (X(i,9,2)-X(i,10,2)) (X(i,9,3)-X(i,10,3))];
  v2=[(X(i,11,1)-X(i,10,1)) (X(i,11,2)-X(i,10,2)) (X(i,11,3)-X(i,10,3))];
  angle2(i)=(atan2(norm(cross(v1,v2)), dot(v1,v2)))*180/pi;

  v1=[(X(i,13,1)-X(i,14,1)) (X(i,13,2)-X(i,14,2)) (X(i,13,3)-X(i,14,3))];
  v2=[(X(i,15,1)-X(i,14,1)) (X(i,15,2)-X(i,14,2)) (X(i,15,3)-X(i,14,3))];
  angle3(i)=(atan2(norm(cross(v1,v2)), dot(v1,v2)))*180/pi;

  v1=[(X(i,17,1)-X(i,18,1)) (X(i,17,2)-X(i,18,2)) (X(i,17,3)-X(i,18,3))];
  v2=[(X(i,19,1)-X(i,18,1)) (X(i,19,2)-X(i,18,2)) (X(i,19,3)-X(i,18,3))];
  angle4(i)=(atan2(norm(cross(v1,v2)), dot(v1,v2)))*180/pi;

  v1=[(X(i,4,1)-X(i,2,1)) (X(i,4,2)-X(i,2,2)) (X(i,4,3)-X(i,2,3))];
```

```
    v2=[(X(i,1,1)-X(i,2,1)) (X(i,1,2)-X(i,2,2)) (X(i,1,3)-X(i,2,3))];
    angle5(i)=(atan2(norm(cross(v1,v2)), dot(v1,v2)))*180/pi;

    v1=[(X(i,5,1)-X(i,13,1)) (X(i,5,2)-X(i,13,2)) (X(i,5,3)-X(i,13,3))];
    v2=[(X(i,14,1)-X(i,13,1)) (X(i,14,2)-X(i,13,2)) (X(i,14,3)-X(i,13,3))];
    angle6(i)=(atan2(norm(cross(v1,v2)), dot(v1,v2)))*180/pi;

    v1=[(X(i,9,1)-X(i,17,1)) (X(i,9,2)-X(i,17,2)) (X(i,9,3)-X(i,17,3))];
    v2=[(X(i,18,1)-X(i,17,1)) (X(i,18,2)-X(i,17,2)) (X(i,18,3)-X(i,17,3))];
    angle7(i)=(atan2(norm(cross(v1,v2)), dot(v1,v2)))*180/pi;
end
```

Sample Run: The Matlab code creates the skeletal image and feature vector as given in Fig. 3 and Fig. 8 correspondingly.

References

1. A. Kendon, *Gesture: Visible Action as Utterance* (Cambridge University Press, Cambridge, 2004
2. J. Han, L. Shao, D. Xu, J. Shotton, Enhanced computer vision with Microsoft kinect sensor: a review. IEEE Trans. Cybern. **43**(5), 1318–1334 (2013)
3. E.E. Stone, M. Skubic, Fall detection in homes of older adults using the Microsoft kinect. IEEE J. Biomed. Health Inf. **19**(1), 290–301 (2015)
4. M.-C. Hu, C.-W. Chen, W.-H. Cheng, C.-H. Chang, J.-H. Lai, J.-L. Wu, Real-Time Human Movement Retrieval and Assessment With Kinect Sensor. IEEE Trans. Cybern. **45**(4), 742–753 (2015)
5. M. Pal, S. Saha, A. Konar, A Fuzzy c means clustering approach for gesture recognition in healthcare, in *Knee*, vol. 1, p. C7
6. S. Saha, M. Pal, A. Konar, D. Bhattacharya, Automatic gesture recognition for health care using relieff and fuzzy kNN, in *Information Systems Design and Intelligent Applications* (Springer, Berlin, 2015), pp. 709–717
7. M. Parajuli, D. Tran, W. Ma, D. Sharma, Senior health monitoring using Kinect, in *2012 Fourth International Conference on Communications and Electronics (ICCE)* (2012), pp. 309–312
8. E. Stone, M. Skubic, Fall detection in homes of older adults using the Microsoft Kinect (2014)
9. X. Yu, L. Wu, Q. Liu, H. Zhou, Children tantrum behaviour analysis based on Kinect sensor, in *2011 Third Chinese Conference on Intelligent Visual Surveillance (IVS)* (2011), pp. 49–52
10. K. Khoshelham, S.O. Elberink, Accuracy and resolution of kinect depth data for indoor mapping applications. Sensors **12**(2), 1437–1454 (2012)
11. D. Pagliari, L. Pinto, Calibration of kinect for Xbox one and comparison between the two generations of Microsoft sensors. Sensors **15**(11), 27569–27589 (2015)
12. S. Roweis, *Levenberg-marquardt optimization*, (Univ. Toronto, Unpubl., 1996)
13. J.J. Moré, The Levenberg-Marquardt algorithm: implementation and theory, in *Numerical Analysis* (Springer, Berlin, 1978), pp. 105–116
14. Y. Zhu, X. Ji, Expected values of functions of fuzzy variables. J. Intell. Fuzzy Syst. **17**(5), 471–478 (2006)

15. M. Oszust, M. Wysocki, Recognition of signed expressions observed by Kinect Sensor, in *2013 10th IEEE International Conference on Advanced Video and Signal Based Surveillance (AVSS)*, (2013), pp. 220–225
16. T.G. Dietterich, Approximate statistical tests for comparing supervised classification learning algorithms. Neural Comput. **10**(7), 1895–1923 (1998)
17. S. García, D. Molina, M. Lozano, F. Herrera, A study on the use of non-parametric tests for analyzing the evolutionary algorithms' behaviour: a case study on the CEC'2005 special session on real parameter optimization. J. Heuristics **15**(6), 617–644 (2009)
18. J. H. Zar, *Biostatistical Analysis* (Pearson Education India, 1999)

Recognizing Subtle Micro-facial Expressions Using Fuzzy Histogram of Optical Flow Orientations and Feature Selection Methods

S. L. Happy and Aurobinda Routray

Abstract Micro-expressions are the subtle and short-lived facial deformations that convey the inner feelings of a person. Automatic recognition of micro-expressions has potential applications in many areas. However, extraction of the appropriate feature, for encoding the subtle movements during the micro-expressions, is a very challenging work. The use of spatial and spatio-temporal features are studied extensively for this problem. However, the face appearance does not change appreciably during a micro-expression. Moreover, the muscle movements are also very small, almost indistinguishable. Rather, these changes possess a temporal pattern. We use the fuzzy histogram of optical flow orientation (FHOFO) features to encode the temporal patterns associated with facial micro-movements. The FHOFO constructs fuzzified angular histograms from the facial movement vectors. The feature descriptors of a micro-expression clip usually possess high dimension and suffer from the curse of dimensionality. To this end, we explore different feature selection methods to reduce the dimension of the descriptor. Experimentally we found that FHOFO achieves significant accuracy on the publicly available databases and its performance is consistently well across the databases.

Keywords Facial micro-expression analysis · Fuzzy histogram · Feature selection

1 Facial Micro-expressions

Communication is an essential part of our lives. In addition to the verbal communication, we often communicate through nonverbal means, such as signs, gestures, postures, etc. Sharing information via e-mail or Internet is very popular these days. We spend a lot of time interacting with digital devices. However, these devices are

S. L. Happy (✉) · A. Routray
Department of Electrical Engineering, Indian Institute of Technology, Kharagpur, India
e-mail: happy@iitkgp.ac.in

A. Routray
e-mail: aroutray@iitkgp.ac.in

W. Pedrycz and S.-M. Chen (eds.), *Computational Intelligence for Pattern Recognition*, Studies in Computational Intelligence 777,
https://doi.org/10.1007/978-3-319-89629-8_13

emotionally blind at this moment. They are inefficient at conveying our affective signals, which are at the heart of successful human relations. To this end, affective computing is an emerging field of study with its focus to develop systems that can recognize, interpret, and emulate human affects [1]. Among the various channels that communicate our emotions, the facial expression is considered to be the most accurate identifier [2]. The facial expressions can be grouped into two categories based on the context and the duration they last, namely: micro-expression and macro-expression.

When a person tries to conceal his affective state, the facial muscles move involuntarily as the reflex to the emotional state. Such nonverbal leakage of genuine feelings through subtle facial movements are termed as facial micro-expressions [3]. Deciphering these micro-expressions can help the doctors to indirectly assess the situation of patients or the police to understand the behavior of people during the criminal investigation. As it is very difficult to fake these actions, recognizing facial micro-expressions has many potential applications in the field of criminal investigation, surveillance, clinical diagnosis, lie detection, etc. However, these micro-expressions are very subtle and short-lived (between 170 and 500 ms [4]) in nature. Even the trained people can recognize only 47% cases in practical situations [5]. Therefore, there is a great need for automated systems for accurate recognition of micro-expressions.

On the other hand, the macro-expressions last on the face for 0.5–4 s that we easily notice and understand. Most of the state-of-art facial expression recognition methods [6, 7] are capable of classifying macro-expressions with significant accuracy on the public databases. However, limited work has been reported on micro-expression classification [6]. Literature suggests the use of high frame rate videos to encode the features of micro-expressions properly. Furthermore, the muscle movements during the micro-expressions are very subtle and almost unrecognizable to naked eyes, which increases the complexity of classification task. The subtleties of micro-expression sequences are demonstrated in Fig. 1. For the convenience of the reader, we summarized the acronyms used in this article under Table 1.

Many spatio-temporal features have been proposed in the literature to encode the texture as well as the temporal changes of the micro-expressions. However, facial texture vary from individual to individual and the change in appearance of facial areas are negligible during a micro-expression. Consequently, only the motion vectors can be used for micro-expression recognition, while ignoring the texture deformations. Here the idea is to obtain a descriptor from the motion vectors that can effectively represent all variations of facial dynamics. Constructing an angular histogram out of the motion vectors can effectively represent the facial local movements. However, the facial movement vectors differ a lot depending on the face pose and the camera angle. Besides, the display of same expression for a particular individual may also vary from instance to instance based on its intensity and other factors. Instead of using hard thresholding to obtain the angular histogram, soft assignment of motion vectors makes the descriptor robust to camera view angle and head poses. Therefore,

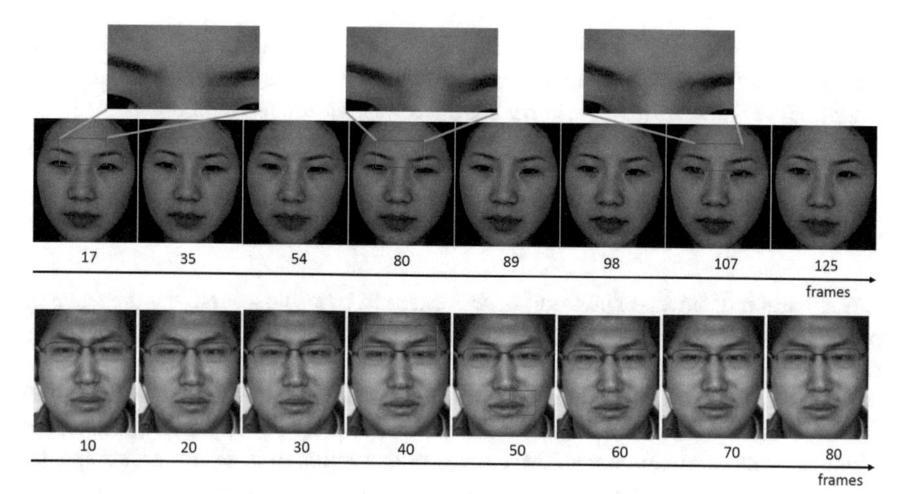

Fig. 1 Example of the frame sequences of micro-expressions in CASME database. The movement is easier to detect when the video is played at low frame rate

Table 1 Abbreviations and their definitions

Abbreviations	Definitions
FACS	Facial action coding system
FDM	Facial dynamic map
FHOFO	Fuzzy histogram of optical flow orientation
FS	Fisher score
FSV	Feature selection via concave minimization
HOG	Histogram of oriented gradient
HOOF	Histogram of oriented optical flow
KNN	K-nearest neighbor
LBP	Local binary pattern
LBP-TOP	Local binary pattern from three orthogonal planes
LDA	Linear discriminant analysis
LOSO	Leave-one-subject-out
LOVO	Leave-one-video-out
LS	Laplace score
OS	Optical strain
PWFP	Pair-wise feature proximity
ROI	Region of interest
SVM	Support vector machine
TIM	Temporal interpolation model

fuzzy membership functions are utilized to assign some values to each histogram bin. Here the fuzzy is introduced in the design of the feature descriptor followed by the classification algorithms to carry out micro-expression recognition.

2 Methods for Recognizing Micro-expressions

Classification of micro-expression is performed in two-folds. First, the presence of micro-expressions needs to be detected from the facial videos. This process of temporal segmentation is also called as micro-expression spotting or detection. The presence of a micro-expression in a long video is marked by the starting and the end frames, also called on-set and off-set frames respectively [8, 9]. Usually, the on-set frame indicates the beginning of facial muscle contraction from a neutral state and the off-set frame indicates the falling back of facial muscles to their usual positions. The frame at which the micro-expression becomes most intense is called as the apex. The apex can be any frame between on-set and off-set frames, not necessarily exactly the middle one between them. Figure 2 illustrates the micro-expression spotting process. These annotations are generally included in the database, with the help from psychologists, for evaluating the automated spotting algorithms. Further, the features are extracted from the segmented part of the video and used for classification purpose.

Researchers use facial motion vectors, the geometric deformations, and gradient histogram descriptors for spotting micro-expressions. In [10], the spotting is carried out by the use of optical strain (OS), which is a method based on the optical flow

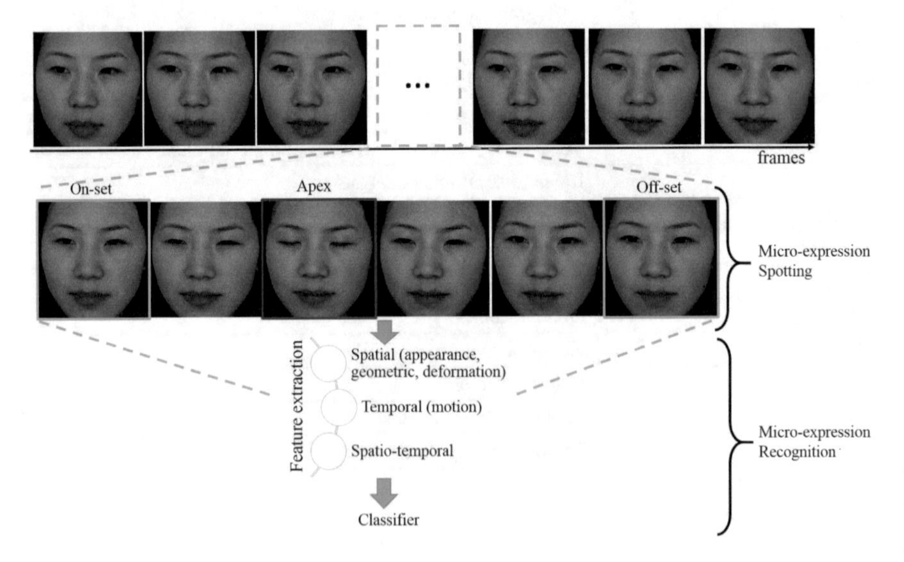

Fig. 2 Stages in classifying micro-expressions

vectors. Optical flow fields describe the movement directions and their strengths, while OS measures the extent of facial muscle deformations at every pixel location. Similar attempts were made in [9] to find out the apex frame in the micro-expression clip for further feature extraction. The peak, on-set, and off-set are detected in [11] using optical flow field and direction continuity obtained from spatio-temporal integration of the motion features across variable sized sliding windows. Wang et al. [12] suggest the use of main directional maximal difference, which is computed using the optical flow vectors between distance frames, for detecting micro-expressions. Active shape model and Procrustes analysis are utilized in [13] for spotting micro-expressions. Polikovsky et al. [14] divided the face into video cubes on the basis of facial action coding system (FACS) and the motion in each region is described by the help of 3-D gradient orientation histogram. They successfully distinguished the onset, apex and offset stages of micro-expressions. Li et al. [15] performed experiments with local binary patterns (LBP) and histogram of optical flow (HOOF), and found LBP to be more efficient at spotting micro-expressions. The use of histogram of oriented gradients (HOG) is reported in [16] for detection of micro-expressions.

After spotting the micro-expressions, the representation of micro-expressions is a challenging task. The recognition performance solely depends upon the discriminating nature of the encoding techniques [6]. The feature extraction techniques broadly fall into three categories, namely: spatial, temporal, and spatio-temporal. Each expression can be associated with a unique facial appearance. Thus, the facial changes, which are also called as spatial features, can be observed in the image plane. These features represent facial geometry, appearance, deformation etc. The relative distance of facial components or the presence of wrinkles, furrows are easily encoded by spatial features, making it a strong candidate for recognizing full-blown expressions. However, most of them fail to represent the subtle behavior associated with micro-expressions as the facial appearance does not change appreciably. In addition to spatial changes, the evolution of facial expressions through time also possess distinct pattern. The temporal features encode the motion and other changes as time progresses. The subtle movements of facial regions during a micro-expression are easily missed if the temporal change (movement of facial muscles as time progresses) is not observed very closely [17]. Thus, the key to success in recognizing micro-expression is the analysis of the temporal dynamics. Some researchers suggest encoding temporal changes, while others report using both spatial and temporal deformations for accurate representation. The spatio-temporal features combine both spatial and temporal properties to represent a video sequence.

The spatial features include LBP, Gabor filter, HOG, shape features, the relative distance between facial components, etc. The eye, eyebrow, and mouth are categorized into several states in [18] based on facial geometry followed by classification of micro-expressions. The use of Gabor filters is reported to be useful in [19, 20]. Lie detection system developed in [21] uses the relative distance variations between facial components as features. The facial landmark points are feature tracked and quantified for micro-expression recognition in [22] based on FACS systems.

Among the temporal features, optical flow based methods are also popular for encoding micro-expressions. OS [23] is a feature derived from optical flow which encodes the rate of change of facial muscle movements during a micro-expression effectively. Liong et al. [24] used the strength of OS as the weights for other features for representing facial micro-movements. The main directional mean optical flow feature is proposed in [25] to efficiently use the optical flow vectors to encode the movement and spatial information for expression classification. Facial dynamic map (FDM) [26] characterizes the dynamics of facial muscles by the principal motion directions of muscle movements.

Many spatio-temporal feature extraction techniques [15, 27, 28] have been proposed in the literature. Among them, the extensions of LBP are the most famous ones. Volume LBP (VLBP) and LBP from three-orthogonal-planes (LBP-TOP) [29] are reported to be powerful spatio-temporal descriptors. LBP-TOP considers three orthogonal planes of a video cube, namely XY-LBP, XT-LBP, and YT-LBP. The LBP features from each of these planes are concatenated to represent the LBP-TOP descriptor. These features are extracted from space-time volume from multiple frames. Computation of VLBP and LBP-TOP are similar to LBP [30], except that the neighbors from consecutive frames are also included in the computation. Mid-level features are generated from LBP-TOP and other features in [31]. In [32], the authors normalized the video cubes and extracted LBP-TOP features in their experiments. LBP-TOP is used in [33], which achieves significant accuracy. The block-based features are well-known for preserving both local and global features.

Local phase quantization from three orthogonal planes [34] is reported to attain superior AU detection. Pyramid of the histogram of oriented gradients with three orthogonal planes [35] is another feature which achieves better performance in recognizing expressions. Spatio-temporal local monogenic binary patterns [36] are reported to have robustness toward illumination variations. Some researchers use 3D gradients [37] as features for representing micro-expressions. Centralized binary patterns from three orthogonal panels (CBP-TOP) [38] is another extension of LBP which is reported to achieve high performance in micro-expression recognition. Spatio-temporal completed local quantization patterns (STCLQP) was proposed in [27] to encode sign, magnitude, and orientation in a codebook. In [39], dual-cross patterns (DCP-TOP) and hot wheel patterns (HWP-TOP) are proposed using three orthogonal planes.

Other than using the feature extraction techniques, some researchers classify micro-expressions by constructing suitable manifold out of the expression clips. The videos are considered as third order tensors and discriminative tensor subspace analysis is used in [40] followed by extreme machine learning for classification purpose. Robust principal component analysis is used on spatio-temporal directional features in [28] for extracting subtle motion features of micro-expressions. Tensor independent color space (TICS) [41] considers the color components independent of each other and uses it as the fourth dimension, thus converting the video clips into 4D tensors. High accuracy of micro-expression recognition is reported using CIELab and CIELuv color spaces.

Deep learning methods are very successful for object detection and recognition purposes. Many researchers use these methods because of their ability to learn the inherent shape and appearance of the object from the training data. However, such algorithms need huge training data for tuning the network parameters. As of now, a few micro-expression databases are available. Moreover, the number of samples in each database is very less, given the exhausting job of manual annotation of micro-expressions from long videos. Lack of sufficient data is a hindrance in using deep learning techniques for micro-expression recognition [42]. Therefore, Patel et al. [42] used transfer learning from objects and facial expressions. In [43], a convolutional neural network (CNN) is used to encode the spatial features, which is further improved by incorporating the temporal characteristic through long short-term memory (LSTM). However, the use of deep learning to recognize dynamic features is relatively new. This is an open area of research which can further flourish for accurate representation of facial dynamics.

Though appearance, geometry and deformation features work well in case of macro-expression, their performance on micro-expression is questionable. Since the face undergoes significant visible changes during macro-expressions, these features encode the prominent facial behavior concerning the expressions. However, facial changes in micro-expressions are very subtle [24, 25], which may not be captured using the traditional feature extractors. Consequently, the features describing facial texture may not be enough and can be ignored for micro-expression recognition. On the contrary, the low amplitude motion vectors resulting from fine facial movements can be effective to represent micro-expressions.

The fuzzy histogram of optical flow orientations (FHOFO) feature, explained in Sect. 3, extracts the motion features effectively from micro-expression clips and explores the effectiveness of motion patterns during a micro-expression. It is an extension of histogram of oriented optical flow (HOOF) features [44] especially to encode the subtle movements of micro-expressions. FHOFO carefully constructs the angular bins from the motion vector directions based on the fuzzy membership function. Moreover, duration and magnitude of micro-expression vary a lot from sample to sample. Therefore, FHOFO uses only the motion direction ignoring the subtle motion magnitudes.

Micro-expression recognition requires analysis of videos captured at a very high frame rate. Though creating a subtle expression database is quite difficult due to complex collecting procedures, a few databases are reported in the literature, namely: SMIC [45], CASME [46], and CASME2 [33]. Recognizing the patterns associated with micro-expression requires the analysis of spatial as well as temporal changes of facial regions. Therefore, the feature descriptor usually has high dimensionality. However, the number of samples available in these databases are quite less. Consequently, the feature space is almost empty [47], and it suffers from the curse of dimensionality. Thus, care must be taken to improve classification performance.

The presence of redundant or noisy features reduces the performance when the sample size is low [48]. In such cases, the feature selection method can be effective to improve the performance of the system. Feature selection methods reduce the dimensionality of the descriptor by selecting the optimal subset of features [49]. Reliable performance for high dimensional data can be achieved with supervised feature selection algorithms with a sufficient number of labeled training data [47]. Ideal selection criteria should evaluate the combinational features to determine the optimal feature subset. The feature selection criteria in the literature [47, 50] either use individual feature based separability or the distance between the points. The feature separability criteria usually use the between and within class variances. Thus, the attributes in a descriptor are evaluated based on the statistical analysis or the performance of each attribute in the classification task. The best performing attributes are then selected for further processing. We demonstrate the use of different feature selection methods to evaluate the goodness of features for encoding micro-expressions.

3 Fuzzy Histogram of Optical Flow Orientations (FHOFO)

The rapid and subtle micro-expressions are difficult to observe in naked eyes. As appreciable facial muscle movements do not happen during micro-expressions, the movement patterns are more important than the magnitude of motion. Hence, the magnitude information can be ignored for feature representation. Moreover, the duration and magnitude of micro-expression vary from sample to sample. Thus, only the accumulated motion directions into different angular histogram bins can be a crucial feature descriptor. As can be understood by the terms, FHOFO extracts an angular histogram out of the motion vectors obtained from optical flow field. It uses only the motion orientations to construct the histogram.

Optical flow is a widely used method in computer vision for estimating the motion in a video. The optical flow vector at each pixel location indicates the pixel level displacement between adjacent frames in terms of direction and magnitude. It assumes the brightness constancy and small continuous motion [51] to estimate the motion between two frames. Assume that v_{j_x} and v_{j_y} are the horizontal and vertical velocity vectors at jth pixel location with intensity I_j. The brightness consistency can be approximated by,

$$\begin{bmatrix} I_{j_x} & I_{j_y} \end{bmatrix} \begin{bmatrix} v_{j_x} \\ v_{j_y} \end{bmatrix} + I_{j_t} = 0 \tag{1}$$

where $\begin{bmatrix} I_{j_x} & I_{j_y} \end{bmatrix}$ represents the spatial gradients and I_{j_t} is the temporal gradient at the jth pixel location. The motion magnitude and orientation pair (ρ_j, θ_j) can be obtained from the optical flow vectors by converting the euclidean to polar coordinates, given by $\left(\rho_j = \sqrt{v_{j_x}^2 + v_{j_y}^2}\right)$ and $(-\pi \leq \theta_j < \pi)$.

FHOFO constructs a histogram out of the optical flow orientations as a global motion descriptor. The conventional way of histogram computation assigns one element to one histogram bin. The overall facial muscle movements for an expression is similar in a broad sense. However, the motion directions of a facial region are not always in the exact direction for similar expressions. Rather, the directions of facial muscle movements vary from person to person. Moreover, the motion directions also change with respect to the camera view angle. Thus, constructing histograms in a conventional way might not produce the desired feature vector. To avoid these limitations, histogram fuzzification assigns some values to all histogram bins based on a

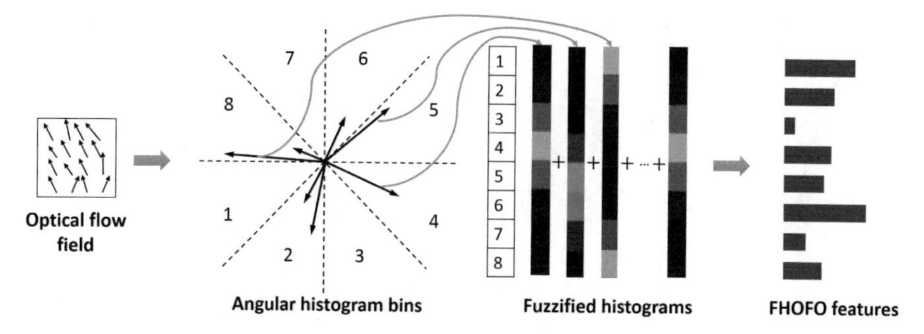

Fig. 3 Computation of FHOFO features

membership function. Thus, each motion vector contributes to multiple angular bins. The illustration of histogram fuzzification is provided in Fig. 3. Using a Gaussian membership function assures that the contribution is high to the closest neighboring bins than the far ones. In addition, it considers the circular continuity of angular histograms at π during fuzzification.

For the computation of FHOFO features, the optical flow vectors with magnitude more than a certain threshold are utilized, which corresponds to the presence of considerable movements on face. Given an image I of size $N \times M$, the FHOFO features, $F(I) = \left[f_1, f_2, \ldots, f_n\right]$, can be computed as

$$f_i = \frac{f_i'}{\sum_{k=1}^{n} f_k'} \tag{2}$$

where

$$f_i' = \sum_{j=1}^{M \times N} \rho_j' \mu_{ij}, \quad \text{and} \quad \rho_j' = \begin{cases} 1, & \text{if } \rho_j > \mathcal{T} \\ 0, & \text{otherwise} \end{cases} \tag{3}$$

Here ρ_j' are the pixel locations where the motion magnitude is greater than threshold $\mathcal{T}$. The contribution of a motion vector at an angle of θ_j to the ith histogram bin is decided by the membership function, given by

$$\mu_{ij} = \exp\left(\frac{-\left(\theta_j - c_i\right)^2}{2\sigma^2}\right) \tag{4}$$

where c_i is the center of ith histogram bin and σ is the variance of the Gaussian membership function. Thus, the significant motion vectors are collected into an angular histogram with soft-assignment of their values into multiple bins. The Gaussian membership function ensures higher contribution to the close histogram bins than the distant ones. Further, the contribution toward different histogram bins can be manipulated using different values of σ. More the value of σ, the motion vectors contribute to more number of bins of the resulting histogram.

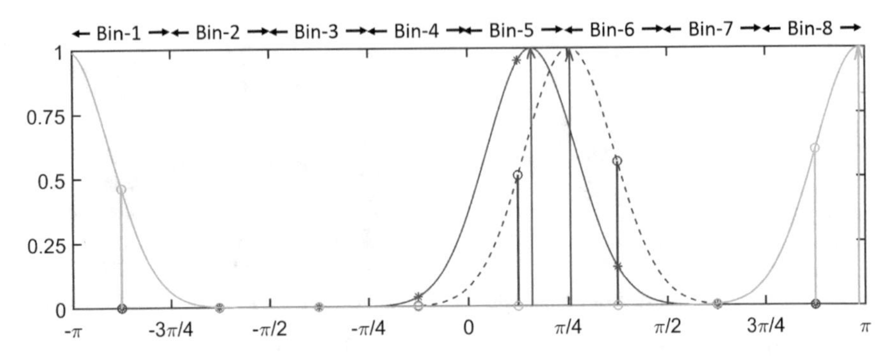

Fig. 4 The Gaussian membership functions used in fuzzification process

Fuzzy histograms are more informative than the traditional histograms, in case of representing motion vectors, which is illustrated in Fig. 4. Consider the instance slightly more than $\pi/4$ with a dotted membership function. This motion vector would be assigned to the 6th bin in case of hard histogram assignment. However, the facial movement directions might change slightly from person to person and with different facial poses. Thus, fuzzy histogram helps to assign the motion directions into multiple bins; thereby close angles contribute to different bins almost equally. The dotted membership function, hence, contributes equally to both bin 5 and 6. Similarly, the movement vector falling in 5th bin contributes to surrounding bins forming a smooth histogram for motion vectors. Moreover, the 8th bin and the 1st bin of the angular histogram are considered to be connected due to circular continuity. The membership function warps at π as $-\pi = \pi$. The contribution of a motion vector in the 8th bin to both 1st and 8th bin is also illustrated in Fig. 4.

As discussed so far, the fuzzy histogram involves the computation of membership values for each motion direction to all the histogram bins. The membership values of each angle to the angular histogram can be computed once and used repeatedly. However, it is impossible to perform for continuous angular values since θ_j can be any real number, and it becomes computationally intensive. Instead, discretization of angular axis is required. Inspired by [52], we first constructed a fine-histogram with a sufficient number of bins. Then the fine histograms were assigned to the coarse ones based on the membership function.

First, the fine histogram was constructed with n' histogram bins in the range $[-\pi, \pi)$. Here n' should be sufficiently large to consider the motion directions in an interval to be the same. The fine-histogram is further mapped to the coarse histogram using the fuzzy membership functions. In this way, the mapping matrix for converting from fine-to-coarse histogram needs to be computed only ones, thereby reducing the computational burden.

This mapping matrix can be represented by M, where the matrix element m_{ij} represents the membership value or the contribution of jth fine histogram bin to the ith coarse angular bin. With n number of coarse histogram bins, we can write,

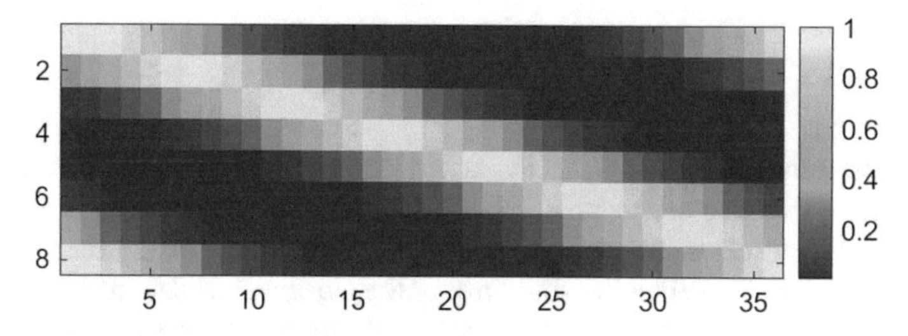

Fig. 5 A membership matrix (M) for $n' = 36$ and $n = 8$

$$F_{n\times 1} = M_{n\times n'} H_{n'\times 1} \tag{5}$$

Figure 5 illustrates a mapping matrix that maps a fine-histogram with 36 bins to a coarse histogram with 8 bins.

In this way, a fuzzy histogram of the motion directions at all pixel locations with motion magnitude above threshold $\mathcal{T}$. The histogram, hence obtained, is further normalized, to sum up to 1. Fuzzy statistics is able to handle the motion vectors during facial expressions better than crisp histograms, thus producing a smooth histogram. Therefore, it improves the facial motion representation and classification performance.

4 Feature Selection

For a classification problem, feature selection aims to select a subset of highly discriminant features that achieves maximal classification accuracy [47]. It removes the redundant features, reduces the risk of over-fitting, and improves the time and space complexity of the data. Thus, the feature selection problem may be formulated as finding m features out of d dimensions that are capable of discriminating the samples of different classes.

Given the samples of two classes, SVM determines the optimal discriminating plane for separating the class samples. Feature selection via concave minimization (FSV) [53] tries to suppress as many components of the normal vector of the separating plane, obtained from SVM, with acceptable performance. Apart from FSV, two widely used filter-based feature selection methods are Fisher Score (FS) and Laplace Score (LS). These methods compute a score for each feature (attribute) of the data and select the best set of attributes based on their individual performance.

Given the data points $x_i \in R^n$ with class labels $y_i \in \{1, 2, \ldots, c\}$, the ith attribute of all the data points can be represented by f^i. FS uses an evaluation criterion which maximizes the between-class variance of the ith attribute (f^i) while minimizing its within-class variance. Thus, the score in FS is given by

$$F(f^i) = \frac{\sum_{k=1}^{c} n_k(\mu_k^i - \mu^i)^2}{\sum_{k=1}^{c} n_k(\sigma_k^i)^2} \tag{6}$$

where μ^i is the overall mean of ith feature, μ_k^i is the ith feature mean of kth class, σ_k^i is the ith feature variance of kth class, c is the number of classes, and n_k is the number of samples of kth class. In contrast, LS uses the similarity of the data points for feature evaluation. Based on the similarity or the closeness of data points, it models the points on a graph structure to reflect the local geometric properties and seeks for the feature set that complies the graph. Thus, it finds the feature set that brings the data points from the same class more closer, while retaining a high global variance. First it computes the similarity score (S_{jk}) between data points x_j and x_k based on any distance measure. Further, LS [54] is calculated by,

$$L(f^i) = \frac{\sum_{j,k}(f^{ij} - f^{ik})^2 S_{jk}}{Var(f^i)} \tag{7}$$

where $Var(\cdot)$ represents the variance. Usually, Euclidean or Mahalanobis distances are used for computing the similarity score for graph construction.

ReliefF is a instance based algorithm which selects instances from similar and dissimilar classes to update the feature weight. It randomly selects an instance (x_l) and determines its two closest instances; one from the same class (near-hit (x_l^+)) and another from dissimilar classes (near-miss (x_l^-)). For L number of random sampling, the score of ith feature is given by

$$R(f^i) = \frac{1}{L}\sum_{l=1}^{m} -(x_l^i - x_l^{+^i})^2 + (x_l^i - x_l^{-^i})^2 \tag{8}$$

Thus, the score of a feature decreases if the feature values are close for similar class samples than the dissimilar class samples.

Pair-wise feature proximity (PWFP) [55] considers a feature to be ‘good’ if the samples of the same class are close along this attribute, while the samples from dissimilar classes are far away. It considers each pair of points to evaluate this property iteratively. The PWFP can be interpreted as sorting the distance between individual attributes of a point pair in ascending order and selecting the first few attributes as the best ones.

$$PWFP(f^i) = \left|\frac{p^i - q^i}{p^i + q^i}\right| \tag{9}$$

Here p^i can be interpreted as the probability of occurrence of ith attribute among the first β number of closest attributes for all possible point pairs. Similarly, q^i can be interpreted as the probability of occurrence of ith feature among the farthest β feature attributes. PWFP has been reported to work well for high dimensional and low sample size data.

5 Micro-expression Recognition Framework

Face registration and its representation is of paramount importance in automated expression recognition. Face registration is the process of aligning faces for appropriate feature extraction. In [56], we demonstrated the superiority of part-based feature extraction to the whole face-based holistic features. A number of other literature [25, 41, 57] also support the idea of extracting features from selective facial areas to improve accuracy. The best way to recognize micro-expressions is by observing the subtle changes around the visually distinctive facial landmarks, such as eyes, mouth corners, etc. In [25], feature extraction was carried out from 36 facial ROIs from all over the face. However, the subtle changes in micro-expression are mostly observable around some of the key facial areas [57]. These ROIs can be located following the occurrence of action units in FACS. Besides, features extracted from multiple facial ROIs are invariant to the slight variations of face poses for near-frontal images. Therefore, we located 36 ROIs based on our previous work [58] for feature extraction. The method proposed in [59] was used for facial landmark detection and the ROIs were determined with reference to these points.

The facial landmark points were detected in the first frame of the video. Assuming negligible head movements during a micro-expression, we cropped the face region from each frame and resized to a resolution of 200×200 pixels. Further, temporal normalization was performed to nullify the effect of variation in micro-expression duration. We used temporal interpolation model (TIM) [32] to interpolate each video sequence into the same number of frames. The face ROIs were extracted from each frame and stacked together to form video cubes. Thus, we obtained 36 video cubes, one for each facial patch, from a micro-expression sequence. The feature descriptors were extracted from each video cube of a sequence and concatenated to represent that sequence. The feature set was further evaluated using different feature selection techniques, and the best subset of features was used for classification.

The summary of the framework used in our work is shown in Fig. 6. First, the face is detected and normalized both spatially and temporally through interpolation. The facial landmarks were detected in the first frame of the sequence, and video cubes were extracted from each patch region. Features were extracted from each facial cubes and concatenated.

6 Experiments and Discussion

6.1 Database Description

CASME database: The CASME database [46] contains 197 micro-expressions clips. Following the methods adopted in [25, 41], they were categorized into positive (9 samples), negative (54 samples), surprise (21 samples) and ‘others’ (113 samples) classes.

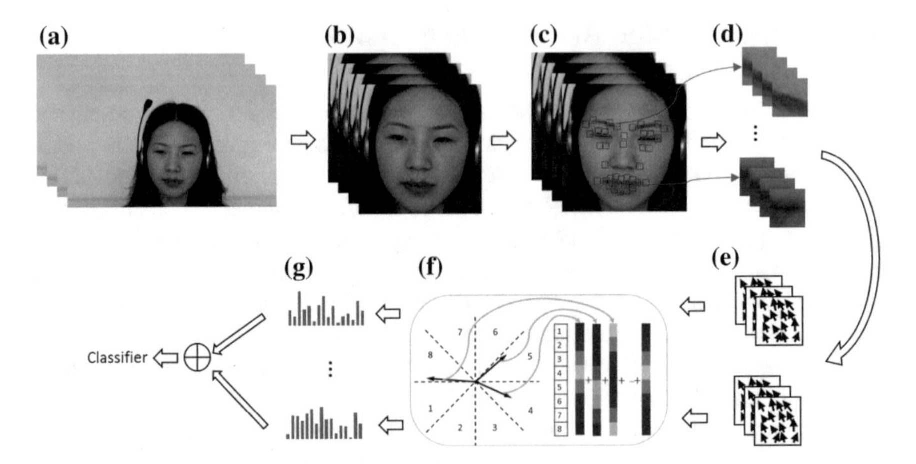

Fig. 6 The block-diagram for micro-expression recognition. **a** Micro-expression frame sequence, **b** Facial region normalization (both spatial and temporal), **c** Localizing active facial patches in the first frame of the sequence, **d** Extracting the video cubes from the face patch regions, **e** Patch-wise optical flow field computation, **f** FHOFO feature extraction from each optical flow field, **g** FHOFO feature representation from all facial patches and concatenation of features

CASME2 database: The CASME 2 database [33] contains 256 micro-expression videos. In our experiments, these micro-expression clips were categorized into positive (32 samples), negative (73 samples), surprise (25 samples), and 'others' (126 samples) classes.

In our experiments, we carried out only micro-expression recognition assuming that the video frames containing micro-expressions are feed to our algorithm. Thus, the video clips starting from onset to offset of micro-expressions are extracted and analyzed for both CASME and CASME2. The number of frames for each video clip is temporally normalized as it varies for each sample.

SMIC Corpus: The high speed (HS) data set is used in our experiments from Spontaneous Micro- expression Corpus (SMIC) [45]. It contains 164 number of micro-expression clips. The class labels included in the database are: positive (51 samples), negative (70 samples) and surprise (43 samples).

6.2 Experimental Protocol and Evaluation Criterion

Data preprocessing is essential for performance improvement. We normalized each micro-expression sequence both spatially and temporally. Some literature suggests the use of different temporal resolution. Moreover, this value varies from database to database. However, a micro-expression recognition system should use the same set of parameters in all cases irrespective of database and duration of sequences. Therefore, we conducted all the experiments using similar parameters for all databases.

Given N number of micro-expression sequences, $D = \{D_1, D_2, \ldots, D_N\}$, we first interpolated each video temporally to a certain number of frames. If the videos are normalized to X frames using TIM, we call it TIM-X. The face was detected, and the ROI locations were computed in the first frame of the normalized sequences. The face ROI was cropped and resized to 200×200 from each frame to form a video cube. This face ROI cube $\mathcal{F}_i$, corresponding to sequence D_i, had a size of $200 \times 200 \times X$. The facial landmarks were detected, and 36 number of active patches were located on the face. Thus, 36 patch cubes $\mathcal{P}_i|_{i=1}^{36}$ were extracted from each $\mathcal{F}_i$. The feature extraction was carried out from the volume cuboids from each ROI location and concatenated afterward.

The uniform LBP-TOP descriptor extracted from one patch cube has the dimension of $59 * 3 = 177$. Thus, the total length of LBP-TOP descriptor for a sequence was $77 \times 36 = 6372$ (36 patches). Similarly, the FHOFO feature computed from the consecutive frames (with TIM-X) has the dimension $(X - 1) * 8$. Thus, the length of FHOFO descriptor was $(X - 1) \times 8 \times 36$. We kept all the parameters used for the ROIs, classifiers, and the feature extraction techniques constant throughout the experiments with all databases. Since the LBP-TOP and FHOFO features possess high dimensionality, different feature selection methods, such as FS, LS, FSV, ReliefF, and PWFP, were used to reduce the dimensionality. The classifier models were trained on the feature vectors obtained from all the training videos. We used three traditional classifiers to access the effectiveness of FHOFO features. We used the K-nearest neighbor classifier (with K = 3), which performs classification based on the closest distance of training samples. Also, linear support vector machine (SVM) and linear discriminant analysis (LDA) were used. The performance of the multi-class classification is reported by using the macro average of F1-scores [60] along with the average accuracy score. The F1-score is a more unbiased performance estimator.

The expression recognition performance can b performed using either leave-one-subject-out (LOSO) or leave-one-video-out (LOVO) cross-validation. In LOSO, one subject's data is considered as test set, whereas the rest data is treated as training data. Thus, this is a subject independent method. Afterwards, this method is repeated for all other subjects, and the average performance is reported. In LOVO, the model is trained with all samples except one, which is treated as the test sample. This process is repeated for all the data. Both LOVO and LOSO cross-validation techniques were performed in our experiments.

6.3 Parameter Selection

As explained earlier, the duration of micro-expression clips in different databases varies a lot. Moreover, the videos, in different databases, are recorded at different frames-per-seconds. All micro-expression clips of a particular class can be assumed to have similar motion patterns, in spite of the varying duration from onset to offset. Therefore, temporal normalized is necessary to bring all the clips to sequences of the same number of frames. To achieve best results, researchers normalized the video

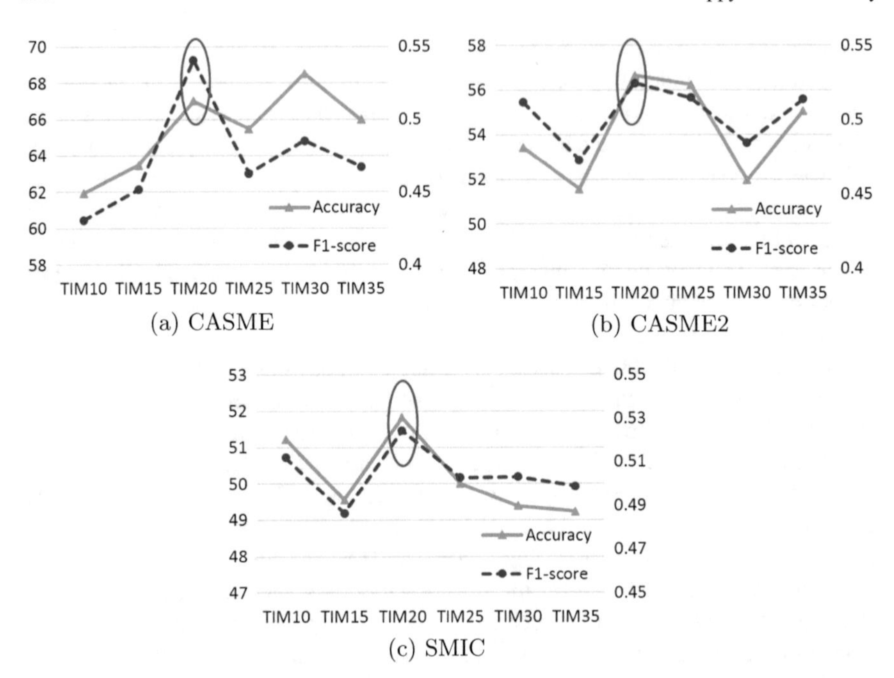

(a) CASME (b) CASME2

(c) SMIC

Fig. 7 Performance of the FHOFO features for different temporal normalization (TIM-X). The primary the secondary vertical axes represent the average recognition accuracy and the F1-score respectively

clips to a different number of frames for different databases. However, we seek for a universal temporal normalization that will achieve good accuracy in most databases. Thus, we performed all the experiments under similar parameter settings. To do so, we first estimated the best set of parameters that would perform well in most cases.

The literature [15, 61–63] suggests normalizing the temporal frame number to 10–20 for best performance in case of SMIC database. However, this number might not be suitable for other databases. Normalizing all videos from different databases to a particular number appears practically appealing for a stable operation. Therefore, we empirically found out the best number for temporal normalization. Figure 7 shows the best accuracies and F1-scores for different TIM-X. It can be observed in Fig. 7 that performance is better with TIM-20 in all databases. Thus, all the video clips of every database were normalized to a temporal length of 20 frames (TIM-20). This indicates that the computers can detect micro-expressions even it is captured in a device with low frame rate, as low as 20 fps. Though it is difficult for us to detect the micro-expressions at a low frame rate, the features used by computers are capable of using a low frame rate video to recognize the micro-expressions. Our claim is supported by the research of Li et al. [15] as they report that the performance does not improve by increasing the temporal resolution. Other literature [32, 62] also support the finding that the facial information captured at low frame rate are adequate for recognizing micro-expressions.

Table 2 The best performance achieved by FHOFO with LOSO cross-validation and TIM-20

	n'	n	σ	Accuracy	F1 score
CASME	48	8	14	67.01	0.5849
CASME2	42	8	12	56.64	0.5248
SMIC	36	6	6	51.83	0.5243

We followed similar procedure and parameter settings in all of our experiments. To this end, we kept all parameters, such as the ROI size, classifiers settings, FHOFO parameters, etc., alike for all databases. The effectiveness of FHOFO depends on the choice of the number of bins in fine and coarse histograms, and the membership function used for fuzzification purpose. Experimentally, we found that the Gaussian membership function is suitable for this task. A grid search was carried out in order to obtain the best set of parameters for FHOFO, i.e., the number of fine and coarse histogram bin (n', n) and the variance of Gaussian membership function (σ).

Table 2 summarizes the best results obtained in different databases for different parameter settings. Linear SVM with LOSO cross-validation was used unless otherwise specified. Empirically, we found that stable performance was obtained with $\sigma = 10$ across the databases. Similarly, the number of fine and coarse histogram bins were decided empirically by varying n' and n for a fixed value of $\sigma = 10$ as shown in Fig. 8. We observed that performance is better for $n' = 36$ for all databases, and lower value of n seemed to fit for our purpose. Thus, we selected $n' = 36$, $n = 8$, $\sigma = 10$, and $\mathcal{T} = 0.0002$ throughout all our experiments with a trade-off of recognition accuracies across the databases.

Table 3 summaries the results obtained with different classifiers and databases for various feature extraction techniques.

6.4 *Performance of FHOFO and Other Methods*

The performance of FHOFO is compared with two widely used feature extraction methods for micro-expression recognition: uniform LBP-TOP and HOOF. Again, the performance of uniform LBP-TOP depends on the parameter selection. Performing a series of experiments, we concluded that the optimal set of parameters for LBP-TOP is the spatial radius of 1 pixel (Rx = Ry = 1) and temporal radius of 3 (Rt = 3). In addition, we observed that the performance of FHOFO is superior to that of the best performance of LBP-TOP with a margin of 10% in CASME2 and 6% in SMIC. Note that the same parameters were used for FHOFO in our experiments for all databases, whereas the parameters were varied for LBP-TOP to obtain the best performance.

Table 3 summaries the micro-expression classification results obtained with LBP-TOP (Rx = Ry = 1 and Rt = 3), HOOF ($n = 8$), and FHOFO ($n' = 36, n = 8$). We can observe that the best performance increased from LBP-TOP to HOOF to

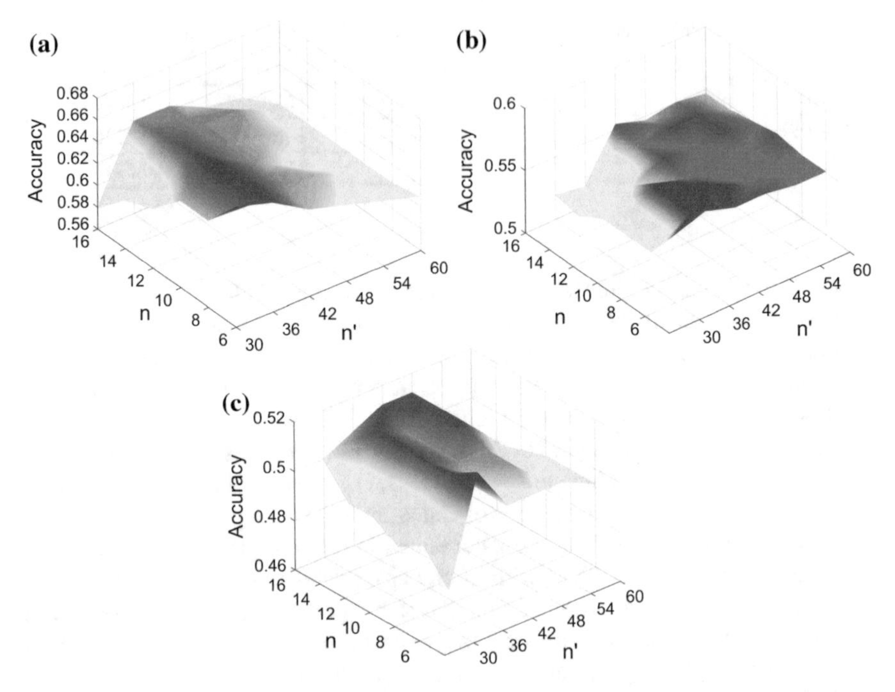

Fig. 8 The performance of FHOFO with varying n' and n (constant $\sigma = 10$) in different databases

FHOFO in two databases: CASME2 and SMIC. However, in CASME database, the performance of LBP-TOP was better than HOOF. However, FHOFO achieved the best performance in CASME with an accuracy of 65.99%. Moreover, the F1 score has improved significantly with FHOFO features. For example, FHOFO achieved an F1 score of 0.54 in CASME, which is better from other features at a high margin. Similar trends can be observed in other databases as well. The performance of linear SVM was found to be consistently well for all databases.

Considering only the linear SVM classifier, we can observe that the performance of FHOFO is superior to that of HOOF and LBP-TOP. The improvement can be observed in both average accuracy and F1 score. For example, in case of CASME database, the accuracy of LBP-TOP (64.46%) and FHOFO (65.99%) are close. However, F1 score has increased from 0.39 to 0.54, which is noteworthy. This can be interpreted as the failure of LBP-TOP in data imbalance scenario. High accuracy with lower F1 score conveys that the class with more samples are more accurately classified compared to the class with less training samples.

Table 4 summaries the performance of FHOFO using LOVO cross-validation. As expected, the performance of LOVO is better than LOSO because of the use of more number of training samples. The system performance reaches up to 71% in CASME and 64% in CASME2 with LOVO cross-validation.

Table 3 LOSO cross-validation results of different feature extraction techniques for different classifiers

(a) Accuracy with LBP-TOP with Rx = Ry = 1 and Rt = 3

Classifiers	CASME		CASME2		SMIC	
	Accuracy (%)	F1 score	Accuracy (%)	F1 score	Accuracy (%)	F1 score
Linear SVM	**64.46**	0.3999	**46.87**	**0.3708**	41.46	0.4258
3 points KNN	61.42	**0.4588**	37.11	0.2731	32.93	0.3401
LDA	58.38	0.3159	44.92	0.3278	**43.90**	**0.4530**

(b) Accuracy with HOOF

Classifiers	CASME		CASME2		SMIC	
	Accuracy (%)	F1 score	Accuracy (%)	F1 score	Accuracy (%)	F1 score
Linear SVM	**57.87**	**0.3654**	51.17	**0.4781**	**47.56**	**0.4852**
3 points KNN	41.62	0.2188	**52.34**	0.4344	35.57	0.3810
LDA	56.35	0.3306	45.70	0.4136	40.85	0.4076

(c) Accuracy with FHOFO

Classifiers	CASME		CASME2		SMIC	
	Accuracy (%)	F1 score	Accuracy (%)	F1 score	Accuracy (%)	F1 score
Linear SVM	**65.99**	**0.5409**	**55.86**	**0.5197**	**51.22**	**0.5182**
3 points KNN	53.30	0.3663	50.39	0.4088	42.68	0.4427
LDA	56.85	0.3787	49.22	0.4426	48.17	0.4808

Table 4 LOVO cross-validation results with FHOFO features

Classifiers	CASME		CASME2		SMIC	
	Accuracy (%)	F1 score	Accuracy (%)	F1 score	Accuracy (%)	F1 score
Linear SVM	**71.57**	**0.5924**	**64.06**	**0.6025**	56.10	0.5536
3 points KNN	57.36	0.4516	55.08	0.4716	50.61	0.5301
LDA	68.02	0.4849	62.50	0.5844	**59.15**	**0.5807**

6.5 Performance of Feature Selection Methods

LBP-TOP with the parameter set Rx = Ry = 1 and Rt = 3 performed the best [58] for all databases when using TIM-20. Similarly, FHOFO performed well with parameters $n' = 36$, $n = 8$, and $\sigma = 10$. Therefore, we used the same set of parameters in our experiments for respective feature extraction techniques. The recognition accuracies for different classifiers are shown in Figs. 9, 10, and 11 for CASME, CASME2 and SMIC databases respectively. It shows the performances of SVM, KNN, and LDA classifiers on the raw feature set compared to the performance of feature selection techniques. In case of feature selection methods, we selected 3000 number of features for all databases and used linear SVM for classification purpose.

As can be observed from Fig. 9, the performance of SVM was better than KNN and LDA. However, the performance improved with the use of feature selection techniques, such as PWFP, LS, FSV, and ReliefF. In CASME, we observed that the LBP-TOP features performed better than FHOFO for LDA and KNN classifiers. However, the performance was improved by SVM. In addition, the use of feature selection methods, such as PWFP and ReliefF, further elevated the recognition accuracy.

It can be observed in Fig. 10 that the performance of FHOFO was better than LBP-TOP in all situations. The performance was further improved with PWFP feature selection. However, the other feature selection methods performed inferior to the raw FHOFO descriptor. The loss of information during the feature selection methods might be the reason behind the performance deterioration. Similar trends can also be observed in case of SMIC database (Fig. 11).

As observed from Figs. 9, 10, and 11, the PWFP feature selection method achieved significant performance among others. Similarly, the overall performance of FHOFO was better than LBP-TOP. Therefore, the accuracies of a different number of selected features are reported to strengthen the proposal.

As can be seen in Fig. 12, the performance of PWFP on FHOFO features is stable for a different number of selected features and across different databases as well. Although the performances of other methods were close to PWFP in Fig. 12a, it still had an advantage at higher number of features. Similarly, FSV leaded in CASME2

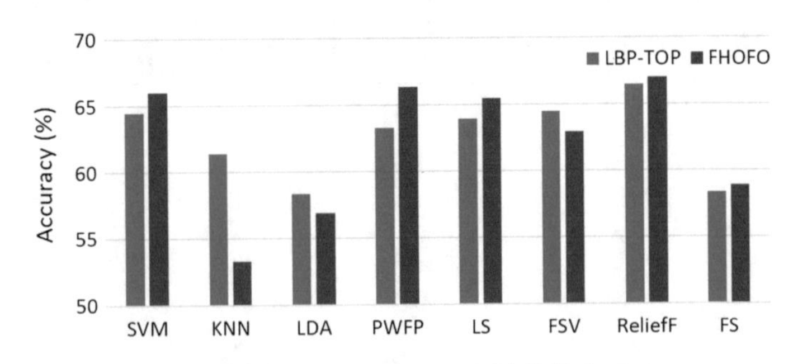

Fig. 9 The recognition accuracy of various classifiers in CASME database

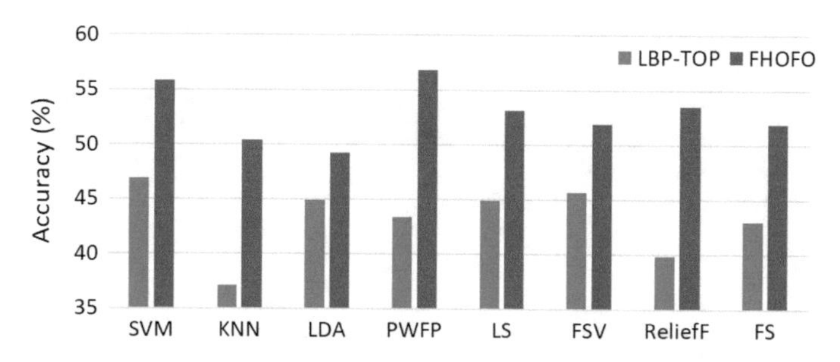

Fig. 10 The recognition accuracy of various classifiers in CASME2 database

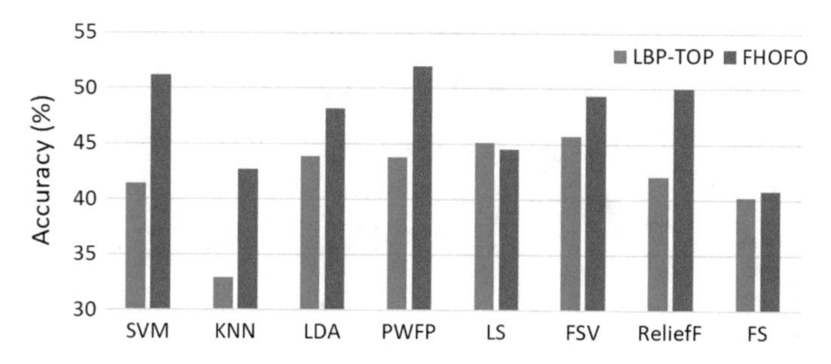

Fig. 11 The recognition accuracy of various classifiers in SMIC database

database at low number of features (Fig. 12b). However, the number of features to be selected should not be changed in practical scenarios. Thus, a feature selection method which works well at different number of selected features is desirable. Comparing with ReliefF, PWFP increased the performance by 4% on an average. Although LS and ReliefF performed well in CASME, the performance of PWFP is stable for all number of selected features.

The performance of different feature selection methods converged in Fig. 12c (for SMIC) when more number of features were selected. However, the accuracy gradually decreased. Thus, we can infer using more number of features increased the noise level. Besides, it is clear from Fig. 12c that FS performed the worst in SMIC. FSV achieved the best performance when the number of features was minimal. However, PWFP provided stable performance throughout.

6.6 Comparison with the State-of-Art Methods

In literature, experiments were carried out with a different number of samples and different numbers of expression categories as well. Moreover, the normalization methods, classifiers, and cross-validation methods vary from literature to literature. Therefore, there is no easy way of comparing our results with literature. The results

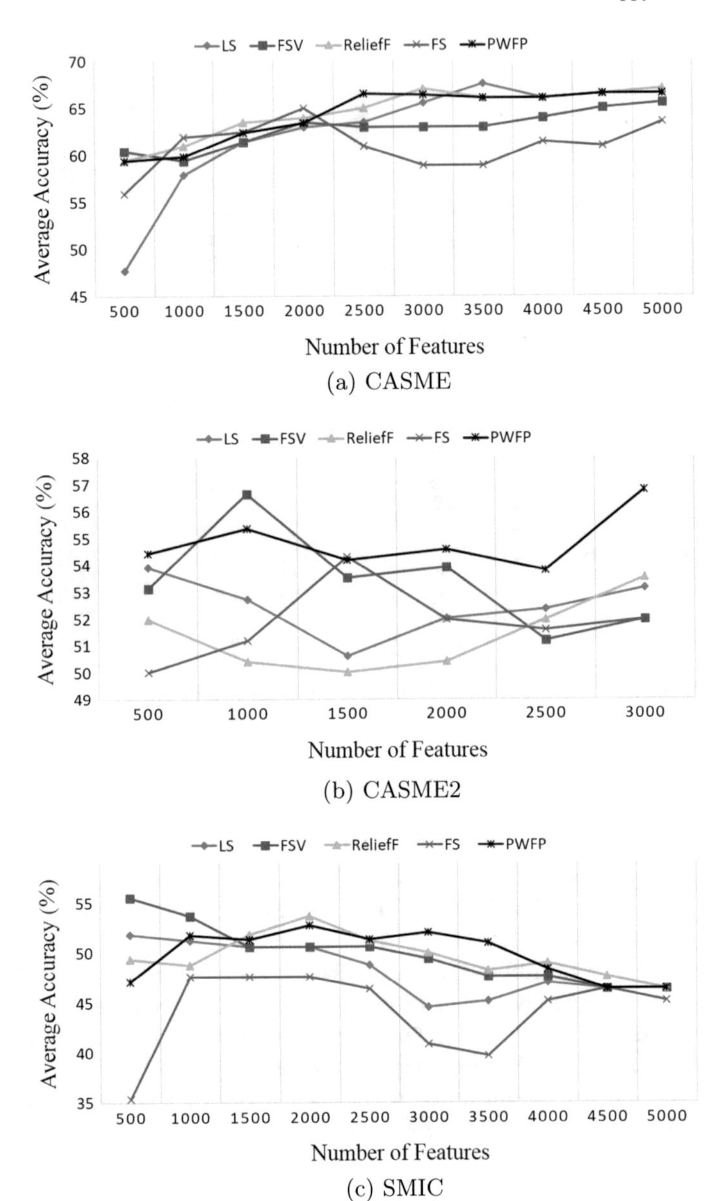

Fig. 12 The performance of FHOFO descriptors in different databases with varying number of selected features

Table 5 Comparison to state-of-the-art in micro-expression recognition

	CASME	CASME2	SMIC
Wang et al. [64]	41.2	38.4	N/A
Yan et al. [65]	61.88	N/A	N/A
Li et al. [45]	N/A	N/A	48.78
Liong et al. [63]	N/A	61.54	52.44
Monogenic Riesz wavelet [66]	N/A	46.15	N/A
OSW-LBP-TOP [24]	N/A	N/A	56.09
Eulerian motion magnification [17]	N/A	51	N/A
TICS [41]	61.86	62.3	N/A
Wang et al. [67]	N/A	45.75	55.49
Transfer learning [42]	N/A	47.3	53.6
STCLQP [27]	57.31	58.39	64.02
FDM [26]	56.14	45.93	54.88
FHOFO	67.01	56.64	51.83
FHOFO with PWFP	67.24	56.78	52.01

reported in of some literature are provided in Table 5 and their performances are compared with the proposed method.

Wang et al. [64] used five expression classes and obtained an accuracy close to 40% in both CASME and CASME2. However, their performance with 4 class classification is not reported. Similarly, Huang et al. [27] reported 58.39% accuracy in CASME2 with five class classification, where as 57.31% accuracy in CASME with four class classification. However, we have considered four classes in our experiments, thus, comparing the performance of our system directly with these methods is not appropriate. Instead we can notice that FHOFO method achieved an accuracy of 67.01% in CASME and 56.64% in CASME2, which are on an average close or better than other methods. This confirms that FHOFO extracts suitable discriminative features from video clips for micro-expression classification.

From Table 5, we can observe that the performance of FHOFO with PWFP is more than 67% in CASME, which is superior to the performance of other methods. TICS [41] achieved 61.86 and 62.3% in CASME and CASME2 respectively. Though its performance was better in CASME2, it underperformed for CASME. In [63], Optical strain features were used along with the weighted version of the same, which achieved an accuracy of 52.44% in SMIC and 61.54% in CASME2. Wang et al. [67] used LBP six intersection points and LBP three mean orthogonal planes for micro-expression recognition. They achieved an accuracy of 45.75 and 55.49% in CASME2 and SMIC respectively. Similarly, FDM [26] characterized the micro movements and performed well in all datasets. Monogenic Riesz wavelet representation [66] with SVM achieved an accuracy of 46.15% in CASME2. Optical strain weighted LBP-TOP (OSW-LBP-TOP) features were proposed in [24] which improved the recognition accuracy to 56.09% in SMIC. As can be observed, the methods in literature achieved good accuracy in one database while compromising the accuracy in another. However, the proposed method achieved above average accuracy in all databases.

Compared to other literature, STCLQP [27] performed the best in SMIC database. On the other hand, Optical strain features in [63] performed the best in CASME2. Though comparing performances directly is very difficult due to various experimental settings, the proposed method outperformed in CASME database and its performance is also significant in other databases as well. FHOFO obtained an accuracy of 67.24% in CASME and 56.78% in CASME2 with PWFP feature selection technique. Moreover, the classification accuracy can be improved by further tuning the classifier parameters. Choosing suitable normalization parameters and FHOFO parameters may further enhance the performance. However, here the idea is to prove the consistency of the system performance across the databases. Therefore, we kept the parameters constant throughout all the experiments with a linear classifier. Thus, the proposed combination of features followed by the feature selection methods proved to be a suitable for micro-expression recognition.

7 Conclusion

Recognizing micro-expressions is challenging due to its rapid and subtle nature. Information from temporal dimension plays a significant role in classifying micro-expressions. The patterns associated with facial muscle movements can be incorporated into the feature representation to improve system performance. The FHOFO feature explained here, is a global motion direction descriptor, constructs an angular histogram out of the optical field flow orientations. It is similar to HOOF features with the soft assignment of weights into neighboring histogram bins. Since the change of facial appearance is negligible during a micro-expression, hence, FHOFO ignores the spatial changes. We also discussed the improvement in performance with the use of feature selection methods. Some feature selection methods performed better compared to KNN and LDA. However, their performance is inferior to the performance of SVM alone, which can be inferred as a loss of information. The use of PWFP with SVM appears to be beneficial in classifying micro-expressions.

The work presented here uses the manual annotation data for on-set and off-set frames to segment the micro-expressions. However, spotting of micro-expression is still a challenging task. Manual annotation of onset, apex and offset frames, during the creation of a database, is time-consuming and subjective in nature. Besides, the databases created in laboratory environments are constrained in terms of facial pose, illumination variation, self-occlusion, and data acquisition quality. Lack of sufficient samples with wide variation is a major obstacle in the evaluation of different feature extraction techniques. Development of a system for fast and reliable in-the-wild performance is a long-term challenge.

Researchers reported different feature extraction techniques for classifying micro-expressions. However, a comprehensive analysis of different feature extraction techniques considering all configurations can be beneficial to figure out the strong and weak points with different features. Feature-level or decision-level fusion of different feature extraction techniques is another open area of research. The duration of micro-expressions varies from sample to sample. Several interpolation techniques are currently being used to normalize the data both spatially and temporally. Systems with ability to analyze multi-resolution images are also desirable. The feature dimensionality usually becomes high to encode the subtle changes. In such cases, the feature selection and dimensionality reduction methods can be investigated extensively to improve the performance of the system.

The ultimate aim of affective computing is to accurately estimate the human affects. Imagine a classroom situation where a student can be under any of the following affective states: frustrated, depressed, angry, bored or confused. All such states can be identified by interpreting speech, facial expressions, gestures, and other physiological signals. The future of affective computing is to recognize all possible human affects from multimodal cues. The integration of affective technologies with consumer level products is still in its infancy. Development of robust algorithms for representation and assimilation of information from multiple cues holds the key to making the science fiction of today, the reality of tomorrow.

Acknowledgments The authors would like to thank Ministry of Human Resource Development, Govt. of India for funding this research.

References

1. R.W. Picard, R. Picard, *Affective Computing*, vol. 252 (MIT Press, Cambridge, 1997)
2. P. Ekman, *Emotions Revealed: Recognizing Faces and Feelings to Improve Communication and Emotional Life* (Times Books, New York, 2003)
3. P. Ekman, Lie catching and microexpressions, *The Philosophy of Deception* (2009), pp. 118–133
4. W.-J. Yan, Q. Wu, J. Liang, Y.-H. Chen, X. Fu, How fast are the leaked facial expressions: the duration of micro-expressions. J. Nonverbal Behavior **37**(4), 217–230 (2013)
5. M. Frank, M. Herbasz, K. Sinuk, A. Keller, C. Nolan, I see how you feel: training laypeople and professionals to recognize fleeting emotions, in *The Annual Meeting of the International Communication Association, New York City* (2009)
6. E. Sariyanidi, H. Gunes, A. Cavallaro, Automatic analysis of facial affect: a survey of registration, representation, and recognition. IEEE Trans. Pattern Anal. Mach. Intell. **37**(6), 1113–1133 (2015)
7. Y. Tian, T. Kanade, J.F. Cohn, Facial expression recognition, *Handbook of Face Recognition* (Springer, Berlin, 2011), pp. 487–519
8. M. Valstar, M. Pantic, Combined support vector machines and hidden Markov models for modeling facial action temporal dynamics, *Human–Computer Interaction* (2007), pp. 118–127
9. S.-T. Liong, J. See, K. Wong, R.C.-W. Phan, Automatic micro-expression recognition from long video using a single spotted apex, in *Asian Conference on Computer Vision* (Springer, 2016), pp. 345–360

10. M. Shreve, J. Brizzi, S. Fefilatyev, T. Luguev, D. Goldgof, S. Sarkar, Automatic expression spotting in videos. Image Vis. Comput. **32**(8), 476–486 (2014)
11. D. Patel, G. Zhao, M. Pietikäinen, Spatiotemporal integration of optical flow vectors for micro-expression detection, in *International Conference on Advanced Concepts for Intelligent Vision Systems* (Springer, 2015), pp. 369–380
12. S.-J. Wang, S. Wu, X. Qian, J. Li, X. Fu, A main directional maximal difference analysis for spotting facial movements from long-term videos. Neurocomputing **230**, 382–389 (2017)
13. Z. Xia, X. Feng, J. Peng, X. Peng, G. Zhao, Spontaneous micro-expression spotting via geometric deformation modeling. Comput. Vis. Image Underst. **147**, 87–94 (2015)
14. S. Polikovsky, Y. Kameda, Facial micro-expression detection in hi-speed video based on facial action coding system (FACS). IEICE Trans. Inf. Syst. **96**(1), 81–92 (2013)
15. X. Li, H. Xiaopeng, A. Moilanen, X. Huang, T. Pfister, G. Zhao, M. Pietikainen, Towards reading hidden emotions: a comparative study of spontaneous micro-expression spotting and recognition methods. IEEE Trans. Affect. Comput. **8**, 29–42 (2017)
16. A.K. Davison, M.H. Yap, C. Lansley, Micro-facial movement detection using individualised baselines and histogram-based descriptors, in *IEEE International Conference on Systems, Man, and Cybernetics (SMC)* (2015), pp. 1864–1869
17. A.C. Le Ngo, Y.-H. Oh, R.C.-W. Phan, J. See, Eulerian emotion magnification for subtle expression recognition, in *IEEE International Conference on Acoustics, Speech and Signal Processing (ICASSP)* (2016), pp. 1243–1247
18. K. Sumi, T. Ueda, Micro-expression recognition for detecting human emotional changes, in *International Conference on Human-Computer Interaction* (Springer, 2016), pp. 60–70
19. H. Zheng, Micro-expression recognition based on 2D Gabor filter and sparse representation. J. Phys.: Conf. Ser. **787**(1), 012013 (2017) (IOP Publishing)
20. Q. Wu, X. Shen, X. Fu, The machine knows what you are hiding: an automatic micro-expression recognition system, in *Affective Computing and Intelligent Interaction* (2011), pp. 152–162
21. M. Owayjan, A. Kashour, N. Al Haddad, M. Fadel, G. Al Souki, The design and development of a lie detection system using facial micro-expressions, in *2012 2nd International Conference on Advances in Computational Tools for Engineering Applications (ACTEA)* (IEEE, 2012), pp. 33–38
22. S. Yao, N. He, H. Zhang, O. Yoshie, Micro-expression recognition by feature points tracking, in *2014 10th International Conference on Communications (COMM)* (IEEE, 2014), pp. 1–4
23. M. Shreve, S. Godavarthy, D. Goldgof, S. Sarkar, Macro-and micro-expression spotting in long videos using spatio-temporal strain, in *2011 IEEE International Conference on Automatic Face and Gesture Recognition and Workshops (FG 2011)* (IEEE, 2011), pp. 51–56
24. S.-T. Liong, J. See, R.C.-W. Phan, A.C. Le Ngo, Y.-H. Oh, K. Wong, Subtle expression recognition using optical strain weighted features, in *Asian Conference on Computer Vision* (Springer, 2014), pp. 644–657
25. Y.-J. Liu, J.-K. Zhang, W.-J. Yan, S.-J. Wang, G. Zhao, X. Fu, A main directional mean optical flow feature for spontaneous micro-expression recognition. IEEE Trans. Affect. Comput. **7**, 299–310 (2016). https://doi.org/10.1109/TAFFC.2015.2485205
26. F. Xu, J. Zhang, J.Z. Wang, Microexpression identification and categorization using a facial dynamics map. IEEE Trans. Affect. Comput. **8**(2), 254–267 (2017)
27. X. Huang, G. Zhao, X. Hong, W. Zheng, M. Pietikäinen, Spontaneous facial micro-expression analysis using spatiotemporal completed local quantized patterns. Neurocomputing **175**, 564–578 (2016)
28. S.-J. Wang, W.-J. Yan, G. Zhao, X. Fu, C.-G. Zhou, Micro-expression recognition using robust principal component analysis and local spatiotemporal directional features, in *Computer Vision-ECCV Workshops* (2014), pp. 325–338
29. G. Zhao, M. Pietikainen, Dynamic texture recognition using local binary patterns with an application to facial expressions. IEEE Trans. Pattern Anal. Mach. Intell. **29**(6), 915–928 (2007)
30. T. Ojala, M. Pietikainen, T. Maenpaa, Multiresolution gray-scale and rotation invariant texture classification with local binary patterns. IEEE Trans. Pattern Anal. Mach. Intell. **24**(7), 971–987 (2002)

31. J. He, J.-F. Hu, X. Lu, W.-S. Zheng, Multi-task mid-level feature learning for micro-expression recognition. Pattern Recognit. **66**, 44–52 (2017)
32. T. Pfister, X. Li, G. Zhao, M. Pietikäinen, Recognising spontaneous facial micro-expressions, in *IEEE International Conference on Computer Vision (ICCV)* (2011), pp. 1449–1456
33. W.-J. Yan, X. Li, S.-J. Wang, G. Zhao, Y.-J. Liu, Y.-H. Chen, X. Fu, CASME II: an improved spontaneous micro-expression database and the baseline evaluation. PloS one **9**(1), e86041 (2014)
34. B. Jiang, M. Valstar, B. Martinez, M. Pantic, A dynamic appearance descriptor approach to facial actions temporal modeling. IEEE Trans. Cybern. **44**(2), 161–174 (2014)
35. X. Fan, T. Tjahjadi, A spatial-temporal framework based on histogram of gradients and optical flow for facial expression recognition in video sequences. Pattern Recognit. **48**(11), 3407–3416 (2015)
36. X. Huang, G. Zhao, W. Zheng, M. Pietikäinen, Spatiotemporal local monogenic binary patterns for facial expression recognition. IEEE Signal Process. Lett. **19**(5), 243–246 (2012)
37. S. Polikovsky, Y. Kameda, Y. Ohta, Facial micro-expression detection in hi-speed video based on facial action coding system (FACS). IEICE Trans. Inf. Syst. **96**(1), 81–92 (2013)
38. Y. Guo, C. Xue, Y. Wang, M. Yu, Micro-expression recognition based on CBP-TOP feature with ELM. Optik-Int. J. Light Electron Opt. **126**(23), 4446–4451 (2015)
39. X. Ben, X. Jia, R. Yan, X. Zhang, W. Meng, Learning effective binary descriptors for micro-expression recognition transferred by macro-information. Pattern Recognit. Lett. (2017)
40. S.-J. Wang, H.-L. Chen, W.-J. Yan, Y.-H. Chen, X. Fu, Face recognition and micro-expression recognition based on discriminant tensor subspace analysis plus extreme learning machine. Neural Process. Lett. **39**(1), 25–43 (2014)
41. S.-J. Wang, W.-J. Yan, X. Li, G. Zhao, C.-G. Zhou, X. Fu, M. Yang, J. Tao, Micro-expression recognition using color spaces. IEEE Trans. Image Process. **24**(12), 6034–6047 (2015)
42. D. Patel, X. Hong, G. Zhao, Selective deep features for micro-expression recognition, in *2016 23rd International Conference on Pattern Recognition (ICPR)* (IEEE, 2016), pp. 2258–2263
43. D.H. Kim, W. Baddar, J. Jang, Y.M. Ro, Multi-objective based spatio-temporal feature representation learning robust to expression intensity variations for facial expression recognition. IEEE Trans. Affect. Comput. (2017)
44. R. Chaudhry, A. Ravichandran, G. Hager, R. Vidal, Histograms of oriented optical flow and Binet-Cauchy kernels on nonlinear dynamical systems for the recognition of human actions, in *IEEE Conference on Computer Vision and Pattern Recognition* (2009), pp. 1932–1939
45. X. Li, T. Pfister, X. Huang, G. Zhao, M. Pietikainen, A spontaneous micro-expression database: inducement, collection and baseline, in *IEEE International Conference and Workshops on Automatic Face and Gesture Recognition (FG)* (2013), pp. 1–6
46. W.-J. Yan, Q. Wu, Y.-J. Liu, S.-J. Wang, X. Fu, CASME database: a dataset of spontaneous micro-expressions collected from neutralized faces, in *IEEE International Conference and Workshops on Automatic Face and Gesture Recognition (FG)* (2013), pp. 1–7
47. J. Tang, S. Alelyani, H. Liu, Feature selection for classification: a review, *Data Classification: Algorithms and Applications* (2014), p. 37
48. Z. Zhao, L. Wang, H. Liu, Efficient spectral feature selection with minimum redundancy, in *24th AAAI Conference on Artificial Intelligence* (2010)
49. L. Yu, H. Liu, Efficient feature selection via analysis of relevance and redundancy. J. Mach. Learn. Res. **5**, 1205–1224 (2004)
50. L. Yin, Y. Ge, K. Xiao, X. Wang, X. Quan, Feature selection for high-dimensional imbalanced data. Neurocomputing **105**, 3–11 (2013)
51. B.D. Lucas, T. Kanade, An iterative image registration technique with an application to stereo vision, in *International Joint Conference on Artificial Intelligence*, vol. 81 (1981), pp. 674–679
52. J. Han, K.-K. Ma, Fuzzy color histogram and its use in color image retrieval. IEEE Trans. Image Process. **11**(8), 944–952 (2002)
53. P.S. Bradley, O.L. Mangasarian, Feature selection via concave minimization and support vector machines, in *International Conference on Machine Learning (ICML)*, vol. 98 (1998), pp. 82–90
54. X. He, D. Cai, P. Niyogi, Laplacian score for feature selection, in *NIPS*, vol. 186 (2005), p. 189

55. S.L. Happy, R. Mohanty, A. Routray, An effective feature selection method based on pair-wise feature proximity for high dimensional low sample size data, in *European Signal Processing Conference (EUSIPCO)* (2017)
56. S.L. Happy, A. Routray, Automatic facial expression recognition using features of salient facial patches. IEEE Trans. Affect. Comput. **6**(1), 1–12 (2015)
57. M. Shreve, S. Godavarthy, D. Goldgof, S. Sarkar, Macro-and micro-expression spotting in long videos using spatio-temporal strain, in *IEEE International Conference on Automatic Face and Gesture Recognition and Workshops* (2011), pp. 51–56
58. S.L. Happy, A. Routray, Fuzzy histogram of optical flow orientations for micro-expression recognition. IEEE Trans. Affect. Comput. (2017)
59. A. Asthana, S. Zafeiriou, G. Tzimiropoulos, S. Cheng, M. Pantic, From pixels to response maps: discriminative image filtering for face alignment in the wild. IEEE Trans. Pattern Anal. Mach. Intell. **37**(6), 1312–1320 (2015)
60. M. Sokolova, G. Lapalme, A systematic analysis of performance measures for classification tasks. Inf. Process. Manag. **45**(4), 427–437 (2009)
61. A.C. Le Ngo, R.C.-W. Phan, J. See, Spontaneous subtle expression recognition: imbalanced databases and solutions, in *Asian Conference on Computer Vision* (Springer, 2014), pp. 33–48
62. X. Huang, S.-J. Wang, G. Zhao, M. Piteikainen, Facial micro-expression recognition using spatiotemporal local binary pattern with integral projection, in *IEEE International Conference on Computer Vision (ICCV) Workshops* (2015), pp. 1–9
63. S.-T. Liong, J. See, R.C.-W. Phan, Y.-H. Oh, A.C. Le Ngo, K. Wong, S.-W. Tan, Spontaneous subtle expression detection and recognition based on facial strain. Signal Process.: Image Commun. **47**, 170–182 (2016)
64. S.-J. Wang, W.-J. Yan, T. Sun, G. Zhao, X. Fu, Sparse tensor canonical correlation analysis for micro-expression recognition. Neurocomputing **214**, 218–232 (2016)
65. W.-J. Yan, S.-J. Wang, Y.-J. Liu, Q. Wu, X. Fu, For micro-expression recognition: database and suggestions. Neurocomputing **136**, 82–87 (2014)
66. Y.-H. Oh, A.C. Le Ngo, J. See, S.-T. Liong, R.C.-W. Phan, H.-C. Ling, Monogenic Riesz wavelet representation for micro-expression recognition, in *IEEE International Conference on Digital Signal Processing (DSP)* (IEEE, 2015), pp. 1237–1241
67. Y. Wang, J. See, R.C.-W. Phan, Y.-H. Oh, Efficient spatio-temporal local binary patterns for spontaneous facial micro-expression recognition. PloS one **10**(5), e0124674 (2015)

Improved Deep Neural Network Object Tracking System for Applications in Home Robotics

Berat A. Erol, Abhijit Majumdar, Jonathan Lwowski, Patrick Benavidez, Paul Rad and Mo Jamshidi

Abstract Robotic navigation in GPS-denied environments requires case specific approaches for controlling a mobile robot to any desired destinations. In general, a nominal path is created in an environment described by a set of distinct objects, in other words such obstacles and landmarks. Intelligent voice assistants or digital assistance devices are increasing their importance in today's smart home. Especially, by the help of fast-growing Internet of Things (IoT) applications. These devices are amassing an ever-growing list of features such as controlling states of connected smart devices, recording tasks, and responding to queries. Assistive robots are the perfect complement to smart voice assistants for providing physical manipulation. A request made by a person can be assigned to the assistive robot by the voice assistant. In this chapter, a new approach for autonomous navigation is presented using pattern recognition and machine learning techniques such as Convolutional Neural Networks to identify markers or objects from images and videos. Computational

B. A. Erol (✉) · A. Majumdar · J. Lwowski · P. Benavidez (✉) · M. Jamshidi
Autonomous Control Engineering (ACE) Laboratories, Department of Electrical and Computer Engineering, The University of Texas at San Antonio, One UTSA Circle, San Antonio, TX 78249, USA
e-mail: Berat.Erol@utsa.edu

P. Benavidez
e-mail: Patrick.Benavidez@utsa.edu; p_b_2003@hotmail.com

A. Majumdar
e-mail: abhijit.g.majumdar@gmail.com

J. Lwowski
e-mail: Jonathan.Lwowski@gmail.com

M. Jamshidi
e-mail: Mo.Jamshidi@utsa.edu

B. A. Erol · P. Benavidez · P. Rad
Open Cloud Institute, The University of Texas at San Antonio, San Antonio, TX 78249, USA
e-mail: Peyman.Najafirad@utsa.edu

P. Rad
Department of Information Systems & Cyber Security, The University of Texas at San Antonio, One UTSA Circle, San Antonio, TX 78249, USA

W. Pedrycz and S.-M. Chen (eds.), *Computational Intelligence for Pattern Recognition*, Studies in Computational Intelligence 777,
https://doi.org/10.1007/978-3-319-89629-8_14

intelligence techniques are implemented along with Robot Operating System and object positioning to navigate towards these objects and markers by using RGB-depth camera. Multiple potential matching objects detected by the robot with deep neural network object detectors will be displayed on a screen installed on the assistive robot to improve and evaluate Human-Robot Interaction (HRI).

Keywords Neural network · Computational intelligence · Simultaneous localization and mapping (SLAM) · Multi object tracking · Deep convolutional neural network (DCNN) · Depth camera · Autonomous navigation · Human robot interaction · Human computer interface · Machine learning · GPS denied environment · Real time implementation

1 Introduction

Improvements on computational intelligence in parallel with assistive robotics, and reinforced applications for object identification engines based on visual sensory readings from RGB-D cameras have increased the accuracy of cooperative task assignments in robotics. Moreover, by implementing pattern recognition fundamentals for object classification, object detection, and computer vision, their impacts on human robot interactions are becoming more crucial than ever. A visual representation of the various fields that pattern recognition has played a major contribution along with their established applications can be seen in Fig. 1. Pattern recognition has especially contributed to applications of intelligent voice assistants and Internet of Things devices, such as Amazon's Echo platforms and Google Home products. This has led to today's smart home applications to decrease the difficult necessity of requiring high computational power, and still can handle the task scheduling requirements easily. These devices are amassing an ever-growing list of features such as controlling states of connected smart devices, playing music, managing alarms, recording tasks, and responding to queries. Potential industrial applications of smart digital assistants are numerous; however, applications are limited due to the digital-only nature of the device. On the other hand, more intelligent assistive robots with higher computational power can loosen this constraint and even make another contribution by providing physical manipulation. A request made by a person can be assigned to the assistive robot by the digital assistant; then, the robot performs its duties while keeping the person updated by verbal feedback. Most of the time the problem lies in this stage and raises a question. Did the robot understand the request correctly and do what the user asked them to do in the right way?

Human-Computer Interaction (HCI) focuses on finding answers for how to efficiently design computers with the latest technology that will provide better user interfaces to make users feel more comfortable with them. As its vision states, HCI interests not only human users and their benefits from a system of computers, but also it investigates the way to use integrated hardware and software platforms. The

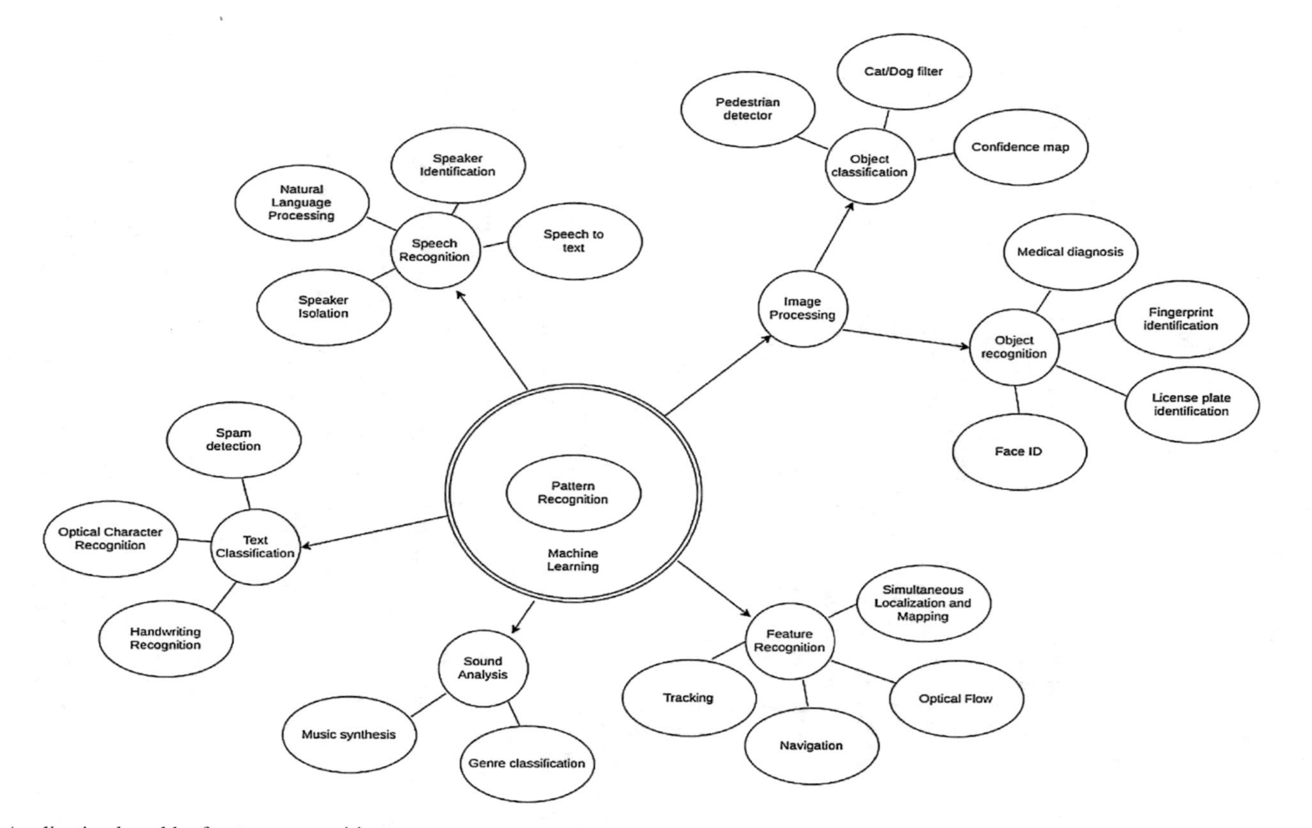

Fig. 1 Application breadth of pattern recognition

goal is to build an interactive relationship between human and machines that are controlled and observed by computers.

Humanoid robots are a well-studied class of robots with the greatest potential in the future to assist with activities of advanced work tasks, and even in a household setting. They are built to perform tasks in a manner that closely follows human form and functionality. Humanoid robots can observe, interact with, and mimic humans. Many have capability to recognize voices, faces, geometric shapes, objects, tools, and environments. They are equipped with very capable sensors, controllers and processing power. Two of the main roadblocks to incorporating humanoid robots are the complexity of robot dynamics and high costs of system components. Hybrids of humanoids and unmanned ground vehicle robots are a popular solution to reduce complexity and the cost of such robotic system. Human-robot interaction (HRI) has increased its importance over the last decades due to the desired conclusion of industrial laws that required higher efficiency and increased productivity. Collaborative working of humans and robots has led to the development of state-of-the-art applications and interfaces for multiple purposes, i.e. robotic arms for the automobile manufacturing lines, unmanned ground vehicles, artificial body parts, and robotic platforms for surgical operations. Since recent developments in the field make robotic systems much more reliable and resilient to changes in the environment, the use of pattern recognition, computational intelligence as a sensory feedback mechanism for a HRI system has become more essential.

Intelligent voice assistants are the central hubs of smart home technologies and are able to perform a variety of tasks. These devices are becoming a must-have for any smart environment applications by featuring list of functionalities, such as controlling the smart thermostats, following a scheduled works, managing smart home observation systems or alarms, and most importantly responding to verbal user queries. What they cannot do is control objects that are not "smart devices". Home-based assistive robots are the perfect candidates for applications requiring physical manipulation. Steep barriers of cost, limited functionality, and relatively slow performance are preventing the adoption of robots in the home.

In this chapter, we present applications of a proposed intelligent object detection and tracking system for improving functionality and performance of home-based robotic systems. The proposed system is applied to two unique problems: real-time sensing for manipulation and improved environmental awareness for mapping and localization.

Real-time sensing and control is an important milestone on the path towards home based robotic systems. The development of Neural Networks in the past decade has led to substantial increases in the performance of such intelligent networks. Neural networks are a group of nodes that are designed to represent neurons in the brain. These nodes are interconnected and simulate the learning process in the brain. Neural networks are a type of supervised machine learning, which requires the neural network to be trained using labeled data [1]. Slightly more advanced than neural networks, convolutional neural networks are more efficient for classifying images [2]. Convolutional neural networks work by using convolutions with various kernels to detect different features such as horizontal and vertical edges. The convolutional layers are then stacked on top of each other to detect more complicated features. The final layers of a convolutional neural network are a normal neural network. This is used to combine the features detected by the convolutional layers together to make a classification decision.

Using Deep Convolutional Neural Networks for images, on the other hand, has provided us with trained systems to be able to detect objects of interest with a very high level of accuracy [2–4]. With faster, real-time tracking algorithms [5], the robot can have a better understanding of the dynamics of its environment, and act more responsively, which is desirable in a human-in-the-loop scenario [6, 7]. The benefit of deep neural networks compared to traditional methods to perform these tasks is their ability to adapt better to the object of interest and reject unwanted noise. This makes it ideal for such systems to be incorporated into different environments (e.g. home, office, factory, etc.) without the need to modify the algorithm to adapt to enable detection. Experimental results for tracking objects for manipulation are detailed in the following sections.

Simultaneous Localization and Mapping (SLAM) is a method in robotics to map and navigate GPS-denied environments, such as a home. GPS-denied localization can be a computational constraint for any autonomous navigating task in an unknown environment. Such environments create a problem for locating objects and performing automation operations while creating and following its map by sensory readings, vision sources and etc. This problem requires a system to understand its environment, identify the objects, and localize them as stated in [8]. Sensor data is mapped to odometry data to determine the correct placement in a map. Once a map is developed, a robot can navigate the map and remember where it has gone. Visual SLAM (VSLAM) is the natural extension with visual inputs from one or more cameras. Image based feature extraction is the main method used in acquiring sensory data in VSLAM. In the extraction process, features such as edges and corners from any object are recorded. In cases of complex, dynamic environments where VSLAM will likely be used, landmark selection becomes a difficult problem given a low level of a priori information on the acquired features. Generated maps will contain transient features from objects that moved or disappeared from the environment. With new methods of image classification, namely convolutional neural networks, landmarks can be selected from the environment based off their known properties. The object detector and tracker presented in this chapter is used to find appropriate landmarks for navigation. Preliminary results for the landmark selection process will be presented in this chapter.

In this chapter, an improved object tracking algorithm is proposed for a home-based HRI system. The home-based HRI system presented in this chapter has capabilities of voice and object recognition and Internet of things compatibilities. Exchanges in the system are those associated with the management of tasks in activities of daily living. The chapter is formatted as follows. Section 2 provides a background on the related work. Section 3 details the proposed HRI smart home system. Section 4 describes the prototype of the proposed system. Section 5 details the experimental results with the system components. Finally, Sect. 6 presents the conclusions.

2 Literature and Related Works

A multi object tracking problem mainly consists of two parts: observation model and tracking [9]. The object identification and tracking algorithm used in this chapter uses a convolutional neural network for the model observation part, and uses that data with established tracking algorithms with modifications to suite to our application. Recent developments in object detection using neural networks offers real time performance, which is essential to the application in hand. Networks like You Only Look Once (YOLO) [10], Single Shot MultiBox Detector (SSD) [11], Faster R-CNN [12], R-FCN [13], OverFeat [14], such real-time performance optimal for application in robotics. However, research indicates that the accuracy of such detections reduces with the increase in detection speed [15].

Once the object to be tracked is determined, it is extracted from the frame, thus enabling a smaller region of pixels to be fed into the tracking algorithms to identify the object in future frames. Such tracking, which use feature matching, color segmentation, edge detection, background subtraction etc. can be performed using algorithms like Kanade-Lucas-Tomasi Feature Tracker (KLT) [16], Extended Lucas-Kanade Tracking [17], Online-boosting Tracking [18], Spatio-Temporal Context Learning [19], Locality Sensitive Histograms [20], TLD: Tracking—Learning—Detection [21], CMT: Clustering of Static-Adaptive Correspondences for Deformable Object Tracking [22], Kernelized Correlation Filters [23]. The performance of these methods can be compared and evaluated through benchmarking tools [24] to figure out which one is optimal for one's application. There are several open source libraries which integrate several of these algorithms to facilitate their use in an application like OpenCV [25] and Modular Tracking Framework [26].

Previous related work by the authors of this chapter follows. A framework for navigation and target tracking system for mobile robot was presented using 3D depth image data and used color image recognition, depth camera data and fuzzy logic to control and navigate the robot [27]. Design of a testbed for Large-Scale autonomous system of vehicles was proposed for localization, navigation and control of multiple networked robotic platforms by using cloud computing in [28]. A real-time cloud-based VSLAM was provided in [29] with enhancements to reduce processing time and storage requirements for a mobile robot. A visual SLAM based cooperative mapping study with cloud back-end proposed the importance of the object identification for the mobile navigation and localization [8]. Design and development of a multi-agent home-based assistive robotic system for the elderly and disabled was provided in [30–32]. Furthermore, a cloud architecture for large scale systems of autonomous vehicles was presented in [33]. A foundation for deep neural network control was provided in [34]. An initial deep vision landmark framework was developed for robot navigation by Puthussery et al. in [35]. This system utilized the Inception V3 engine to classify image frames into trained object classes. The most probable detected class was recorded along with positional information relative to the robot in a map. After mapping, objects selected for further inspection were approached by the robot. In this chapter, we utilize a deep neural network based object detector, which has the capability to detect multiple classes per frame with bounding boxes identifying the detected objects. The use of a real-time object detector greatly improves map resolution, classification throughput, and data acquisition time in the mapping process.

Along with the multi-object tracking algorithms, this chapter also uses unsupervised learning approaches. These unsupervised learning approaches are used for various functions such as object ownership association. The traditional unsupervised learning approaches include k-means [36] and fuzzy c-means clustering [37]. However, in recent years these clustering algorithms have been improved due to some of the traditional methods drawbacks. One such drawback is that both k-means and fuzzy c-means need to know the number of clusters beforehand. In many situations, the number of clusters are unknown. Many different methods have since been developed to remove this constraint. For example, Ester et al. developed Density-based spatial clustering of applications with noise (DBSCAN), which groups together closely packed points [38]. One of the largest advantages of DBSCAN is that it does not require the a priori knowledge of the number of clusters. Another issue associated with k-means and fuzzy c-means clustering is that for large numbers of points, the runtime can be very slow. The typical implementation of k-means has a complexity of $O(\mathrm{N}(\mathrm{D}+\mathrm{K}))$, where N is the number of points, D is the number of dimensions, and K is the number of centroids [39]. Fuzzy c-means is even slower than k-means, with the typical time complexity of $\mathrm{O}(\mathrm{NK}^2\mathrm{D})$ [40]. Although these algorithms can run slow with large numbers of points, some advances have been made to improve this. For example, Kolen and Hutcheson were able to reduce the time complexity of fuzzy c-means down to O(NKD) by removing the need to store a large matrix during iterations, which is significantly faster [41]. Arthur et al. reduced the time complexity of k-means down to $\mathrm{O}(log\mathrm{K})$ by initializing the cluster centers by using points in the dataset that are further away from each other in a probabilistic manner [39, 42].

3 Proposed System

The proposed system is comprised of components which implement the following process. An elderly user makes a request to the voice assistant for a retrieval type task to be completed. In this case, the item to be retrieved is a drink. The task is broken down into its components: action(s), location(s) and object(s). In this example, a robot is tasked to inspect an object. On its way to interact with it, various objects are detected, tracked and mapped. Once candidates for the selected object are detected, a catalog is created for the user to verify. The user provides input to the robot, or voice assistant with a camera, via facial expressions to express satisfaction with the actions of the assistant. Emotion levels are used to select the closest match to the desired output of the system. The robot then completes the task utilizing its physical manipulation capabilities.

3.1 Vision-Based Object Detection and Mapping

Identification and tracking of unique objects requires the following steps. First the object should be detected and classified. Once detected, the object should be tracked frame by frame to ensure that duplicate results are not recorded. Estimates of the objects position are recorded into a map using estimates of the robot pose and properties of the camera. Further detail on each step is provided below.

3.1.1 Multi-object Detection

We use a generalized Convolutional Neural Network (CNN) in our system to perform multi-object detection using an RGB frame captured by a camera on the robot. With the recent development and availability of powerful mobile computers with multi-processing capabilities like the CUDA-cores, we are able to process these frames in real time speeds, to detect multiple objects in a single forward pass of the network. This enables us to use CNN for real time applications like SLAM. The CNN architecture is inspired from open-sourced projects [10–14], the initial layers of which are pre-trained as object classifiers using available datasets of common objects [43–45]. The latter layers of such networks are trained to maximize the Intersection-Over-Union (IOU) of the most likely objects detected in the frame with the bounding box of these objects, also available as supplement to the datasets. We extend these algorithms by adding a higher level of abstracted computer intelligence. For the networks, we use the pre-trained models which are available for most of the networks

3.1.2 Object Ownership Clustering

Sometimes the user may present the robot with ambiguity such as the task of getting the user "their" glasses. This can be an ambiguous task because "their" is a pronoun meaning that the glasses belong to them. "Their" does not provide the robot with any physical description of the glasses, which could cause a problem if multiple people wear glasses in the household. Therefore, ownership of objects could be a very important attribute to consider. For example, if the robot were to see two pairs of glasses, initially the ownership of each pair is unknown so the robot will have to ask the user which glasses is "theirs". Once the robot can determine the ownership of each pair of glasses, the identifying physical descriptors of the glasses can be saved into a database. Since it is very likely that the glasses are placed next to other objects that belong to the owner of the glasses, the robot can assume that the nearby objects also have a possibility of belonging to the glasses owner. To allow the robot to make these assumptions, clustering algorithms can be used.

Once the ownership of an item is verified by the robot, the robot can utilize a clustering algorithm. This algorithm will cluster objects together based on Euclidean

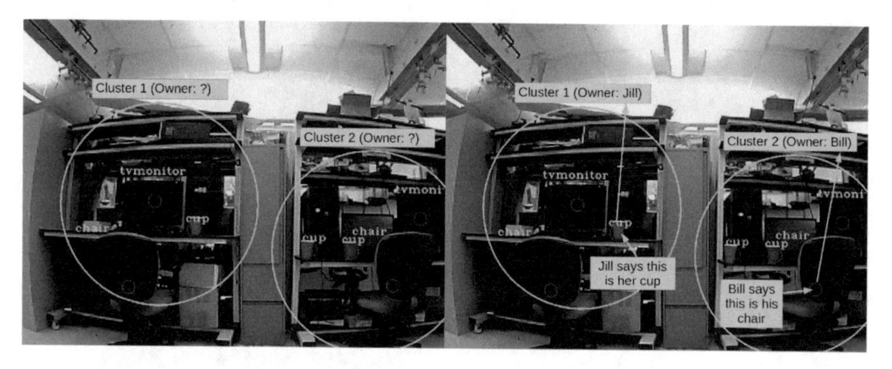

Fig. 2 Example scenario of how clustering can be used to solve ambiguity

distance. Any objects that are in the same cluster as the verified objects can be stored into a database as objects that possibly belong to the owner of the verified object. Now if the user provides the robot with another ambiguous request, the robot can use the objects that were in the same cluster as the verified object to solve the ambiguity. In the same scenario as before, also seen in Fig. 2, if Jill asks the robot to get "her" cup, and there are multiple pairs of glasses, the robot will not know which cup belong to Jill. The robot will then ask Jill to verify which cup is hers. Once Jill responds, the robot will then proceed to get that cup. As the robot is getting the cup, the robot will cluster the objects near the cup. Since Jill's chair and monitor are near her cup, the robot will cluster them into the same cluster as the cup. The robot will then store the ownership of the chair and monitor as having a high probably of belonging to Jill. Now the robot brings Jill's cup back to Jill, and then Jill requests the robot to get her chair. Normally this would be another ambiguous request, but since the robot now knows that the chair was near the cup, the chair has a high probability of being Jill's. This allows the robot to be able to immediately go and get the chair and bring it back to Jill. Since there is still a small chance that the chair is not Jill's chair, the robot will still verify with Jill to ensure that it is actually her chair. A flowchart of the algorithm that the robot can use to solve ownership ambiguity can be seen in Fig. 3.

3.1.3 Object Tracking

Though neural network based multi object detector performs very well as an object tracking neural network, it fails to distinguish between multiple instances of the same type of object that it is detecting. To identify objects for the purpose of automation, we need to track an object. This means to be able to distinguish between two similar objects that might be detected by the multi object detector.

The camera frame will be processed by the multi object detector to provide the location and classification of objects in the frame, while smaller Region-of Interest (ROI) will be selected within the frame based on the same location information to detected features on the object, hence assigning uniqueness to the object. When

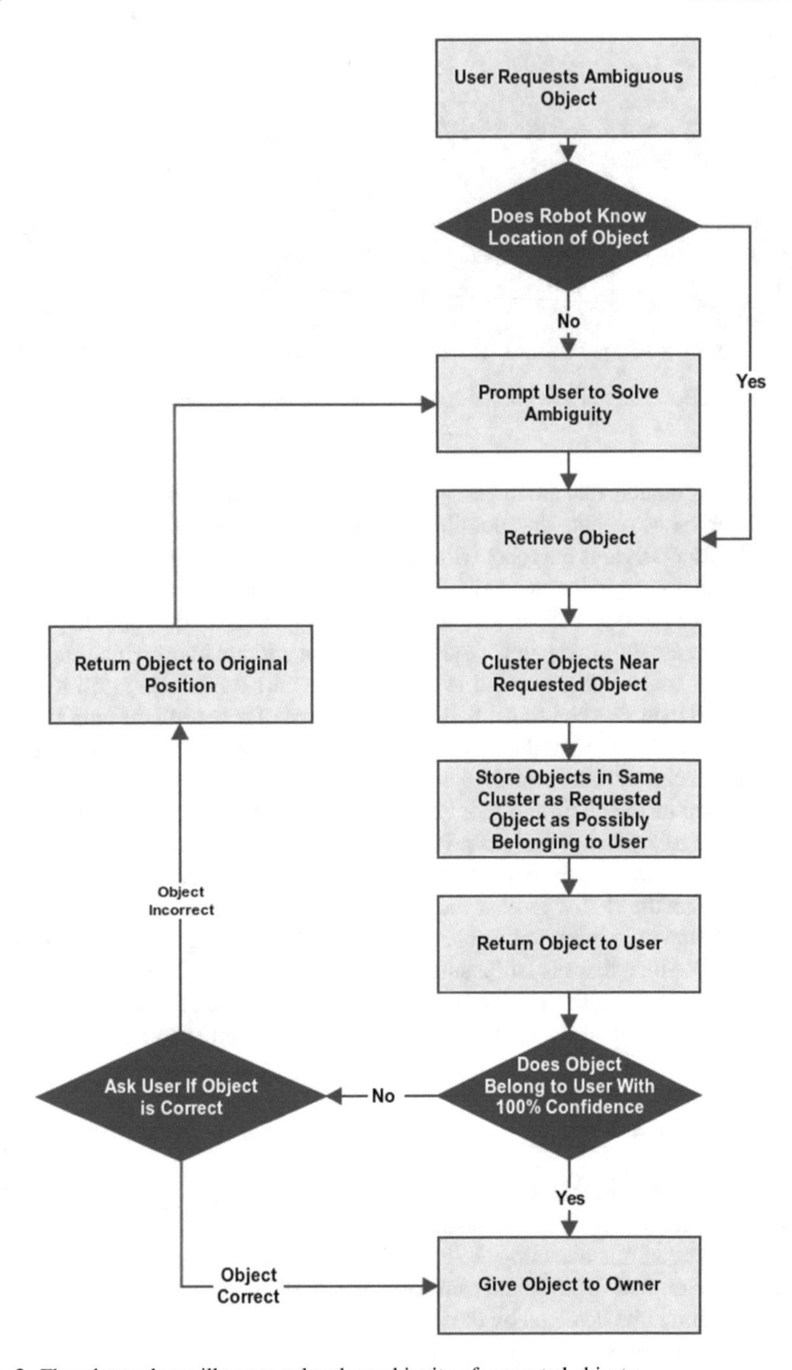

Fig. 3 Flowchart robot will use to solve the ambiguity of requested objects

needed, feature matching is used to solve the problem of ambiguities when two or more objects of the same class overlap, or one object goes out of the camera field of view.

For the purpose of experiments, the number of reliable features to track may be set arbitrarily, and refined to a more experimentally tested decision in the test iterations. A feature management algorithm is utilized to decide on the actions to take in case of loss of features mapped and the minimum number of features required to have a reliable tracking of the object. It is observed that while detecting features of an object, the count of the features may not suffice the need for reliably distinguishing similar objects. To overcome this issue, the algorithm guides the robot to move towards a particular object, once detected, until it has enough features in its feature map. These features of the object are recorded into the memory of the robot along with a picture of the object for reference. A library of all the similar objects and their associated features are stored and then queried with the user to find out which one is of interest. Once a selection is made by the user, other features may be discarded, while the features of the selected unambiguous object is used to track the robot back to the object.

3.1.4 Object Mapping

Positioning of the robot with respect to the environment is important since we need the robot to find its position back to the user and the detected objects. Similarly localizing the detected objects with respect to the environment is also important to plan a path for the robot to maneuver to the object. The kinematic model for the proposed robot's locomotion is a combination of a differential drive kinematics and serial manipulator kinematics. To simplify system development for this current research, the humanoid torso robot was assumed to remain in a fixed pose.

Mapping of the robot and the environment is performed using a combination of different sensors on the robot. Every sensor has different kind of error associated with them. For example, the odometer on the robot is prone to error due to slippage. To overcome this problem, we use an Extended Kalman Filter (EKF) algorithm to perform a sensor fusion between the positions obtained from the odometer, Inertial Measurement Units (IMU), the visual odometry reported by the camera using Simultaneous Localization and Mapping (SLAM) techniques. The use of Robot Operating System (ROS) packages enables us to perform such sensor fusion with minimal effort.

Traditional SLAM algorithms using feature detection are complemented with using multi object detection as reference points on the map to localize the robot. The objects detected by the system act as landmarks in the mapping process. Another EKF is applied to this system to provide a filtered map and localization of the robot. Mapping of the objects is performed once the objects of interest are detected and features are selected and stored. The locations of the objects are stored alongside the features and the picture of the object. With the use of modified neural network based object tracking, we could hand pick a certain category of objects that may offer more remarkable features to distinguish between other features. Since these features are conglomerated into objects detected by the algorithm, hence the matching process is less intensive than traditional SLAM, and at the same time provide more error correction from similar features, which is a large problem in the field of SLAM.

An important part of integrating an HRI into a system is for the robot to learn. A key way to do this for a task assigned robot is to remember the choice made by the user. As an implementation example, the first time the robot is asked to locate a bottle, and the user selects a specific bottle from a list of bottles the robot found, the specific bottle is stored along with its features, location and a picture for reference in the robot memory. Later, if the user instructs the robot to find the same object, the first guess that the robot makes is of the stores bottle in its memory. The user may want a different bottle and deny the robot, however there is a higher possibility that the user may want the same object again, which in turn improves the confidence of the robot with the user, and hence may make the robot more reliable. This is also useful in saving time for the robot to look for an object that it had already looked for earlier.

3.1.5 Object Database Creation

This section defines the creation of a virtual database of the objects detected, tracked and mapped by the robot. As described earlier, various real time multi-object detector algorithms may be used to track the current position of an object in the frame of the image frame, as observed by the robot. Once the presence of N object is confirmed, each of them are compared to the existing database to find if any of these objects are already in the library. In order to check if the object being recognized is already in the library, it takes into consideration various attributes about the detected object which includes color, features, ownership information, location and last access of the same object. Such attributes about each recognized object is compared with the corresponding attributes of all objects in the database, to compute a confidence level as shown in Fig. 4. This confidence level is then used to determine among three possible actions to be taken:

- If the object is already present in the database, affirm the presence of the same object in the image frame and determine if other action need to be performed on the object, for example pick up the object.
- If the confidence of the same object being present in the database is above a certain threshold but not high enough, compare to the most probable object in the database and update the object attributes in the database with the observed attributes.
- In case the confidence of the object being among the ones in the database is lower than a set threshold, add the object and its corresponding attributes as a new object in the database.

It should be noted that the attributes are weighted, when evaluating the confidence levels. This is because, certain attributes may be more reliable than others. For example, if the robot detects a bottle in its image frame, being an object that can be moved

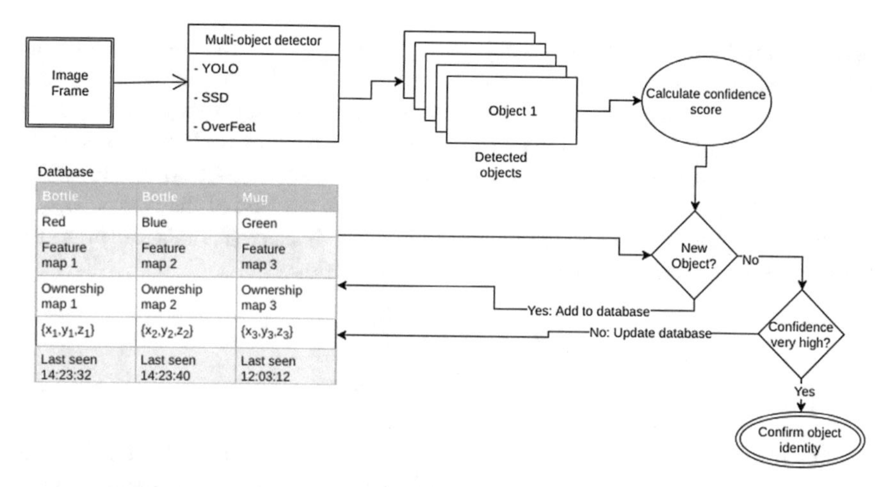

Fig. 4 Object database query, creation and update

since the last observation of the object, we assign a low weight on the location of the object while determining the confidence of the object being the same bottle in the database.

On the other hand, the features recorded for the bottle is assigned a higher weight, since a higher score on feature matching is more accurate indication of the same object being view by the robot image frame. Since we already define the objects that the object tracker can recognize, we also pre-define the weights associated with the different attributes of the corresponding object.

We can hence formulate, Eq. 1, a weighted average confidence level calculation of a detected object as:

$$C_n = \vec{W}\vec{A}$$
$$\vec{W} = \left\{w_c, w_f, w_o, w_l, w_t\right\} \text{ and } \vec{A} = \left\{s_c, s_f, s_o, s_l, s_t\right\}^T \quad (1)$$

where, $C_n \rightarrow$ Confidence level of object n,

$w_c, w_f, w_o, w_l, w_t \rightarrow$ Pre-defined weight vector for different attributes: color, feature matching, ownership, location and last access time, respectfully,

$s_c, s_f, s_o, s_l, s_t \rightarrow$ the attribute scores for color matching, feature matching, ownership, location and last access time.

3.1.6 Determining Optimal Action Sequence

One important decision the robot needs to make, is the sequence of the actions to be taken when multiple commands are requested by the user. This can be difficult because there are a lot of variables that can be considered. To simplify this process, we assume that the robot can only retrieve one object at a time, some of the objects locations are known, some of the objects locations are unknown, and the user has placed a higher priority on some objects versus others. To determine the optimal action sequence, the robot will first receive multiple requests from the user. Using machine learning algorithms, the robot will to decide whether each request is a low, medium or high priority request. The robot will sort the actions with known locations based on the cost calculated using Eq. 2, where C is the cost of the action, D_{RO} is the distance from the robot to the object, D_{OU} is the distance from the object to the user, and P is the predicted user's priority for that action.

$$C = \frac{D_{RO} + D_{Ou}}{P}$$
$$P_{low} < P_{medium} < P_{high} \tag{2}$$

If there are not any objects with known locations, the robot will search until it finds an object. The robot will then proceed to retrieve the first item, while simultaneously searching for objects with unknown locations. If the robot finds an unknown object, the robot will rerun the auction algorithm and determine the new optimal action sequence. If the robot does not find any unknown objects, then the robot will grab the object and return it to the user while still searching for unknown objects. The robot will repeat this process until all the actions have been completed. A flowchart describing this algorithm in a high level can be seen in Fig. 5.

3.1.7 Avoiding an Obstacle in the Environment

Once the robot has decided the optimal sequence of actions to take, the robot needs to successfully travel to the object to retrieve them. While traveling to the objects, the robot may encounter obstacles. These obstacles need to be avoided in order to ensure the safety of the robot and to not cause damage to the household. To avoid these obstacles, the already onboard camera can be utilized. Using the vision based obstacle avoidance algorithms such as the reactive vision only sliding mode controller developed by [46]. The robot can use its front facing camera to avoid obstacles, while still moving towards the object of interest. The results of the reactive vision based obstacle avoidance algorithm can be seen in Fig. 6.

4 Prototyping Robotic Smart Home System

4.1 Robotic System Hardware

The assistive robot used in this system is a hybrid of an unmanned ground vehicle and a humanoid robot, Fig. 7. For the humanoid portion of the hybrid machine, the humanoid robot torso is used which is a 3D printed open source robot from the

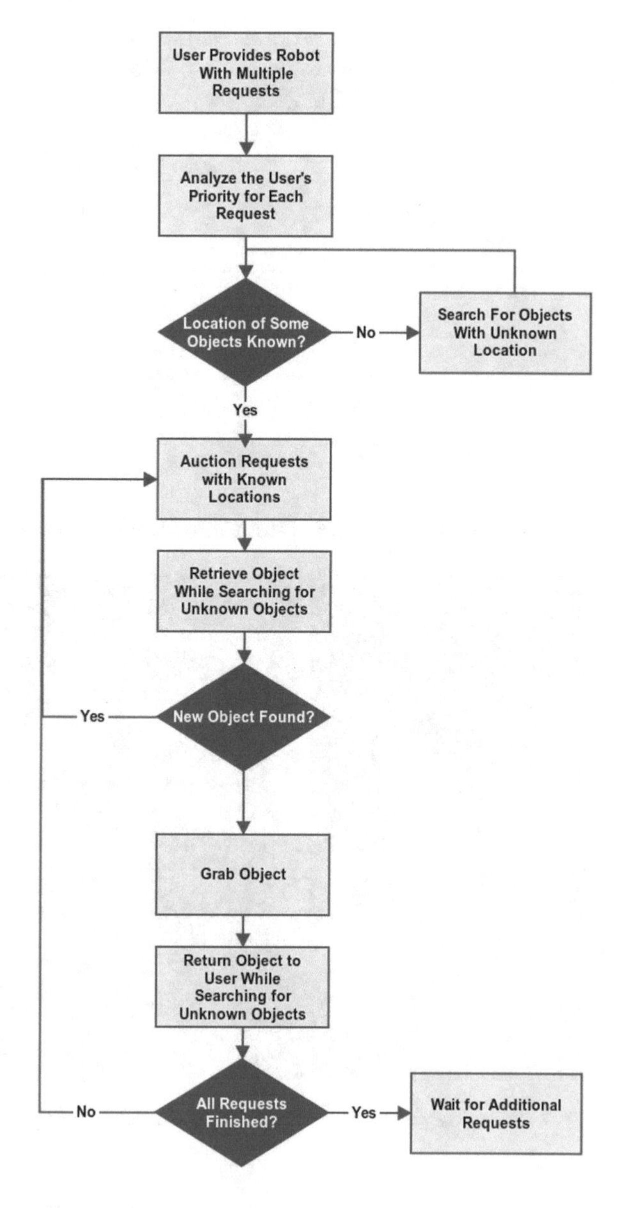

Fig. 5 Flowchart robot will use when the user requests multiple actions

torso up. The humanoid torso was combined with a Kobuki Turtlebot 2 research platform from YujinRobot later on as summarized in Fig. 8, which is an unmanned ground vehicle. The Turtlebot2 was selected due to its customizable capability and open source software. The rover is equipped with a Yujin Robot Kobuki base, a 14.8 V Lithium-Ion battery, and a Hardkernel ODROID XU4 minicomputer. The ODROID XU4 minicomputer was selected as the embedded computer for the UGVs. It features a Samsung Exynos 5422 octa-core CPU, 2 GB DDR3 RAM, USB 2.0/3.0

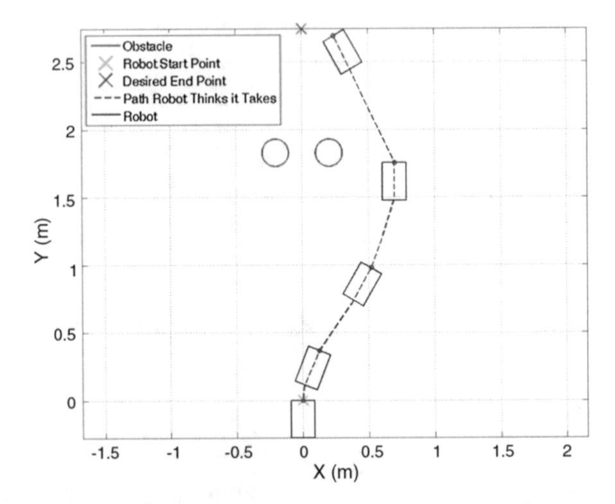

Fig. 6 Results of the vision based obstacle avoidance algorithm developed by Lwowski et al.

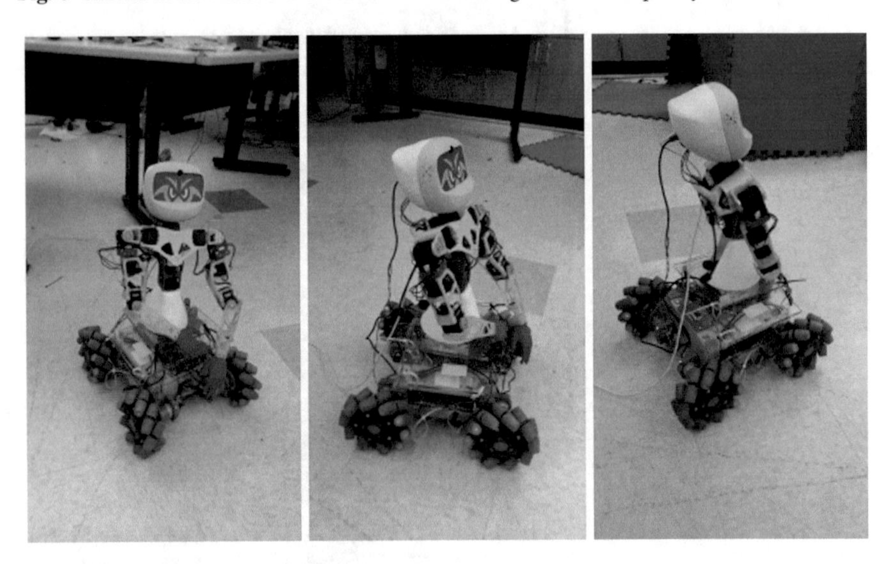

Fig. 7 Preliminary low-cost prototype for hybrid 3D printed mobile assistive robotic system

and a 64 GB eMMC card for storage. A Meanwell DC-DC converter was connected to the Kobuki's 12 V 5A output to supply power for the ODROID XU4 (5 V/4A requirement).

In addition, the rovers come with cliff sensors (left, center, right), wheel drop sensors (left, right), a single axis gyro and motor overload protection. The hybrid robotic platform is compatible with the Robot Operating System (ROS) by extensions of APIs supplied for both research platforms. To obtain better directional awareness, a BOSCH BNO055 Inertial Measuring Unit was added to provide absolute orientation

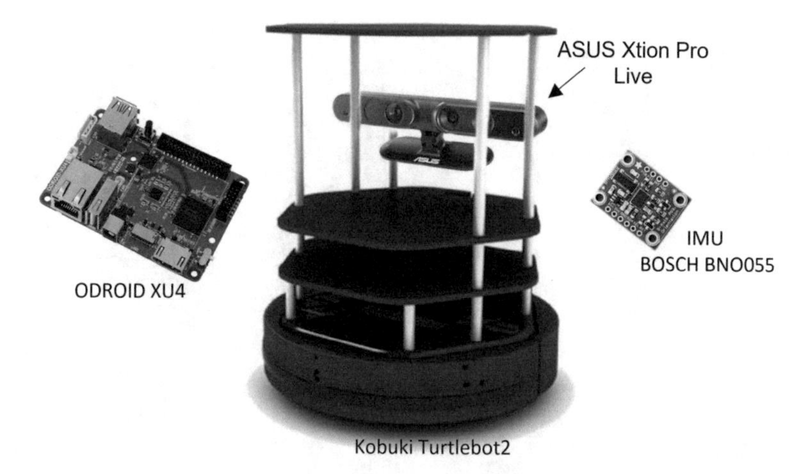

Fig. 8 Kobuki Turtlebot 2 has been chosen and modified with ASUS Xtion Pro Live RGB-D camera, powered with ODROID XU4microcontroller and BOSCH BND055 IMU installed

to the system. This IMU integrates multiple sensors to obtain a stable absolute output: a triaxial 14-bit accelerometer, a triaxial 16-bit gyroscope and triaxial magnetometer.

The humanoid robot torso has been mounted on the Kobuki Turtlebot 2 robot as shown in Fig. 9. A camera mounted in the head of humanoid robot torso is used to detect objects in the environment. Control of the hybrid robot is performed using ROS. The *Kobuki_ROS* package handles control of the base robot. A custom designed head unit was designed to support addition of a five-inch touchscreen LCD display (ODROID-VU5), monocular camera, stereo speakers for synthesized auditory feedback, and a microphone for obtaining commands from the user.

An overview of the prototype system is provided below in Fig. 10. The smart home system includes interfaces for voice, vision, cloud-based computation, and robotic platforms. In this section, preparation of robotic hardware, the HRI, object detection and tracking algorithms and the control loop are discussed.

4.1.1 Human-Robot Interface

Visual Interface

A software interface was developed for the new humanoid robot torso's head unit using pyqt3 for the graphics and ROS for the interface to the data. A ROS software package ace poppy hri display was developed for this work. Inputs to this package are the desired target, the robot state, and text to display. A question and answer game, used in HMI, is implemented by the robot and user of the system. An example of the HRI is displayed on the humanoid robot torso head unit display in Fig. 11. The example shows the user request "check the plant" to the robot, a response "is

Fig. 9 Redesigned prototype of the system with the mobile platform and the torso robot is installed performing object identification and navigation tasks

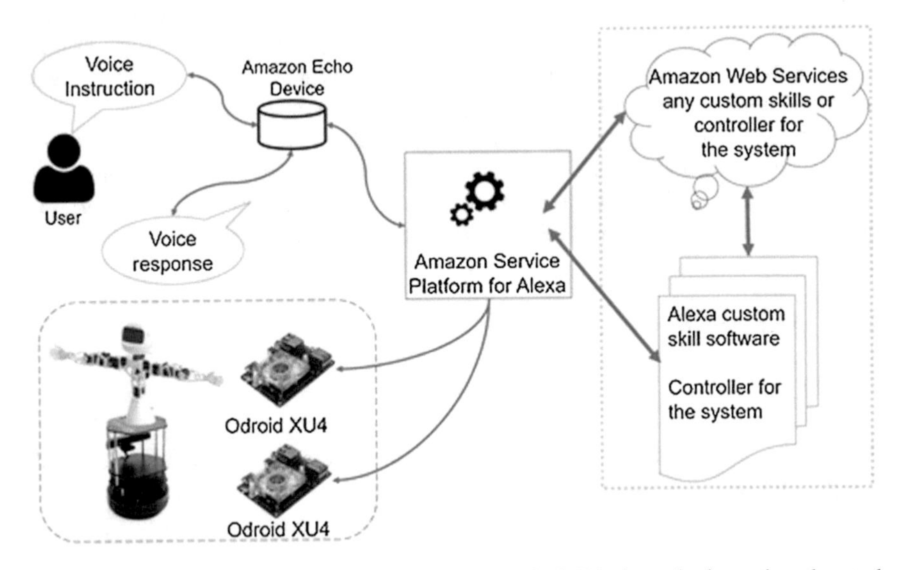

Fig. 10 System representation included simple flow chart for IoT device and voice-activated control

Fig. 11 Question and answer type HRI for a selected cropped output of the database displaying a plant

this what you wanted me to check?", and a picture of the item that the robot checked. Stylistically, the text of the question and answer HRI is like the voice user interface (VUI) de facto standards used in voice assistants. This was implemented like VUI as voice is used to provide the robot tasks and is shown in Fig. 10.

Auditory Interface

Auditory commands are provided to the robot using a home voice assistant, in this case an Amazon Echo Dot. Auditory responses from the robot are generated using a combination of Linux programs *espeak* for synthesis of words into a WAV file and *aplay* for playing back the WAV file. The option to use a WAV file was selected for a combination of reasons. Most important of all, *espeak* tends to connect to the audio service jack-server slowly or fails intermittently, where *aplay* plays back the audio almost instantaneously and consistently. The second reason is to maintain a history of responses to the user for quality purposes. As a synthesizer, *espeak* tends to be limited in its ability to pronounce certain words and people's names. There is a need to sometimes break a word into its phonetic components to synthesize it correctly.

4.2 Object Detection and Tracking

4.2.1 Modification for Compatibility with Video Source

Darknet, the software package for YOLO, was installed on a high-performance desktop computer with an Intel i5 and a NVIDIA GTX-1080 with 8 GB of memory. The desktop computer is the detection engine for the robot platform. The source code for the neural network was modified to allow connections to the camera feed through Wi-Fi. The Robot Operating System (ROS) was used to provide Wi-Fi interface to the camera feed over a protocol similar to TCP called TCPROS. To be compatible with ROS, use of a branched version of darknet written in C++ was necessary. This software branch contains the modifications necessary to generate a shared library file libdarknet-cpp-shared.so and arapaho, a C++ API to the library. ROS packages ace_arapaho and ace_arapaho_msgs were developed by the authors to use the arapaho API. These packages provide the capability to publish the identified objects with labels, timestamps, and the relevant region of interest bounds of the image. An additional input to the package is the object filter list.

Objects in the list are filtered out from the reported identifications. Outputs of the ace_arapaho ROS node are passed to a feature tracking package, developed by the authors, called ace_object_tracker. Inputs to this package are parameters from the motion of the robot, parameters of the camera, and the image ROIs from the multi object detector. This package develops initial models of the detected objects from the inputs provided to it. These models are used to uniquely identify the incoming data as belonging to a unique object.

4.2.2 Feature-Matching Enhanced Object Detector

It is important to note that even though the output of neural network based multi object detectors resemble tracking, it is just a multi-object detection algorithm which works at a very high throughput. Though it is able to detect the location of an object within the image frame, it does not track objects as individual items. Hence in a frame with more than one instance of the same type of object, for example two different bottles, will be tracked as the object bottle, irrespective of their differences. This ambiguity is problematic in cases where the user wants the system to find a specific object. To resolve this issue, we combine traditional object tracking methods with neural network multi object detectors. The system uses multi object detectors to detect objects in the frame, which are then passed onto feature tracking algorithms. Features are then selected from within the bounding box provided by the object detectors for a particular object. These features are then used to identify a specific object when such ambiguity arise.

4.2.3 Confidence Gradient Tracking

Another common issue when using such multi object detectors is the lack of reliable detection in every frame. Since they process every image individually, isolated from the previous image sequence, it often causes alternating loss and detection of objects in the scene. Another problem while using such algorithms is the detection of false positives. We use a method of confidence gradient tracking to overcome both these issues, to achieve a reliable tracking. Our algorithm uses a complimentary filter to smoothen the detection confidence level of a particular object being tracked. The gradient of this filtered confidence level is monitored by the system. While the robot is moving in a particular direction, if the filtered confidence gradient is positive and the filtered confidence level builds up to a set threshold of confidence level, the existence of the object is confirmed. This confirmation of the object in the scene initiates the feature tracking algorithm, which starts to record features of the object, while the robot is moving towards the object. The algorithm instructs the robot to keep moving towards the object until the number of recorded reliable features for the object matches a pre-defined minimum number. Implementation of the algorithm is explained in more detail in Sect. 5.

5 Experimental Results

5.1 Processing Rate for Object Detection and Tracking

Initial performance tests were executed with a direct USB 2.0 connection from the desktop computer to the camera onboard the robot. In this configuration images are processed at about 27 FPS, which is similar to the camera frame throughput. Further tests utilized images transmitted over TCPROS on a Wi-Fi IEEE 802.11 N connection from the robot to the desktop computer. In this configuration, we were able to process image frames around 5 FPS. The drastic reduction in processing rate is solely due to the transmission of raw image data over Wi-Fi. Compressed image streams will be examined in the future to achieve a higher data rate over Wi-Fi. Selected outputs of this result are displayed in Fig. 12 that show both some correct and incorrect detections. For a couple examples, a cardboard box is covered in white paper is labeled a sink, and a large cabinet is classified as a refrigerator. Items that were detected more reliably (e.g. clocks, bottles, chairs, TV monitors, etc.) were selected as target objects for the experiment.

Fig. 12 Selected outputs of YOLO demonstrating correct and incorrect classifications of objects in an image frame

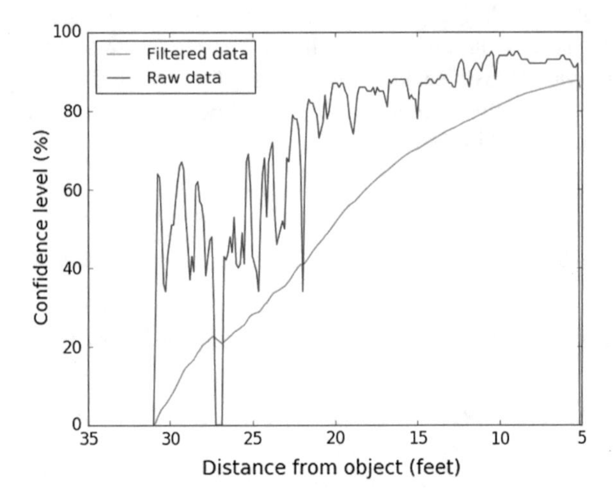

Fig. 13 Detection confidence analysis with respect to distance to object

5.2 *Confidence Gradient Tracking*

Figure 13 shows a plot of the raw and filtered confidence level of a chair detected, as the camera frame moves closer to the chair. As can be observed from the figure, the raw confidence outputs of a particular object being tracked by the multi object detector is noisy. However, the filtered data shows a general increasing trend of confidence signifying a true positive detection of the object. To recognize this behavior, our algorithm differentiates the filtered confidence level. This differential is used to indicate the increasing or falling nature of the confidence level of a detected object. However, as observed in Fig. 13, there are regions where the algorithm either fails to detect the object or the detection is a false positive.

For this reason, certain regions of the differential are negative. On the other hand, for our algorithm to work, we would want to detect a negative differential slope only

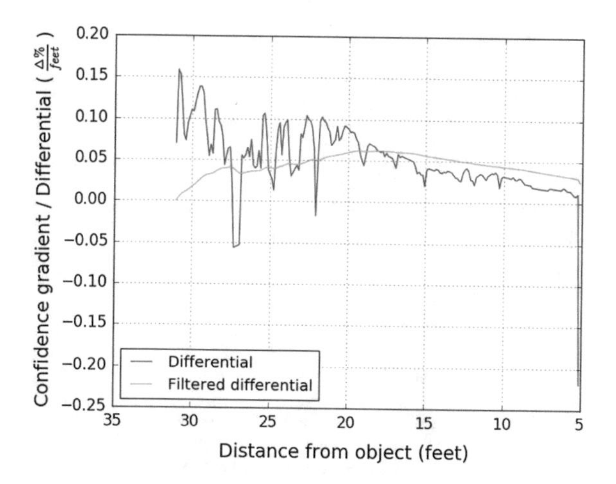

Fig. 14 Filtered differential of the confidence levels with respect to distance to object

when the object for detection is not in the frame. In order to detect a true positive in cases where the object is not detected for few frames, we filter the differential plot as shown in Fig. 14.

If the filtered differential remains positive while the robot is moved towards the object, a true positive of the detected object is established. The filtered confidence data is monitored while the robot is moving towards the object to assign a Region of Interest (ROI) for feature recognition and storage, for matching and identification later. It should be noted that differentiating the original unfiltered confidence levels and then filtering them, generates an output which radically alternates between the positive and negative. This is an expected behavior for a noisy signal.

An example of the differential of the raw confidence levels can be seen in Fig. 15. However, such differential signal cannot be used to distinguish between false positive and a true positive, using the previously explained algorithm. As a result, we use a filtered confidence levels before differentiating the data.

Experimental tests were performed on 3 different objects—person, chair and bottle for three instances each. The algorithm described above was implemented in each case and the results obtained are tabulated in Table 1. Two thresholds were used to determine a true positive. The first one, set to 50% confidence was used to trigger the robot to turn towards the object and move towards it. The next threshold of 60% was used to declare an ROI to start tracking features of the object. The filtered differential was monitored to make sure that the results are not false positives.

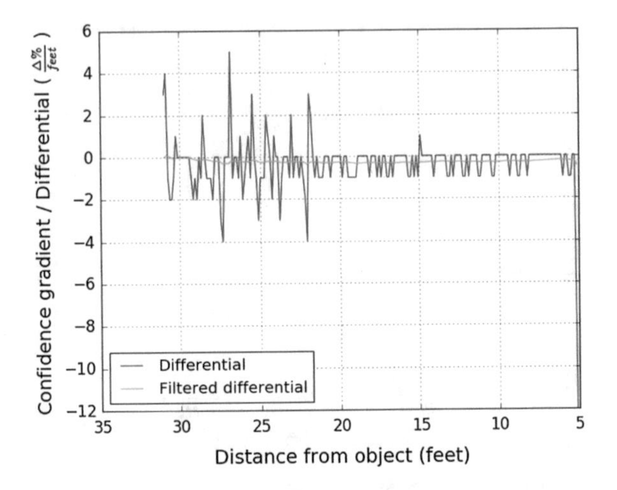

Fig. 15 Differential of unfiltered confidence levels: Alternates rapidly across zero level

Table 1 Confidence gradient tracking (Threshold (true positive/feature recording) = 50%/60%)

Object tracked	Distance from object with true positive affirmation (Feet)	Line of sight angle offset (°)	Maximum features matched for tracking capability	Tracking possible?
Bottle 1	12.77	10	730	Yes
Chair 1	20.26	0	382	Yes
Person 1	20.66	5	127	No
Chair 2	17.18	10	312	Yes
Chair 3	18	0	302	Yes
Bottle 2	5.69	5	N/A	No
Bottle 3	9.55	0	779	Yes
Person 2	20.94	15	911	Yes
Person 3	14.62	10	630	Yes

6 Conclusions

Robotic navigation in GPS-denied environments highly depended on specific approaches for locomotion and navigation tasks. Improvements on computational intelligence tools and pattern recognition approaches, along with reinforced learning applications for object identification engines based on improved RGB-D cameras, have increased the accuracy of cooperative multi-tasking assignments in robotics. Pattern recognition and machine learning techniques, such as Convolutional Neural Networks to identify markers or objects from images and videos, improved the performance in the autonomous navigation and localization experiments.

The use of robotics in the home environment is a very complicated scenario with lots of problems. In this chapter, many problems such as object detection and tracking, object ownership, object mapping, object database creation, determining optimal action sequence, and obstacle avoidance have been addressed. Solving these problems are one of the necessary steps into creating a robust fully functioning home robotic assistant that could be used in our everyday lives. In the future, we plan to integrate all of these decouple systems together, to create a more complete robotic assistant for the home environment. This robotic assistant could then be tested in many different situations in order to gather data and improve our algorithms. These tests would also help us identify new problems that will need to be solved in the future to make the system even more robust and useful.

Acknowledgements The authors would like to acknowledge the support from Air Force Research Laboratory and OSD for sponsoring this research under agreement number FA8750-15-2-0116. The U.S. Government is authorized to reproduce and distribute reprints for Governmental purposes notwithstanding any copyright notation thereon. The views and conclusions contained herein are those of the authors and should not be interpreted as necessarily representing the official policies or endorsements, either expressed or implied, of Air Force Research Laboratory, OSD, or the U.S. Government. The work partially supported by the Open Cloud Institute at The University of Texas at San Antonio.

References

1. S.B. Kotsiantis, I. Zaharakis, P. Pintelas, Supervised machine learning: A review of classification techniques, 2007, 3–24
2. A. Krizhevsky, I. Sutskever, G.E. Hinton, Imagenet classification with deep convolutional neural networks," in *Advances in neural information processing systems*, 2012, pp. 1097–1105
3. C. Szegedy, W. Liu, Y. Jia, P. Sermanet, S. Reed, D. Anguelov, D. Erhan, V. Vanhoucke, A. Rabinovich, Going deeper with convolutions, in *Proceedings of the IEEE Conference on Computer Vision and Pattern Recognition*, 2015, pp. 1–9
4. C. Szegedy, V. Vanhoucke, S. Ioffe, J. Shlens, Z. Wojna, Rethinking the inception architecture for computer vision, in *Proceedings of the IEEE Conference on Computer Vision and Pattern Recognition*, 2016, pp. 2818–2826
5. A. Karpathy, What I learned from competing against a convnet on imagenet, 2014, http://karpathy.github.io/2014/09/02/whati-learned-from-competing-against-a-convnet-on-imagenet
6. J. Redmon, S. Divvala, R. Girshick, A. Farhadi, You only look once: Unified, real-time object detection, in *Proceedings of the IEEE Conference on Computer Vision and Pattern Recognition*, 2016, pp. 779–788
7. W. Liu, D. Anguelov, D. Erhan, C. Szegedy, S. Reed, C.-Y. Fu, A.C. Berg, Ssd: Single shot multibox detector, in *European Conference on Computer Vision* (Springer, 2016), pp. 21–37
8. B.A. Erol, S. Vaishnav, J.D. Labrado, P. Benavidez, M. Jamshidi, Cloud-based control and vslam through cooperative mapping and localization, in *World Automation Congress (WAC)* (IEEE, 2016), pp. 1–6
9. L. Fan, Z. Wang, B. Cail, C. Tao, Z. Zhang, Y. Wang, S. Li, F. Huang, S. Fu, F. Zhang, A survey on multiple object tracking algorithm, in *2016 IEEE International Conference on Information and Automation (ICIA)* (IEEE, 2016), pp. 1855–1862
10. J. Redmon, A. Farhadi, Yolo9000: better, faster, stronger, 2016, arXiv:1612.08242

11. W. Liu, D. Anguelov, D. Erhan, C. Szegedy, S.E. Reed, C. Fu, A.C. Berg, SSD: single shot multibox detector. CoRR, abs/1512.02325 (2015). http://arxiv.org/abs/1512.02325
12. S. Ren, K. He, R.B. Girshick, J. Sun, Faster R-CNN: towards real-time object detection with region proposal networks. CoRR abs/1506.01497 (2015). http://arxiv.org/abs/1506.01497
13. J. Dai, Y. Li, K. He, J. Sun, R-FCN: object detection via region-based fully convolutional networks CoRR. abs/1605.06409 (2016). http://arxiv.org/abs/1605.06409
14. P. Sermanet, D. Eigen, X. Zhang, M. Mathieu, R. Fergus, Y. LeCun, Overfeat: Integrated recognition, localization and detection using convolutional networks. CoRR abs/1312.6229 (2013). http://arxiv.org/abs/1312.6229
15. J. Huang, V. Rathod, C. Sun, M. Zhu, A. Korattikara, A. Fathi, I. Fischer, Z. Wojna, Y. Song, S. Guadarrama, K. Murphy, Speed/accuracy trade-offs for modern convolutional object detectors. CoRR abs/1611.10012 (2016). http://arxiv.org/abs/1611.10012
16. C. Tomasi, T. Kanade, Detection and tracking of point features. Int. J. Comput. Vision (Tech. Rep.) (1991)
17. S. Oron, A. Bar-Hille, S. Avidan, Extended lucas-kanade tracking, in *European Conference on Computer Vision* (Springer, Cham, 2014), pp. 142–156
18. H. Hu, B. Ma, Y. Wu, W. Ma, K. Xie, Kernel regression based online boosting tracking. J. Inf. Sci. Eng. **31**(1), 267–282 (2015)
19. K. Zhang, L. Zhang, Q. Liu, D. Zhang, M.H. Yang, Fast visual tracking via dense spatio-temporal context learning, in *European Conference on Computer Vision* (Springer, Cham, 2014), pp. 127–141
20. S. He, Q. Yang, R.W. Lau, J. Wang, M.H. Yang, Visual tracking via locality sensitive histograms, in *Proceedings of the IEEE Conference on Computer Vision and Pattern Recognition*, 2013, pp. 2427–2434
21. Z. Kalal, K. Mikolajczyk, J. Matas, Tracking-learning-detection. IEEE Trans. Pattern Anal. Mach. Intell. **34**(7), 1409–1422 (2012). https://doi.org/10.1109/TPAMI.2011.239
22. G. Nebehay, R. Pflugfelder, *Clustering of static-adaptive correspondences for deformable object tracking* (Comput. Vision Pattern Recognit., IEEE, 2015)
23. J.F. Henriques, R. Caseiro, P. Martins, J. Batista, High-speed tracking with kernelized correlation filters. CoRR abs/1404.7584, 2014. http://arxiv.org/abs/1404.7584
24. Y. Wu, J. Lim, M.-H. Yang, Online object tracking: A benchmark, in *IEEE Conference on Computer Vision and Pattern Recognition (CVPR)*, 2013
25. Open source computer vision library, https://github.com/itseez/opencv
26. A. Singh, M. Jagersand, Modular tracking framework: a unified approach to registration based tracking. CoRR. abs/1602.09130 (2016). http://arxiv.org/abs/1602.09130
27. P. Benavidez, M. Jamshidi, Mobile robot navigation and target tracking system," in *2011 6th International Conference on System of Systems Engineering (SoSE)* (IEEE, 2011), pp. 299–304
28. J.D. Labrado, B.A. Erol, J. Ortiz, P. Benavidez, M. Jamshidi, B. Champion, Proposed testbed for the modeling and control of a system of autonomous vehicles, in *2016 11th IEEE System of Systems Engineering Conference (SoSE)*, 2016, pp. 1–6
29. P. Benavidez, M. Muppidi, P. Rad, J.J. Prevost, M. Jamshidi, L. Brown, Cloud-based realtime robotic visual slam, in *2015 9th Annual IEEE International Systems Conference (SysCon)* (IEEE, 2015), pp. 773–777
30. P. Benavidez, *Low-cost Home Multi-robot Rehabilitation System for the Disabled Population* (Google, 2015)
31. P. Benavidez, M. Kumar, S. Agaian, M. Jamshidi, Design of a home multi-robot system for the elderly and disabled, in *2015 10th System of Systems Engineering Conference (SoSE)*, 2015, pp. 392–397
32. P. Benavidez, M. Kumar, B. Erol, M. Jamshidi, S. Agaian, Software interface design for home-based assistive multi-robot system, in *2015 10th System of Systems Engineering Conference (SoSE)* (IEEE, 2015), pp. 404–409
33. S.A. Miratabzadeh, N. Gallardo, N. Gamez, K. Haradi, A. R. Puthussery, P. Rad, M. Jamshidi, Cloud robotics: A software architecture: For heterogeneous large-scale autonomous robots, in *World Automation Congress (WAC)*, (IEEE, 2016), pp. 1–6

34. M. Roopaei, P. Rad, M. Jamshidi, Deep learning control for complex and large scale cloud systems, in *Intelligent Automation & Soft Computing*, 2017, pp. 1–3
35. A.R. Puthussery, K. Haradi, M. Jamshidi, A deep vision landmark framework for robot navigation, in *2017 12th IEEE System of Systems Engineering Conference (SoSE)*, 2017, pp. 1–6
36. S. Lloyd, Least squares quantization in pcm. IEEE Trans. Inf. Theory **28**(2), 129–137 (1982)
37. J.C. Dunn, A fuzzy relative of the isodata process and its use in detecting compact well-separated clusters. J. Cybern. **3**(3), 32–57 (1973)
38. M. Ester, H.-P. Kriegel, J. Sander, X. Xu, A density-based algorithm for discovering clusters a density-based algorithm for discovering clusters in large spatial databases with noise, in *Proceedings of the Second International Conference on Knowledge Discovery and Data Mining* (AAAI Press, 1996), pp. 226–231
39. X. Jin, J. Han, *K-Means Clustering* (Springer, US, 2010), pp. 563–564
40. A. Stetco, X.-J. Zeng, J. Keane, Fuzzy c-means++: Fuzzy c-means with effective seeding initialization. Expert Syst. Appl. **42**(21), 7541–7548 (2015)
41. J.F. Kolen, T. Hutcheson, Reducing the time complexity of the fuzzy c-means algorithm. IEEE Trans. Fuzzy Syst. **10**(2), 263–267 (2002)
42. D. Arthur, S. Vassilvitskii, K-means++: The advantages of careful seeding, in *Proceedings of the Eighteenth Annual ACM-SIAM on Discrete Algorithms, SODA*, 2007, pp. 1027–1035
43. T. Lin, M. Maire, S.J. Belongie et al., Microsoft COCO: common objects in context. CoRR abs/1405.0312, 2014. http://arxiv.org/abs/1405.0312
44. M. Everingham, L. Van Gool, C.K.I. Williams, J. Winn, A. Zisserman, The pascal visual object classes (voc) challenge. Int. J. Comput. Vision **88**(2), 303–338 (2010)
45. J. Deng, W. Dong, R. Socher, L.-J. Li, K. Li, L. Fei-Fei, ImageNet: A Large-Scale Hierarchical Image Database, in *CVPR09*, 2009
46. J. Lwowski, L. Sun, R.M. Saavedra, R. Sharma, D. Pack, A reactive bearing angle only obstacle avoidance technique for unmanned ground vehicles. J. Autom. Control Res. **1**, 31–37 (2014). http://dx.doi.org/10.11159/jacr.2014.004

Low Cost Parkinson's Disease Early Detection and Classification Based on Voice and Electromyography Signal

Farika T. Putri, Mochammad Ariyanto, Wahyu Caesarendra, Rifky Ismail, Kharisma Agung Pambudi and Elta Diah Pasmanasari

Abstract Parkinson's disease (PD) is one of the health problems concerning for elderly population. Manageable symptom is an important thing for Parkinson's sufferer in order to be independent enough to do daily activities. As a solution to Parkinson's early detection method, this research purpose is to develop a low cost diagnostic tool for PD which inexpensive yet accurate and easy to use by neurologist, enriching and giving new insight for neurologist about voice and electromyography (EMG) signal analysis result. It can be very useful for PD clinical evaluation and spreading awareness about PD as well as the important of early diagnose to citizen. Parkinson's detection method in this research uses pattern recognition method, the first step is initiated with voice and EMG data acquisition. Second step is feature extraction using five features for voice and EMG signal. The last step is classification using Adaptive Neuro-Fuzzy Inference System (ANFIS) and Artificial Neural network (ANN) methods. The pattern recognition of PD is divided in two sections, the first is for two class classification, and the second is four stage classification based on Hughes Scale which commonly used in Indonesia as PD diagnose guideline. Based on the results, voice method classification has higher accuracy than EMG classification because the feature for voice is a good feature which can well classified the voice data. Voice data sampling rate is higher than EMG data sampling rate which means voice data

F. T. Putri · M. Ariyanto (✉) · W. Caesarendra · R. Ismail · K. A. Pambudi
Center for Biomechanics, Biomaterial, Biomechatronics, Biosignal Processing (CBIOM3S), Diponegoro University, Semarang, Indonesia
e-mail: ari_janto5@yahoo.co.id

F. T. Putri
e-mail: farikatonoputri@gmail.com

W. Caesarendra
School of Mechanical and Aerospace Engineering, Nanyang Technological University, Singapore, Singapore
e-mail: wcaesarendra@ntu.edu.sg

E. D. Pasmanasari
Department of Neurology, Kariadi General Hospital & Faculty of Medicine, Diponegoro University, Semarang, Indonesia
e-mail: eltadiah@gmail.com

W. Pedrycz and S.-M. Chen (eds.), *Computational Intelligence for Pattern Recognition*, Studies in Computational Intelligence 777,
https://doi.org/10.1007/978-3-319-89629-8_15

recording has more data each second than EMG data. Two class classification has higher accuracy than four class classification both in ANN and ANFIS. Based on the four class classification results in both of voice and EMG signals using ANN and ANFIS, the probable class has the lowest accuracy of all classes.

Keywords Parkinson's disease · Pattern recognition · Voice · EMG · ANN ANFIS

1 Introduction

1.1 Parkinson's Disease (PD)

Parkinson's Disease (PD) was first described by Dr. James Parkinson in a book entitled "An Essay on the Shaking Palsy" published in 1817. For several centuries later known as shaking palsy and in the medical terms known by the name of paralysis agitans. Along with its development later called Parkinson's Disease to commemorate the services of Dr. James Parkinson as the first person to explain the disease [1].

Many studies have been done by experts related to PD but until now not yet known exactly what causes of PD. The Parkinson's Institute team in California, USA conducted a study by interviewing 519 PD patients and 511 healthy people associated with occupational history and exposure to toxins experienced including pesticides and solvent fluids [2]. The results show that those who work in education, agriculture, health workers and welders are not directly related to PD. Researchers found 8.5 percent of PD patients are people who are often exposed to pesticides. The results of this study proves that there is a relationship between pesticides with PD, not just fertilizers but other chemicals [2]. The National Institute of Neurological Disorders and Stroke (NINDS) identifies several possible causes of PD, among others [3]:

- Premature aging, premature aging of brain cells (neurons) allows one of the causes of PD
- Oxidative damage, free radicals are very unstable and potentially damaging molecules produced through normal chemical reactions in the body
- Environmental toxins, both external and internal toxins that can damage the cells (neurons) of the body
- Genetic influence, 10 to 15 percent of PD patients have close relatives who also experience PD symptoms

PD shows some symptoms in patients, among others as follows [3, 4]:

- Tremor at rest
  Tremor at rest is one of the typical symptoms of PD so that PD is often called shaking palsy. The shaking that occurs during this break can be seen on the hands,

arms, legs, jaws and faces. Tremor generally begins from the hand and for the early stages of tremor occurs on one side of the body only.

- Rigidity
  Rigidity is the stiffness that occurs in the limbs. The human muscle basically has opposite muscle sections. The basic principle of movement is when one muscle is activated and the opposite muscle is relaxed. Rigidity will occur when the muscles are moving and the opposite muscles are both in active state because the signals from the brain are disrupted.
- Bradykinesia
  Bradykinesia is taken from the Greek word meaning "slow motion". Bradykinesia can be seen from facial expressions such as masks (hypomimia) and decreased frequency of eye blinks, delay in moving and decreased fine motor coordination such as not being able to button clothes or cut meat.
- Gait (walking posture)
  The running posture will be slightly disturbed at the beginning. Disturbed walking posture can be seen from the arm that does not move when the patient is walking. When walking will experience freezing (silent in place) for a few moments.
- The problem of balance and posture instability
  Instability posture makes PD patients have a hunched posture. Impaired balance and coordination cause PD patients to fall easily and injury.

1.2 Classification of Parkinson's Disease (PD)

Parkinson's disease has a stage of development of symptoms. Each stage is measured on a widely used scale. Several common scales are used to determine the stage of PD development in patients. UPDRS (Unified Parkinson's Disease Rating Score), Hoehn and Yahr Scales, and Schwab and England of Daily Living Scales are the most commonly used scales in PD staging. Each scale reflects the burden of PD disease and how far the impact of symptoms on the patient. This scale is particularly useful for defining disease progression as well as appropriate treatments for patients.

The UPDRS scale is the most widely used scale for determining PD stadium in patients. The UPDRS consists of 44 sections where each section has an assessment of 0–4. The number 0 indicates that the patient is a healthy person and the number 4 indicates the patient has symptoms of PD [5, 6].

Schwab and England of Daily Living Scales is one of the scales to assess PD stadium. The scale consists of 100% which patients completely independent and able to perform all tasks without obstacles, difficulties and slowness. Basically normal to 0% which patients depends entirely [7].

Hoehn and Yahr Scales consists of two, namely Hoehn and Yahr Scales and Modified Hoehn and Yahr Scales. Hoehn and Yahr Scales are used as guidelines in determining the stages of PD patients. Hoehn and Yahr Scales consists of stage 1 to stage 5. Stage 1 is when only the one side of the body affected and Stage 5 when the patients lying in bed and or wheelchair unless assisted [7]. While Modified Hoehn and Yahr Scales consists of Stage 0 to Stage 5 where Stage 0 means there is no sign of disease and Stage 5 means the patients lying in bed and or wheelchair unless assisted [7].

PD examination in Indonesia is conducted clinically. Dr. Kariadi General Hospital distinguishes PD into three criteria according to Hughes criteria, as follows [8]:

- Possible
  Expressed as possible if there is one of the main symptoms of tremor at rest, rigidity, bradykinesia and failure of postural reflexes.
- Probable
  It is stated probable when there is a combination of two main symptoms (including postural reflex failure) or alternatively asymmetric rigidity or asymmetric bradykinesia.
- Definite
  It states defitinite if there is a combination of three of the four symptoms or two symptoms with one symmetrical symptom.

2 Data Acquisition

The first method in pattern recognition is data acquisition. Data acquisition can be described as the process of physical phenomenon using sensors and computer. This pattern recognition application on PD detection use two kinds of data acquisition: voice and Electromyography (EMG) data acquisition. The data acquisition process is conducted in two group of research participants, healthy participants and PD participants. There are some regulations associated with human data collection. Ethical clearance and informed consent must be fulfilled before data acquisition is conducted. Ethical clearance is an ethical appropriation given by research ethic committee for a research which involving human and animals. Ethical clearance approval for this research is obtained from Medical faculty, Diponegoro University in Semarang, Indonesia

2.1 Voice

Voice data acquisition process begins with the recording equipment preparation. Unidirectional microphone is chosen as part of recording device. Unidirectional microphone is a microphone which receives voice from one direction. Unidirectional type microphone selection aims to minimize noise from the environment. The selected recording device is Yamada multifunctional microphone DM-Q6000 and it can save the voice recorded in mp3 format through external storage like a flash disk. Figure 1 shows used unidirectional microphone and Yamada multifunctional recording device. The selected sample rate in this voice acquisition is 44 kHz.

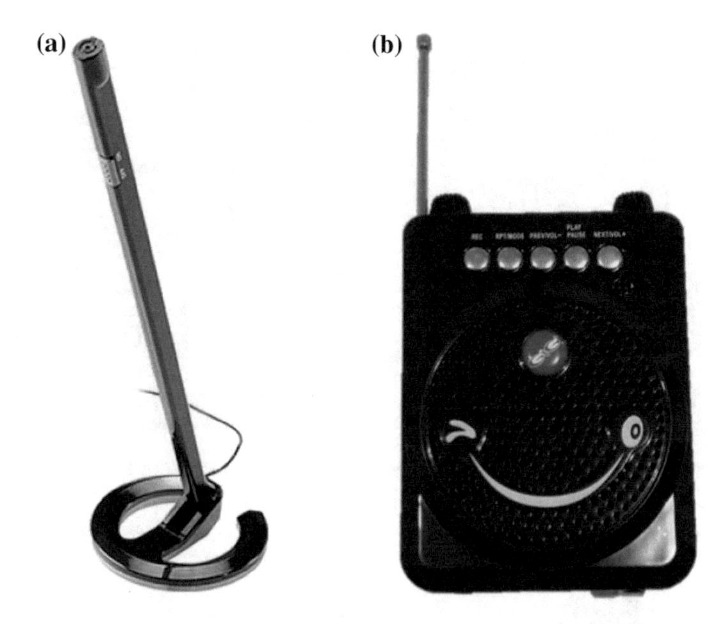

Fig. 1 Voice recording data acquisition tool. **a** Unidirectional microphone, and **b** Yamada multi-functional microphone DM-Q6000

Parkinson's patients have distinctive symptoms that are different from healthy people. In patients with Parkinson's there is a change of voice that comes out of the larynx. The change is the hoarseness of the vocal letters [5]. Based on these considerations, then the procedure data retrieval is done by recording the patient's vocal voice and healthy participants [5, 9, 10]. Voice data acquisition procedure is taken as follows:

- Research participants speak "aaaa ......" with a stable pitch for about 10 s [5, 9, 10]. When talking try positioning the mouth close to the microphone.
- Data retrieval is conducted in six times.

Figure 2 below is a picture of the process of taking voice signals in Parkinson's patients and healthy people. The voice signal is saved in .mp3 format. The format is selected because it can store the voice signal for 44 kHz with less byte and the .mp3 format also can be operated easily in MATLAB environment. Images will be blurred to protect privacy and patient confidentiality. The example of acquired voice signal from healthy and study participant with PD can be shown in Fig. 3.

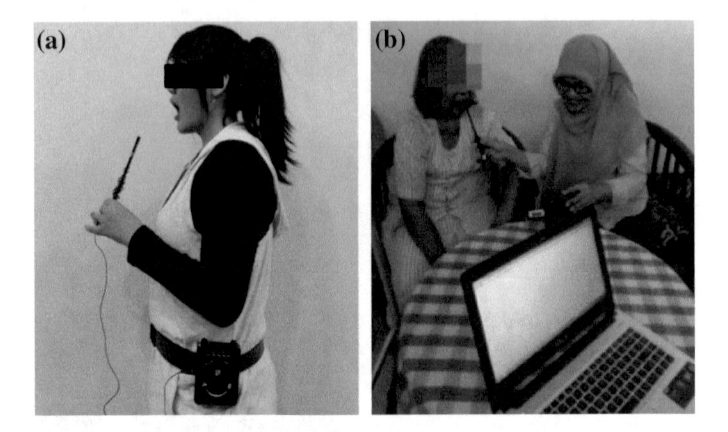

Fig. 2 Participants voice data acquisition process, **a** Healthy participant, **b** Participant with PD

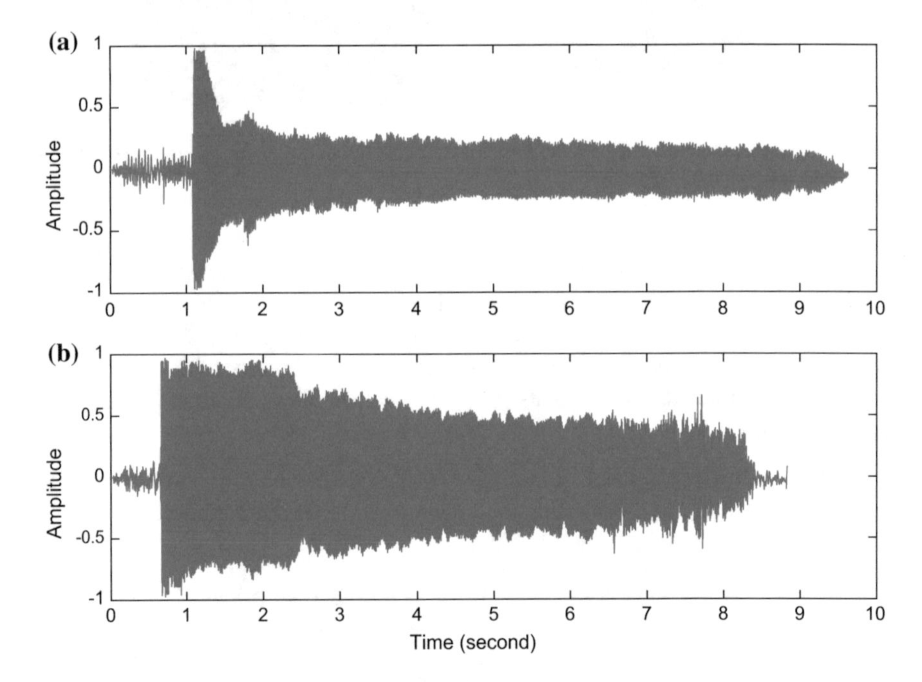

Fig. 3 Participant's voice signals, **a** Healthy Participant, **b** Participant with PD

2.2 Electromyography (EMG)

Gait data acquisition is conducted using BITalino EMG sensor. Figure 4 shows the BITalino plugged kit sensor. BITaino plugged kit consists of data acquisition component as main board, sensors consisting of EMG sensors, ECG (electrocardiography), EDA (electrodermal activity), accelerometer and LUX sensor (luminous sensor), 500

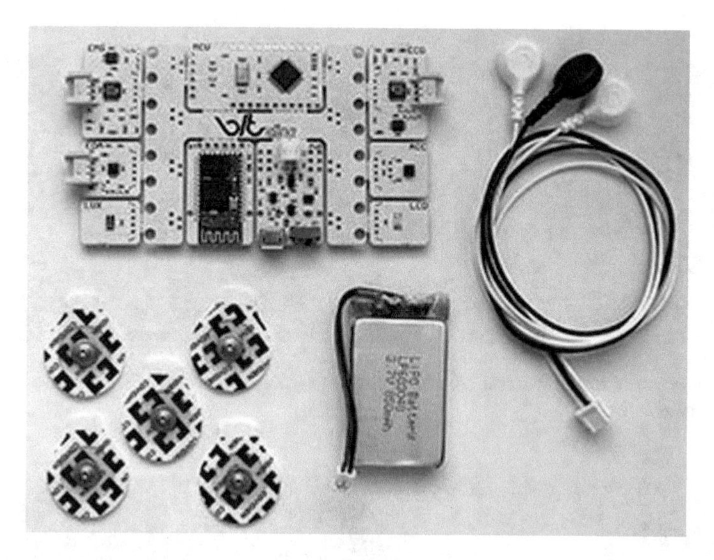

Fig. 4 BITalino plugged kit

mAh battery, main connection cable, cable pad for EMG sensor and cable pad for ECG and EDA sensors. The sample rate on BITalino plugged kit that can be selected is 1, 10, 100 and 1000 Hz.

BITalino plugged kit uses bluetooth 3.0 as wireless interface with a computer and a maximum capture distance of 10 m. The selected sample rate is 1000 Hz. BITalino can connect with MATLAB software and open access software: OpenSignals as data acquisition tool. This study uses OpenSignals as a software for storing the gait signal data. The most noticeable symptoms in Parkinson's patients are tremor and stiffening of the muscles of the limbs. With these considerations then gait signal data acquisition is conducted as follows [11]:

- The EMG sensor is mounted on the *Tibialis Anterior*, *Gastrocnemius Medialis,* and *Gastrocnemius Lateralis*. Installation EMG sensor is based on SENIAM guidelines [11, 12]. Figure 5 shows the EMG sensor position attachment.
- First data collection task is the study participants sat down and tapped the fingertips and heels on the floor alternately for 20 s. Data recording is conducted 3 times.
- Second data collection task is participants sitting and raising feet approximately 10 cm from the floor and twisted the ankle for 20 s. Data recording is conducted 3 times.

Figure 6 shows gait data recording with EMG sensors on the study participants. For the sake of privacy and the participant's secrecy, the face part is intentionally not shown. The example of acquired gait signal using EMG signal for healthy and study participant with PD can be shown in Fig. 7.

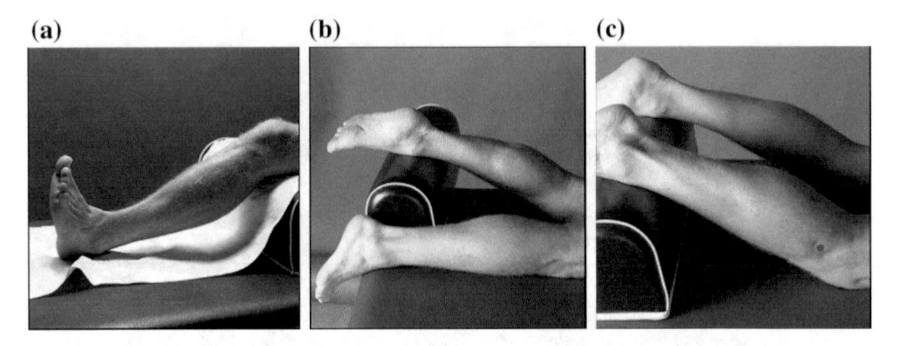

Fig. 5 EMG sensor attachment on *Tibialis Anterior* (**a**), *Gastrocnemius Medialis* (**b**) and *Gastrocnemius Lateralis* (**c**) [12]

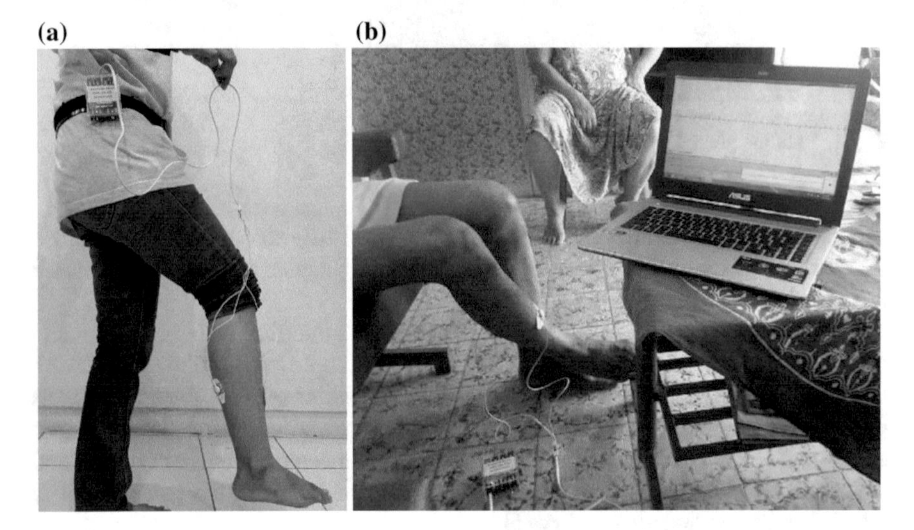

Fig. 6 Participants gait data acquisition process, **a** Healthy Participant, **b** Study participant with PD

3 Feature Calculation

In this study, the pattern recognition in PD staging is conducted using voice and EMG signal. The previous study [5, 13, 14] used 22 voice features. In this study, 22 voice features are selected into five features as presented in Sect. 3.1. All of the voice features are calculated in frequency domain. EMG signals use five selected features as discussed in Sect. 3.2. The EMG features consist of three features in time domain and two features in frequency domain. In the previous study, the five features of EMG have high accuracy on finger motion classification in five class pattern recognition [15].

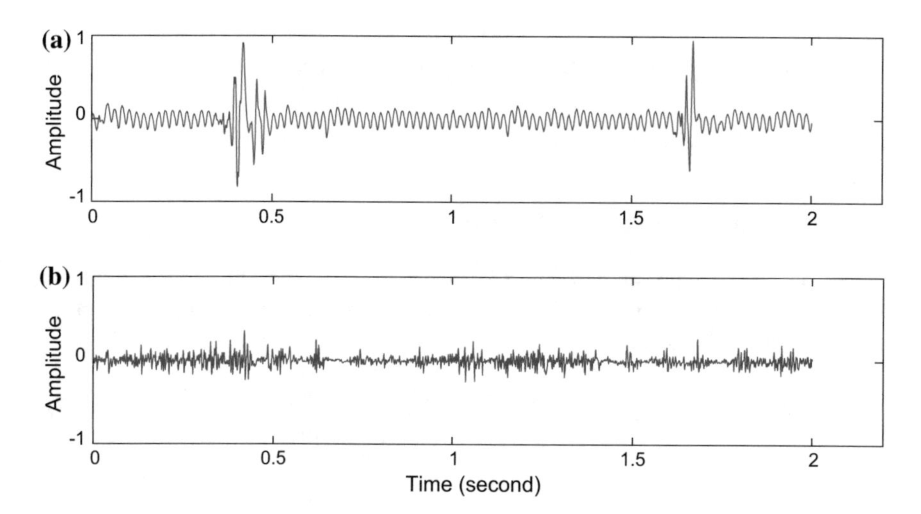

Fig. 7 Participant's gait signals using EMG, **a** Healthy Participant, **b** Participant with PD

3.1 Voice Features

a. Multi-Dimensional Voice Program/MDVP (F0)

F0 is the average value of the fundamental frequency. F0 estimation is important for voice signal characterization. Gender and age affect estimated value F0. In addition to gender and the estimated age of F0 as well influenced by emotional state while speaking, while talking through phone, incoming voice interference as background and indication when the speaker is in a state of alcohol. But the factors is only small and negligible [4]. Some researchers develop algorithms for value estimation F0 which is called Pitch Detection Algorithm (PDA). Generally the PDA has three stages of preprocessing, identification possible values of F0 and post-processing [4]. Identification of the algorithm include DYPSA (Dynamic Programming Projected Phase-Slope Algorithm), PRAAT, SHRP and SWIPE (Sawtooth Waveform Inspired Pitch Estimator) [4].

b. Local Jitter

Jitter serves to identify disorders and small irregularities which occurs in the period from cycle to cycle [4]. Equation (1) is the equation for local jitter [16].

$$Jitter\ local = \frac{jitter\,(\text{second})}{mean\,period} \tag{1}$$

where:

$$Mean\,period = \sum\nolimits_{i=1}^{N} \frac{T_i}{N}\ (\text{second}) \tag{2}$$

$$Jitter\ second = \frac{\sum_{i=2}^{N} |T_i - T_{i-1}|}{(N-1)}\ (\text{second}) \tag{3}$$

T_i = Period interval-i (second)
N = Sum of interval period

c. Period Perturbation Quotient 5 (PPQ5)

PPQ5 is the average value of the difference between the periods at a given interval with the four closest periods divided by the average period value [16]. Equation (4) is used to calculate PPQ5 [16].

$$PPQ5 = \frac{Absolute\ PPQ5\ (\text{second})}{Mean\ period\ (\text{second})} \tag{4}$$

where:

$$Absolute\ PPQ5 = \frac{\sum_{i=3}^{N-2} |T_i - (T_{i-2} + T_{i-1} + T_i + T_{i+1} + T_{i+2})/5|}{N-4} \tag{5}$$

$$mean\ period = \sum_{i=1}^{N} \frac{T_i}{N} \tag{6}$$

d. Recurrence Period Density Entropy (RPDE)

RPDE is a feature that can be used to determine the period deviation on a repeating signal [4]. The vocal cords have the ability to maintain vocal stability while oscillating. In patients with PD there is a larger period deviation than healthy people when the vocal cords oscillate. The following Eq. (7) is the equation for RPDE.

$$RPDE = \frac{\sum_{i}^{T\max} p(i) \ln(p(i))}{\ln(T\max)} \tag{7}$$

e. Pitch Period Entropy (PPE)

PPE is a value that describes the subject's deviation or incompetence in maintaining the stability of the voice tone. PPE is a feature developed by a professor from MIT named Max A. Little [5].

3.2 EMG Features

a. Integrated EMG (IEMG)
IEMG is one of the EMG signal features that belong to the time domain feature. IEMG is commonly used as early detection of EMG signal interpretation for clinical areas [17, 18]. The following Eq. (8) is an equation for the IEMG feature.

$$IEMG = \sum_{i=1}^{N} |x_i| \tag{8}$$

where:

$N =$ number of EMG data
$i =$ ith EMG data
$x_i =$ raw ith of EMG data

b. Mean Absolute Value (MAV)
MAV is one of the EMG signal features that is included in the time domain feature. MAV is often used to interpret EMG signals. The following Eq. (9) is the equation for MAV.

$$MAV = \frac{1}{N} \sum_{i=1}^{N} |x_i| \tag{9}$$

c. Variance of EMG (VAR)
VAR is one of the EMG signal features that is included in the time domain feature. VAR is defined as the mean value of the square of the deviation value for the EMG signal. The following Eq. (10) is the equation for VAR [19].

$$VAR = \frac{1}{N-1} \sum_{i=1}^{N} x_i^2 \tag{10}$$

d. Mean Frequency (MNF)
Mean frequency (MNF) is an average frequency which is computed as sum of product of the signal power spectrum and the frequency divided by total sum of the spectrum intensity [20]. It can be calculated as expressed in (11).

$$MNF = \frac{\sum_{j=1}^{M} f_i P_j}{\sum_{j=1}^{M} P_j} \tag{11}$$

where f_j is spectrum frequency at frequency bin j, P_j is the signal power spectrum at frequency bin j, and M is length of the frequency.

e. Mean Power *(MNP)*
Mean power is defined as the average value of EMG signal in frequency domain. It can be expressed as in Eq. (12).

$$MNP = \frac{\sum_{j=1}^{M} P_j}{M} \tag{12}$$

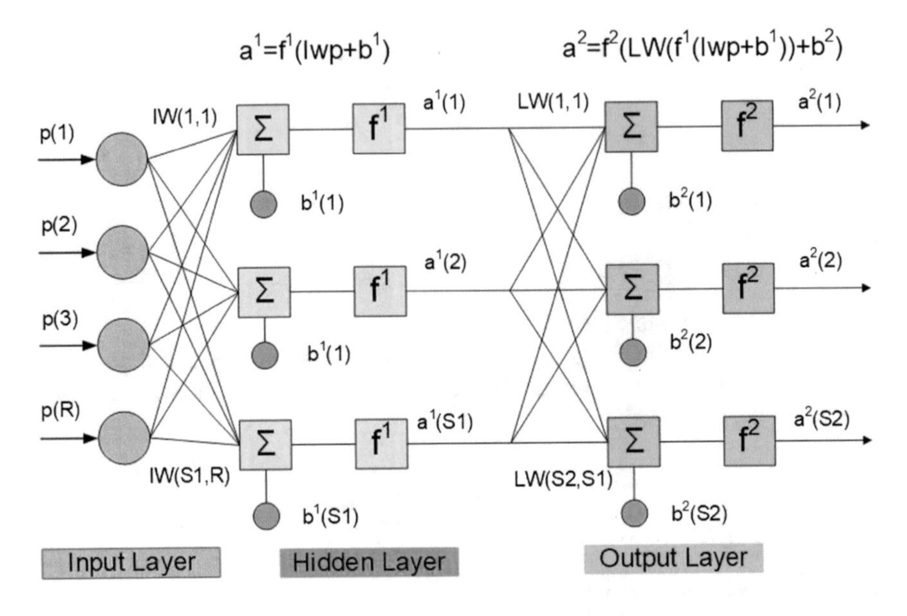

Fig. 8 Commonly used neural network structure

4 Pattern Recognition

4.1 Artificial Neural Network

In this section, the pattern recognition for voice and EMG features is classified using the neural network as can be shown in Fig. 8. The standard network that is used for pattern recognition method consists of input layer, hidden layer, and output layer. The first output neuron in the hidden layer can be expressed as in (13).

$$a^1 = f^1(IWp + b^1) \tag{13}$$

where a^1 is output vector in hidden layer, p is an n-length input vector, IW is input weight matrix, f^1 is transfer function of hidden layer, and b^1 is the bias vector of hidden layer. In the Fig. 8, R indicates the number of elements in input vector, while S1 and S2 denote the number of neuron in hidden layer and output layer respectively.

The first output neuron in the output layer as expressed in (14)

$$a^2 = f^2(LW(f^1(IWp + b^1)) + b^2) \tag{14}$$

where a^2 is output vector in output layer, LW is output layer weight matrix, f^2 is transfer function of the output layer, and b^2 is the bias vector of the output layer.

The Levenberg-Marquardt training algorithm is used in this study. It was designed to approach second-order training speed without having to compute the Hessian matrix. As typical training feedforward networks, the performance function of this training algorithm has the form of a sum of squares, and the Hessian matrix can be approximated using Eq. (15).

$$H = J^T J \tag{15}$$

and the gradient can be calculated as

$$g = J^T e \tag{16}$$

where J is the Jacobian matrix that contains first derivatives of the network errors with respect to the weights and biases, and e is a vector of network errors.

The Levenberg-Marquardt training algorithm uses Eq. (17) to approximate the Hessian matrix

$$x_{k+1} = x_k - \left[J^T J + \mu I\right]^{-1} J^T e \tag{17}$$

When the scalar μ is zero, the Eq. (17) uses the approximated Hessian matrix. When μ is large, this becomes gradient descent with a small step size.

In this research study, Mean Square Error (MSE) is utilized in ANN for classification. The MSE measures the magnitude of the forecast errors as shown in (18). Better model will show the smaller values of MSE.

$$mse_{error} = \frac{\sum (y_1 - y_2)^2}{m} \tag{18}$$

where y_1 is the real output in classification, y_2 is the output from ANN classification, and m is the total number of samples in the classification.

In two class classification of PD, the used ANN has five input features and has two class classification results for healthy and PD. While In four class classification of PD, the ANN has five input features and four class classification results using Hughes scale for healthy, possible, probable, and definite. The training algorithm employs Levenberg-Marquardt training algorithm. The neural network has one hidden layer and 25 neurons in hidden layer both in two class and four class classifications. Finally, the proposed ANN's structure can be presented in Fig. 9. Hyperbolic tangent sigmoid transfer function is used in hidden layer and soft max transfer function is employed in output layer.

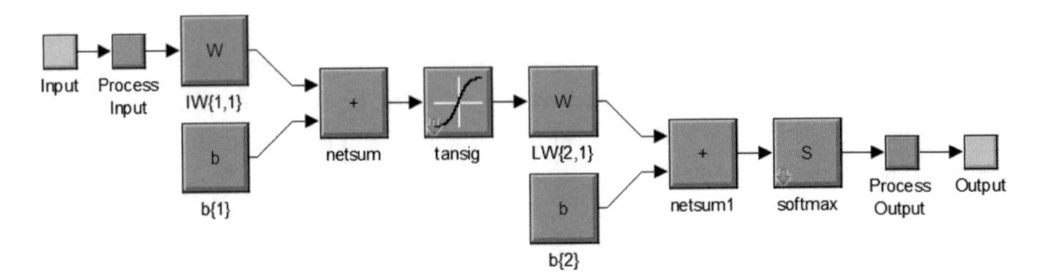

Fig. 9 Utilized ANN input and output for PD pattern recognition

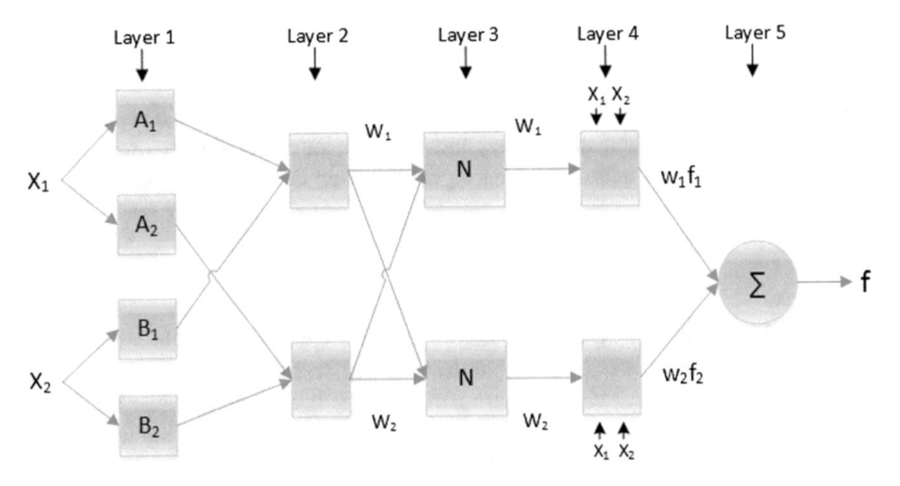

Fig. 10 Typical ANFIS architecture with first-order Sugeno fuzzy model

4.2 Adaptive Neuro Fuzzy Inference System

The adaptive neuro-fuzzy inference system (ANFIS) is a kind of neuro-fuzzy classifier method which integrates the neural network's adaptive capability and the fuzzy logic qualitative approach [21]. The common rule set with two fuzzy if-then rules for a first-order Sugeno fuzzy model can be written as follows

Rule 1: If x is A1 and y is B1, then f1 = p1x + q1y + r1;
Rule 2: If x is A2 and y is B2, then f2 = p2x + q2y + r2;

where p1, p2, q1, q2, r1 and r2 are linear parameters, and A1, A2, B1 and B2 are nonlinear parameters.

The ANFIS architecture consists of five layers as depicted in Fig. 10. The layers in ANFIS can be described as follows

Layer 1: All the nodes are adaptive nodes. The output of the *i*th node in layer 1 is denoted as $O_{1,i}$. The outputs of this layer are the fuzzy membership function of the inputs that can be expressed as in Eqs. (19) and (20)

$$O_{1,i} = \mu_{Ai}(x), \quad for\, i = 1, 2 \tag{19}$$

$$O_{1,i} = \mu_{Bi}(y), \quad for\, i = 3, 4 \tag{20}$$

where x (or y) is the input to nodes i and Ai (or Bi–2) generating a linguistic label coupled with the node. In this study, the membership function for A (or B) can be any parameterized as a gaussian membership function as written in (21)

$$\mu A_i(x) = \exp\left\{-\frac{1}{2}\frac{(x - c_i)^2}{\sigma_i^2}\right\} \tag{21}$$

where (c_i, σ_i) are the parameter set.

Layer 2: This layer perform as a simple multiplier. Each node in this layer calculates the firing strengths of each rule via multiplying the incoming signals and sends the product out. The outputs of this layer can be expressed as in (22)

$$O_{2,i} = w_i = \mu_{Ai}(x)\mu_{Bi}(y), \quad i = 3, 4 \tag{22}$$

Layer 3: The nodes are also fixed nodes, indicate a normalization role to the firing strengths from the previous layer. The ith node of this layer calculates the ratio of the ith rule's firing strength to the sum of all rules' firing strengths as in (23)

$$O_{3,i} = \bar{w}_i = \frac{w_i}{\sum_{i=1}^{2} w_i} = \frac{w_i}{w_1 + w_2} \quad i = 1, 2 \tag{23}$$

Layer 4: This nodes are adaptive nodes. Parameters in layer 4 will be referred to as consequent parameters. The output of each node in this layer is the product of the normalized firing strength and a first order polynomial. The output of this layer can be express in (24)

$$O_{4,i} = \bar{w}_i f_i = \bar{w}_i(p_i x + q_i y + r_i) \tag{24}$$

where $\bar{w}_i$ is a normalized firing strength from layer 3.

Layer 5: There is only single fixed node. This node performs the summation of all incoming signals from layer 4. The output of layer 5 is summarized as in (25)

$$O_{5,1} = \sum_i \bar{w}_i f_i = \frac{\sum_i w_i f_i}{\sum_i w_i} \tag{25}$$

In this study, the utilized ANFIS structure can be seen in Fig. 11. The structure has 243 rules. The layer 1 uses gaussian membership function. Each of input has three gaussian membership functions. The layer 4 uses linear output function. For the training algorithm, hybrid learning is selected. The used Fuzzy operators are product

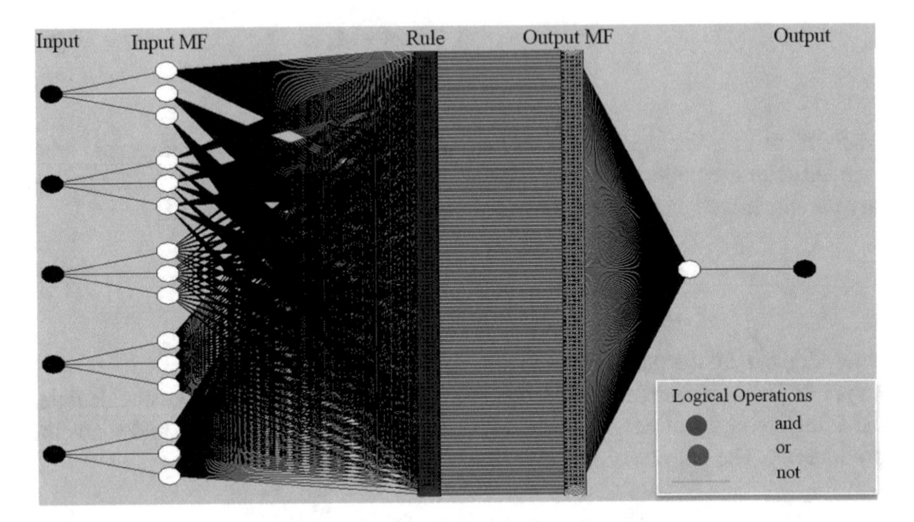

Fig. 11 Proposed ANFIS architecture with first-order Sugeno fuzzy model for PD classification

for And method and probabilistic OR for Or method. The selected implication and aggregation method are minimum and maximum method. Weighted average of all rule outputs method is used for defuzzification process.

5 Result

All of the involved study participants are PD patients from dr. Kariadi general hospital in Semarang. Eight healthy participants and 15 study participants with PD are involved in this study. Table 1 shows the list of study participants. There are 23 study participants who join in the research, 15 participants with PD both male and female with age range from 39 years old to 80 years old, and 8 healthy participants both male and female with age range from 52 years old to 75 years old. Classification process in this research used total 126 data for voice signals and 69 data for EMG signals. Voice and EMG signal data have been collected from 8 healthy participant and 15 people with PD participants. Healthy participants recorded their voice six times per person. Voice signal have been collected from PD patients consist of 10 patients with six time recording each, 3 patients with four recording each and 2 patients with 3 recording each. There are different data recording for PD patients because some of patients struggling to produce their voice to be recorded. EMG signals have been collected three times each for both healthy and PD participants.

For the convenience and confidentiality of the study participants, the participant name's identity is kept secret. Unfortunately, the data acquisition for EMG signals is less than the data from voice signals because of the condition on the study participants with PD. Few PD participants can be acquired using EMG signals depending

Table 1 List of data capturing study participants

Parkinson participants				Healthy participants		
Study participants	Sex	Age (years old)	Status	Study participants	Sex	Age (years old)
S 1	Male	68	Definite	S 16	Female	71
S 2	Male	53	Possible	S 17	Female	53
S 3	Male	59	Possible	S 18	Female	75
S 4	Female	79	Definite	S 19	Male	70
S 5	Female	66	Definite	S 20	Female	55
S 6	Female	39	Probable	S 21	Female	70
S 7	Female	58	Possible	S 22	Male	52
S 8	Female	54	Probable	S 23	Male	60
S 9	Male	66	Probable			
S 10	Male	80	Probable			
S 11	Male	76	Probable			
S 12	Male	70	Definite			
S 13	Male	68	Probable			
S 14	Male	68	Definite			
S 15	Male	72	Definite			

on his/her health condition. The classification results are presented in performance measures i.e. accuracy, precision, recall, F1 score, and Cohen's kappa. The resulted value of Kappa is characterized based on the reference [22].

5.1 *Classification Result of Artificial Neural Network Method*

Classification using ANN method recognizes the terms of training and testing. Training in ANN is a dataset that serves to adjust the weighting and biases in the ANN method. Testing in ANN is a dataset that serves to test the final solution to confirm the strength and toughness of the ANN classifier. The dataset selected for training and testing process is randomly assigned by the ANN. The dataset is selected based on a percentage of 70% of the data for training, and 30% for testing. The ANN classification results for this study consist of training and testing results for two class classification using voice signals, four class classification using voice signals, two class classification using EMG signals and four class classification using EMG signals.

Table 2 Accuracy of voice signals for two classes using ANN in testing

Actual class	Classification result	
	Healthy	PD
Healthy	12	5
PD	0	21
Accuracy (%)	70.59	100
Total accuracy (%)	86.84	

Table 3 Performance measures of voice signals for two classes using ANN in testing

Performance	Values
Accuracy	0.8684
Precision	1
Recall (Sensitivity)	0.7059
F1 score	0.8276
Kappa	0.7262

5.1.1 Classification Results of Voice Signals for Two Classes

The data for training and testing consists of 126 voice record data from healthy and participants with PD. The data are divided into two sets, 88 for training and 38 for testing. The data for training and testing are selected randomly. For training data, the data are divided in two sets i.e. 31 for healthy class and 57 for participants with PD.

Table 2 shows the accuracy of the testing result with the ANN of the voice signal for two classes. The dataset included in the first and second class is class with healthy and participants with PD randomly selected 17 and 21 data respectively. Accuracy testing for healthy and PD is 70.59 and 100%. The overall testing accuracy for the two classes is 86.84%. The performance measures of the classification results can be summarized in Table 3. Based on the Table 3, F1 score and Kappa are 0.8276 and 0.7262. Based on the Kappa value, the classifier has substantial agreement.

Table 4 shows the overall accuracy (training and testing) of the ANN classifier for two classes. The target for the first class of healthy class shows an accuracy of 89.58%. The target class for the second class is the participants with PD showing an accuracy of 100%. The overall accuracy result is 96.03%. The performance measures of the overall classification can be summarized as in Table 5. It indicates that the ANN classifier has both high precision and high recall. The resulted Kappa shows that the classifier has almost perfect agreement.

5.1.2 Classification Results of Voice Signals for Four Classes

In four class classification, the data for training and testing consists of 126 voice record data from 48 healthy and 78 PD. The overall data are divided into two sets, 88 for training and 38 for testing. For training data, the data is divided in four sets

Table 4 Accuracy of voice signals for two classes using ANN in overall

Target class	Classification result	
	Healthy	PD
Healthy	43	5
PD	0	78
Accuracy (%)	89.58	100
Total accuracy (%)	96.03	

Table 5 Performance measures of voice signals for two classes using ANN in overall

Performance	Values
Accuracy	0.9603
Precision	1
Recall (Sensitivity)	0.8958
F1 score	0.9451
Kappa	0.9141

Table 6 Accuracy of voice signals for four classes using ANN in testing

Actual class	Classification Result			
	Healthy	Possible	Probable	Definite
Healthy	9	1	3	1
Possible	0	1	0	3
Probable	1	4	4	1
Definite	0	3	2	5
Accuracy (%)	64.29	25	40	50
Total accuracy (%)	50			

i.e. 34 for healthy, 11 for possible, 20 for probable, and 23 for definite. The data for training and testing are selected randomly.

Table 6 shows the ANN test results for four class classification of voice signals. The dataset for testing is divided in four sets i.e. healthy, possible, probable, and definite randomly selected from 14, 4, 10, and 10 data respectively. The accuracy of each class is 64.29% for healthy, 25% for possible, 40% for probable, and 50% for definite. The overall testing accuracy for the four classes is 50%. The performance measures of classifier in overall can be summarized in Table 7. Based on the resulted Kappa, the classification result has fair agreement.

Table 8 reveals the ANN classification overall (training and testing) results for four class classification of voice signals. The overall accuracy of each class is 87.5% for healthy, 80% for possible, 70% for probable, and 84.85% for definite. The overall accuracy for the four classes is 81.75%.

The performance measures of the overall classification for four class classification are summarized as in Table 9. Based on the resulted Kappa, it shows that the classifier has moderate agreement. Healthy class has the highest accuracy of all classes.

Table 7 Performance measures of voice signals for four classes using ANN in testing

Performance	Healthy	Possible	Probable	Definite	Total
Accuracy	0.6429	0.2500	0.4000	0.5000	0.5000
Precision	0.9	0.1111	0.4444	0.5000	0.4889
Recall (Sensitivity)	0.6429	0.2500	0.4000	0.5000	0.4482
F1 score	0.7500	0.1538	0.4211	0.5000	0.4562
Kappa	0.4265	0.6999	0.5805	0.5534	0.2500

Table 8 Accuracy of voice signals for four classes using ANN in overall

Target class	Classification result			
	Healthy	Possible	Probable	Definite
Healthy	42	1	4	1
Possible	0	12	0	3
Probable	2	5	21	2
Definite	0	3	2	28
Accuracy (%)	87.5	80	70	84.85
Total accuracy (%)	81.75			

Table 9 Performance measures of voice signals for four classes using ANN in overall

Performance	Healthy	Possible	Probable	Definite	Total
Accuracy	0.8750	0.8000	0.7000	0.8485	0.8175
Precision	0.9546	0.5714	0.7778	0.8235	0.7818
Recall (Sensitivity)	0.8750	0.8000	0.7000	0.8485	0.8059
F1 score	0.9130	0.6667	0.7368	0.8358	0.7881
Kappa	0.3039	0.7281	0.5796	0.4981	0.5132

5.1.3 Classification Results of EMG Signals for Two Classes

The data for training and testing consists of 69 EMG signal record data from healthy and participants with PD. The data are divided into two sets, 48 for training and 21 for testing. The data for training and testing are selected randomly in EMG signal classification. For training data, the data is divided in two sets i.e. 16 for healthy class and 32 for participants with PD.

Table 10 shows the accuracy of the testing result of the EMG signal for two classes. The dataset in the first and second class is a class with healthy and participants with PD randomly selected 8 and 13 data respectively from testing data. Accuracy testing for healthy and PD is 87.5% and 84.62%. The overall testing accuracy for the two classes using EMG signals is 85.71%. The performance measures of the classification

Table 10 Accuracy of EMG signals for two classes using ANN in testing

Actual class	Classification result	
	Healthy	PD
Healthy	7	1
PD	2	11
Accuracy (%)	87.5	84.62
Total accuracy (%)	85.71	

Table 11 Performance measures of EMG signals for two classes using ANN in testing

Performance	Values
Accuracy	0.8571
Precision	0.7778
Recall (Sensitivity)	0.8750
F1 score	0.8235
Kappa	0.7042

Table 12 Accuracy of EMG signals for two classes using ANN in overall

Target class	Classification result	
	Healthy	PD
Healthy	22	2
PD	3	42
Accuracy (%)	91.67	93.33
Total accuracy (%)	92.75	

Table 13 Performance measures of EMG signals for two classes using ANN in overall

Performance	Values
Accuracy	0.9275
Precision	0.8800
Recall (Sensitivity)	0.9167
F1 score	0.8980
Kappa	0.8418

results in two class classification can be shown in Table 11. Based on the Kappa value in Table 11, the classifier has substantial agreement.

The overall accuracy (training and testing) of the ANN for two classes can be shown in Table 12. The overall accuracy for healthy class and participants with PD is 91.67 and 93.33% respectively. The overall accuracy result is 92.75%. The performance measures of the overall classification can be summarized in Table 13. Based on F1 Score, It indicates that the classifier has both high precision and high recall. The resulted Kappa in overall classification result shows that the classifier has almost perfect agreement.

Table 14 Accuracy of EMG signals for four classes using ANN in testing

Target class	Classification result			
	Healthy	Possible	Probable	Definite
Healthy	3	1	1	3
Possible	1	1	0	1
Probable	1	1	0	2
Definite	1	0	1	4
Accuracy (%)	37.5	33.33	0	66.67
Total accuracy (%)	38.1			

Table 15 Performance measures of EMG signals for four classes using ANN in testing

Performance	Healthy	Possible	Probable	Definite	Total
Accuracy	0.3750	0.3333	0	0.6667	0.3810
Precision	0.5000	0.3333	0	0.4000	0.3083
Recall (Sensitivity)	0.3750	0.3333	0	0.6667	0.3438
F1 score	0.4286	0.3333	NaN	0.5000	NaN
Kappa	0.4762	0.7429	0.7506	0.3950	0.3942

5.1.4 Classification Results of EMG Signals for Four Classes

In four class classification using EMG signals, the data for training and testing consists of 69 voice record data from 24 healthy and 45 participants with PD. The data are divided into two sets, 48 for training and 21 for testing. For training data, the data are divided in four sets i.e. 16 for healthy, 6 for possible, 11 for probable, and 15 for definite. The data for training and testing are selected randomly.

Table 14 shows the ANN classification testing results for four class classification of EMG signals. The dataset for testing is divided in four sets i.e. healthy, possible, probable, and definite randomly selected from 8, 3, 4, and 6 data respectively. The testing accuracy of each class is 37.5% for healthy, 33.33% for possible, 0% for probable, and 66.67% for definite. The overall testing accuracy for the four classes is 38.1%. The performance measures of classifier in testing are presented in Table 15. F1 Score has the value of NaN (Not a Number). It indicates that the classifier has poor precision and poor recall. Based on the resulted Kappa, the classification result has fair agreement.

The overall accuracy result of the EMG signal classification is shown in Table 16. The overall accuracy result for the ANN classification is 76.81%. The lowest accuracy is probable class. The performance measures of classifier in overall are presented in Table 17. Based on the resulted Kappa, the overall classification result has fair agreement in four class classification using EMG signals.

Table 16 Accuracy of EMG signals for four classes using ANN in overall

Target class	Classification result			
	Healthy	Possible	Probable	Definite
Healthy	19	1	1	3
Possible	1	6	0	2
Probable	1	2	9	3
Definite	1	0	1	19
Accuracy (%)	79.17	66.67	60	90.48
Total accuracy (%)	76.81			

Table 17 Performance measures of EMG signals for four classes using ANN in overall

Performance	Healthy	Possible	Probable	Definite	Total
Accuracy	0.7917	0.6667	0.6000	0.9047	0.7681
Precision	0.8636	0.6667	0.8182	0.7037	0.7630
Recall (Sensitivity)	0.7917	0.6667	0.6000	0.9048	0.7408
F1 score	0.8261	0.6667	0.6923	0.7917	0.7442
Kappa	0.3861	0.7516	0.6464	0.3687	0.3816

5.2 *Classification Result of Adaptive Neuro-Fuzzy Inference System (ANFIS)*

The dataset selected for the training and testing process is randomly assigned. The dataset is selected based on a percentage of about 70% of the data for training, and about 30% for the testing. ANFIS classification results consist of training and testing results for two class classification using voice signals, four class classification using voice signals, two class classification using EMG signals and four class classification using EMG signals.

5.2.1 Classification Results of Voice Signals for Two Classes

The data for training and testing consists of 126 voice record data from 48 healthy and 78 with PD. The data are divided into two sets, 88 for training and 38 for testing. The data for training and testing are selected randomly. For training data, the data are divided in two sets i.e. 29 for healthy class and 59 for PD. Table 18 shows the accuracy of the testing result of voice signal for two classes. The dataset in the first and second class is a class with healthy and participants with PD randomly selected 19 and 19 data respectively. Accuracy testing for healthy and PD is 100% and 100% respectively. The overall testing accuracy is 100%. The performance measures of the classification results can be summarized in Table 19. Based on the Table 19, F1 score

Table 18 Accuracy of voice signals for two classes using ANFIS in testing

Actual class	Classification result	
	Healthy	PD
Healthy	19	0
PD	0	19
Accuracy (%)	100	100
Total accuracy (%)	100	

Table 19 Performance of voice signals for two classes using ANFIS in testing

Performance	Values
Accuracy	1
Precision	1
Recall (Sensitivity)	1
F1 score	1
Kappa	1

Table 20 Accuracy of voice signals for two classes using ANFIS in overall

Actual class	Classification result	
	Healthy	PD
Healthy	48	0
PD	0	78
Accuracy (%)	100	100
Total accuracy (%)	100	

Table 21 Performance measures of voice signals for two classes using ANFIS in overall

Performance	Values
Accuracy	1
Precision	1
Recall (Sensitivity)	1
F1 score	1
Kappa	1

and Kappa are 1 and 1. It indicates that the classification results has best precision and best recall. Based on the Kappa value, the classifier has perfect agreement.

Table 20 shows the overall accuracy (training and testing) of the ANFIS for two classes. The overall accuracy of both healthy and PD class is 100%. The overall accuracy result is 100%. The performance measures of the overall classification can be summarized as in Table 21. It indicates that the classifier has both best precision and best recall. The resulted Kappa show that the classifier has perfect agreement in overall two class classification using voice signals.

Table 22 Accuracy of voice signals for four classes using ANFIS in testing

Actual class	Classification result			
	Healthy	Possible	Probable	Definite
Healthy	15	0	0	0
Possible	0	1	1	3
Probable	2	2	1	3
Definite	0	0	1	8
Accuracy (%)	100	20	12.5	88.89
Total accuracy (%)	67.57			

Table 23 Performance measures of voice signals for four classes using ANFIS in testing

Performance	Healthy	Possible	Probable	Definite	Total
Accuracy	1	0.200	0.125	0.8889	0.6757
Precision	0.8824	0.3333	0.33	0.5714	0.5301
Recall (Sensitivity)	1	0.2000	0.125	0.8889	0.5535
F1 score	0.9375	0.2500	0.1818	0.6957	0.5162
Kappa	0.1720	0.8003	0.7305	0.4418	0.1351

5.2.2 Classification Results of Voice Signals for Four Classes

In four class classification using ANFIS, the data for training and testing consist of 126 voice record data from 48 healthy and 78 PD. The data are divided into two sets, 89 for training and 37 for testing. For training data, the data are divided in four sets i.e. 33 for healthy, 10 for possible, 22 for probable, and 24 for definite. The data for training and testing are selected randomly.

The classification test result for four class classification using ANFIS of EMG signals is presented in Table 22. The dataset for testing is divided in four sets i.e. healthy, possible, probable, and definite randomly selected from 15, 5, 8, and 9 data respectively. The testing accuracy of each class is 100% for healthy, 20% for possible, 12.5% for probable, and 88.89% for definite. The overall testing accuracy for the four classes is 67.57%. The performance measures of classifier in testing can be summarized in Table 23. Based on the resulted Kappa, the classification result has slight agreement.

Table 24 reveals the classification results in overall (training and testing) for four class classification of voice signals. The overall accuracy of each class is 100% for healthy, 73.33% for possible, 76.67% for probable, and 96.97% for definite. The overall accuracy for the four classes is 90.48%. The performance measures of the overall classification for four classes can be presented in Table 25. Based on the resulted Kappa, it shows that the classifier has substantial agreement.

Table 24 Accuracy of voice signals for four classes using ANFIS in overall

Actual class	Classification result			
	Healthy	Possible	Probable	Definite
Healthy	48	0	0	0
Possible	0	11	1	3
Probable	2	2	23	3
Definite	0	0	1	32
Accuracy (%)	100	73.33	76.67	96.97
Total accuracy (%)	90.48			

Table 25 Performance measures of voice signals for four classes using ANFIS in overall

Performance	Healthy	Possible	Probable	Definite	Total
Accuracy	1	0.7333	0.7667	0.9697	0.9048
Precision	0.9600	0.8462	0.9200	0.8421	0.8921
Recall (Sensitivity)	1	0.7333	0.7667	0.9697	0.8674
F1 score	0.9796	0.7857	0.8364	0.9014	0.8758
Kappa	0.2320	0.7834	0.5816	0.4564	0.7460

Table 26 Accuracy of EMG signals for two classes using ANFIS in overall

Actual class	Classification result	
	Healthy	PD
Healthy	2	4
PD	3	10
Accuracy (%)	33.33	76.92
Total accuracy (%)	63.16	

5.2.3 Classification Results of EMG Signals for Two Classes

The training and testing data consist of 69 EMG signal data from healthy and PD. The data are divided into two sets, 50 for training and 19 for testing. For training data, the data is divided in two sets i.e. 18 for healthy class and 32 for participants with PD. Table 26 shows the accuracy of testing result of the EMG signals for two classes. The dataset in the first and second class are a class with healthy and participants with PD randomly selected 6 and 13 data respectively from testing data. The overall testing accuracy for the two classes is 63.16%. The performance measures can be shown in Table 27. Based on the Kappa value, the classifier has slight agreement in testing.

The overall accuracy for two classes is presented in Table 28. The overall accuracy for healthy class and participants with PD class is 70.83% and 91.11% respectively. The overall accuracy result is 84.06%. The performance measures of the overall classification can be summarized in Table 29. The resulted Kappa on overall classification result shows that the classifier has substantial agreement.

Table 27 Performance measures of EMG signals for two classes using ANFIS in testing

Performance	Values
Accuracy	0.6316
Precision	0.4000
Recall (Sensitivity)	0.3333
F1 score	0.3636
Kappa	0.1074

Table 28 Accuracy of EMG signals for two classes using ANFIS in overall

Actual class	Classification result	
	Healthy	PD
Healthy	17	7
PD	4	41
Accuracy (%)	70.83	91.11
Total accuracy (%)	84.06	

Table 29 Performance measures of EMG signals for two classes using ANFIS in overall

Performance	Values
Accuracy	0.8406
Precision	0.8095
Recall (Sensitivity)	0.7083
F1 score	0.7556
Kappa	0.6381

5.2.4 Classification Results of EMG Signals for Four Classes

The training and testing data consist of 69 voice record data from 24 healthy and 45 PD. The data are divided into two sets, 47 for training and 22 for testing. For training data, the data are divided in four sets i.e. 16 for healthy, 6 for possible, 10 for probable, and 15 for definite. Table 30 shows the accuracy of test results for four class classification of EMG signals. The dataset for testing is divided in four sets i.e. healthy, possible, probable, and definite randomly selected from 8, 3, 5, and 6 data respectively. The overall testing accuracy for the four classes is 31.82%. The performance measures of classifier in testing are shown in Table 31. F1 score has the value of NaN (Not a Number). It indicates that the classifier has poor precision and poor recall. Based on the Kappa value, the classification result has moderate agreement.

The overall accuracy result of the EMG signal classification using ANFIS is presented in Table 32. The overall accuracy result for the ANFIS is 75% for healthy, 66.67% for possible, 40% for probable, and 61.91% for definite. The lowest accuracy is probable class. The performance measures of classifier in overall can be presented in Table 33. Based on the Kappa value, the overall classification result has slight agreement in four class classification using EMG signals.

Table 30 Accuracy of EMG signals for four classes using ANFIS in testing

Actual class	Classification result			
	Healthy	Possible	Probable	Definite
Healthy	4	1	0	3
Possible	0	2	1	0
Probable	4	0	0	1
Definite	2	3	0	1
Accuracy (%)	50	66.67	0	16.67
Total accuracy (%)	31.82			

Table 31 Performance measures of EMG signals for four classes using ANFIS in testing

Performance	Healthy	Possible	Probable	Definite	Total
Accuracy	0.5000	0.6667	0	0.1667	0.3182
Precision	0.4000	0.3333	0	0.2000	0.2333
Recall (Sensitivity)	0.5714	0.7895	0.9412	0.7500	0.3333
F1 score	0.4444	0.4444	NaN	0.1818	NaN
Kappa	0.4040	0.6318	0.7479	0.6061	0.4500

Table 32 Accuracy of EMG signals for four classes using ANFIS in overall

Actual class	Classification result			
	Healthy	Possible	Probable	Definite
Healthy	18	1	2	3
Possible	0	6	2	1
Probable	6	2	6	1
Definite	2	6	0	13
Accuracy (%)	75	66.67	40	61.91
Total accuracy (%)	62.32			

Table 33 Performance measures of EMG signals for four classes using ANFIS in overall

Performance	Healthy	Possible	Probable	Definite	Total
Accuracy	0.7500	0.6667	0.4000	0.6191	0.6232
Precision	0.6923	0.4000	0.6000	0.7222	0.6036
Recall (Sensitivity)	0.7500	0.6667	0.4000	0.6191	0.6089
F1 score	0.7200	0.5000	0.4800	0.6667	0.5917
Kappa	0.3743	0.6812	0.6722	0.4992	0.0048

6 Conclusion and Future Works

There are some conclusions that can be concluded from the research:

- Two class classification has higher accuracy than four class classification both in neural network and adaptive fuzzy-inference system.
- Voice method classification has higher accuracy than EMG classification because the feature for voice is a good feature which can well classified the voice data. Voice data sampling rate is higher than EMG data sampling rate which means voice data recording has more data each second than EMG data.
- EMG signal classification has less accuracy because there is a lot of noise in the EMG Sensor and it has one channel with low sampling rate i.e. 1000 Hz.
- Based on the four class classification results in both of voice and EMG signals using ANN and ANFIS, the probable class has the lowest accuracy of all.

To increase the accuracy of pattern recognition method, it can be conducted by using higher sampling rate up to 100 kHz and more channel in EMG sensor. The low accuracy in four class classifications especially in testing can caused by wrong staging PD of patient. For example when the PD study participants met the neurologist, he/she has been diagnosed with staging possible, but when the researcher met the study participant, the staging of PD become probable or definite when his/her healthy condition become worse. When this study participant's signal is used, it can give miss classification and decrease the accuracy in PD pattern recognition.

Future research can be conducted by adding a new method for PD detection: hand tremor method. Based on the research, almost all PD participants have a tremor in their hands as sign of PD symptom. With hand tremor detection method, there is a hope that this PD detection tool will be very accurate. The accuracy result between voice, gait EMG and hand tremor can be combined in order to achieve higher accuracy.

References

1. L.I. Golbe, M.H. Mark, J.I. Sage, *Parkinson's Disease Handbook* (The American Parkinson Disease Association Inc., 2010)
2. Ikatan Apoteker Indonesia. (2015) www.ikatanapotekerindonesia.net/site-feature/36-community-pharmacy/scientific-publications-on-health/1185-Apa-Itu-Parkinson.html
3. National Institute of Neurological Disorder and Stroke (NINDS). (2015) www.ninds.nih.gov/disorders/parkinson_disease/detail_parkinson_disease.html
4. A. Tsanas, Accurate Telemonitoring of Parkinson's Disease Symptom Severity Using Nonlinear Speech Signal Processing and Statistical Machine Learning. Doctoral Dissertation Biomedical Engineering Department, University of Oxford, St. Cross College, United Kingdom (2012)
5. M.A. Little, P.E. McSharry, E.J. Hunter, J. Spielman, O.L. Ramig, Suitability of dysphonia measurements for telemonitoring of parkinson disease. IEEE Trans. Biomed. Eng. **56**(4), 1015–1022 (2009)
6. W. Poewe, M.B. Stern, S. Fahn, P. Martinez-Martin, *UPDRS. Chicago.* (Movement Disorder Society, America, 2008)

7. P. Martinez-Martin, *Guide to Assessment Scales in Parkinson's Disease*. Springer Healthcare (2014)
8. T. Syamsudin, G. Dewanto, dan Subagya: Buku Panduan Tatalaksana Penyakit Parkinson dan Gangguan Gerak Lainnya. Indonesia: Perhimpunan Dokter Spesialis Saraf Indonesia (PERDOSSI) (2013)
9. C. Azad, S. Jain, K. Vijay, Design and analysis of data mining based prediction model for parkinson's disease. Int. J. Comput. Sci. Eng. **3**(3), 181–189 (2014)
10. M. Shahbakhti, D. Taherifar, Linear and non-linear speech features for detection of parkinson's disease. In: Proceeding of Biomedical Engineering International Conference (BMEiCON-2013) (2013), pp. 1–3
11. P. Kuegler, S. Member, C. Jaremenko, J. Schalachelzki, Automatic recognition of parkinson disease using surface electromyography during standarized gait tests. In: Proceeding of 2013 35th Annual International Conference of the IEEE Engineering in Medicine and Biology Society (EMBC) (2013), pp. 5781–5784
12. D.F. Stegeman, H.J. Hermens, Standards for Surface Electromyography: the European Project: Surface EMG for Non-invasive Assessment of Muscles (SENIAM). (2015). www.seniam.org
13. W. Caesarendra, M. Ariyanto, J.D. Setiawan, M. Arozi, C.R. Chang, A pattern recognition method for stage classification of Parkinson's disease utilizing voice features. In: Proceeding of International Conference on Biomedical Engineering and Sciences (IECBES) (2014), pp. 87–92
14. W. Caesarendra, F.T. Putri, M. Ariyanto, J.D. Setiawan, Pattern recognition methods for multi stage classification of parkinson's disease utilizing voice features. In: Proceeding of Advanced Intelligent Mechatronics (AIM), pp. 802–807 (2015)
15. M. Ariyanto, W. Caesarendra, A.R. Winoto, Finger movement pattern recognition method using artificial neural network based on electromyography (EMG) sensor. In: Proceeding of International Conference on Automation, Cognitive Science, Optics, Micro Electro-¬Mechanical System, and Information Technology, Bandung, Indonesia (2015), pp. 12–17
16. M. Kumar, Classification of Parkinson's Disease using Multipass LVQ, Logistic Model Tree, K Star for Audio Data Set. Master Thesis Computer Engineering Department, Dalarna University, Falun, Sweden (2011)
17. D. Tkach, H. Huang, T. Kuiken, A study of stability of time-domain features for electromyography pattern recognition. J. Neuro Eng. Rehabil. **7**(21), 1–13 (2010)
18. M. Zardoshti-Kermani, B.C. Wheeler, K. Badie, R.M. Hashemi, EMG feature evaluation for movement control of upper extremity prostheses. IEEE Trans. Rehabil. Eng. **3**, 324–333 (1995)
19. S.H. Park, S.P. Lee, EMG pattern recognition based on artificial intelligence techniques. IEEE Trans. Rehabil. Eng. **6**(4), 400–405 (1998)
20. M.A. Oskoei, H. Hu, Support vector machine based classification scheme for myoelectric control applied to upper limb. IEEE Trans. Biomed. Eng. **55**(8), 1956–1965 (2008)
21. J.S.R. Jang, C.T. Sun, E. Mizutani, *Neuro-Fuzzy and Soft Computing* (Prentice Hall, Englewood Cliffs, 1997)
22. J.R. Landis, G.G. Koch, The measurement of observer agreement for categorical data. Biometrics **33**(1), 159–174 (1977)

Index

W. Pedrycz and S.-M. Chen (eds.), *Computational Intelligence for Pattern Recognition*, Studies in Computational Intelligence 777,
https://doi.org/10.1007/978-3-319-89629-8

Printed in the United States
By Bookmasters